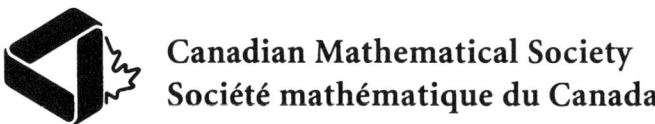

Canadian Mathematical Society
Société mathématique du Canada

Editors-in-Chief
Rédacteurs-en-chef
Jonathan Borwein
Peter Borwein

Springer
New York
Berlin
Heidelberg
Hong Kong
London
Milan
Paris
Tokyo

CMS Books in Mathematics
Ouvrages de mathématiques de la SMC

1   HERMAN/KUČERA/ŠIMŠA   Equations and Inequalities

2   ARNOLD   Abelian Groups and Representations of Finite Partially Ordered Sets

3   BORWEIN/LEWIS   Convex Analysis and Nonlinear Optimization

4   LEVIN/LUBINSKY   Orthogonal Polynomials for Exponential Weights

5   KANE   Reflection Groups and Invariant Theory

6   PHILLIPS   Two Millennia of Mathematics

7   DEUTSCH   Best Approximation in Inner Product Spaces

8   FABIAN ET AL.   Functional Analysis and Infinite-Dimensional Geometry

9   KŘÍŽEK/LUCA/SOMER   17 Lectures on Fermat Numbers

10 BORWEIN   Computational Excursions in Analysis and Number Theory

11 REED/SALES (Editors)   Recent Advances in Algorithms and Combinatorics

12 HERMAN/KUČERA/ŠIMŠA   Counting and Configurations

13 NAZARETH   Differentiable Optimization and Equation Solving

14 PHILLIPS   Interpolation and Approximation by Polynomials

15 BEN-ISRAEL/GREVILLE   Generalized Inverses, Second Edition

16 ZHAO   Dynamical Systems in Population Biology

17 GÖPFERT ET AL.   Variational Methods in Partially Ordered Spaces

Alfred Göpfert   Christiane Tammer
Hassan Riahi   Constantin Zălinescu

Variational Methods in Partially Ordered Spaces

With 13 Illustrations

Springer

Alfred Göpfert
Department of Mathematics and
    Computer Science
Martin-Luther-University
    Halle Wittenberg
Halle D-06099
Germany
goepfert@mathematik.uni-halle.de

Christiane Tammer
Department of Mathematics and
    Computer Science
Martin-Luther-University
    Halle-Wittenberg
Halle D-06099
Germany
tammer@mathematik.uni-halle.de

Hassan Riahi
Faculté des Sciences Semlalia,
    Mathematiques
Université Cadi Ayyad
B.P. S 15
Marrakech 40000
Morocco
h-riahi@ucam.ac.ma

Constantin Zălinescu
Faculty of Mathematics
University of "Al. I. Cuza" Iasi
Bd. Copou, Nr. 11
700506 Ia\c si
Romania
zalinesc@uaic.ro

Editors-in-Chief
Rédacteurs-en-chef
Jonathan Borwein
Peter Borwein
Centre for Experimental and Constructive Mathematics
Department of Mathematics and Statistics
Simon Fraser University
Burnaby, British Columbia V5A 1S6
Canada
cbs-editors@cms.math.ca

Mathematics Subject Classification (2000): 00A71, 93A30, 65K10

Library of Congress Cataloging-in-Publication Data
Variational methods in partially ordered spaces / Alfred Göpfert ... [et al.].
        p. cm. — (CMS books in mathematics ; 17)
    Includes bibliographical references and index.
    ISBN 0-387-00452-1 (alk. paper)
        1. Partially ordered spaces.   2. Vector spaces.   3. Mathematical optimization.   I. Göpfert,
A. (Alfred)   II. Series.
    QA611.3 V37   2003
    512′.52—dc21   2003044935

ISBN 0-387-00452-1   Printed on acid-free paper.

Printed in the United States of America.

9 8 7 6 5 4 3 2 1   SPIN 10911884

Typesetting: Pages created by the authors using the Springer LaTeX2e macro package (w/svmono.cls).

www.springer-ny.com

Springer-Verlag   New York Berlin Heidelberg
A member of BertelsmannSpringer Science+Business Media GmbH

Preface

In mathematical modeling of processes occurring in industrial systems, logistics, management science, operations research, networks, and control theory one often encounters optimization problems involving more than one objective function, so that multiobjective optimization (or vector optimization, initiated by W. Pareto) has received new impetus. The growing interest in multiobjective problems, both from the theoretical point of view and as it concerns applications to real problems, asks for a general scheme that embraces several existing developments and stimulates new ones. With this book we intend to give direct access to new results and new applications of this quickly growing field.

Mathematical Background

In particular, we discuss basic tools of partially ordered spaces and apply them to variational methods in nonlinear analysis and for optimization problems; i.e., we present the relevant functional analysis for our presentations, especially separation theorems for not necessarily convex sets, which are important for the characterization of solutions, for the proof of existence results and optimality conditions in multicriteria optimization. We use the optimality conditions in order to derive numerical algorithms for special classes of vector optimization problems.

Purpose of This Book

We believe that our book will be of interest to graduate students in mathematics, economics, and engineering as well as researchers in pure and applied mathematics, economics, engineering, geography, and town planning. A sound knowledge of linear algebra and introductory real analysis should provide readers with sufficient background for this book.

On the one hand, the book has the character of a monograph, because the authors use many of their own results and applications; on the other hand, it is a textbook, because we would like to present in a sense a state of the art of the field in an understandable, useful, and teachable way.

Organization

Firstly, we shall give some simple examples to show which kinds of problems can be handled with the methods of the book. Then the three main chapters follow.

In the first of them we deal with connections between order structures and topological structures of sets, give a new nonconvex separation theorem, which is very useful for scalarization, and study different properties of multi-functions.

The second of them contains our results concerning the theory of multi-criteria optimization and equilibrium problems directly. Approximate efficiency, scalarization, new existence results with respect to different order relations, well-posedness, sensitivity, duality, and optimality conditions with respect to general vector optimization problems and a big section on minimal point theorems belong to this chapter as well as new results of vector equilibrium problems and their applications to vector-valued variational inequalities.

Those new theoretical results are applied in the last chapter of the book in order to construct numerical algorithms, especially proximal-point algorithms and geometrical algorithms based on duality assertions. It is possible to use the special structure of several classes of multicriteria optimization problems (location problems, approximation problems, fractional programming problems, multicriteria control problems) for deriving optimality conditions and corresponding algorithms. We discuss concrete applications (approximation problems, location problems in town planning, multicriteria equilibrium problems, fractional programming) with solution procedures and in some cases with corresponding software. Here and in the whole book there are examples to illustrate the results or to check stated conditions. The chapters are followed by a list of references, a list of symbols and a big enough index.

The book was written by four authors, we wrote it together, and it was at every time stimulating and profitable to consider the problems from different sides. A. Göpfert, Chr. Tammer, and C. Zălinescu contributed to Sections 1.1, 2.1–2.3, 3.1, and 3.10; C. Zălinescu wrote Sections 2.4–2.7 and 3.2–3.6; H. Riahi wrote Sections 3.8, 3.9, 4.2.4, and 4.2.5; and A. Göpfert and Chr. Tammer wrote Sections 1.2–1.6, 3.7, 3.11, 4.1, 4.2.1–4.2.3, and 4.3–4.6.

Acknowledgments

We are grateful to Manfred Goebel, Sylke Sauter, Kristin Winkler, and Eugen Zălinescu for numerous suggestions for technically improving the manuscript. Each author is grateful to her/his coauthors for their wonderful typing of the manuscript. We would like to thank an anonymous referee for his/her remarks. We are happy to publish this monograph in the series *Canadian Mathematical Society Series of Monographs and Advanced Texts* by Springer.

<table>
<tr><td>Halle, Germany</td><td>Alfred Göpfert</td></tr>
<tr><td>Halle, Germany</td><td>Christiane Tammer</td></tr>
<tr><td>Marrakech, Morocco</td><td>Hassan Riahi</td></tr>
<tr><td>Iaşi, Romania</td><td>Constantin Zălinescu</td></tr>
</table>

January 2003

Contents

Preface v

List of Figures xiii

1 Examples 1
  1.1 Cones in Vector Spaces 1
  1.2 Equilibrium Problems 3
  1.3 Location Problems in Town Planning 6
  1.4 Multicriteria Control Problems 8
  1.5 Multicriteria Fractional Programming Problems 10
  1.6 Stochastic Efficiency in a Set 11

2 Functional Analysis over Cones 13
  2.1 Order Structures 13
  2.2 Functional Analysis and Convexity 26
  2.3 Separation Theorems for Not Necessarily Convex Sets 39
  2.4 Convexity Notions for Sets and Multifunctions 45
  2.5 Continuity Notions for Multifunctions 51
  2.6 Continuity Properties of Multifunctions Under Convexity
      Assumptions 70
  2.7 Tangent Cones and Differentiability of Multifunctions 73

3 Optimization in Partially Ordered Spaces 81
  3.1 Solution Concepts 81
     3.1.1 Approximate Minimality 81
     3.1.2 A General Scalarization Method 84
  3.2 Existence Results for Efficient Points 86
     3.2.1 Preliminary Notions and Results Concerning
           Transitive Relations 87
     3.2.2 Existence of Maximal Elements with Respect to
           Transitive Relations 89

    3.2.3 Existence of Efficient Points with Respect to Cones 94
    3.2.4 Types of Convex Cones and Compactness with
          Respect to Cones 101
    3.2.5 Classification of Existence Results for Efficient Points .. 104
    3.2.6 Some Density and Connectedness Results 110
3.3 Continuity Properties with Respect to a Scalarization
    Parameter 115
3.4 Well-Posedness of Vector Optimization Problems 119
3.5 Continuity Properties 122
    3.5.1 Continuity Properties of Optimal-Value Multifunctions . 122
    3.5.2 Continuity Properties for the Optimal Multifunction
          in the Case of Moving Cones 135
    3.5.3 Continuity Properties for the Solution Multifunction ... 138
3.6 Sensitivity of Vector Optimization Problems 141
3.7 Duality 154
    3.7.1 Duality Without Scalarization 155
    3.7.2 Duality by Scalarization 159
    3.7.3 Duality for Approximation Problems 162
3.8 Vector Equilibrium Problems and Related Topics 168
    3.8.1 Vector Equilibrium Problems 169
    3.8.2 General Vector Monotonicity 171
    3.8.3 Existence of Vector Equilibria by Use of the
          Generalized KKM Lemma 172
    3.8.4 Existence by Scalarization of Vector Equilibrium
          Problems 174
    3.8.5 Some Knowledge About the Assumptions 177
    3.8.6 Some Particular Cases 180
    3.8.7 Mixed Vector Equilibrium Problems 183
3.9 Applications to Vector Variational Inequalities 186
    3.9.1 Vector Variational-Like Inequalities 186
    3.9.2 Perturbed Vector Variational Inequalities 188
    3.9.3 Hemivariational Inequality Systems 189
    3.9.4 Vector Complementarity Problems 192
    3.9.5 Application to Vector Optimization Problems 194
    3.9.6 Minimax Theorem for Vector-Valued Mappings 196
3.10 Minimal-Point Theorems in Product Spaces and
    Corresponding Variational Principles 197
    3.10.1 Not Authentic Minimal-Point Theorems 199
    3.10.2 Authentic Minimal-Point Theorems 202
    3.10.3 Minimal-Point Theorems and Gauge Techniques 205
    3.10.4 Minimal-Point Theorems and Cone-Valued Metrics 209
    3.10.5 Fixed Point Theorems of Kirk–Caristi Type 212
3.11 Optimality Conditions 213
    3.11.1 Lagrange Multipliers and Saddle Point Assertions 213
    3.11.2 ε-Saddle Point Assertions 218

4   Applications 229
  4.1   Approximation Problems 229
     4.1.1   General Approximation Problems 229
     4.1.2   Finite-dimensional Approximation Problems 235
     4.1.3 L_p-Approximation Problems 241
     4.1.4   Example: The Inverse Stefan Problem................ 242
  4.2   Solution Procedures 244
     4.2.1   A Proximal-Point Algorithm for Real-Valued Control
          Approximation Problems 244
     4.2.2   Computer Programs for the Application of the
          Proximal-Point Algorithm 259
     4.2.3   An Interactive Algorithm for the Vector Control
          Approximation Problem........................... 263
     4.2.4   Proximal Algorithms for Vector Equilibrium Problems . 266
     4.2.5   Relaxation and Penalization 277
  4.3   Location Problems...................................... 282
     4.3.1   Formulation of the Problem 282
     4.3.2   An Algorithm for the Multicriteria Location Problem .. 285
     4.3.3   A Mathematica Program for Solving the Multicriteria
          Location Problem 286
     4.3.4   Comparison of Alternatives 287
     4.3.5   Application to a Problem of Town Planning 289
  4.4   Multicriteria Fractional Programming 294
     4.4.1   Solution Concepts 294
     4.4.2   Generalized Dinkelbach Transformation 297
     4.4.3   Possibilities for a Solution Approach 300
  4.5   Multicriteria Control Problems 303
     4.5.1   The Formulation of the Problem 303
     4.5.2   An ε-Minimum Principle for Multicriteria Optimal
          Control Problems 304
     4.5.3   A Multicriteria Stochastic Control Problem 309
  4.6   Stochastic Efficiency in a Set........................... 316

List of Abbreviations 319

References 325

Index 345

List of Figures

1.3.1 The set of efficient elements of the multicriteria location
   problem (P) with the maximum norm   8
1.3.2 The set of efficient points of the multicriteria location problem
   (P) with the Lebesgue norm instead of the maximum norm   8

2.3.1 Level sets of the functional φ in (2.23). 42

3.1.1 The set of approximately efficient elements (where the distance
   between it and the set of efficient elements is unbounded). 82
3.1.2 The set of approximately efficient elements with a bigger cone C. 83

4.2.1 Application of Algorithm II in a C^{++} computer program. 260
4.2.2 Solutions x_0^1, x_0^2, and x_0^3 of the location problem generated by
   the proximal-point algorithm choosing different weights α_i
   $(i = 1, \ldots, n)$. 262
4.3.3 The set of efficient elements of the multicriteria location
   problem (P) with the maximum norm. 287
4.3.4 The set of efficient points of the multicriteria location problem
   (P) with the Lebesgue norm instead of the maximum norm. . . . 288
4.3.5 Comparison of the alternatives x^1, x^2, x^3 with respect to the
   criteria C_1, C_2, C_3, C_4, C_5. 290
4.3.6 The set of efficient points of the multicriteria location problem
   (P_L). 291
4.3.7 The set of efficient points $\mathcal{X}_{\text{Eff}}^{\text{Leb}}$ of (P_L) with the Lebesgue
   norm instead of the maximum norm. 292
4.3.8 Radar chart for the alternatives x^1, x^2, and x^3. 293

1

Examples

1.1 Cones in Vector Spaces

Vector optimization in partially ordered spaces requires, among other things, that one studies properties of cones; as arguments recall:

- Conic approximation of sets, e.g., in order to prove necessary optimality conditions; think of tangent cones, generated cones . . . .
- Cone-valued mappings; think of the subdifferential of indicator functions, which coincide with the normal cone, used in order to state optimality conditions for optimization problems with nonfunctional constraints.
- Order relations in linear spaces induced by cones; think of Pareto efficiency and the consideration of smaller or larger cones in order to vary the set of efficient points.
- Cones in order to represent inequalities; think of side restrictions or vector variational inequalities.
- Cones in connection with some theoretical procedures; think of Phelps's cone related to Ekeland's variational principle in normed spaces, minimal points with respect to cone orderings.
- Dual cones or polar cones; think of dual assertions and the bipolar theorem.

A cone in a vector space is an algebraic notion, but its use in theory and applications of optimization demands that one considers cones in topological vector spaces and studies, besides algebraic properties such as pointedness and convexity, also analytical ones such as those of being closed, normal, Daniell, nuclear, or having a base with special properties. A short overview of such qualities can be found at the end of this section; the corresponding proofs are given by hints to references or to later sections. The examples of cones given below show interesting results on cones in infinite-dimensional vector spaces, which are important in vector optimization and control theory.

Example 1.1.1. Let X be a normed vector space with $\dim X = \infty$, and let $x' : X \to \mathbb{R}$ be a linear but not continuous function. Then the cone

$$C = \{0\} \cup \{x \mid x'(x) > 0\}$$

has a base $B = \{x \in C \mid x'(x) = 1\}$, but $0 \in \operatorname{cl} B$, since B is dense in C. For bases see Section 2.2.

Example 1.1.2. Let X be a locally convex space and C a convex cone in X. Then

$$C^\# \neq \emptyset \Rightarrow \operatorname{cl} C \cap (-C) = \{0\} \Rightarrow C \text{ is pointed,}$$

where

$$C^\# := \{x^* \in X^* \mid x^*(x) > 0 \; \forall x \in C \setminus \{0\}\}.$$

The first implication becomes an equivalence if $\dim X < \infty$. The converse of the second implication is not true, even if $X = \mathbb{R}^2$; $C = ((0,\infty) \times \mathbb{R}) \cup (\{0\} \times [0,\infty))$ is a counterexample.

To prove the first implication consider $u^* \in C^\#$. Assume that there exists $\bar{c} \in \operatorname{cl} C \cap (-C)$, $\bar{c} \neq 0$. Then there exists a sequence $(c^n)_{n \in \mathbb{N}} \subset C$ such that $(c^n) \to \bar{c}$. Because $u^*(c^n) \geq 0$, we have that $u^*(\bar{c}) = \lim u^*(c^n) \geq 0$, which contradicts the fact that $u^* \in C^\#$ and $\bar{c} \in -C \setminus \{0\}$.

Assume $\dim X < \infty$; let us prove the converse of the first implication. Let $C \neq \{0\}$ (if $C = \{0\}$ then $C^\# = X^*$) and suppose that $\operatorname{cl} C \cap (-C) = \{0\}$. Then

$$0 \notin \operatorname{raint} C^+ \subset C^\#. \tag{1.1}$$

Indeed, if $0 \in \operatorname{raint} C^+$, then C^+ is a linear subspace, and consequently $C^{++} = \operatorname{cl} C$ is a linear subspace. This implies that $\operatorname{cl} C \cap (-C) = -C \neq \{0\}$, a contradiction. Now let $\bar{x}^* \in \operatorname{raint} C^+$. Assume that $\bar{x}^* \notin C^\#$; then there exists $\bar{x} \in C \setminus \{0\}$ such that $\bar{x}^*(\bar{x}) = 0$. Let $x^* \in C^+$. Then there is $\lambda > 0$ such that $(1+\lambda)\bar{x}^* - \lambda x^* \in C^+$. So $((1+\lambda)\bar{x}^* - \lambda x^*)(\bar{x}) = -\lambda x^*(\bar{x}) \geq 0$, whence $x^*(\bar{x}) \leq 0$; it follows that $-\bar{x} \in C^{++} = \operatorname{cl} C$. It follows that $\operatorname{cl} C \cap (-C) \neq \{0\}$, a contradiction. Taking into account $\dim X < \infty$, (1.1) gives $C^\# \neq \emptyset$, since $\operatorname{raint} C^+$ is nonempty (recall that every nonempty finite-dimensional convex set has a nonempty relative algebraic interior; see [168, p. 9]).

Example 1.1.3. The convex cone C in a Banach space X has the **angle property** if for some $\varepsilon \in (0,1]$ and $x^* \in X^* \setminus \{0\}$ we have $C \subset \{x \in X \mid x^*(x) \geq \varepsilon \|x^*\| \cdot \|x\|\}$. It follows that $x^* \in C^\#$. Since for $X = \mathbb{R}^n = X^*$ the last inequality means $\cos(x^*, x) \geq \varepsilon$, it is clear where the name "angle property" comes from. The class of convex cones with the angle property is very large (for all $\varepsilon \in (0,1)$ and $x^* \in X^* \setminus \{0\}$ the set $\{x \in X \mid x^*(x) \geq \varepsilon \|x^*\| \cdot \|x\|\}$ is a closed convex cone with the angle property and nonempty interior). In fact, in normed spaces, a convex cone has the angle property iff it is well-based. So, the cone $\mathbb{R}^n_+$ in $\mathbb{R}^n$ has the angle property (with $x^* = (1,1,\ldots,1) \in \mathbb{R}^n$ and $\varepsilon = 2^{-1/2}$), but the ordinary order cone $\ell_2^+ \subset \ell_2$ does not have it.

Indeed, if for some $x^* \in \ell_2 \setminus \{0\}$ and some $\varepsilon > 0$, $\ell_2^+ \subset \{x \in \ell_2 \mid x^*(x) \geq \varepsilon \|x^*\| \cdot \|x\|\}$, then $x_n^* \geq \varepsilon \|x^*\|$ (because $e_n = (0,\ldots,1,\ldots) \in \ell_2^+$), whence the contradiction $\varepsilon \|x^*\| \leq 0$ (since $(x_n^*) \to 0$ for $n \to \infty$).

Overview of Several Properties of Cones

Let X be a Banach space, C and K proper (i.e., $\{0\} \neq C, K \neq X$) convex cones in X, K^+ the continuous dual cone of K, and

$$K^\# := \{x^* \in K^+ \mid x^*(x) > 0 \ \forall x \in K \setminus \{0\}\}$$

the quasi-interior of K^+. Then the relations below hold. For more relationships among different kinds of cones and spaces look at Section 3.2; for cones with base see Section 2.2.

$$
\begin{array}{ccccc}
& & K \text{ has compact base} & \Longrightarrow & K \text{ is Daniell} \\
& & \Downarrow \ \Uparrow \ {\scriptstyle K = \overline{K}} & & \\
& & {\scriptstyle X = \mathbb{R}^n} & & \\
K \text{ has angle property} & \Longleftrightarrow & K \text{ is well-based} & \Longleftrightarrow & \mathrm{int}\, K^+ \neq \emptyset \\
& & \Downarrow & & \\
\exists C \ : \ K \setminus \{0\} \subset \mathrm{int}\, C & \Longleftrightarrow & K \text{ is based} & \Longleftrightarrow & K^\# \neq \emptyset \\
& & \Downarrow \ \Uparrow \ {\scriptstyle K = \overline{K}} & & \\
& & {\scriptstyle X \text{ separable}} & & \\
K \text{ well-based} & \overset{K = \overline{K}, \ X = \mathbb{R}^n}{\Longleftarrow} & K \text{ pointed} & \Longleftarrow & K^+ - K^+ = X^* \\
\Downarrow & & \Uparrow & & \Updownarrow \\
\overline{K} \text{ is normal} & \Longleftrightarrow & K \text{ is normal} & \Longrightarrow & K \text{ is } w\text{-normal.}
\end{array}
$$

1.2 Equilibrium Problems

Let us consider a common scalar optimization problem

$$\varphi(x) \to \min \text{ s.t. } x \in \mathcal{B}, \tag{1.2}$$

where $\mathcal{B}$ is a given nonempty set in a space X and $\varphi : \mathcal{B} \to \mathbb{R}$ a given function. Let $\overline{x} \in \mathcal{B}$ be a solution of (1.2); that is,

$$\varphi(\overline{x}) \leq \varphi(y) \ \forall y \in \mathcal{B}. \tag{1.3}$$

Then, setting $f(x, y) := \varphi(x) - \varphi(y)$ for $x, y \in \mathcal{B}$, $\overline{x}$ solves also the problem

$$\text{find } \overline{x} \in \mathcal{B} \text{ such that } f(\overline{x}, y) \leq 0 \ \forall y \in \mathcal{B}. \tag{1.4}$$

For given $\mathcal{B}$ and $f : \mathcal{B} \times \mathcal{B} \to \mathbb{R}$, a problem of the kind (1.4) is called an **equilibrium problem** and $\mathcal{B}$ its feasible set. A large number of quite different problems can be subsumed under the class of equilibrium problems as, e.g., saddle point problems, Nash equilibria in noncooperative games, complementarity problems, variational inequalities, and fixed point problems. To sketch the last one, consider X a Hilbert space and $T : \mathcal{B} \to \mathcal{B}$ a given mapping. With $f(x, y) := (x - Tx \,|\, x - y)$ we have that $\bar{x} \in \mathcal{B}$ is a fixed point of T (i.e., $T\bar{x} = \bar{x}$) if and only if $\bar{x}$ satisfies $f(\bar{x}, y) \leq 0 \ \forall y \in \mathcal{B}$. Indeed, if $\bar{x}$ is an equilibrium point, then taking $\bar{y} := T\bar{x}$, we have

$$0 \geq (\bar{x} - T\bar{x} \,|\, \bar{x} - T\bar{x}) = \|\bar{x} - T\bar{x}\|^2,$$

and so $\bar{x} = T\bar{x}$ as claimed. The other direction of the assertion is obvious.

There are powerful results that ensure the existence of a solution of the equilibrium problem (1.4); one of the most famous is **Fan's theorem**; see [111]:

Theorem 1.2.1. *Let $\mathcal{B}$ be a compact convex subset of the Hausdorff locally convex space (H.l.c.s.) X and let $f : \mathcal{B} \times \mathcal{B} \to \mathbb{R}$ be a function satisfying*

$\forall y \in \mathcal{B} : x \to f(x, y)$ *is lower semicontinuous,*
$\forall x \in \mathcal{B} : y \to f(x, y)$ *is concave,*
$\forall y \in \mathcal{B} : f(y, y) \leq 0.$

Then there is $\bar{x} \in \mathcal{B}$ with $f(\bar{x}, y) \leq 0$ for every $y \in \mathcal{B}$.

Since we mainly deal with multicriteria problems, we now take $\mathcal{B} \subset X$ as above and $\varphi : \mathcal{B} \to Y$, where Y is the Euclidean space $\mathbb{R}^n$ ($n > 1$) or any locally convex space (l.c.s.). In Y we consider a convex pointed cone C, for instance $C = \mathbb{R}^n_+$ if $Y = \mathbb{R}^n$. Then, as usual and in accordance with (1.3), $\bar{x} \in \mathcal{B}$ solves the multicriteria problem $\varphi(x) \to \min, x \in \mathcal{B}$, if

$$\forall y \in \mathcal{B} : \varphi(y) \notin \varphi(\bar{x}) - (C \setminus \{0\}). \tag{1.5}$$

Sometimes, if the cone C has a nonempty interior (e.g., $C = \mathbb{R}^n_+$ when $X = \mathbb{R}^n$), one looks for so-called weak solutions of multicriteria problems replacing (1.5) by

$$\forall y \in \mathcal{B} : \varphi(y) \notin \varphi(\bar{x}) - \operatorname{int} C. \tag{1.6}$$

As above, we introduce $f(x, y) := \varphi(x) - \varphi(y)$ for $x, y \in \mathcal{B}$, and so in accordance with (1.4) we get the following **vector equilibrium problem**: find $\bar{x} \in \mathcal{B}$ (sometimes called vector equilibrium) with the property

$$\forall y \in \mathcal{B} : f(\bar{x}, y) \notin C \setminus \{0\}, \tag{1.7}$$

or, considering weak solutions of multicriteria problems, find $\bar{x} \in \mathcal{B}$ (sometimes called **weak vector equilibrium**) with the property

$$\forall y \in \mathcal{B} : f(\bar{x}, y) \notin \operatorname{int} C. \tag{1.8}$$

Looking for existence theorems for solutions of (1.7) we mention Sections 3.8, 3.9, 4.2.4, and 4.2.5, which are devoted to vector-valued equilibrium problems.

Vector equilibria play an important role in mathematical economics, e.g., if one deals with **traffic control**. As an example we describe an extension of **Wardrop's principle** for weak traffic equilibria to the case in which the route **flows** have to satisfy the travel demands between origin–destination pairs and route capacity restrictions, and the travel cost function is a mapping.

Let us consider a **traffic** (or **transportation**) **network** (N, L, P), where N denotes a finite set of nodes, L a finite set of links, $P \subseteq N \times N$ a set of origin–destination pairs, and $|P|$ the cardinality of P. We denote by

- $d \in \mathbb{R}^{|P|}$ the travel demand vector with $d_p > 0$ for all $p \in P$;
- $\mathbb{R}$ a finite set of routes (paths) between the origin–destination pairs such that for each $p \in P$, the set $R(p)$ of routes in $\mathbb{R}$ connecting p is nonempty;
- Q the pair–route incidence matrix ($Q_{pr} = 1$ if $r \in R(p)$, $Q_{pr} = 0$ otherwise); i.e., Q is a matrix of type $(|P|, |R|)$.

We suppose that the demand vector is given on $\mathbb{R}^{|P|}$. Then we introduce the set K of traffic flow vectors (path flow vectors), where $K := \{v \in \mathbb{R}^{|R|} \mid 0 \le v_r \ \forall r \in R, \ Qv = d\}$. Note that, in more detailed form, the condition $Qv = d$ states that for all $p \in P$, $\sum_{r \in R(p)} v_r = d_p$. The set K is the feasible set for the desired vector equilibrium problem.

In order to formulate the vector equilibrium problem, a valuation of the traffic flow vectors is necessary. Therefore, let Y be a topological vector space partially ordered by a suitable convex, pointed cone C with int $C \ne \emptyset$ and F a travel cost function, which assigns to each route flow vector $v \in \mathbb{R}^{|R|}$ a vector of marginal travel costs $F(v) \in L(\mathbb{R}^{|R|}, Y)$, where $L(\mathbb{R}^{|R|}, Y)$ is the set of all linear operators from $\mathbb{R}^{|R|}$ into Y.

The **weak traffic equilibrium problem** we describe consists in finding $u \in K$ such that (in accordance to (1.4))

$$\phi(u, v) := -F(u)(v - u) \notin \text{int } C \ \forall v \in K, \tag{1.9}$$

recalling that in the scalar case $Y = \mathbb{R}^1$, the product $F(u)v$ is a price for v.

Any route flow $u \in K$ that satisfies (1.9) is called a **weak equilibrium flow**. It has been shown that u is a weak equilibrium flow if and only if u satisfies a generalized vector form of **Wardrop's principle**, namely, for all origin–destination $p \in P$ and all routes $s, t \in \mathbb{R}$ connecting p, i.e., $s, t \in R(p)$,

$$F(u)_s - F(u)_t \in -\text{int } C \Rightarrow u_s = 0. \tag{1.10}$$

To explain $F(u)_s$ consider $u \in \mathbb{R}^{|R|}$; then $u = (u_1, \ldots, u_s, \ldots, u_{|R|})$, and for $v = (0, \ldots, 0, u_s, 0, \ldots 0)$ it is $F(u)_s = F(u)v$.

Condition (1.10), due to its decomposed (user-oriented) form, is often more practical than the original definition (1.9), since it deals with pairs p and paths $s, t \in R(p)$ directly: When the traffic flow is in vector equilibrium, users choose only Pareto-optimal or vector minimum paths to travel on. In the case

$Y = \mathbb{R}^m, C = \mathbb{R}^m_+$ it is easily seen that (1.10) follows from (1.9). Let $u \in K$ satisfy (1.9); take $p \in P$ and $s, t \in R(p)$. Choose a flow $\overline{v}$ such that

$$\overline{v}_\lambda = \begin{cases} u_\lambda & \text{if } \lambda \neq t, \lambda \neq s, \\ 0 & \text{if } \lambda = s, \\ u_t + u_s & \text{if } \lambda = t, \end{cases} \quad (\lambda = 1, 2, \dots, |R|).$$

Then $\overline{v}_\lambda \in K$, since $\sum_{1 \leq \lambda \leq |R|} \overline{v}_\lambda = \sum_{1 \leq \lambda \leq |R|} u_\lambda = d_p$. Consequently, from (1.9), writing $F(u)$ as a matrix $(F_{\mu\lambda}) \in \mathbb{R}^{m \times |R|}$, we have $\forall \mu = 1, \dots, m$,

$$\sum_{1 \leq \lambda \leq |R|} F_{\mu\lambda}(\overline{v}_\lambda - u_\lambda) = F_{\mu t}(u_s) + F_{\mu s}(-u_s) = (F_{\mu t} - F_{\mu s})(u_s) \not< 0.$$

But from the first condition in (1.10) it is $F_{\mu s} - F_{\mu t} < 0 \; \forall \mu$, so $u_s = 0$. For further references see [86, 88, 134, 226, 254, 370, 382]. For other applications of multicriteria decision-making to economic problems see [15].

1.3 Location Problems in Town Planning

Urban development is connected with conflicting requirements of areas for dwelling, traffic, disposal of waste, recovery, trade, and others. Using methods of location theory may be one way of supporting urban planning to determine the **best location** for a special new layout or for arrangements. For references see e.g. [102, 101, 152, 156, 171, 173, 174, 170, 219, 305, 368, 369, 371].

In our investigations we consider the special situation in East German towns. One of the actual main problems of town planning is the traffic problem, due to the extremely high increase of motorized individual traffic in recent years. The lack of parking space is a part of the traffic problem. This is typical for many newly built residential areas in East Germany. Such a residential area is Halle-Silberhöhe, which was built at the beginning of the 1980s. In this district 5, 6, and 11-story blocks dominate. In our example we consider two residential sections, which count about 9300 inhabitants. This area has a size of 800 m × 1000 m. There exist 1750 parking facilities, representing a shortage of 1950. The impact is that many inhabitants park their cars on green areas.

One way to solve this problem of **inadequate parking facilities** is to build multistory parking garages. Now, the problem is to find the **best location** for such a multistory garages.

It would be possible to formulate our problem as a **real-valued location problem (Fermat–Weber problem)**, which is the problem to determine a location x of a new facility such that the weighted sum of distances between n given facilities a^i $(i = 1, \dots, n)$ and x is minimal. Using this approach it is very difficult to say how the weights λ_i $(i = 1, \dots, n)$ are to be chosen. Another difficulty may arise if the solution of the corresponding optimal location is not practically useful. Then we need new weights, and again we don't know how to choose the weights.

So the following approach is of interest: We formulate the problem as a **vector-valued (or synonymously vector or multicriteria) location problem**

$$(P) \qquad \begin{pmatrix} \|x - a^1\|_{\max} \\ \|x - a^2\|_{\max} \\ \cdots \\ \|x - a^n\|_{\max} \end{pmatrix} \longrightarrow v - \min_{x \in \mathbb{R}^2},$$

where $x, a^i \in \mathbb{R}^2$ $(i = 1, \ldots, n)$,

$$\|x\|_{\max} = \max\{|\, x_1\,|, |\, x_2\,|\},$$

and "$v - \min_{x \in \mathbb{R}^2}$" means that we study the problem of determining the set of efficient points of an objective function $f : X \longrightarrow \mathbb{R}^n$ with respect to a cone $C \subset \mathbb{R}^n$:

$$\mathrm{Eff}(f[X], C) := \{f(x) \mid x \in X, \quad f[X] \cap (f(x) - (C \setminus \{0\})) = \emptyset\}.$$

Remark 1.3.1. For applications in town planning it is important that we can choose different norms in the formulation of (P). The decision which of the norms will be used depends on the course of the roads in the city or in the district or on other influences coming from the practical background of the planning problem.

In the following example we study the problem (P) with $C = \mathbb{R}^n_+$, where $\mathbb{R}^n_+$ denotes the usual ordering cone in n-dimensional Euclidean space.

In Section 4.3 we consider a location problem in town planning, formulate a multicriteria location problem, derive optimality conditions, and present several algorithms for solving multicriteria location problems and the corresponding computer programs. It is well known that the set of solutions in vector optimization (set of efficient elements) may be large, and so we will carry out a comparison of alternatives by using a graphical representation.

For the problem in town planning mentioned above we fix the given points $a^1 = (-1.5, 3.5)$, $a^2 = (1, 3)$, $a^3 = (1, 0)$, $a^4 = (-3, -2)$, $a^5 = (3.5, -1.5)$, $a^6 = (2, 2)$, $a^7 = (-2, 2)$, $a^8 = (4, 1)$, $a^9 = (-3, 2)$.

If the decision-maker prefers the maximum norm, we get the **solution set of the multicriteria location problem** (P) as shown in Figure 1.3.1.

But if the decision-maker prefers the dual norm to the maximum norm (this norm is called the Lebesgue norm), the **solution set** has the form shown in Figure 1.3.2

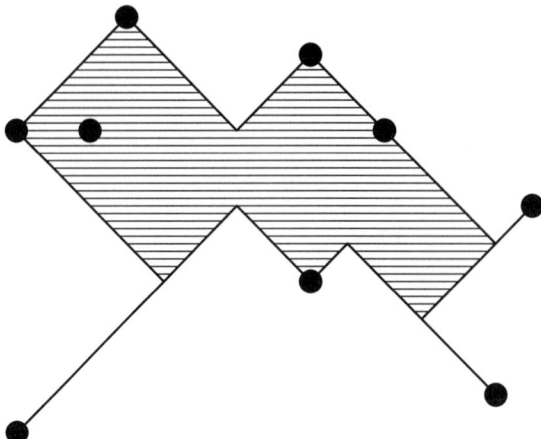

Figure 1.3.1. The set of efficient elements of the multicriteria location problem (P) with the maximum norm

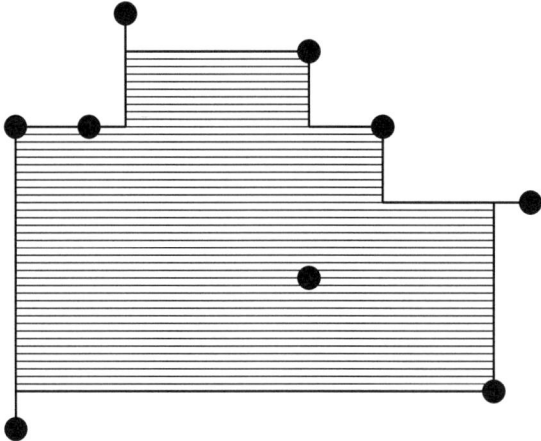

Figure 1.3.2. The set of efficient points of the multicriteria location problem (P) with the Lebesgue norm instead of the maximum norm

1.4 Multicriteria Control Problems

In control theory often one has the problem to minimize more than one objective function, for instance, a cost functional as well as the distance between the final state and a given point.

To realize this task usually one takes as objective function a weighted sum of the different objectives. However, the more natural way would be to study the set of efficient points of a vector optimization problem with the given objective functions. It is well known that the weighted sum is only a special surrogate problem to find efficient points, which has the disadvantage that in the nonconvex case one cannot find all efficient elements in this way.

In order to formulate a multicriteria control problem we introduce a system of differential equations that describe the motion of the controlled system. Let $x(t)$ be an n-dimensional phase vector that characterizes the state of the controlled system at time t. Furthermore, let $u(t)$ be an m-dimensional vector that characterizes the controlling action realized at time t. In particular, consider in a time interval $0 \leq t \leq T$ the system of ordinary differential equations with initial condition

$$\left. \begin{aligned} \frac{dx}{dt}(t) &= \varphi(t, x(t), u(t)), \\ x(0) &= x_0 \in \mathbb{R}^n. \end{aligned} \right\} \tag{1.11}$$

It is assumed that the controlling actions $u(t)$ satisfy the restrictions

$$u(t) \quad \in \quad U \subset \mathbb{R}^m.$$

The vector x is often called the **state**, and u is called the **control** of (1.11); a pair (x, u) satisfying (1.11) and the control restriction is called an **admissible solution** (or **process**) of (1.11).

Under additional assumptions it is possible to ensure the existence of a solution (x, u) of (1.11) on the whole time interval $[0, T]$ or at least almost everywhere on $[0, T]$ (compare Section 4.5).

Introducing the criteria or performance space, the objective function $f : (X \times U) \longrightarrow (Y, C_Y)$, where Y is a linear space, and $C_Y \subset Y$ is a proper convex cone, we have as **multicriteria optimal control problem** (also called multiobjective or vector-valued or vector optimal control problem)

(P): Find some control $\bar{u}$ such that the corresponding trajectory $\bar{x}$ satisfies

$$f(x, u) \notin f(\bar{x}, \bar{u}) - (C_Y \setminus \{0\})$$

for all solution pairs (x, u) of (1.11). The pair $(\bar{x}, \bar{u})$ is then called an optimal process of (P).

If we study the problem to minimize the distance f_1 of the final state $x(T)$ of the system (1.11) and a given point as well as a cost functional $\int_0^T \Phi(t, x(t), u(t))dt$ by a control u, we have a special case of (P) with

$$f(x, u) = \begin{pmatrix} f_1(x(T)) \\ \int_0^T \Phi(t, x(t), u(t))dt \end{pmatrix}.$$

In the following we explain that **cooperative differential games** are special cases of the multicriteria control problem (P). We consider a game with $n \geq 2$ players and define

$$Y := Y_1 \times Y_2 \times \cdots \times Y_n,$$

the product of the criteria spaces of each of the n players,

$$C_Y := C_{Y_1} \times C_{Y_2} \times \cdots \times C_{Y_n},$$

the product of n proper convex cones,

$$f(x, u) := (f_1(x, u), \ldots, f_n(x, u)),$$

the vector of the loss functions of the n players,

$$U := U_1 \times U_2 \times \cdots \times U_n,$$

the product of the control sets of the n players, and define

$$u := (u_1, \ldots, u_n).$$

The player j tries to minimize his cost function f_j (the utility or profit function is $-f_j$) with respect to the partial order induced by the cone C_{Y_j} influencing the system (1.11) by means of the function $u_j \in U_j$. And "cooperative game" means that a state is considered optimal if no player can reduce his costs without increasing the costs of another player. Such a situation is possible because each cost function f_j depends on the same control tuple u. The optimal process gives the **Pareto minimum** of the cost function $f(x, u)$.

It is well known that it is difficult to show the existence of optimal (or efficient) controls of (P), whereas **suboptimal controls** exist under very weak conditions. So it is important to derive some assertions for suboptimal controls. This is done in Section 4.5. There an application of a variational principle for vector optimization problems yields an ε-minimum principle for (P), which is (for $\varepsilon = 0$) closely related to Pontryagin's minimum principle (cf. Section 4.5).

1.5 Multicriteria Fractional Programming Problems

Many problems in economics can be formulated as fractional programming problems. In the papers of Hirche ([166]), Schaible ([315]), Schaible and Ibaraki ([316]) the following **real-valued fractional programming problem** is considered:

$$\varphi(x) := a_1^T x + \frac{a_2^T x}{b_2^T x} \longrightarrow \max_{x \in \mathcal{A}},$$

where $x \in \mathbb{R}^n$, $a_1, a_2, b_2 \in \mathbb{R}_+^n$, $a_1^T x, a_2^T x, \ldots$ are scalar products, $b_2^T x \neq 0$, and $\mathcal{A} = \{x \in \mathbb{R}^n \mid Bx - b \in -K\}$. The cone K is a convex cone in $\mathbb{R}^m$. The objective function $a_1^T x$ describes the production scope, and $\frac{a_2^T x}{b_2^T x}$ describes the viability.

From the practical as well as from the mathematical point of view it is more useful to formulate the problem as a **multicriteria fractional programming problem**:

$$f(x) = \left(\frac{a_1^T x}{1}, \ \frac{a_2^T x}{b_2^T x} \right) \longrightarrow v - \max_{x \in \mathcal{A}}.$$

Applying the **Dinkelbach transformation**, one gets a useful surrogate parametric optimization problem and the corresponding algorithms:

$$f_\lambda(x) = \left(a_1^T x - \lambda_1, \ a_2^T x - \lambda_2 b_2^T x \right) \longrightarrow v - \max_{x \in \mathcal{A}},$$

where $\lambda \in \mathbb{R}^2$.

In Section 4.4 we discuss possibilities to handle this transformed vector optimization problem by means of parametric optimization. We then derive a three-level dialogue algorithm in order to solve the transformed problem.

1.6 Stochastic Efficiency in a Set

Uncertainty is the key ingredient in many decision problems. Financial planning, cancer screening, and airline scheduling are just examples of areas in which ignoring uncertainty may lead to inferior or simply wrong decisions. There are many ways to model uncertainty; one that has proved particularly fruitful is to use probabilistic models.

Two methods are frequently used for modeling choice among uncertainty prospects: stochastic dominance (Ogryczak and Ruszczynski [282], [283], Whitmore and Findlay ([377]); Levy ([236]) and mean-risk analysis (Markowitz ([251]).

The **stochastic dominance** is based on an axiomatic model of risk-averse preferences: It leads to conclusions that are consistent with the axioms. Unfortunately, the stochastic dominance approach does not provide us with a simple computational recipe; it is, in fact, a multiple criteria model with a continuum of criteria.

The **mean-risk approach** quantifies the problem in a lucid form of only two criteria:

- The mean, representing the expected outcome;
- The risk, a scalar measure of the variability of outcomes.

The mean-risk model is appealing to decision-makers and allows a simple trade-off analysis, analytical or geometrical. On the other hand, mean-risk approaches are not capable of modeling the gamut of risk-averse preferences. Moreover, for typical dispersion statistics used as risk measures, the mean-risk approach may lead to inferior conclusions (compare Section 4.6).

The seminal portfolio optimization model of Markowitz ([249]) uses the variance as the risk measure in the mean-risk analysis, which results in a formulation of a quadratic programming model. Since then, many authors have pointed out that the mean-variance model is, in general, not consistent with stochastic dominance rules. The use of the semivariance rather than variance as the risk measure was already suggested by Markowitz ([250]) himself.

2

Functional Analysis over Cones

2.1 Order Structures

As seen in the introduction, we are concerned with certain sets M with order structures. In the sequel we give the basic definitions.

As usual, when M is a nonempty set, $M \times M$ is the set of ordered pairs of elements of M:

$$M \times M := \{(x_1, x_2) \mid x_1, x_2 \in M\}.$$

Definition 2.1.1. *Let M be a nonempty set and $\mathcal{R}$ a nonempty subset of $M \times M$. Then $\mathcal{R}$ is called a **binary relation** or an **order structure on** M, and $(M, \mathcal{R})$ is called a set M with **order structure** $\mathcal{R}$. The fact that $(x_1, x_2) \in \mathcal{R}$ will be denoted by $x_1 \mathcal{R} x_2$. We say that $\mathcal{R}$ is*

(a) ***reflexive*** *if $\forall x \in M : x \mathcal{R} x$,*
(b) ***transitive*** *if $\forall x_1, x_2, x_3 \in M : x_1 \mathcal{R} x_2, x_2 \mathcal{R} x_3 \Rightarrow x_1 \mathcal{R} x_3$,*
(c) ***antisymmetric*** *if $\forall x_1, x_2 \in M : x_1 \mathcal{R} x_2, x_2 \mathcal{R} x_1 \Rightarrow x_1 = x_2$.*

*$\mathcal{R}$ is called a **preorder** on M if $\mathcal{R}$ satisfies (b) and a **partial order** on M if $\mathcal{R}$ satisfies (a), (b), and (c). In both cases the fact that $(x_1, x_2) \in \mathcal{R}$ is denoted by $x_1 \leq_{\mathcal{R}} x_2$, or simply $x_1 \leq x_2$ if there is no risk of confusion. The binary relation $\mathcal{R}$ is called a **linear** or **total order** if $\mathcal{R}$ is a partial order and*

(d) *for all $x_1, x_2 \in M$ either $x_1 \leq_{\mathcal{R}} x_2$ or $x_2 \leq_{\mathcal{R}} x_1$,*

*i.e., any two elements of M are **comparable**. Finally, if each nonempty subset M' of M has a first element x' (meaning that $x' \in M'$ and $x' \leq_{\mathcal{R}} x \ \forall x \in M'$), then M is called **well-ordered**.[1]*

---

[1] Recall Zermelo's theorem: For every nonempty set M there exists a partial order $\mathcal{R}$ on M such that $(M, \mathcal{R})$ is well-ordered.

An important example of a relation is

$$\Delta_M := \{(x, x) \mid x \in M\}.$$

It is obvious that Δ_M is reflexive, transitive, and antisymmetric, but (d) is satisfied only when M is a singleton. Recall that the **inverse** of the relation $\mathcal{R} \subset M \times M$ is the relation

$$\mathcal{R}^{-1} := \{(x_1, x_2) \in M \times M \mid (x_2, x_1) \in \mathcal{R}\};$$

if $\mathcal{S}$ is another relation on M, the **composition** of $\mathcal{R}$ and $\mathcal{S}$ is the relation

$$\mathcal{S} \circ \mathcal{R} := \{(x_1, x_3) \mid \exists x_2 \in M \ : \ (x_1, x_2) \in \mathcal{R}, \ (x_2, x_3) \in \mathcal{S}\}.$$

Using this notation, the conditions (a), (b), (c), and (d) are equivalent to $\Delta_M \subset \mathcal{R}$, $\mathcal{R} \circ \mathcal{R} \subset \mathcal{R}$, $\mathcal{R} \cap \mathcal{R}^{-1} \subset \Delta_M$ and $\mathcal{R} \cup \mathcal{R}^{-1} = M \times M$, respectively.

Definition 2.1.2. *Let $\mathcal{R}$ be an order structure on the nonempty set M and let $M_0 \subset M$ be nonempty. The element $x_0 \in M_0$ is called a **maximal (minimal) element** of M_0 **relative to** $\mathcal{R}$ if*

$$\forall x \in M_0 \ : \ x_0 \mathcal{R} x \Rightarrow x \mathcal{R} x_0 \quad (\forall x \in M_0 \ : \ x \mathcal{R} x_0 \Rightarrow x_0 \mathcal{R} x). \qquad (2.1)$$

The class of maximal (minimal) elements of M_0 with respect to (w.r.t. for short) $\mathcal{R}$ is denoted by $\mathrm{Max}(M_0, \mathcal{R})$ *($\mathrm{Min}(M_0, \mathcal{R})$).*

Note that x_0 is a maximal element of M_0 w.r.t. $\mathcal{R}$ iff x_0 is a minimal element of M_0 w.r.t. $\mathcal{R}^{-1}$, and so $\mathrm{Max}(M_0, \mathcal{R}) = \mathrm{Min}(M_0, \mathcal{R}^{-1})$.

Remark 2.1.3. If the order structure $\mathcal{R}$ in Definition 2.1.2 is antisymmetric, then $x_0 \in M_0$ is maximal (minimal) iff

$$\forall x \in M_0 \ : \ x_0 \mathcal{R} x \Rightarrow x = x_0 \quad (\forall x \in M_0 \ : \ x \mathcal{R} x_0 \Rightarrow x_0 = x). \qquad (2.2)$$

Partial orders $\mathcal{R}$ play an important role for introducing a solution concept in multicriteria optimization (compare Chapter 3 and 4). In these chapters we denote, as usual, sets of maximal or minimal elements by $\mathrm{Eff}_{\mathrm{Max}}(M_0, \mathcal{R})$ and $\mathrm{Eff}_{\mathrm{Min}}(M_0, \mathcal{R})$, respectively. Furthermore, these sets are called sets of (maximal- or minimal-) **efficient points** of M_0 with respect to $\mathcal{R}$. For brevity we sometimes leave out maximal, minimal, Max or Min.

Remark 2.1.4. When $\mathcal{R}$ is an order structure on M and $\emptyset \neq M_0 \subset M$, then $\mathcal{R}_0 := \mathcal{R} \cap (M_0 \times M_0)$ is an order structure on M_0; in such a situation the set M_0 will always be endowed with the order structure $\mathcal{R}_0$ if not stated explicitly otherwise. If $\mathcal{R}$ is a preorder (partial order, linear order) on M, then $\mathcal{R}_0$ is a preorder (partial order, linear order) on M_0. So x_0 is a maximal (minimal) element of M_0 relative to $\mathcal{R}$ iff x_0 is a maximal (minimal) element of M_0 relative to $\mathcal{R}_0$.

Example 2.1.5. (1) Let X be a nonempty set and let $M := \mathcal{P}(X)$ be the class of subsets of X. The binary relation $\mathcal{R} := \{(A, B) \in M \times M \mid A \subset B\}$ is a partial order on M; if X has at least two elements, then $\mathcal{R}$ is not a linear order.

(2) Let $\mathbb{N}$ be the set of nonnegative integers and

$$\mathcal{R}_\mathbb{N} := \{(n_1, n_2) \in \mathbb{N} \times \mathbb{N} \mid \exists p \in \mathbb{N} : n_2 = n_1 + p\}.$$

Then $\mathbb{N}$ is well-ordered by $\mathcal{R}_\mathbb{N}$. Of course, $\mathcal{R}_\mathbb{N}$ defines the usual order relation on $\mathbb{N}$, and $n_1 \mathcal{R}_\mathbb{N} n_2$ will always be denoted by $n_1 \leq n_2$ or, equivalently, $n_2 \geq n_1$.

(3) Let $\mathbb{R}$ be the set of real numbers and let $\mathbb{R}_+ := [0, \infty[$ be the set of nonnegative real numbers. The usual order structure on $\mathbb{R}$ is defined by

$$\mathcal{R}_1 := \{(x_1, x_2) \in \mathbb{R} \times \mathbb{R} \mid \exists y \in \mathbb{R}_+ : x_2 = x_1 + y\}.$$

Then $\mathcal{R}_1$ is a linear order on $\mathbb{R}$, but $\mathbb{R}$ is not well-ordered by $\mathcal{R}_1$. As usual, the fact $x_1 \mathcal{R}_1 x_2$ will always be denoted by $x_1 \leq x_2$ or, equivalently, $x_2 \geq x_1$.

(4) Let $n \in \mathbb{N}$, $n \geq 2$. Consider the binary relation $\mathcal{R}_n$ on $\mathbb{R}^n$ defined by

$$\mathcal{R}_n := \{(x, y) \in \mathbb{R}^n \times \mathbb{R}^n \mid \forall i \in \overline{1, n} : x_i \leq y_i\},$$

where $x = (x_1, \ldots, x_n)$, $y = (y_1, \ldots, y_n)$ and $\overline{1, n} := \{i \in \mathbb{N} \mid 1 \leq i \leq n\}$. Then $\mathcal{R}_n$ is a partial order on $\mathbb{R}^n$, but $\mathcal{R}_n$ is not a linear order; for example, the elements e_1 and e_2 are not comparable w.r.t. $\mathcal{R}_n$, where $e_i := (0, \ldots, 0, 1, 0, \ldots, 0) \in \mathbb{R}^n$ (all components of e_i being 0 excepting the ith one, which is 1).

Remark 2.1.6. Notice that every well-ordered subset W of $\mathbb{R}$ (endowed with its usual partial order recalled above) is at most countable. Indeed, every element $y \in W$, excepting the greatest element w of W (if it exists), has a successor $s(y) \in W$; of course, if $y, y' \in W$, $y < y'$, then $s(y) \leq y'$. So, fixing $q_y \in \mathbb{Q}$ such that $y < q_y < s(y)$ for $y \in W \setminus \{w\}$, we get an injective function from $W \setminus \{w\}$ into $\mathbb{Q}$, and so W is at most countable.

Even when $\mathcal{R}$ is a partial order on M, a nonempty subset M_0 of M may have zero, one, or several maximal elements, but if $\mathcal{R}$ is a linear order, then every subset has at most one maximal (minimal) element.

Definition 2.1.7. *Let $\emptyset \neq M_0 \subset M$ and let $\mathcal{R}$ be an order structure on M. We say that M_0 is **lower (upper) bounded** (w.r.t. $\mathcal{R}$) if there exists $a \in M$ such that $a\mathcal{R}x$ ($x\mathcal{R}a$) for every $x \in M_0$; a is called a **lower (upper) bound** of M_0 (w.r.t. $\mathcal{R}$). If, moreover, $\mathcal{R}$ is a partial order, we say that $a \in M$ is the **infimum (supremum)** of M_0 if a is a lower (upper) bound of M_0 and for any lower (upper) bound a' of M_0 we have that $a'\mathcal{R}a$ ($a\mathcal{R}a'$).*

An important problem in vector optimization is the existence of maximal elements w.r.t. order structures. Of fundamental importance in this sense is **Zorn's lemma** (or Zorn's axiom).

Axiom 2.1.8 (Zorn) *Let $(M, \leq)$ be a reflexively preordered set. If every nonempty totally ordered subset of M is upper bounded, then M has maximal elements.*

Very often, order structures are defined on subsets of linear spaces. We will consider only **real** linear spaces, so later on we shall omit the word real.

Definition 2.1.9. *Let X be a nonempty set. We say that X is a **(real) linear space** or a **set equipped with linear structure** if an addition (that is, a mapping $+ : X \times X \to X$) and a multiplication by scalars (that is, a mapping $\cdot : \mathbb{R} \times X \to X$) are defined satisfying the following conditions:*

(a) $\forall x, y, z \in X \; : \; (x + y) + z = x + (y + z)$ *(associativity)*,
(b) $\forall x, y \in X \; : \; x + y = y + x$ *(commutativity)*,
(c) $\exists 0 \in X, \; \forall x \in X \; : \; x + 0 = x$ *(null element)*,
(d) $\forall x \in X, \; \exists x' \in X \; : \; x + x' = 0$; *we write $x' = -x$,*
(e) $\forall x, y \in X, \; \forall \lambda \in \mathbb{R} \; : \; \lambda(x + y) = \lambda x + \lambda y$,
(f) $\forall x \in X, \; \forall \lambda, \mu \in \mathbb{R} \; : \; (\lambda + \mu)x = \lambda x + \mu x$,
(g) $\forall x \in X, \; \forall \lambda, \mu \in \mathbb{R} \; : \; \lambda(\mu x) = (\lambda \mu)x$,
(h) $\forall x \in X \; : \; 1x = x$ *(unity element)*.

A linear space is often called a **vector space**. Defining $x + y := (x_1 + y_1, \ldots, x_n + y_n)$ and $\lambda x := (\lambda x_1, \ldots, \lambda x_n)$ for $x, y \in \mathbb{R}^n$ and $\lambda \in \mathbb{R}$, we have that $\mathbb{R}^n$ is a linear space.

In the sequel we shall use the following notation: Let A, B be nonempty subsets of a linear space X, $\emptyset \neq \Gamma \subset \mathbb{R}$, $x \in X$, and $\lambda \in \mathbb{R}$; then

$$A + B := \{x + y \mid x \in A, \, y \in B\}, \quad A + \emptyset := \emptyset + A := \emptyset, \quad x + A := \{x\} + A,$$

and

$$\Gamma \cdot A := \{\gamma x \mid \gamma \in \Gamma, x \in A\}, \quad \lambda A = \{\lambda\} \cdot A, \quad \Gamma \cdot \emptyset = \emptyset.$$

In particular, $A - B = A + (-1)B = \{x - y \mid x \in A, \, y \in B\}$. If, moreover, I is a nonempty set and $\emptyset \neq A_i \subset X$ for every $i \in I$, we have

$$\lambda(A + B) = \lambda A + \lambda B, \quad \lambda\left(\bigcup_{i \in I} A_i\right) = \bigcup_{i \in I} \lambda A_i, \quad \lambda\left(\bigcap_{i \in I} A_i\right) = \bigcap_{i \in I} \lambda A_i.$$

Let X be a linear space and $\emptyset \neq X_0 \subset X$; recall that X_0 is a **linear subspace** of X if $\lambda x + \mu y \in X_0$ for all $x, y \in X_0$ and $\lambda, \mu \in \mathbb{R}$. This shows that X_0 is stable w.r.t. addition and multiplication by scalars, and X_0 endowed with these operations is itself a linear space. Also recall that the nonempty set $M \subset X$ is **affine** (or a **linear manifold**) if $\lambda x + (1 - \lambda)y \in M$ for all

$x, y \in M$ and $\lambda \in \mathbb{R}$; it is known that M is affine iff there exist $x_0 \in X$ and X_0 a linear subspace of X such that $M = x_0 + X_0$ (X_0 being called the parallel linear subspace to M). The nonempty set C of X is called **convex** if $[x, y] := \{\lambda x + (1 - \lambda)y \mid \lambda \in [0, 1]\} \subset C$ for all $x, y \in C$. By convention the empty set $\emptyset$ is considered to be affine and convex. It is obvious that a linear subspace is affine and an affine set is convex; moreover, any intersection of linear subspaces, affine sets, or convex sets is a linear subspace, an affine set, or a convex set, respectively. These properties give us the possibility to introduce the **linear hull**, the **affine hull**, and the **convex hull** of a nonempty set $A \subset X$ as being, respectively,

$$\operatorname{lin} A := \bigcap \{Y \subset X \mid A \subset Y, \ Y \text{ linear subspace}\},$$

$$\operatorname{aff} A := \bigcap \{M \subset X \mid A \subset M, \ M \text{ linear manifold}\},$$

$$\operatorname{conv} A := \bigcap \{C \subset X \mid A \subset C, \ C \text{ convex set}\}.$$

It is obvious (and well known) that for $X = \mathbb{R}^n$ and $\mathcal{R} = \mathcal{R}_n$ (from Example 2.1.5 (4)) one has

$$\forall x_1, x_2 \in X, \ \forall \lambda \in \mathbb{R} \ : \ x_1 \mathcal{R} x_2, \ 0 \leq \lambda \Rightarrow \lambda x_1 \mathcal{R} \lambda x_2, \tag{2.3}$$

$$\forall x_1, x_2, x \in X \ : \ x_1 \mathcal{R} x_2 \Rightarrow (x_1 + x) \mathcal{R}(x_2 + x). \tag{2.4}$$

It is simple to give examples of relations satisfying (2.4). In fact, a nonempty relation $\mathcal{R}$ on the linear space X satisfies (2.4) if and only if there exists $\emptyset \neq D \subset X$ such that $\mathcal{R} = \mathcal{R}_D$, where

$$\mathcal{R}_D := \{(x_1, x_2) \in X \times X \mid x_2 - x_1 \in D\}.$$

Moreover, $\mathcal{R}_D$ is reflexive if and only if $0 \in D$, and $\mathcal{R}_D$ is transitive if and only if $D + D \subset D$.

Definition 2.1.10. *Let $\mathcal{R}$ be an order structure on the linear space X; we say that $\mathcal{R}$ is **compatible** with the linear structure of X if (2.3) and (2.4) hold.*

Fortunately, there exists a large class of relations $\mathcal{R}$ on a linear space X that are compatible with the linear structure of X: This is the class of relations determined by cones of X.

Definition 2.1.11. *A nonempty set $C \subset X$ is a **cone** if*

(a) $\forall x \in C, \ \forall \lambda \in \mathbb{R}_+ \ : \ \lambda x \in C.$

Of course, if C is a cone, then $0 \in C$. The cone C is called:
(b) **convex** *if $\forall x_1, x_2 \in C \ : \ x_1 + x_2 \in C$,*
(c) **nontrivial** *or **proper** if $C \neq \{0\}$ and $C \neq X$,*
(d) **reproducing** *if $C - C = X$,*
(e) **pointed** *if $C \cap (-C) = \{0\}$.*

Note that the cone C satisfies condition (b) in the definition above if and only if C is a convex set.

Example 2.1.12. (1) Let

$$\mathbb{R}^n_+ := \{x \in \mathbb{R}^n \mid x_i \geq 0 \; \forall i \in \overline{1,n}\} = \{x \in \mathbb{R}^n \mid (0,x) \in \mathcal{R}_n\}. \tag{2.5}$$

$\mathbb{R}^n_+$ is obviously a cone in the linear space $\mathbb{R}^n$, which fulfills all the conditions of Definition 2.1.11.

(2)   Let $C[0,1]$ be the linear space of all real functions defined and continuous on the interval $[0,1] \subset \mathbb{R}$; addition and multiplication by scalars are defined, naturally, by

$$(x+y)(t) = x(t) + x(t), \quad (\lambda x)(t) = \lambda x(t) \quad \forall t \in [0,1]$$

for $x, y \in C[0,1]$ and $\lambda \in \mathbb{R}$. Then

$$C_+[0,1] := \{x \in C[0,1] \mid x(t) \geq 0 \; \forall t \in [0,1]\} \tag{2.6}$$

is a convex, nontrivial, pointed, and reproducing cone in $C[0,1]$. Note that the set

$$Q := \{x \in C_+[0,1] \mid x \text{ is nondecreasing}\} \tag{2.7}$$

is also a convex, nontrivial, and pointed cone in the space $C[0,1]$, but it doesn't satisfy condition (d) from Definition 2.1.11: $Q - Q$ is the proper linear subspace of all functions with bounded variation of $C[0,1]$.

(3)   Consider the set $C \subset \mathbb{R}^n$ defined by

$$C := \{x = (x_1, \ldots, x_n)^T \in \mathbb{R}^n \mid x_1 > 0, \text{ or}$$
$$x_1 = 0, \; x_2 > 0, \text{ or}$$
$$\cdots$$
$$x_1 = \cdots = x_{n-1} = 0, \; x_n > 0, \text{ or}$$
$$x = 0\};$$

C satisfies all the conditions of Definition 2.1.11.

With the help of cones we can characterize compatibility between linear and order structures:

Theorem 2.1.13. *Let X be a linear space and let C be a cone in X. Then the relation*

$$\mathcal{R}_C := \{(x_1, x_2) \in X \times X \mid x_2 - x_1 \in C\} \tag{2.8}$$

is reflexive and satisfies (2.3) and (2.4). Moreover, C is convex if and only if $\mathcal{R}_C$ is transitive, and, respectively, C is pointed if and only if $\mathcal{R}_C$ is antisymmetric. Conversely, if $\mathcal{R}$ is a reflexive relation on X satisfying (2.3) and (2.4), then $C := \{x \in X \mid 0\mathcal{R}x\}$ is a cone and $\mathcal{R} = \mathcal{R}_C$.

PROOF. Let us denote the fact that $x_1 \mathcal{R}_C x_2$ by $x_1 \leq x_2$.

(i) $\mathcal{R}_C$ is reflexive because $0 \in C$.

(ii) $\mathcal{R}_C$ satisfies (2.3); indeed, if $x_1, x_2 \in X$ and $\lambda \in \mathbb{R}_+$ then

$$x_1 \leq x_2 \underset{(2.8)}{\Longrightarrow} x_2 - x_1 \in C \underset{\text{Def. 2.1.11(a)}}{\Longrightarrow} \lambda(x_2 - x_1) \in C$$

$$\underset{\text{Def. 2.1.9(e)}}{\Longrightarrow} \lambda x_2 - \lambda x_1 \in C \underset{(2.8)}{\Longrightarrow} \lambda x_1 \leq \lambda x_2.$$

(iii) $\mathcal{R}_C$ satisfies (2.4); indeed, if $x_1, x_2, x \in X$ then

$$x_1 \leq x_2 \underset{(2.8)}{\Longrightarrow} x_2 - x_1 \in C \underset{\text{Def. 2.1.9}}{\Longrightarrow} (x_2 + x) - (x_1 + x) \in C \underset{(2.8)}{\Longrightarrow} x_1 + x \leq x_2 + x.$$

(iv) If C is convex, then $\mathcal{R}_C$ is transitive:

$$x_1 \leq x_2, \ x_2 \leq x_3 \underset{(2.8)}{\Longrightarrow} x_2 - x_1 \in C, \ x_3 - x_2 \in C \underset{\text{Def. 2.1.11(b)}}{\Longrightarrow} x_3 - x_1 \in C$$

$$\underset{(2.8)}{\Longrightarrow} x_1 \leq x_3.$$

Conversely, assume that $\mathcal{R}_C$ is transitive. Then

$$x_1, x_2 \in C \underset{(2.8)}{\Longrightarrow} 0 \leq x_1, \ x_1 \leq x_1 + x_2 \underset{\text{Def. 2.1.1(b)}}{\Longrightarrow} 0 \leq x_1 + x_2 \underset{(2.8)}{\Longrightarrow} x_1 + x_2 \in C.$$

(v) If C is pointed, then $\mathcal{R}_C$ is antisymmetric:

$$x_1 \leq x_2, \ x_2 \leq x_1 \underset{(2.8)}{\Longrightarrow} x_2 - x_1 \in C \cap (-C) = \{0\} \Longrightarrow x_2 = x_1.$$

Conversely, assume that $\mathcal{R}_C$ is antisymmetric. Then

$$x \in C \cap (-C) \underset{(2.8)}{\Longrightarrow} 0 \leq x, \ x \leq 0 \underset{\text{Def. 2.1.1(c)}}{\Longrightarrow} x = 0.$$

(vi) Let $\mathcal{R}$ be a reflexive compatible order structure in X and consider

$$C := \{x \in X \mid 0\mathcal{R}x\}. \tag{2.9}$$

C is really a cone. Indeed, let $\lambda \in \mathbb{R}_+$ and $x \in C$; then $0\mathcal{R}x$. Since $\mathcal{R}$ fulfills (2.3), we obtain that $\lambda 0 \mathcal{R} \lambda x$, and so $\lambda x \in C$. Moreover,

$$x_1 \mathcal{R} x_2 \underset{(2.4)}{\Longleftrightarrow} (x_1 - x_1)\mathcal{R}(x_2 - x_1) \underset{(2.9)}{\Longleftrightarrow} x_2 - x_1 \in C \underset{(2.8)}{\Longleftrightarrow} x_1 \mathcal{R}_C x_2.$$

Therefore $\mathcal{R} = \mathcal{R}_C$.   □

The preceding theorem shows that when $\emptyset \neq C \subset X$, the relation $\mathcal{R}_C$ defined by (2.8) is a reflexive preorder iff C is a convex cone, and $\mathcal{R}_C$ is a partial order iff C is a pointed convex cone.

Also note that $\mathcal{R}_{\mathbb{R}_+^n} = \mathcal{R}_n$ (defined in Example 2.1.5 (4)), while the relation $\mathcal{R}_C$ with $C \subset \mathbb{R}^n$ defined in Example 2.1.12 (3) is a linear order, called the **lexicographic order** on $\mathbb{R}^n$.

The cone $\mathbb{R}^n_+ \subset \mathbb{R}^n$ (defined in (2.5)) has an interesting property: Taking $B := \{x \in \mathbb{R}^n_+ \mid \sum_{i=1}^n x_i = 1\}$, for every $x \in \mathbb{R}^n_+ \setminus \{0\}$ there exist unique elements $b \in B$ and $\lambda > 0$ such that $x = \lambda b$. Indeed, just take $\lambda := x_1 + \cdots + x_n$ (> 0) and $b := \lambda^{-1} x$. Taking into account this example, we introduce the following definition.

Definition 2.1.14. *Let X be a linear space and C a nontrivial convex cone in X. A nonempty convex subset B of C is called a **base** for C if each nonzero element $x \in C$ has a unique representation of the form $x = \lambda b$ with $\lambda > 0$ and $b \in B$.*

Note that if B is a base of the nontrivial convex cone C, then $0 \notin B$. Indeed, in the contrary case, taking $b \in B \setminus \{0\}$ we have that $b = 1 \cdot b = 2 \cdot (\frac{1}{2}b)$ with $b, \frac{1}{2}b \in B$, contradicting the uniqueness of the representation of $b \in C \setminus \{0\}$.

Theorem 2.1.15. *Let C be a nontrivial convex cone in the linear space X and let $B \subset X$ be a convex set. The following assertions are equivalent:*

(i) *B is a base of C;*
(ii) *$C = \mathbb{R}_+ B$ and $0 \notin \operatorname{aff} B$;*
(iii) *there exists a linear functional $\varphi : X \to \mathbb{R}$ such that $\varphi(x) > 0$ for every $x \in C \setminus \{0\}$ and $B = \{x \in C \mid \varphi(x) = 1\}$.*

PROOF. (i) $\Rightarrow$ (ii) Let B be a base for C; then, by Definition 2.1.14, $C = \mathbb{R}_+ B$. Since B is convex, $\operatorname{aff} B = \{\mu b + (1 - \mu)b' \mid b, b' \in B, \ \mu \in \mathbb{R}\}$. Assume that $0 \in \operatorname{aff} B$, and so $0 = \mu b + (1 - \mu)b'$ for some $b, b' \in B$, $\mu \in \mathbb{R}$; since $0 \notin B$, $\mu \notin [0, 1]$. Therefore there exist $\mu_0 > 1$, $b_0, b'_0 \in B$ such that $\mu_0 b_0 = (\mu_0 - 1)b'_0 \in C$, contradicting the definition of the base. Hence $0 \notin \operatorname{aff} B$.

(ii) $\Rightarrow$ (iii) Assume that $C = \mathbb{R}_+ B$ and $0 \notin \operatorname{aff} B$. Consider $b_0 \in B$ and $X_0 := \operatorname{aff} B - b_0$. Then X_0 is a linear subspace of X and $b_0 \notin X_0$. Let $L_0 \subset X_0$ be a base of X_0. Then $L_0 \cup \{b_0\}$ is linearly independent; let us complete it to a base L of X. There exists a unique linear function $\varphi : X \to \mathbb{R}$ such that $\varphi(x) = 0$ for any $x \in L \setminus \{b_0\}$ and $\varphi(b_0) = 1$. Since $\operatorname{aff} B = b_0 + X_0$, we have that $\varphi(x) = 1$ for every $x \in \operatorname{aff} B$, and so $B \subset \{x \in C \mid \varphi(x) = 1\}$. Conversely, let $x \in C$ be such that $\varphi(x) = 1$. Then $x = tb$ for some $t > 0$ and $b \in B$. It follows that $1 = \varphi(x) = t\varphi(b) = t$, and so $x \in B$.

(iii) $\Rightarrow$ (i) Assume that $\varphi : X \to \mathbb{R}$ is linear, $\varphi(x) > 0$ for every $x \in C \setminus \{0\}$, and $B = \{x \in C \mid \varphi(x) = 1\}$. Consider $x \in C \setminus \{0\}$ and take $t := \varphi(x) > 0$ and $b := t^{-1}x$; of course, $x = tb$. Because $b \in C$ and $\varphi(b) = 1$ we have that $b \in B$. Assume that $x = t'b'$ for some $t' > 0$ and $b' \in B$. Then $t = \varphi(x) = t'\varphi(b') = t'$, whence $b = b'$. Hence every nonnull element x of C has a unique representation tb with $t > 0$ and $b \in B$; i.e., B is a base of C. $\qquad\square$

Often it is useful to have also a topological structure on the set under consideration.

Definition 2.1.16. *Let X be a nonempty set. A **topology** τ on X is a family of subsets of X satisfying the following conditions:*

(a) *every union of sets of τ belongs to τ,*
(b) *every finite intersection of sets of τ belongs to τ,*
(c) *the empty set $\emptyset$ and the whole set X belong to τ.*

The elements of τ are called **open** sets, and X equipped with τ is called a **topological space** and is denoted by (X, τ).

The dual notion for open set is that of closed set. So, the subset $A \subset (X, \tau)$ is **closed** if $X \setminus A$ is open; it follows easily that every intersection of closed sets is closed and every finite union of closed sets is closed. The **interior** and the **closure** of the subset A of the topological space (X, τ) are defined, respectively, by

$$\text{int } A := \bigcup \{D \subset X \mid D \subset A, \ D \text{ open}\},$$

$$\text{cl } A := \overline{A} := \bigcap \{F \subset X \mid A \subset F, \ F \text{ closed}\}.$$

It is obvious that $\text{int } A$ is open and $\text{cl } A$ is closed.

Important examples of topological spaces are furnished by metric spaces. Recall that a **metric** on a nonempty set X is a mapping $\rho : X \times X \to \mathbb{R}_+$ having the properties

$$\forall x, y \in X : \rho(x, y) = 0 \Leftrightarrow x = y,$$
$$\forall x, y \in X : \rho(x, y) = \rho(y, x),$$
$$\forall x, y, z \in X : \rho(x, z) \leq \rho(x, y) + \rho(y, z).$$

The set X equipped with the metric ρ is called a **metric space** and is denoted by (X, ρ).

Having the metric space (X, ρ), $x \in X$ and $r > 0$, consider the (closed) ball of center x and radius r :

$$B(x, r) := B_\rho(x, r) := \{y \in X \mid \rho(x, y) \leq r\}.$$

Define

$$\tau_\rho := \{\emptyset\} \cup \{D \subset X \mid \forall x \in D, \ \exists r > 0 \ : \ B(x, r) \subset D\}.$$

It is easy to verify that τ_ρ is a topology on X, and so X becomes a topological space; a metric space (X, ρ) will always be endowed with the topology τ_ρ if not stated explicitly otherwise. Note that $B(x, r)$ is a closed set, while the open ball $\{y \in X \mid \rho(x, y) < r\}$ is an open set.

Taking $X = \mathbb{R}^n$ and $\rho(x, y)$ the usual Euclidean distance between $x \in \mathbb{R}^n$ and $y \in \mathbb{R}^n$, ρ is a metric on $\mathbb{R}^n$, and the open sets w.r.t. ρ are the usual open sets of $\mathbb{R}^n$. As in $\mathbb{R}^n$, in an arbitrary topological space (X, τ) we define other topological notions, that of neighborhood being very useful.

Definition 2.1.17. *Let (X, τ) be a topological space and $x \in X$. The subset U of X is a **neighborhood** of x (relative to τ) if there exists an open set $O \in \tau$ such that $x \in O \subset U$. The class of all neighborhoods of x will be denoted by $\mathcal{N}_\tau(x)$, or simply $\mathcal{N}(x)$. A subset $\mathcal{B}(x)$ of $\mathcal{N}_\tau(x)$ is called a **base of neighborhoods** (filter base) of x relative to τ if for every $U \in \mathcal{N}_\tau(x)$ there exists $V \in \mathcal{B}(x)$ such that $V \subset U$. One says that (X, τ) is **first-countable** if every element $x \in X$ has an (at most) countable neighborhood base.*

For the topological space (X, τ) and a neighborhood base $\mathcal{B}(x)$ of x for every $x \in X$, the family $\{\mathcal{B}(x) \mid x \in X\}$ has the following properties:

(NB1) $\forall x \in X, \forall U \in \mathcal{B}(x) : x \in U,$

(NB2) $\forall x \in X, \forall U_1, U_2 \in \mathcal{B}(x), \exists U_3 \in \mathcal{B}(x) : U_3 \subset U_1 \cap U_2,$

(NB3) $\forall x \in X, \forall U \in \mathcal{B}(x), \exists V \in \mathcal{B}(x), \forall y \in V, \exists W \in \mathcal{B}(y) : W \subset U.$

Conversely, if for every $x \in X$ one has a nonempty family $\mathcal{B}(x) \subset \mathcal{P}(X)$ such that $\{\mathcal{B}(x) \mid x \in X\}$ satisfies the conditions (NB1)–(NB3), then there exists a unique topology τ on X such that $\mathcal{B}(x)$ is a neighborhood base for x, for every $x \in X$; moreover,

$$\forall x \in X : \mathcal{N}_\tau(x) = \{U \subset X \mid \exists V \in \mathcal{B}(x) : V \subset U\}.$$

In fact, the topology (X, ρ) of a metric space is constructed in this manner: The family $\{\mathcal{B}_0(x) \mid x \in X\}$ with $\mathcal{B}_0(x) := \{B_\rho(x, 1/n) \mid n \in \mathbb{N} \setminus \{0\}\}$ satisfies the conditions (NB1)–(NB3) above; the topology defined by $\{\mathcal{B}_0(x) \mid x \in X\}$ is nothing else but the topology τ_ρ defined above. Since $\mathcal{B}_0(x)$ is countable for every $x \in X$, we obtain that every metric space is first-countable.

As seen above (property (NB2)), taking $\mathcal{B}(x)$ a neighborhood base of $x \in (X, \tau)$ and considering the relation $\succeq$ on $\mathcal{B}(x)$ defined by $U_1 \succeq U_2$ iff $U_1 \subset U_2$, then for all $U_1, U_2 \in \mathcal{B}(x)$ there exists $U_3 \in \mathcal{B}(x)$ such that $U_3 \succeq U_1$ and $U_3 \succeq U_2$. This is the prototype of a directed set. So, we say that the nonempty set I is **directed** by $\succeq$ if $\succeq$ is a partial order on I and

$$\forall i_1, i_2 \in I, \exists i_3 \in I : i_3 \succeq i_1 \text{ and } i_3 \succeq i_2.$$

Note that if $(I', \succeq')$ and $(I'', \succeq'')$ are directed sets, then $I' \times I''$ is directed by the relation $\succeq$ defined by $(i', i'') \succeq (j', j'')$ iff $i' \succeq' j'$ and $i'' \succeq'' j''$. Another important example of directed set is $\mathbb{N}$ endowed with its usual order $\geq$. A subset J of a directed set $(I, \succeq)$ is called **cofinal** if for every $i \in I$ there exists $j \in J$ such that $j \succeq i$; it follows immediately that $(J, \succeq)$ is directed if J is a cofinal subset of I. The above discussion shows that $(\mathcal{B}(x), \succeq)$ is directed if $\mathcal{B}(x)$ is a neighborhood base for $x \in (X, \tau)$. This consideration is the base of the following definition.

Definition 2.1.18. *A mapping $\varphi : I \to X$, where $(I, \succeq)$ is a directed set, is called a **net** or a **generalized sequence** of X; generally, the net φ is denoted by $(x_i)_{i \in I}$, where $x_i := \varphi(i)$. When X is equipped with a topology τ we say that the net $(x_i)_{i \in I} \subset X$ **converges** to $x \in X$ if*

$$\forall V \in \mathcal{N}_\tau(x), \ \exists i_V \in I, \ \forall i \succeq i_V \ : \ x_i \in V.$$

We denote this fact by $(x_i) \overset{\tau}{\to} x$ or simply $(x_i) \to x$, and x is called a **limit** of (x_i).

Lemma 2.1.19. *Let* (X, τ) *be a topological space. Then every convergent net of* X *has a unique limit if and only if*

$$\forall x, y \in X, \ x \neq y, \ \exists U \in \mathcal{N}_\tau(x), \ V \in \mathcal{N}_\tau(y) \ : \ U \cap V = \emptyset, \qquad (2.10)$$

i.e., (X, τ) *is* **Hausdorff**.

PROOF. Suppose that (2.10) does not hold. Then there exist $x, y \in X, \ x \neq y$, such that

$$\forall U \in \mathcal{N}_\tau(x), \ \forall V \in \mathcal{N}_\tau(y), \ \exists x_{U,V} \in U \cap V.$$

As seen above, $I := \mathcal{N}_\tau(x) \times \mathcal{N}_\tau(y)$ is directed, and so $(x_{U,V})_{(U,V) \in I}$ is a net in X. It is obvious that $(x_{U,V}) \overset{\tau}{\to} x$ and $(x_{U,V}) \overset{\tau}{\to} y$, which shows that the limit of nets is not unique in this case.

Suppose now that (2.10) holds and assume that the net $(x_i)_{i \in I} \subset X$ converges to x and y with $x \neq y$. Take $U \in \mathcal{N}_\tau(x)$ and $V \in \mathcal{N}_\tau(y)$ such that $U \cap V = \emptyset$. Then there exist $i_U, i_V \in I$ such that $x_i \in U$ for $i \succeq i_U$ and $x_i \in V$ for $i \succeq i_V$. Because I is directed, there exists $i_0 \in I$ such that $i_0 \succeq i_U$ and $i_0 \succeq i_V$. We thus obtain the contradiction that $x_{i_0} \in U \cap V$. The proof is complete. $\qquad\square$

A useful notion is that of subnet; so, we say that the net $(y_k)_{k \in K}$ is a **subnet** of the net $(x_i)_{i \in I}$ if there exists a mapping $\psi : (K, \succ) \to (I, \succeq)$ such that $y_k = x_{\psi(k)}$ for every $k \in K$ and for every $i \in I$ there exists $k_i \in K$ such that $\psi(k) \succeq i$ for $k \succ k_i$. If J is a cofinal subset of the directed set $(I, \succeq)$, then $(x_i)_{i \in J}$ is a subnet of the net $(x_i)_{i \in I}$. With the help of nets and subnets it is possible to characterize several topological notions as the closure of a set, the cluster points, the compact subsets, and the continuity of mappings between topological spaces in the same manner as these notions are characterized in $\mathbb{R}^k$ by using sequences and subsequences. For example, a subset A of the topological space (X, τ) is closed if and only if $x \in A$ whenever the net $(x_i)_{i \in I} \subset A$ converges to $x \in X$, but generally, taking sequences is not sufficient; we say that A is **sequentially closed** if $x \in A$ whenever the sequence $(x_n)_{n \in \mathbb{N}} \subset A$ converges to $x \in X$.

Recall that the function $f : (X, \tau) \to (Y, \sigma)$ is **continuous at** $x \in X$ if

$$\forall V \in \mathcal{N}_\sigma(f(x)), \ \exists U \in \mathcal{N}_\tau(x), \ \forall x' \in U \ : \ f(x') \in V;$$

f is **continuous** if f is continuous at every $x \in X$. When X is a linear space and τ is a topology on X, it is important to have compatibility between the linear and topological structures of X.

Definition 2.1.20. *Let X be a linear space endowed with a topology τ. We say that (X, τ) is a **topological linear space** or **topological vector space** (t.l.s. or t.v.s. for short) if both operations on X (the addition and the multiplication by scalars) are continuous; in this case τ is called a **linear topology** on X.*

Since these operations are defined on product spaces, we recall that for two topological spaces (X_1, τ_1) and (X_2, τ_2), there exists a unique topology on $X_1 \times X_2$, denoted by $\tau_1 \times \tau_2$, with the property that

$$\mathcal{B}(x_1, x_2) := \{U_1 \times U_2 \mid U_1 \in \mathcal{N}_{\tau_1}(x_1), \ U_2 \in \mathcal{N}_{\tau_2}(x_2)\}$$

is a neighborhood base of (x_1, x_2) w.r.t. $\tau_1 \times \tau_2$ for every $(x_1, x_2) \in X_1 \times X_2$; $\tau_1 \times \tau_2$ is called the **product topology** on $X_1 \times X_2$. Of course, in Definition 2.1.20 the topology on $X \times X$ is $\tau \times \tau$, and the topology on $\mathbb{R} \times X$ is $\tau_0 \times \tau$, where τ_0 is the usual topology of $\mathbb{R}$. It is easy to see that when (X, τ) is a topological linear space, $a \in X$ and $\lambda \in \mathbb{R} \setminus \{0\}$, the mappings $T_a, H_\lambda : X \to X$ defined by $T_a(x) = a + x$, $H_\lambda(x) := \lambda x$, are bijective and continuous with continuous inverses i.e., they are **homeomorphisms**). It follows, that $V \in \mathcal{N}_\tau(0)$ iff $a + V \in \mathcal{N}_\tau(a)$ for every $a \in X$ and every neighborhood of 0 is absorbing ($A \subset X$ is **absorbing** if for every $x \in X$ there exists $\delta > 0$ such that $[-\delta, \delta] \cdot x \subset A$, or equivalently, for every $x \in X$ there exists $\delta > 0$ such that $[0, \delta] \cdot x \subset A$). When X is a t.v.s., the class of all neighborhoods of $0 \in X$ will be denoted also by $\mathcal{N}_X$. It is well known that the origin 0 of a t.v.s. (X, τ) has a neighborhood base formed by balanced absorbing sets ($\emptyset \neq A \subset X$ is **balanced** if $[-1, 1] \cdot A = A$). Taking into account the formula

$$\mathrm{cl}\, A = \bigcap \{A + V \mid V \in \mathcal{B}\} \tag{2.11}$$

for the closure of a subset A of the t.l.s. (X, τ), where $\mathcal{B}$ is a neighborhood base for 0, one obtains that the origin of a topological linear space has a neighborhood base formed by closed, balanced, and absorbing sets.

Note also that a topological linear space (X, τ) is Hausdorff iff for every $x \in X \setminus \{0\}$ there exists $V \in \mathcal{N}_X$ such that $x \notin V$, or equivalently (see (2.11)), the set $\{0\}$ is closed.

Further examples of topological linear spaces are given in the next section.

We discuss now the connections between **topology** and **order**. Contrary to the definition of an ordered linear space (i.e., a linear space endowed with a compatible reflexive preorder), the definition of an ordered topological linear space does not require any direct relation to exist between the order and the topology involved. However, because a compatible reflexive preorder on a linear space is defined by a convex cone, generally one asks that the cone defining the order be closed, have nonempty interior, or be normal. Before giving the definition of a normal cone let us say that the nonempty set A of the linear space X is **full** with respect to the convex cone $C \subset X$ if $A = [A]_C$, where

$$[A]_C := (A + C) \cap (A - C);$$

note that $[A]_C$ is full w.r.t. C for every set $A \subset X$.

Definition 2.1.21. *Let (X, τ) be a t.v.s. and let $C \subset X$ be a convex cone. Then C is called **normal** (relative to τ) if the origin $0 \in X$ has a neighborhood base formed by full sets w.r.t. C.*

In the next result we give several characterizations of normal cones in a topological linear space.

Theorem 2.1.22. *Let (X, τ) be a topological linear space and let $C \subset X$ be a convex cone. Then the following statements are equivalent:*

(i) *C is normal,*
(ii) *$\forall V \in \mathcal{N}_X, \exists U \in \mathcal{N}_X \; : \; [U]_C \subset V$,*
(iii) *for all nets $(x_i)_{i \in I}, (y_i)_{i \in I} \subset X$ such that $0 \leq_C x_i \leq_C y_i$ for every $i \in I$ one has $(y_i) \to 0 \Rightarrow (x_i) \to 0$,*
(iv) *$\operatorname{cl} C$ is normal.*

PROOF. Recall that $x \leq_C y$ means $y - x \in C$. (i) $\Leftrightarrow$ (ii) is obvious from the definition of the normal cone.
   (ii) $\Rightarrow$ (iii) Let $V \in \mathcal{N}_X$; then there exists $U \in \mathcal{N}_X$ such that $[U]_C \subset V$. Since $(y_i) \to 0$, there exists $i_U \in I$ such that $y_i \in U$ for $i \succeq i_U$. Then $x_i \in C \cap (y_i - C) \subset (U + C) \cap (U - C) = [U]_C \subset V$ for $i \succeq i_U$, and so $(x_i) \to 0$. (iii) $\Rightarrow$ (ii) Assume, by contradiction, that (ii) does not hold. Then there exists $V_0 \in \mathcal{N}_X$ such that $[U]_C \not\subset V_0$ for every $U \in \mathcal{N}_X$. Hence, for every $U \in \mathcal{N}_X$ there exist $y_U, y_U' \in U$ and $x_U, x_U' \in C$ such that $z_U := y_U + x_U = y_U' + x_U' \notin V_0$. Of course, $(y_U), (y_U') \to 0$, and so $(y_U' - y_U) \to 0$, too. Since $0 \leq_C x_U \leq_C y_U' - y_U$, by hypothesis we obtain that $(z_U) \to 0$. This contradicts the fact that $z_U \notin V_0$ for every $U \in \mathcal{N}_X$. (iv) $\Rightarrow$ (ii) is obvious because $A \subset [A]_C \subset [A]_{\operatorname{cl} C}$ for $\emptyset \neq A \subset X$. (ii) $\Rightarrow$ (iv) Let $V \in \mathcal{N}_X$; then there exists $U \in \mathcal{N}_X$ such that $[U]_C \subset V$. By the continuity of "$+$", there exists $U_0 \in \mathcal{N}_X$ such that $U_0 + U_0 \subset U$. Using (2.11), we obtain that

$$[U_0]_{\operatorname{cl} C} = (U_0 + \operatorname{cl} C) \cap (U_0 - \operatorname{cl} C) \subset (U_0 + U_0 + C) \cap (U_0 - (U_0 + C))$$
$$\subset (U + C) \cap (U - C) = [U]_C \subset V.$$

Therefore (ii) holds. The proof is complete. $\square$

Corollary 2.1.23. *Let (X, τ) be a Hausdorff t.v.s. and let $C \subset X$ be a convex cone. If C is normal, then $\operatorname{cl} C$ is pointed, and so C is pointed, too.*

PROOF. Indeed, if $x \in \operatorname{cl} C \cap (-\operatorname{cl} C)$, then $x \in (\{0\} + \operatorname{cl} C) \cap (\{0\} - \operatorname{cl} C) \subset (U + \operatorname{cl} C) \cap (U - \operatorname{cl} C) = [U]_{\operatorname{cl} C}$ for every $U \in \mathcal{N}_X$. Since $\operatorname{cl} C$ is normal (by the preceding theorem), the family $\{[U]_{\operatorname{cl} C} \mid U \in \mathcal{N}_X\}$ is a neighborhood base of 0. Because X is Hausdorff (see also (2.11)), $x = 0$. $\square$

In the next section we shall give several characterizations of normal cones in locally convex spaces.
   It is well known that a real nonincreasing and bounded below sequence (x_n) converges to $\inf\{x_n \mid n \in \mathbb{N}\}$.

Let (X, τ) be a Hausdorff t.v.s. partially ordered by the convex cone C; we say that the net $(x_i)_{i \in I} \subset X$ is nonincreasing if

$$\forall i, j \in I \ : \ j \succeq i \Rightarrow x_j \leq_C x_i. \qquad (2.12)$$

Let $\emptyset \neq A \subset X$; we say that A is **lower bounded with respect to** C if A is lower bounded with respect to $\mathcal{R}_C$ (see Definition 2.1.7). Similarly, $a \in X$ is a lower bound (infimum) of A w.r.t. C if a is so for $\mathcal{R}_C$. Hence $a \in X$ is a lower bound of A w.r.t. C if $a \leq_C x$ for every $x \in A$; a is the infimum of A w.r.t. C if a is a lower bound and for any lower bound a' of A we have that $a' \leq_C a$. The infimum of A w.r.t. C will be denoted by $\inf_C A$ when it exists.

Proposition 2.1.24. *Let (X, τ) be a Hausdorff t.v.s. partially ordered by the closed convex cone C. If the net $(x_i)_{i \in I} \subset X$ is nonincreasing and convergent to $x \in X$, then $\{x_i \mid i \in I\}$ is bounded below and $x = \inf\{x_i \mid i \in I\}$.*

PROOF. Let $A := \{x_i \mid i \in I\}$ and fix $i_0 \in I$. Then $x_i \in x_{i_0} - C$ for $i \succeq i_0$, and so $x \in \mathrm{cl}(x_{i_0} - C) = x_{i_0} - C$. Therefore $x \leq_C x_{i_0}$ for every $i_0 \in I$. Hence x is a lower bound of A. Let $x' \in X$ be an arbitrary lower bound of A. Then $x_i \in x' + C$ for every $i \in I$, and so $x \in \mathrm{cl}(x' + C) = x' + C$. Therefore $x = \inf_C A$. $\qquad \square$

In ordered topological linear spaces the result concerning bounded monotone sequences recalled above is not generally true. Consider the linear space ℓ^∞ of all bounded sequences $x = (x^k)_{k \geq 1} \subset \mathbb{R}$ endowed with the norm $\|x\| = \sup\{|x^k| \mid k = 1, 2, \ldots\}$. In ℓ^∞ we consider the "usual" partial order generated by the cone $\ell_+^\infty := \{x \in l^\infty \mid x^k \geq 0 \ \forall k \geq 1\}$; ℓ_+^∞ is a pointed closed convex cone (even reproducing and with nonempty interior).

Example 2.1.25. (Peressini [294, p. 91]) The sequence $(x_n)_{n \geq 1} \subset \ell^\infty$, defined by (for n fixed)

$$x_n^k = \begin{cases} -1 \text{ if } 1 \leq k \leq n, \\ 0 \quad \text{if } k > n, \end{cases}$$

is nonincreasing w.r.t. C, and $\inf\{x_n \mid n \geq 1\} = e' := -e$ where $e = (1, 1, 1, \ldots) \in \ell^\infty$. But $\|x_n - e'\| = 1$ for every $n \geq 1$. Therefore $(x_n)_{n \geq 1}$ does not converge to its infimum.

Sometimes a cone C that partially orders a Hausdorff linear topological space (X, τ) is said to be **Daniell** if any nonincreasing net having a lower bound τ-converges to its infimum (see Jahn [195, p. 29], Luc [244, p. 47], Borwein [38]).

2.2 Functional Analysis and Convexity

An important class of topological linear spaces (t.l.s.) is that of locally convex spaces. Usually this class of t.l.s. is introduced in two equivalent manners. The simplest is that the t.l.s. (X, τ) is a **locally convex space** (l.c.s. for short) if the origin $0 \in X$ has a neighborhood base formed by convex sets. It follows then that the origin has a neighborhood base formed by balanced convex sets; these sets may be taken to be all closed or all open.

Before discussing the second way of introducing locally convex spaces, let X be a linear space and $p : X \to \mathbb{R}$; we say that:

- p is **subadditive** if $\forall x, y \in X \; : \; p(x + y) \leq p(x) + p(y)$;
- p is **positively homogeneous** if $\forall x \in X, \forall \lambda \in \mathbb{R}_+ \; : \; p(\lambda x) = \lambda p(x)$;
- p is **symmetric** if $\forall x \in X \; : \; p(-x) = p(x)$;
- p is **sublinear** if p is subadditive and positively homogeneous;
- p is a **seminorm** if p is sublinear and symmetric; note that in this case $p(x) \geq 0$ for every $x \in X$. Indeed, $0 = p(0) \leq p(x) + p(-x) = 2p(x)$;
- p is a **norm** if p is a seminorm and $p(x) = 0 \Leftrightarrow x = 0$.

Consider X a linear space and $\mathcal{P}$ a nonempty family of seminorms on X. For every nonempty finite subset $P \subset \mathcal{P}$, $x \in X$, and $\varepsilon > 0$ let

$$V(x; P, \varepsilon) := \{y \in X \mid p(y - x) \leq \varepsilon \; \forall p \in P\}.$$

It is obvious that $V(x; P, \varepsilon)$ is a convex set and $V(x; P, \varepsilon) = x + V(0; P, \varepsilon)$; moreover, $V(0; P, \varepsilon)$ is also symmetric (and so is balanced) and absorbing. The family $\{\mathcal{B}(x) \mid x \in X\}$, where

$$\mathcal{B}(x) := \{V(x; P, \varepsilon) \mid \varepsilon > 0, \; \emptyset \neq P \subset \mathcal{P}, \; P \text{ finite}\},$$

satisfies conditions (NB1)–(NB3) (see Section 2.1). Therefore there exists a unique topology $\tau_{\mathcal{P}}$ on X such that $\mathcal{B}(x)$ is a neighborhood base of x w.r.t. $\tau_{\mathcal{P}}$ for every $x \in X$. It is easy to show that $(X, \tau_{\mathcal{P}})$ is a topological linear space, and so $(X, \tau_{\mathcal{P}})$ is a locally convex space, denoted in the sequel by $(X, \mathcal{P})$. In order to show that every l.c.s. can be introduced by using a family of seminorms, let us define the Minkowski functional.

Consider A an absorbing subset of the linear space X; the **Minkowski functional** associated with A is the mapping

$$p_A : X \to \mathbb{R}, \quad p_A(x) := \inf\{t \geq 0 \mid x \in tA\}.$$

In the next result we collect several properties of the Minkowski functional. Recall that the **algebraic interior**, or **core**, of $A \subset X$ is the set $\operatorname{aint} A := A^i := \{a \in X \mid A - a \text{ is absorbing}\}$; of course, $A^i \subset A$.

Lemma 2.2.1. *Let X be a linear space; let $A \subset B \subset X$ be such that A is absorbing and $\lambda > 0$. Then p_A is positively homogeneous, $p_A \geq p_B$, and $p_{\lambda A} = \lambda^{-1} p_A$; if A is symmetric, that is, $A = -A$, then p_A is also symmetric. Assume that A is convex; then p_A is sublinear and*

$$A^i = \{x \in X \mid p_A(x) < 1\} \subset A \subset \{x \in X \mid p_A(x) \leq 1\}. \tag{2.13}$$

If $C \subset X$ is a convex cone and $A = [A]_C$, then

$$0 \leq_C x \leq_C y \Rightarrow p_A(x) \leq p_A(y), \tag{2.14}$$

*i.e., p_A is **monotone** w.r.t. C.*

PROOF. The fact that p_A is positively homogeneous and the relations $p_A \geq p_B$, $p_{\lambda A} = \lambda^{-1} p_A$, and $p_A(-x) = p_A(x)$ if A is symmetric are obvious.

Assume that A is convex and take $x, y \in X$. Then for $t, s \in \mathbb{R}_+$ such that $p_A(x) < t$ and $p_A(y) < s$ there exist $t' \in [0, t]$ and $s' \in [0, s]$ such that $x \in t'A$ and $y \in s'A$. It follows that $x + y \in t'A + s'A = (t' + s')A$, and so $p_A(x + y) \leq t' + s' \leq t + s$. Letting $t \to p_A(x)$ and $s \to p_A(y)$, we obtain that $p_A(x + y) \leq p_A(x) + p_A(y)$. Therefore p_A is sublinear.

The inclusions

$$\{x \in X \mid p_A(x) < 1\} \subset A \subset \{x \in X \mid p_A(x) \leq 1\} \tag{2.15}$$

are obvious. Let $a \in A^i$; since $A - a$ is absorbing, there exists $\delta > 0$ such that $a + \delta a \in A$, i.e., $a \in (1 + \delta)^{-1}A$. Hence $p_A(a) \leq (1 + \delta)^{-1} < 1$. Now let $a \in X$ be such that $p_A(a) < 1$ and consider $x \in X$. Since $p_A(a + tx) \leq p_A(a) + t p_A(x)$ for $t \geq 0$, there exists $\delta > 0$ such that $p_A(a + tx) < 1$ for $t \in [0, \delta]$. From (2.15) we obtain that $a + tx \in A$ for $t \in [0, \delta]$, and so $a \in A^i$. Therefore (2.13) holds.

To prove (2.14) consider $x, y \in X$ such that $0 \leq_C x \leq_C y$ and take $t > 0$ such that $y \in tA$, i.e., $y = ta$ with $a \in A$. Then $x = y - (y - x) = t[a - t^{-1}(y - x)] \in t(A - C)$ and $x = t(0 + t^{-1}x) \in t(A + C)$. So $x \in t[A]_C = tA$. Hence $p_A(x) \leq p_A(y)$. □

The next theorem shows that by the second procedure one obtains all the locally convex spaces.

Theorem 2.2.2. *Let (X, τ) be a locally convex space. Then there exists a nonempty family $\mathcal{P}$ of seminorms on X such that $\tau = \tau_{\mathcal{P}}$.*

PROOF. Let $\mathcal{B}$ be a neighborhood base of $0 \in X$ formed by balanced convex subsets of X. Consider $\mathcal{P} := \{p_U \mid U \in \mathcal{B}\}$. By Lemma 2.2.1 every element of $\mathcal{P}$ is a seminorm. In order to show that $\tau = \tau_{\mathcal{P}}$, consider first $V \in \mathcal{N}_X$. Then there exists $U \in \mathcal{B}$ such that $U \subset V$. Because $V(0; \{p_U\}, 1) \subset U \subset V$, it follows that $V \in \mathcal{N}_{\tau_{\mathcal{P}}}(0)$. Conversely, let $V \in \mathcal{N}_{\tau_{\mathcal{P}}}(0)$. Then there exist $U_1, \ldots, U_n \in \mathcal{B}$ and $\varepsilon > 0$ such that $\{x \in X \mid p_{U_i}(x) \leq \varepsilon \ \forall i \in \overline{1, n}\} \subset V$. Since $\mathcal{B}$ is a neighborhood base (see (NB2) in Section 2.1), there exists $U \in \mathcal{B}$ such that $U \subset U_1 \cap \cdots \cap U_n$, and so $p_U \geq p_{U_i}$ for every $i \in \overline{1, n}$. It follows that $\{x \in X \mid p_U(x) \leq \varepsilon\} \subset V$, whence by (2.13), $\varepsilon U \subset V$. Since $\varepsilon U \in \mathcal{N}_X$, $V \in \mathcal{N}_X$, too. Therefore $\tau = \tau_{\mathcal{P}}$. □

It is useful to note that the Minkowski functional p_U associated with a convex neighborhood U of the origin of the l.c.s. X is continuous and

$$\operatorname{int} U = \{x \in X \mid p_U(x) < 1\}, \quad \operatorname{cl} U = \{x \in X \mid p_U(x) \leq 1\}. \tag{2.16}$$

In particular, every seminorm $p \in \mathcal{P}$ is $\tau_{\mathcal{P}}$-continuous on the l.c.s. $(X, \mathcal{P})$. The proof of the preceding theorem also shows that the locally convex space

(X, τ) is first-countable if and only if the topology τ is defined by an (at most) countable family of seminorms. One can show that a Hausdorff and first-countable l.c.s. is **metrizable**, i.e., that there exists a metric ρ (even invariant under translations) such that $\tau = \tau_\rho$. It is easy to show that the l.c.s. $(X, \mathcal{P})$ is Hausdorff if and only if for every $x \in X \setminus \{0\}$ there exists $p \in \mathcal{P}$ such that $p(x) > 0$. Using this characterization and the Hahn–Banach theorem below, one obtains a remarkable property of locally convex spaces;, that is, the continuous dual space of the nontrivial $(X \neq \{0\})$ Hausdorff l.c.s. $(X, \mathcal{P})$,

$$X^* := \{x^* : X \to \mathbb{R} \mid x^* \text{ is linear and continuous}\},$$

does not reduce to $\{0\}$; more precisely, for every $x \in X \setminus \{0\}$ there exists $x^* \in X^*$ such that $\langle x, x^* \rangle := x^*(x) > 0$. Recall that the linear functional $\varphi : (X, \mathcal{P}) \to \mathbb{R}$ is continuous if and only if there exist $M > 0$ and $p_1, \ldots, p_k \in \mathcal{P}$ such that

$$\varphi(x) \leq M \max\{p_1(x), \ldots, p_k(x)\} \quad \forall\, x \in X.$$

We give now several **examples** of **locally convex spaces**.

Example 2.2.3. (1) Let $\|\cdot\| : X \to \mathbb{R}$ be a norm on the linear space X. Taking $\mathcal{P} := \{\|\cdot\|\}$, the space $(X, \|\cdot\|) := (X, \mathcal{P})$ is called a **normed space**. The normed space $(X, \|\cdot\|)$ is also a metric space, where the metric ρ is defined by $\rho(x, y) := \|x - y\|$, and so is a Hausdorff first-countable topological space. Important examples of normed spaces are:

(a) $\ell^\infty := \{(x_n)_{n \geq 1} \subset \mathbb{R} \mid (x_n)_{n \geq 1} \text{ is bounded}\}$ with the norm $\|(x_n)_{n \geq 1}\|_\infty := \sup\{|x_n| \mid n \geq 1\}$ and its subspaces $\mathfrak{c} := \{(x_n)_{n \geq 1} \subset \mathbb{R} \mid (x_n)_{n \geq 1} \text{ is convergent}\}$ and $\mathfrak{c}_0 = \{(x_n)_{n \geq 1} \subset \mathbb{R} \mid (x_n) \to 0\}$ with the same norm;
(b) $\ell^p := \{(x_n)_{n \geq 1} \subset \mathbb{R} \mid \sum_{n \geq 1} |x_n|^p < \infty\}$, for $p \in [1, \infty[$, with the norm

$$\|(x_n)_{n \geq 1}\|_p := \left(\sum_{n \geq 1} |x_n|^p\right)^{1/p};$$

(c) $C[0,1]$ with the Chebyshev norm $\|x\| := \max_{t \in [0,1]} |x(t)|$;
(d) $L^p(\Omega) := \{x : \Omega \to \mathbb{R} \mid x \text{ is measurable and } \int_\Omega |x(t)|^p \, dt < \infty\}$ with the norm $\|x\|_p := \left(\int_\Omega |x(t)|^p \, dt\right)^{1/p}$, where $p \in [1, \infty[$ and $\Omega \subset \mathbb{R}^n$ is an open set endowed with the Lebesgue measure (in fact, one identifies two functions if they coincide almost everywhere).

The normed space $(X, \|\cdot\|)$ is a **Banach space** if any Cauchy sequence $(x_n)_{n \geq 1} \subset X$ is convergent ((x_n) is Cauchy if $\forall\, \varepsilon > 0$, $\exists\, n_\varepsilon \in \mathbb{N}$, $\forall\, n, m \geq n_\varepsilon : \|x_n - x_m\| < \varepsilon$); the normed spaces mentioned in (a)–(d) are Banach spaces.
    The Banach space $(X, \|\cdot\|)$ is a **Hilbert space** when the norm $\|\cdot\|$ is from an inner product or, equivalently, it satisfies the parallelogram law: $\|x + y\|^2 + \|x - y\|^2 = 2(\|x\|^2 + \|y\|^2)$ for all $x, y \in X$.
    (2) Let $(X, \mathcal{P})$ be a locally convex space and let X_0 be a linear subspace of X. Then $p|_{X_0}$ is a seminorm on X_0 for every $p \in \mathcal{P}$. Taking $\mathcal{P}_0 := \{p|_{X_0} \mid p \in \mathcal{P}\}$, the space $(X_0, \mathcal{P}_0)$ becomes a locally convex space; the topology $\tau_{\mathcal{P}_0}$ is the **trace** of the topology $\tau_\mathcal{P}$ on X_0. If $(X_1, \mathcal{P}_1)$ and $(X_2, \mathcal{P}_2)$ are locally convex

spaces, then $(X_1 \times X_2, \mathcal{P})$ is an l.c.s., where $\mathcal{P} := \{p_1 \times p_2 \mid p_1 \in \mathcal{P}_1, \ p_2 \in \mathcal{P}_2\}$ and $p_1 \times p_2 : X_1 \times X_2 \to \mathbb{R}$, $p_1 \times p_2(x_1, x_2) := \max\{p_1(x_1), p_2(x_2)\}$; the topology $\tau_{\mathcal{P}}$ is exactly $\tau_{\mathcal{P}_1} \times \tau_{\mathcal{P}_2}$.

(3) Let X, Y be two linear spaces and let $\Phi : X \times Y \to \mathbb{R}$ be a separating bilinear form, i.e., $\Phi(x, \cdot)$ and $\Phi(\cdot, y)$ are linear on Y and X for all $x \in X$ and $y \in Y$, respectively, and $\forall\, x \in X \setminus \{0\}$, $\exists\, y \in Y$ such that $\Phi(x, y) \neq 0$, and $\forall\, y \in Y \setminus \{0\}$, $\exists\, x \in X$ such that $\Phi(x, y) \neq 0$. It is obvious that $p_y : X \to \mathbb{R}$, $p_y(x) := |\Phi(x, y)|$, and $p_x : Y \to \mathbb{R}$, $p_x(y) := |\Phi(x, y)|$, are seminorms for all $y \in Y$ and $x \in X$. Taking $\mathcal{P}_X := \{p_y \mid y \in Y\}$, the space $(X, \mathcal{P}_X)$ is a Hausdorff l.c.s.; similarly, taking $\mathcal{P}_Y := \{p_x \mid x \in X\}$, $(Y, \mathcal{P}_Y)$ is a Hausdorff l.c.s. One says that (X, Y, Φ) is a **dual pair**; the topology $\tau_{\mathcal{P}_X}$ is denoted by $\sigma(X, Y)$, while the topology $\tau_{\mathcal{P}_Y}$ is denoted by $\sigma(Y, X)$.

(4) Let (X, τ) be a Hausdorff locally convex space and X^* its topological dual. Then $\Phi : X \times X^* \to \mathbb{R}$ defined by $\Phi(x, x^*) := x^*(x) = \langle x, x^* \rangle$ is a separating bilinear form (see the discussion about X^* before the statement of this example). The topology $w := \sigma(X, X^*)$ is called the **weak topology** of X, while the topology $w^* := \sigma(X^*, X)$ is called the **weak* topology** of X^*; the name weak topology for w is motivated by the fact that $w \subset \tau$, i.e., w is weaker than τ (this means that there are fewer open sets w.r.t. w than open sets w.r.t. τ). It is useful to mention that the topological dual of (X, w) is X^*, while the topological dual of (X^*, w^*) is X (in the sense that for every $\varphi \in (X^*, w^*)^*$ there exists a unique $x_\varphi \in X$ such that $\varphi(x^*) = \langle x_\varphi, x^* \rangle$ for all $x^* \in X^*$; moreover, the mapping $\varphi \in X^* \mapsto x_\varphi \in X$ is an isomorphism of linear spaces. Recall that two locally convex topologies τ_1 and τ_2 on the linear space X are called **compatible** if (X, τ_1) and (X, τ_2) have the same topological dual; the finest topology σ on the locally convex space (X, τ) that is compatible with τ is called the **Mackey topology** of X (such a topology always exists!).

(5) Let $(X, \|\cdot\|)$ be a normed space. Then X^* is also a normed space, and with the norm (always) defined by $\|x^*\| := \sup\{\langle x, x^* \rangle \mid \|x\| \leq 1\}$; $(X^*, \|\cdot\|)$ is even a Banach space. As in (3), on X one has also the weak topology w, while on X^* one has the topology w^*; one has that $w = \tau_{\|\cdot\|}$ iff X is finite-dimensional. Since $(X^*, \|\cdot\|)$ is a normed space, on X^* one has also the topology $\sigma(X^*, X^{**})$; note that $w^* \subset \sigma(X^*, X^{**}) \subset \tau_{\|\cdot\|}$. Taking $x \in X$, the mapping $\phi_x : (X^*, \|\cdot\|) \to \mathbb{R}$, $\phi_x(x^*) := x^*(x)$, is linear and continuous; moreover, $\|\phi_x\| = \|x\|$. The mapping $J_X : (X, \|\cdot\|) \to (X^{**}, \|\cdot\|)$ defined by $J_X(x) := \phi_x$ is a linear operator having the property that $\|J_X(x)\| = \|x\|$ for every $x \in X$. One says that $(X, \|\cdot\|)$ is **reflexive** if J_X is onto. Because $(X^{**}, \|\cdot\|)$ is a Banach space, every reflexive normed space is a Banach space. One gets that $w^* = \sigma(X^*, X^{**})$ iff $(X, \|\cdot\|)$ is a reflexive Banach space. Among the usual Banach spaces mentioned in (1), those that are reflexive are ℓ^p and $L^p(\Omega)$ for $p \in\,]1, \infty[$.

An important result in the theory of locally convex spaces is the **Alaoglu–Bourbaki** theorem: If U is a neighborhood of the origin 0 of the Hausdorff l.c.s. (X, τ), then the polar set U^0 of U is w^*-compact (i.e., every net $(x_i^*)_{i \in I} \subset U^0$ contains a subnet $(x_{\psi(k)}^*)_{k \in K}$ converging to $x^* \in U^0$), where the **polar set** of $\emptyset \neq A \subset X$ is $A^0 := \{x^* \mid \langle x, x^* \rangle \geq -1 \ \forall x \in A\}$; A^0 is, obviously, w^*-closed, convex and contains $0 \in X^*$. If $(X, \|\cdot\|)$ is a normed space and $U_X := \{x \in X \mid \|x\| \leq 1\}$ is the closed unit ball of X, then $(U_X)^0 = U_{X^*}$. Therefore U_{X^*} is w^*-compact. It follows that the normed space $(X, \|\cdot\|)$ is reflexive iff U_X is w-compact. Another famous characterization of reflexive Banach spaces is due to R.C. James: $(X, \|\cdot\|)$ is reflexive iff any $x^* \in X^*$ attains its supremum on U_X.

The dual cone and its quasi-interior are defined analogously to polar sets.

Definition 2.2.4. *Let C be a cone in the l.c.s. X. Then*

$$C^+ := \{x^* \in X^* \mid \langle x, x^* \rangle \geq 0 \ \forall x \in C\}$$

*is called the **dual cone** of C, while*

$$C^\# := \{x^* \in C^+ \mid \langle x, x^* \rangle > 0 \ \forall x \in C \setminus \{0\}\}$$

*is called the **quasi-interior** of C^+.*

Obviously, C^+ is a w^*-closed convex cone as the intersection of a family of closed half-spaces.

As mentioned above, a fundamental tool in functional analysis is the (algebraic) **Hahn–Banach** theorem.

Theorem 2.2.5. *Let X be a linear space, $X_0 \subset X$ a linear subspace, $p : X \to \mathbb{R}$ a sublinear functional, and $\varphi_0 : X_0 \to \mathbb{R}$ a linear functional such that $\varphi_0(x) \leq p(x)$ for every $x \in X_0$. Then there exists a linear functional $\varphi : X \to \mathbb{R}$ such that $\varphi|_{X_0} = \varphi_0$ and $\varphi(x) \leq p(x)$ for every $x \in X$.*

The proof of this result can be found in any book on functional analysis, so it is omitted; just note that for the proof one uses the Zorn axiom mentioned in the first section.

If $(X, \|\cdot\|)$ is a normed space, X_0 is a linear subspace of X, and $\varphi_0 \in (X_0, \|\cdot\|)^*$, then $p := \|\varphi_0\| \cdot \|\cdot\|$ is sublinear, and $\varphi_0(x) \leq p(x)$ for every $x \in X_0$. By the Hahn–Banach theorem, there exists a linear functional $\varphi : X \to \mathbb{R}$ such that $\varphi|_{X_0} = \varphi_0$ and $\varphi(x) \leq p(x)$ for every $x \in X$. It follows that φ is continuous, and so $\varphi \in X^*$, and $\|\varphi\| = \|\varphi_0\|$. If $(X, \mathcal{P})$ is an l.c.s., a similar procedure is possible.

Theorem 2.2.6. *Let $(X, \mathcal{P})$ be a locally convex space, $X_0 \subset X$ a linear subspace, and $\varphi_0 : X_0 \to \mathbb{R}$ a continuous linear functional. Then there exists $\varphi \in X^*$ such that $\varphi|_{X_0} = \varphi_0$.*

PROOF. Because φ_0 is a continuous linear functional, there exist $M > 0$ and $p_1, \ldots, p_k \in \mathcal{P}$ such that $\varphi_0(x) \leq M \cdot \max\{p_1(x), \ldots, p_k(x)\}$ for every $x \in X_0$.

Since $p := M \cdot \max\{p_1, \ldots, p_k\}$ is a seminorm, there exists $\varphi : X \to \mathbb{R}$ such that $\varphi|_{X_0} = \varphi_0$ and $\varphi(x) \leq p(x)$ for every $x \in X$. The last inequality shows that φ is continuous, and so $\varphi \in X^*$. $\qquad\square$

Another important application of the Hahn–Banach theorem is to the **separation of convex sets**. The following result is met in the literature as the **geometric form of the Hahn–Banach theorem**.

Theorem 2.2.7. Let (X, τ) be a locally convex space, $A \subset X$ a convex set with nonempty interior and $x_0 \in X \setminus \operatorname{int} A$. Then there exists $x^* \in X^* \setminus \{0\}$ such that

$$\langle x, x^* \rangle \leq \langle x_0, x^* \rangle \quad \forall\, x \in A. \tag{2.17}$$

PROOF. Replacing, if necessary, A by $A - a$ with $a \in \operatorname{int} A$, we assume that $0 \in \operatorname{int} A$. Thus A is an absorbing convex set, and so, by Lemma 2.2.1, the Minkowski functional p_A is sublinear. Because $x_0 \notin \operatorname{int} A$ we have that $p_A(x_0) \geq 1$ (see (2.16)). Let $X_0 := \operatorname{lin}\{x_0\} = \mathbb{R}x_0$ and $\varphi_0 : X_0 \to \mathbb{R}$ defined by $\varphi_0(tx_0) := tp_A(x_0)$ for every $t \in \mathbb{R}$. Since $\varphi_0(-x_0) = -p_A(x_0) \leq p_A(-x_0)$, we obtain that $\varphi_0(x) \leq p_A(x)$ for every $x \in X_0$. Using the Hahn–Banach theorem, there exists a linear functional $\varphi : X \to \mathbb{R}$ such that $\varphi|_{X_0} = \varphi_0$ and $\varphi(x) \leq p(x)$ for every $x \in X$. The last inequality shows that φ is continuous. Denoting φ by x^* and taking into account (2.16), we obtain that

$$\langle x, x^* \rangle \leq p_A(x) \leq 1 \leq p_A(x_0) = \varphi_0(x_0) = \langle x_0, x^* \rangle \quad \forall\, x \in A,$$

and so the conclusion holds. $\qquad\square$

Another useful variant of the separation theorem is given in the next result.

Theorem 2.2.8. Let (X, τ) be a locally convex space, $A \subset X$ a nonempty closed convex set, and $x_0 \in X \setminus A$. Then there exists a continuous linear functional $x^* \in X^*$ such that

$$\sup\{\langle x, x^* \rangle \mid x \in A\} < \langle x_0, x^* \rangle. \tag{2.18}$$

PROOF. Since $x_0 \notin A = \operatorname{cl} A$, from (2.11) we get the existence of an open convex neighborhood U of 0 such that $x_0 \notin A + U$. Since $A + U$ is convex and open, by the preceding theorem there exists $x^* \in X^* \setminus \{0\}$ such that

$$\langle x_0, x^* \rangle \geq \sup\{\langle x, x^* \rangle + \langle u, x^* \rangle \mid x \in A,\ u \in U\}$$
$$= \sup\{\langle x, x^* \rangle \mid x \in A\} + \sup\{\langle u, x^* \rangle \mid u \in U\}.$$

But $\sup\{\langle u, x^* \rangle \mid u \in U\} > 0$ (otherwise, $x^* = 0$, because U is absorbing), and so (2.18) holds. $\qquad\square$

Using the preceding theorem in the topological dual X^* of the l.c.s. X endowed with the weak* topology we obtain that

$$(A \cap B)^0 = \overline{\operatorname{conv}}(A^0 \cup B^0) = \operatorname{cl}\left(\bigcup\nolimits_{\lambda \in [0,1]} (\lambda A^0 + (1 - \lambda)B^0)\right) \tag{2.19}$$

for all closed convex subsets A, B of X with $0 \in A \cap B$, where $\overline{\mathrm{conv}}\, E := \overline{\mathrm{conv}\, E}$ is the **closed convex hull** of the subset E of the topological vector space (t.v.s.) Y.

The last property of the Minkowski functional in Lemma 2.2.1 turns out to be useful in characterizing normal cones in locally convex spaces.

Proposition 2.2.9. *Let (X, τ) be an l.c.s. and $C \subset X$ a convex cone.*

(i) *C is normal if and only if there exists a family $\mathcal{P}$ of monotone seminorms w.r.t. C such $\tau = \tau_\mathcal{P}$.*

(ii) *If C is normal, then C^+ is reproducing; i.e., $X^* = C^+ - C^+$.*

(iii) *C is w-normal if and only if C^+ is reproducing.*

PROOF. (i) Assume that C is normal; since X is an l.c.s., there exists a base $\mathcal{B}$ of symmetric convex and full neighborhoods of $0 \in X$. Let $\mathcal{P} = \{p_U \mid U \in \mathcal{B}\}$. By (the proof of) Theorem 2.2.2, $\tau = \tau_\mathcal{P}$. From Lemma 2.2.1 we have that every p_U is monotone w.r.t. C for every $U \in \mathcal{B}$. Conversely, suppose that the topology τ is determined by the family $\mathcal{P}$ of monotone seminorms w.r.t. C. Let $V \in \mathcal{N}_X$. Then there exist $\varepsilon > 0$ and $p_1, \ldots, p_n \in \mathcal{P}$ such that

$$\{x \in X \mid p_i(x) \le \varepsilon \ \forall i \in \overline{1, n}\} \subset V.$$

Consider $U = \{x \in X \mid p_i(x) \le \varepsilon/3 \ \forall i \in \overline{1,n}\} \subset V$ and take $y \in [U]_C$. It follows that $y = x_1 + c_1 = x_2 - c_2$ with $x_1, x_2 \in U$ and $c_1, c_2 \in C$. Since $0 \le_C c_1 = x_2 - x_1 - c_2 \le_C x_2 - x_1$, we obtain that

$$p_i(c_1) \le p_i(x_2 - x_1) \le p_i(x_2) + p_i(x_1) \le 2\varepsilon/3$$

for $1 \le i \le n$, whence

$$p_i(y) = p_i(x_1 + c_1) \le p_i(x_1) + p_i(c_1) \le \varepsilon/3 + 2\varepsilon/3 = \varepsilon$$

for $1 \le i \le n$. Therefore $y \in V$. Hence $[U]_C \subset V$. From Theorem 2.1.22 we obtain that C is normal.

(ii) Let $x^* \in X^*$; then $U := \{x \in X \mid \langle x, x^* \rangle \ge -1\} \in \mathcal{N}_X$, and so, by Theorem 2.1.22, there exists a convex neighborhood V of $0 \in X$ such that $(V + C) \cap (V - C) \subset U$. Therefore

$$x^* \in \left((V + C) \cap (V - C)\right)^0$$

$$= \overline{\mathrm{conv}}\left((V + C)^0 \cup (V - C)^0\right) = \mathrm{conv}\left((V + C)^0 \cup (V - C)^0\right)$$

$$= \bigcup_{\lambda \in [0,1]} \left(\lambda(V + C)^0 + (1 - \lambda)(V - C)^0\right)$$

$$\subset \bigcup_{\lambda \in [0,1]} \left(\lambda C^0 + (1 - \lambda)(-C)^0\right) \subset (C^+ - C^+).$$

We have used the formulae $\mathrm{cl}\left((V + C) \cap (V - C)\right) = \mathrm{cl}(V + C) \cap \mathrm{cl}(V - C)$, (2.19), and the fact that $(V + C)^0$ and $(V - C)^0$ are w^*-compact by the Alaoglu–Bourbaki theorem. Therefore C^+ is reproducing.

(iii) If C is w-normal, by (ii) we obtain that C^+ is reproducing. Suppose now that C^+ is reproducing and the nets $(x_i)_{i \in I}, (y_i)_{i \in I} \subset X$ are such that $0 \leq_C x_i \leq_C y_i$ for all $i \in I$ and $(y_i)_{i \in I} \xrightarrow{w} 0$. Let $x^* \in X$. Since C^+ is reproducing, $x^* = x_1^* - x_2^*$ with $x_1^*, x_2^* \in C^+$. It follows that $0 \leq \langle x_i, x_1^* \rangle \leq \langle y_i, x_1^* \rangle$ for every $i \in I$, and so $(\langle x_i, x_1^* \rangle) \to 0$. Similarly, $(\langle x_i, x_2^* \rangle) \to 0$, and so $(\langle x_i, x^* \rangle) \to 0$. Therefore $(x_i)_{i \in I} \xrightarrow{w} 0$. By Theorem 2.1.22 we obtain that C is w-normal. $\square$

The statements (ii) and (iii) of the preceding proposition can be found in Peressini [294, Prop. 2.1.21] Some further characterizations of normal cones are possible, if we restrict to normed spaces. We recall that the subset B of the topological linear space (X, τ) is **bounded** if for every $V \in \mathcal{N}_X$ there exists $\lambda > 0$ such that $\lambda B \subset V$; if $(X, \mathcal{P})$ is an l.c.s., the set B is bounded (in the sense of the preceding definition) if $p(B)$ is bounded in $\mathbb{R}$ for every $p \in \mathcal{P}$.

Theorem 2.2.10. *Let $(X, \|\cdot\|)$ be a normed vector space and $C \subset X$ a convex cone. The following statements are equivalent:*

(i) *C is normal;*
(ii) *$[U_X]_C$ is bounded;*
(iii) *there exists an equivalent monotone (w.r.t. C) norm $\|\cdot\|_0$;*
(iv) *there exists $\alpha > 0$ such that $\|y\| \geq \alpha \|x\|$ whenever $x, y \in X$ with $0 \leq_C x \leq_C y$;*
(v) *there exists $\beta > 0$ such that $\|x + y\| \geq \beta$ whenever $x, y \in C$ with $\|x\| = \|y\| = 1$.*

PROOF. (i) $\Rightarrow$ (ii) Let $V := U_X \in \mathcal{N}_{\|\cdot\|}(0)$. Using Theorem 2.1.22, there exists $r > 0$ such that $[rU_X]_C = r[U_X]_C \subset U_X$. Therefore $[U_X]_C$ is bounded.

(ii) $\Rightarrow$ (i) is immediate: $\{[rU_X]_C \mid r > 0\}$ is a base of neighborhoods of 0.

(ii) $\Rightarrow$ (iii) There exists $r > 0$ such that

$$rU_X \subset [rU_X]_C =: U \subset U_X.$$

It follows that

$$\|\cdot\| = p_{U_X} \leq p_U \leq p_{rU_X} = r^{-1} p_{U_X} = r^{-1} \|\cdot\|.$$

Since U is a convex, symmetric, and full neighborhood, from the above inequalities we get that $\|\cdot\|_0 := p_U$ is a monotone norm, equivalent to $\|\cdot\|$. (iii) $\Rightarrow$ (iv) There exist $r_1, r_2 > 0$ such that

$$\forall x \in X \; : \; r_1 \|x\| \leq \|x\|_0 \leq r_2 \|x\|.$$

Let $0 \leq_C x \leq_C y$. Then

$$r_1 \|x\| \leq \|x\|_0 \leq \|y\|_0 \leq r_2 \|y\|,$$

whence $\|y\| \geq \alpha \|x\|$, where $\alpha = r_1/r_2 > 0$. (iv) $\Rightarrow$ (ii) Let $y \in [U_X]_C$. Then $y = x_1 + c_1 = x_2 - c_2$ with $x_1, x_2 \in U_X$ and $c_1, c_2 \in C$. It follows that $0 \leq_C c_1 = x_2 - x_1 - c_2 \leq_C x_2 - x_1$, whence $\alpha \|c_1\| \leq \|x_2 - x_1\| \leq 2$. Therefore $\|y\| = \|x_1 + c_1\| \leq \|x_1\| + \|c_1\| \leq 1 + 2/\alpha$. Hence $[U_X]_C$ is bounded.

(iv) $\Rightarrow$ (v) Let $x, y \in C$ with $\|x\| = \|y\| = 1$. Then $0 \leq_C x \leq_C x + y$, whence $\|x + y\| \geq \alpha \|x\| = \alpha$. (v) $\Rightarrow$ (iv) Suppose that (iv) does not hold. Then there exist $(x_n), (y_n) \subset X$ such that $0 \leq_C x_n \leq_C y_n$ and $\|y_n\| < n^{-1} \|x_n\|$ for every $n \in \mathbb{N}^*$. We may suppose that $\|x_n\| = 1$ for every n. Consider $u_n = (y_n - x_n)/\|y_n - x_n\| \in C$. Of course, $\|u_n\| = 1$. From the hypothesis we have that $\|x_n + u_n\| \geq \beta$ for every n. On the other hand,

$$x_n + u_n = x_n + \frac{y_n - x_n}{\|y_n - x_n\|} = \left(1 - \frac{1}{\|y_n - x_n\|}\right) x_n + \frac{1}{\|y_n - x_n\|} y_n. \quad (2.20)$$

But

$$1 - 1/n \leq 1 - \|y_n\| \leq \|x_n - y_n\| \leq 1 + \|y_n\| \leq 1 + 1/n,$$

whence $(\|y_n - x_n\|) \to 1$. From (2.20) we get the contradiction $(x_n + u_n) \to 0$. $\square$

In finite-dimensional spaces the normal cones are recognized easily.

Corollary 2.2.11. *Let (X, τ) be a finite-dimensional Hausdorff t.l.s. and $C \subset X$ a convex cone. Then C is normal if and only if $\mathrm{cl}\, C$ is pointed.*

PROOF. The necessity is valid without the assumption $\dim X < \infty$, as seen in Corollary 2.1.23.

Suppose that $\mathrm{cl}\, C$ is pointed. It is well known that all the Hausdorff linear topologies on a finite-dimensional vector space coincide. Therefore we may suppose that X is $\mathbb{R}^n$ ($n = \dim X$) endowed with the Euclidean norm: $\|(x_1, \ldots, x_n)\| := \left(x_1^2 + \cdots + x_n^2\right)^{1/2}$. Since the mapping $X \times X \ni (x, y) \longmapsto x + y \in X = \mathbb{R}^n$ is continuous and $S = \{(x, y) \in \mathrm{cl}\, C \times \mathrm{cl}\, C \mid \|x\| = \|y\| = 1\}$ is compact, there exist $(\overline{x}, \overline{y}) \in S$ such that $\beta := \|\overline{x} + \overline{y}\| \leq \|x + y\|$ for all $(x, y) \in S$. Because $\mathrm{cl}\, C$ is pointed, $\beta > 0$. Therefore condition (v) of the preceding theorem is satisfied. It follows that C is normal. $\square$

Let X be a Hausdorff l.c.s. and $C \subset X$ a nontrivial convex cone having a base B; then $0 \notin B$. If $0 \notin \mathrm{cl}\, B$, Theorem 2.2.8 gives an element $x_0^* \in X^*$ such that $\langle b, x_0^* \rangle > 0$ for every $b \in \mathrm{cl}\, B$. Since every $x \in C \setminus \{0\}$ has a (unique) representation $x = \lambda b$ with $\lambda > 0$ and $b \in B$, we get $\langle x, x_0^* \rangle = \lambda \langle b, x_0^* \rangle > 0$, and so $x_0^* \in C^\#$. Conversely, if $x_0^* \in C^\#$, the set $B := \{x \in C \mid \langle x, x_0^* \rangle = 1\}$ is a base of C with $0 \notin \mathrm{cl}\, B$ (since x_0^* is continuous). So we have proved the next result.

Theorem 2.2.12. *Let X be a Hausdorff l.c.s. and $C \subset X$ a nontrivial convex cone. Then C has a base B with $0 \notin \mathrm{cl}\, B$ if and only if $C^\# \neq \emptyset$.*

A classical theorem is that of Krein and Rutman: If C is a nontrivial closed convex and pointed cone of a separable normed space X, then $C^\# \neq \emptyset$. So each such cone has a base. Now we assume that $\operatorname{int} C^+ \neq \emptyset$ (the interior being taken w.r.t. a linear topology compatible with w^*). Of course, $\operatorname{int} C^+ \subset C^\#$ (equality holds if $\operatorname{int} C^+ \neq \emptyset$). Then Theorem 2.2.12 can be refined, as shown in [199, Th. 3.8.4].

Theorem 2.2.13. *Let X be a Hausdorff l.c.s. and $C \subset X$ a nontrivial closed convex cone. Then C has a closed and **bounded** base if and only if $\operatorname{int} C^+ \neq \emptyset$, where the interior is considered w.r.t. the strong topology on X^*.*

Recall that the **strong topology** on X^* is that topology for which the family $\{B^0 \mid B \text{ is bounded}\}$ is a base of neighborhoods of the origin of X^*.

Many times, the existence of a base for the convex cone C is too weak. In the following definition we strengthen this condition.

Definition 2.2.14. *Let X be a Hausdorff t.v.s. and $C \subset X$ a nontrivial convex cone.*

(i) *We say that C is **based** if there exists a convex set B such that $C = \mathbb{R}_+ B$ and $0 \notin \operatorname{cl} B$.*

(ii) *We say that C is **well-based** if there exists a bounded convex set B such that $C = \mathbb{R}_+ B$ and $0 \notin \operatorname{cl} B$.*

(iii) *Let the topology of X be defined by the family $\mathcal{P}$ of seminorms. We say that C is **supernormal** or **nuclear** if for every $p \in \mathcal{P}$ there exists $x^* \in X^*$ such that $p(x) \leq \langle x, x^* \rangle$ for all $x \in C$; of course, $x^* \in C^+$ in this case.*

When X is an H.l.c.s., in Definition 2.2.14 (i), (ii) one may ask the uniqueness of the representation of any $x \in C \setminus \{0\}$ as λb with $\lambda > 0$ and $b \in B$; that is, B is also a base in the sense of Definition 2.1.14, but this assumption is not very useful in what follows. Moreover, when X is not an l.c.s., it is not certain that C has a base (in the sense of Definition 2.1.14) when C is based or well-based. Sometimes a convex set $B \subset X$ such that $C = \mathbb{R}_+ B$ and $0 \notin \operatorname{cl} B$ is called also a base of C. When $C = \mathbb{R}_+ B$ with B a bounded convex set and $0 \notin \operatorname{cl} B$, then $\operatorname{cl} C = \mathbb{R}_+ \cdot \operatorname{cl} B$; hence, for a well-based closed convex cone C, the set B in Definition 2.2.14(ii) can be chosen to be closed. Moreover, it follows that C is well-based iff $\operatorname{cl} C$ is well-based.

The notion of nuclear cones in Definition 2.2.14(iii) was introduced by Isac in [175]. Isac [175, 177, 178] and Postolică [303, 304] gave several examples of supernormal cones. From the definition of supernormal cones it is obvious that C is supernormal iff $\operatorname{cl} C$ is supernormal. When $(X, \|\cdot\|)$ is a normed space, the convex cone C is supernormal iff there exists $x^* \in X^*$ such that

$$C \subset \{x \in X \mid \|x\| \leq \langle x, x^* \rangle\}, \qquad (2.21)$$

that is, iff C has the **angle property** in the sense of Cesari and Suryanarayana [60, Def. 4.2] (in fact, this is equivalent to their definition). Of course, x^* from (2.21) is in $C^\#$.

Proposition 2.2.15. (Isac [175]) *Let $(X, \mathcal{P})$ be an H.l.c.s. and $C \subset X$ a proper convex cone. Then*

$$C \text{ well-based } \Rightarrow C \text{ supernormal } \Rightarrow C \text{ normal.}$$

If X is a normed space, then C supernormal $\Rightarrow C$ well-based.

PROOF. Suppose that C is well-based and B is a bounded base of C. Since $0 \notin \operatorname{cl} B$, there exists $\overline{x}^* \in X^*$ such that $1 \leq \langle x, \overline{x}^* \rangle$ for every $x \in B$. Let $p \in \mathcal{P}$; since B is bounded, there exists $\mu \in \mathbb{R}_+$ such that $p(x) \leq \mu$ for every $x \in B$. It follows that $p(x) \leq \langle x, \mu \overline{x}^* \rangle$ for every $x \in C$, and so C is supernormal.

Suppose now that C is supernormal and consider the nets $(x_i)_{i \in I}, (y_i)_{i \in I} \subset X$ such that $0 \leq_C x_i \leq_C y_i$ for all $i \in I$ and $(y_i)_{i \in I} \to 0$. Let $p \in \mathcal{P}$. By hypothesis, there exists $x^* \in C^+$ such that $p(x) \leq \langle x, x^* \rangle$ for all $x \in C$. It follows that $p(x_i) \leq \langle x_i, x^* \rangle \leq \langle y_i, x^* \rangle$ for every $i \in I$. As $(y_i)_{i \in I} \to 0$, $(y_i)_{i \in I} \xrightarrow{w} 0$, and so $(\langle y_i, x^* \rangle)_{i \in I} \to 0$. It follows that $(p(x_i))_{i \in I} \to 0$, and so $(x_i)_{i \in I} \to 0$. Using Theorem 2.1.22 we obtain that C is normal.

Assume now that X is a normed space and C is supernormal. Then there exists $x^* \in C^+$ such that $\|x\| \leq \langle x, x^* \rangle$ for all $x \in C$. Take $B := \{x \in C \mid \langle x, x^* \rangle = 1\}$; it is obvious that B is a bounded base of C. □

The proof above shows that C satisfies a stronger condition than normality when C is supernormal; more exactly, if C is supernormal, then

$$0 \leq_C x_i \leq_C y_i \ \forall i \in I, \ (y_i)_{i \in I} \xrightarrow{w} 0 \Rightarrow (x_i)_{i \in I} \to 0.$$

In particular, for every net $(x_i)_{i \in I} \subset C$ we have $(x_i)_{i \in I} \xrightarrow{w} 0 \Rightarrow (x_i)_{i \in I} \to 0$. Also note (see Isac [177]) that

$$C \text{ is } w\text{-supernormal } \Leftrightarrow C \text{ is } w\text{-normal } \Leftrightarrow C^+ - C^+ = X^*.$$

The last equivalence is stated in Proposition 2.2.9(iii). The implication $\Rightarrow$ of the first equivalence follows from the preceding proposition. Assume that C is w-normal. Let $y^* \in X^*$; since $X^* = C^+ - C^+$, there exist $y_1^*, y_2^* \in C^+$ with $y^* = y_1^* - y_2^*$. Consider $x^* := y_1^* + y_2^* \in C^+$. It follows that $|\langle x, y^* \rangle| \leq |\langle x, y_1^* \rangle| + |\langle x, y_2^* \rangle| = \langle x, y_1^* \rangle + \langle x, y_2^* \rangle = \langle x, x^* \rangle$ for all $x \in C$, and so C is w-supernormal.

Example 2.2.16. (1) The convex cone $C \subset \mathbb{R}^n$ is well-based if and only if $\operatorname{cl} C$ is pointed; hence C is well-based iff C is normal. Indeed, assume that $\operatorname{cl} C$ is pointed and take $S := \{x \in \operatorname{cl} C \mid \|x\| = 1\}$. Because S is compact, $\operatorname{conv} S$ is also compact. But $0 \notin \operatorname{conv} S$; otherwise, $0 = \lambda_1 x_1 + \cdots + \lambda_k x_k$ with $\lambda_i > 0$, $\sum_{i=1}^k \lambda_i = 1$, and $x_i \in S$ for $i \in \overline{1, k}$ ($k \geq 2$). This implies the contradiction $0 \neq -\lambda_k x_k \in \operatorname{cl} C$. It is obvious that $B := C \cap \operatorname{conv} S$ is a bounded convex base of C.

(2) In any nontrivial normed space $(X, \|\cdot\|)$ there exist well-based convex cones, even with nonempty interior: $C := \mathbb{R}_+(x + rU_X)$, where $x \in X \setminus \{0\}$ and $0 < r < \|x\|$.

(3) Among the classical Banach spaces (mentioned in Example 2.2.3) their usual positive cones are well-based only in ℓ^1 and $L^1(\Omega)$.

Lemma 2.2.17. ([385, Prop. 5]) *Let X be a Hausdorff l.c.s., X^* its topological dual, and let C be a closed proper convex cone in X with $\operatorname{int} C \neq \emptyset$. If $\bar{x} \in \operatorname{int} C$, then the set*

$$B = \{x^* \in C^+ \mid \langle \bar{x}, x^* \rangle = 1\}$$

is a weak*-compact base for C^+.

PROOF. Since $\bar{x} \in \operatorname{int} C$, the set $V := C - \bar{x}$ is a neighborhood of $0 \in X$. Therefore its polar set

$$
\begin{aligned}
V^0 &:= \{x^* \in X^* \mid \langle x, x^* \rangle \geq -1 \,\forall\, x \in C\} \\
&= \{x^* \in X^* \mid \langle c, x^* \rangle \geq \langle \bar{x}, x^* \rangle - 1 \,\forall\, c \in C\} \\
&= \{x^* \in C^+ \mid \langle \bar{x}, x^* \rangle \leq 1\}
\end{aligned}
$$

is weak*-compact, by the Alaoglu–Bourbaki theorem. Consequently, B is a weak*-closed subset of V^0 and so weak*-compact. Therefore B is a weak*-compact convex set not containing the origin. Let $x^* \in C^+ \setminus \{0\}$. Then $\gamma := \langle \bar{x}, x^* \rangle > 0$; otherwise, x^* is nonnegative on the neighborhood V of $0 \in X$, and so $x^* = 0$. It follows that $\gamma^{-1} x^* \in B$, which proves that B is a weak*-compact base of C^+. □

Finally, we prove a characterization of well-based cones in normed vector spaces.

Proposition 2.2.18. ([143, Prop. 3]) *Let Y be a normed vector space and $C \subset Y$ a proper convex cone. Then C is well-based if and only if there exist $c^0 \in C$ and $y^* \in C^+$ such that $\langle c^0, y^* \rangle > 0$ and*

$$C \cap S_Y \subset c^0 + \{y \in Y \mid \langle y, y^* \rangle > 0\},$$

where $S_Y = \{y \in Y \mid \|y\| = 1\}$ is the unit sphere in Y.

PROOF. Suppose first that C is well-based with bounded base S; therefore $0 \notin \operatorname{cl} S$ and $C = [0, \infty[\cdot S$. Then there exists $y^* \in Y^*$ such that $1 \leq \langle y, y^* \rangle$ for all $y \in S$. Consider $\widetilde{S} := \{y \in C \mid \langle y, y^* \rangle = 1\}$. It follows that $\widetilde{S}$ is a base of C; moreover, since $\widetilde{S} \subset [0, 1] \cdot S$, $\widetilde{S}$ is also bounded. Taking $c^1 \in C \setminus \{0\}$ we have $C \cap S_Y \subset \lambda c^1 + B_+$ for some $\lambda > 0$, where $B_+ = \{y \in Y \mid \langle y, y^* \rangle > 0\}$. Otherwise,

$$\forall\, n \in \mathbb{N}^*,\ \exists\, c_n \in C \cap S_Y\ :\quad c_n \notin \tfrac{1}{n} c^1 + B_+.$$

Therefore, $\langle c_n, y^* \rangle \leq \tfrac{1}{n}\langle c^1, y^* \rangle$ for every $n \geq 1$. But because $\widetilde{S}$ is a base, $c_n = \lambda_n b_n$ with $\lambda_n > 0$ and $b_n \in \widetilde{S}$; it follows that $1 = \|c_n\| = \lambda_n \|b_n\| \leq \lambda_n M$ with $M > 0$ (because $\widetilde{S}$ is bounded). Therefore

$$M^{-1} \leq \lambda_n = \langle \lambda_n b_n, y^* \rangle = \langle c_n, y^* \rangle \leq n^{-1} \langle c^1, y^* \rangle \qquad \forall\, n \in \mathbb{N}^*,$$

whence $M^{-1} \leq 0$, a contradiction. Thus there exists $\lambda > 0$ such that $C \cap S_Y \subset \lambda c^1 + B_+$. Taking $c^0 := \lambda c^1$ the conclusion follows.

Suppose now that $C \cap S_Y \subset c^0 + B_+$ for some $c^0 \in C$ and $y^* \in C^+$ with $\langle c^0, y^* \rangle =: \gamma > 0$, where B_+ is defined as above. Consider $S = \{y \in C \mid \langle y, y^* \rangle = 1\}$. Let $c \in C \setminus \{0\}$; then $\|c\|^{-1} c = c^0 + y$ for some $y \in B_+$. It follows that $\langle c, y^* \rangle > \gamma \|c\| > 0$; therefore $y^* \in C^{\#}$, and so $c \in]0, \infty[\cdot S$. Since $\mathrm{cl}\, S \subset \{y \in X \mid \langle y, y^* \rangle = 1\}$, we have that S is a base of C. Now let $y \in S$ ($\subset C$). Then $\|y\|^{-1} y \in C \cap S_Y$. There exists $z \in B_+$ such that $\|y\|^{-1} y = c^0 + z$. We get

$$1 = \langle y, y^* \rangle = \|y\| \langle c^0 + z, y^* \rangle \geq \gamma \|y\|,$$

whence $\|y\| \leq \gamma^{-1}$. Therefore S is bounded, and so C is well-based. $\qquad\square$

2.3 Separation Theorems for Not Necessarily Convex Sets

Throughout this section Y is a t.v.s. We shall use some usual notions and notation from convex analysis. So, having the function $f : Y \to \overline{\mathbb{R}}$, its **domain** and **epigraph** are defined, respectively, by

$$\mathrm{dom}\, f := \{y \in Y \mid f(y) < \infty\}, \quad \mathrm{epi}\, f := \{(y, t) \in Y \times \mathbb{R} \mid f(y) \leq t\};$$

f is said to be **convex** if $\mathrm{epi}\, f$ is a convex set, and f is said to be **proper** if $\mathrm{dom}\, f \neq \emptyset$ and f does not take the value $-\infty$. Of course, f is **lower semicontinuous** if $\mathrm{epi}\, f$ is closed.

The aim of this section is to find a suitable functional $\varphi : Y \to \mathbb{R}$ and conditions such that two given nonempty sets A and D can be separated by φ. Provided that D contains the rays generated by $k^0 \in Y \setminus \{0\}$, i.e.,

$$D + [0, \infty) \cdot k^0 \subset D, \tag{2.22}$$

we move $-D$ along this ray and consider the set

$$D' := \{(y, t) \in Y \times \mathbb{R} \mid y \in tk^0 - D\}.$$

The assumption on D shows that D' is of **epigraph type**; i.e., if $(y, t) \in D'$ and $t' \geq t$, then $(y, t') \in D'$. Indeed, if $y \in tk^0 - D$ and $t' \geq t$, since $tk^0 - D = t'k^0 - [D + (t' - t)k^0] \subset t'k^0 - D$, we obtain that $(y, t') \in D'$. Also observe that $D' = T^{-1}(D)$, where $T : Y \times \mathbb{R} \to Y$ is the continuous linear operator defined by $T(y, t) := tk^0 - y$. So, if D is closed (convex, cone), then D' is closed (convex, cone). Since D' is of epigraph type, we associate with D and k^0 the function $\varphi := \varphi_{D,k^0} : Y \to \overline{\mathbb{R}}$ defined by

$$\varphi(y) := \inf\{t \in \mathbb{R} \mid (y, t) \in D'\} = \inf\{t \in \mathbb{R} \mid y \in tk^0 - D\}. \tag{2.23}$$

An illustration of the function φ is furnished by Figure 2.3.1. Obviously, the domain of φ is the set $\mathbb{R}k^0 - D$ and $D' \subset \mathrm{epi}\,\varphi \subset \mathrm{cl}\,D'$, from which it follows that if D is closed, we have that $D' = \mathrm{epi}\,\varphi$, and so φ is a lower semicontinuous (l.s.c.) function.

In the next results we collect several useful properties of $\varphi = \varphi_{D,k^0}$.

Theorem 2.3.1. *Let $D \subset Y$ be a closed proper set and $k^0 \in Y$ be such that (2.22) holds. Then φ is l.s.c., $\mathrm{dom}\,\varphi = \mathbb{R}k^0 - D$,*

$$\{y \in Y \mid \varphi(y) \le \lambda\} = \lambda k^0 - D \quad \forall\ \lambda \in \mathbb{R}, \tag{2.24}$$

and

$$\varphi(y + \lambda k^0) = \varphi(y) + \lambda \quad \forall\,y \in Y,\ \forall \lambda \in \mathbb{R}. \tag{2.25}$$

Moreover,

(a) *φ is convex if and only if D is convex; $\varphi(\lambda y) = \lambda\varphi(y)$ for all $\lambda > 0$ and $y \in Y$ if and only if D is a cone.*

(b) *φ is proper if and only if D does not contain lines parallel to k^0, i.e.,*

$$\forall\,y \in Y,\ \exists t \in \mathbb{R}\ :\ y + tk^0 \notin D. \tag{2.26}$$

(c) *φ is finite-valued if and only if D does not contain lines parallel to k^0 and*

$$\mathbb{R}k^0 - D = Y. \tag{2.27}$$

(d) *Let $B \subset Y$; φ is B-**monotone** (i.e., $y_2 - y_1 \in B \Rightarrow \varphi(y_1) \le \varphi(y_2)$) if and only if $D + B \subset D$.*

(e) *φ is subadditive if and only if $D + D \subset D$.*

Suppose, furthermore, that

$$D + (0, \infty) \cdot k^0 \subset \mathrm{int}\,D. \tag{2.28}$$

Then

(f) *φ is continuous and*

$$\{y \in Y \mid \varphi(y) < \lambda\} = \lambda k^0 - \mathrm{int}\,D, \quad \forall\ \lambda \in \mathbb{R}, \tag{2.29}$$

$$\{y \in Y \mid \varphi(y) = \lambda\} = \lambda k^0 - \mathrm{bd}\,D, \quad \forall\ \lambda \in \mathbb{R}. \tag{2.30}$$

(g) *If φ is proper, then φ is B-monotone $\Leftrightarrow D + B \subset D \Leftrightarrow \mathrm{bd}\,D + B \subset D$. Moreover, if φ is finite-valued, then φ **strictly B-monotone** (i.e., $y_2 - y_1 \in B \setminus \{0\} \Rightarrow \varphi(y_1) < \varphi(y_2)$) $\Leftrightarrow D + (B \setminus \{0\}) \subset \mathrm{int}\,D \Leftrightarrow \mathrm{bd}\,D + (B \setminus \{0\}) \subset \mathrm{int}\,D$.*

(h) *Assume that φ is proper; then φ is subadditive $\Leftrightarrow D + D \subset D \Leftrightarrow \mathrm{bd}\,D + \mathrm{bd}\,D \subset D$.*

PROOF. We have already observed that dom $\varphi = \mathbb{R}k^0 - D$ and φ is l.s.c. when D is closed. From the definition of φ the inclusion $\supset$ in (2.24) is obvious, while the converse inclusion is immediate, taking into account the closedness of D. Formula (2.25) follows easily from (2.24).

(a) Since the operator T defined above is onto and epi $\varphi = T^{-1}(D)$, we have that epi φ is convex (cone) if and only if $D = T(\text{epi }\varphi)$ is so. The conclusion follows.

(b) We have $\varphi(y) = -\infty \Leftrightarrow y \in tk^0 - D$ for every $t \in \mathbb{R} \Leftrightarrow \{y + tk^0 \mid t \in \mathbb{R}\} \subset D$. The conclusion follows.

(c) The conclusion follows from (b) and the fact that dom $\varphi = \mathbb{R}k^0 - D$.

(d) Suppose first that $D + B \subset D$ and take $y_1, y_2 \in Y$ with $y_2 - y_1 \in B$. Let $t \in \mathbb{R}$ be such that $y_2 \in tk^0 - D$. Then $y_1 \in y_2 - B \subset tk^0 - (D + B) \subset tk^0 - D$, and so $\varphi(y_1) \leq t$. Hence $\varphi(y_1) \leq \varphi(y_2)$. Assume now that φ is B-monotone and take $y \in D$ and $b \in B$. From (2.24) we have that $\varphi(-y) \leq 0$. Since $(-y) - (-y - b) \in B$, we obtain that $\varphi(-y - b) \leq \varphi(-y) \leq 0$, and so, using again (2.24), we obtain that $-y - b \in -D$, i.e., $y + b \in D$.

(e) Suppose first that $D + D \subset D$ and take $y_1, y_2 \in Y$. Let $t_i \in \mathbb{R}$ be such that $y_i \in t_i k^0 - D$ for $i \in \{1,2\}$. Then $y_1 + y_2 \in (t_1 + t_2)k^0 - (D + D) \subset (t_1 + t_2)k^0 - D$, and so $\varphi(y_1 + y_2) \leq t_1 + t_2$. It follows that $\varphi(y_1 + y_2) \leq \varphi(y_1) + \varphi(y_2)$. Assume now that φ is subadditive and take $y_1, y_2 \in D$. From (2.24) we have that $\varphi(-y_1), \varphi(-y_2) \leq 0$. Since φ is subadditive, we obtain that $\varphi(-y_1 - y_2) \leq \varphi(-y_1) + \varphi(-y_2) \leq 0$, and so, using again (2.24), we obtain that $-y_1 - y_2 \in -D$, i.e. $y_1 + y_2 \in D$.

Suppose now that (2.28) holds.

(f) Let $\lambda \in \mathbb{R}$ and take $y \in \lambda k^0 - \text{int } D$. Since $\lambda k^0 - y \in \text{int } D$, there exists $\varepsilon > 0$ such that $\lambda k^0 - y - \varepsilon k^0 \in \text{int } D \subset D$. Therefore $\varphi(y) \leq \lambda - \varepsilon < \lambda$, which shows that the inclusion $\supset$ always holds in (2.29) when int $D \neq \emptyset$. Let $\lambda \in \mathbb{R}$ and $y \in Y$ be such that $\varphi(y) < \lambda$. There exists $t \in \mathbb{R}$, $t < \lambda$, such that $y \in tk^0 - D$. It follows that $y \in \lambda k^0 - (D + (\lambda - t)k^0) \subset \lambda k^0 - \text{int } D$. Therefore (2.29) holds, and so φ is upper semicontinuous. Because φ is also lower semicontinuous, we have that φ is continuous. From (2.24) and (2.29) we obtain immediately that (2.30) holds.

(g) Let us prove the second part, the first one being similar (and partially proved in (d)). So, let φ be finite-valued.

Assume that φ is strictly B-monotone and take $y \in D$ and $b \in B \setminus \{0\}$. From (2.24) we have that $\varphi(-y) \leq 0$, and so, by hypothesis, $\varphi(-y - b) < 0$. Using (2.29) we obtain that $y + b \in \text{int } D$. Assume now that bd $D + (B \setminus \{0\}) \subset \text{int } D$. Consider $y_1, y_2 \in Y$ with $y_2 - y_1 \in B \setminus \{0\}$. From (2.30) we have that $y_2 \in \varphi(y_2)k^0 - \text{bd } D$, and so $y_1 \in \varphi(y_2)k^0 - (\text{bd } D + (B \setminus \{0\})) \subset \varphi(y_2)k^0 - \text{int } D$. From (2.29) we obtain that $\varphi(y_1) < \varphi(y_2)$. The remaining implication is obvious.

(h) Let φ be proper. One has to prove bd $D + \text{bd } D \subset D \Rightarrow \varphi$ is subadditive. Consider $y_1, y_2 \in Y$. If $\{y_1, y_2\} \not\subset \text{dom }\varphi$, there is nothing to prove; hence let $y_1, y_2 \in \text{dom }\varphi$. Then, by (2.30), $y_i \in \varphi(y_i)k^0 - \text{bd } D$ for $i \in \{1,2\}$,

and so $y_1 + y_2 \in (\varphi(y_1) + \varphi(y_2))k^0 - (\text{bd } D + \text{bd } D) \subset (\varphi(y_1) + \varphi(y_2))k^0 - D$.
Therefore $\varphi(y_1 + y_2) \leq \varphi(y_1) + \varphi(y_2)$. $\square$

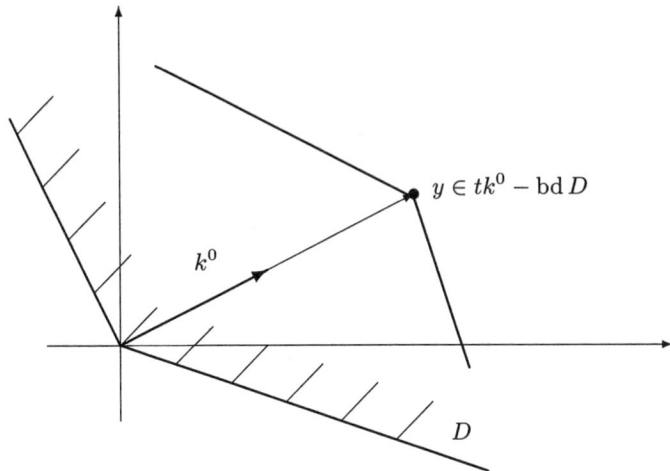

Figure 2.3.1. Level sets of the functional φ in (2.23).

Remark 2.3.2. The conditions (2.22), (2.26), (2.27), and (2.28) are invariant under translations of D, but the conditions $D + D \subset D$ and bd $D + $ bd $D \subset D$ are not. Note also that $\mathbb{R} \subset \text{Im } \varphi_{D,k^0}$ if φ_{D,k^0} is finite somewhere.

Related to condition (2.28) we have the following remark.

Remark 2.3.3. If for $D \subset Y$ and $k^0 \in Y$ one has that if cl $D + (0, \infty) \cdot k^0 \subset \text{int } D$ then cl(int D) = cl D and int(cl D) = int D.

Indeed, if $y \in D$, then $y + n^{-1}k^0 \in \text{int } D$ for any $n \in \mathbb{N}^*$, and so $y \in$ cl(int D); this proves the first equality. Now let $y \in \text{int}(\text{cl } D)$; then there exists $t > 0$ such that $y - tk^0 \in \text{cl } D$. Using the hypothesis we obtain that $y \in \text{int } D$. Therefore the second equality holds, too.

It is obvious that (2.28) $\Rightarrow$ (2.22). Other relations among several conditions used in Theorem 2.3.1 are established in the next result.

Proposition 2.3.4. *Let $D \subset Y$ be a closed proper set and $k^0 \in Y$.*

(i) *If there exists a cone $C \subset Y$ such that $k^0 \in \text{int } C$ and $D + \text{int } C \subset D$, then (2.26), (2.27), and (2.28) hold.*

(ii) *If D is convex, $\operatorname{int} D \neq \emptyset$, and (2.22), (2.27) are satisfied, then (2.26) and (2.28) hold.*

In particular, if the hypotheses of (i) *or* (ii) *hold, then φ_{D,k^0} is finite-valued and continuous, being also convex in case* (ii).

PROOF. (i) Let $y \in Y$. Since $k^0 \in \operatorname{int} C$, $\operatorname{int} C - k^0$ is a neighborhood of 0, and so there exists $t > 0$ such that $ty \in \operatorname{int} C - k^0$. It follows that $y \in \operatorname{int} C - (0,\infty)k^0$. Therefore

$$C + \mathbb{R}k^0 = C - (0,\infty) \cdot k^0 = \operatorname{int} C + \mathbb{R}k^0 = \operatorname{int} C - (0,\infty) \cdot k^0 = Y.$$

Taking $y_0 \in D$, from the inclusion $D + \operatorname{int} C \subset D$, we have that

$$D + \mathbb{R}k^0 \supset y_0 + \operatorname{int} C + \mathbb{R}k^0 = y_0 + Y = Y;$$

i.e., (2.27) holds. Suppose that the line $\mathbb{R}k^0 + y$ is contained in D; then $Y = y + \mathbb{R}k^0 + \operatorname{int} C \subset D + \operatorname{int} C \subset D$, contradicting the properness of D. Since $D + (0,\infty) \cdot k^0 \subset D + \operatorname{int} C \subset D$, it is obvious that (2.28) holds, too.

(ii) Let us show that (2.28) holds in our hypotheses. In the contrary case there exist $y_0 \in D$ and $t_0 \in (0,\infty)$ such that $y_0 + t_0 k^0 \notin \operatorname{int} D$. Since D is convex, by a separation theorem, there exists $y^* \in Y^* \setminus \{0\}$ such that

$$\langle y_0 + t_0 k^0, y^* \rangle \le \langle y, y^* \rangle \quad \forall y \in D.$$

From (2.22) we obtain that $\langle y_0 + t_0 k^0, y^* \rangle \le \langle y_0 + tk^0, y^* \rangle$ for every $t \ge 0$. Since $t_0 > 0$, it follows that $\langle k^0, y^* \rangle = 0$, and so

$$\langle y_0, y^* \rangle \le \langle y + tk^0, y^* \rangle \quad \forall y \in D, \ \forall t \in \mathbb{R}.$$

From (2.27) we obtain that $\langle y_0, y^* \rangle \le \langle y, y^* \rangle$ for all $y \in Y$, which shows that $y^* = 0$. This contradiction proves that (2.28) holds.

Assume now that $y + \mathbb{R}k^0 \subset D$ for some $y \in Y$. Let $d \in D$ and $t \in \mathbb{R}$. Since D is convex, for every $n \in \mathbb{N}^*$ we have that $\frac{n-1}{n}d + \frac{1}{n}(y + tnk^0) \in D$. Taking the limit we obtain that $d + tk^0 \in \operatorname{cl} D = D$. Therefore, using also (2.27), we get the contradiction $Y = D + \mathbb{R}k^0 \subset D$.

Because in both cases conditions (2.26), (2.27), and (2.28) hold, from Theorem 2.3.1(c,f) we have that φ is finite-valued and continuous; moreover, φ is convex in case (ii), D being so. $\square$

Using the preceding result we obtain the following important particular case of Theorem 2.3.1.

Corollary 2.3.5. *Let $C \subset Y$ be a proper closed convex cone and $k^0 \in \operatorname{int} C$. Then*

$$\varphi : Y \to \mathbb{R}, \qquad \varphi(y) := \inf\{t \in \mathbb{R} \mid y \in tk^0 - C\}$$

is a well-defined continuous sublinear function such that for every $\lambda \in \mathbb{R}$,

$$\{y \in Y \mid \varphi(y) \le \lambda\} = \lambda k^0 - C, \quad \{y \in Y \mid \varphi(y) < \lambda\} = \lambda k^0 - \operatorname{int} C.$$

Moreover, φ is strictly $\operatorname{int} C$-monotone.

PROOF. Just take $D = C$ in Theorem 2.3.1 and use Proposition 2.3.4(ii). For the last part note that $C + \operatorname{int} C = \operatorname{int} C$.   □

Now all preliminaries are done, and we can prove the following nonconvex separation theorem.

Theorem 2.3.6. Nonconvex Separation Theorem. *Let $D \subset Y$ be a closed proper set with nonempty interior, $A \subset Y$ a nonempty set such that $A \cap (-\operatorname{int} D) = \emptyset$ and $k^0 \in Y$. Assume that one of the following two conditions holds:*

(i) there exists a cone $C \subset Y$ such that $k^0 \in \operatorname{int} C$ and $D + \operatorname{int} C \subset D$;
(ii) D is convex, (2.22) and (2.27) are satisfied.

Then φ_{D,k^0} is a finite-valued continuous function such that

$$\varphi_{D,k^0}(x) \geq 0 > \varphi_{D,k^0}(-y) \quad \forall x \in A, \; \forall y \in \operatorname{int} D; \qquad (2.31)$$

moreover, $\varphi_{D,k^0}(x) > 0$ for every $x \in \operatorname{int} A$.

PROOF. By Proposition 2.3.4, φ_{D,k^0} is a finite-valued continuous function. By Theorem 2.3.1(f) we have that $-\operatorname{int} D = \{y \in Y \mid \varphi_{D,k^0}(y) < 0\}$, and so (2.31) obviously holds. It is evident that in our conditions $\operatorname{int} A \cap (-\operatorname{int} D) = \emptyset$, whence $\operatorname{int} A \cap \left(-\operatorname{cl}(\operatorname{int} D)\right) = \emptyset$. From Remark 2.23 we obtain that $\operatorname{int} A \cap (-D) = \emptyset$. Using now (2.24) we obtain that $\varphi_{D,k^0}(x) > 0$ for every $x \in \operatorname{int} A$.   □

Of course, if we impose additional conditions on D, we have additional properties of the separating functional φ_{D,k^0} (see Theorem 2.3.1).

As we observed in Proposition 2.3.4, when condition (i) of the preceding theorem holds, condition (2.27) holds, too. When D is a pointed convex cone the converse implication is valid if one replaces the interior with the algebraic interior. To be more precise, we have the following result.

Proposition 2.3.7. *Let $D \subset Y$ be a convex cone such that $D \neq -D$ (i.e., D is not a linear subspace) and $k^0 \in Y$. Then $D + \mathbb{R}k^0 = Y$ if and only if $\{k^0, -k^0\} \cap D^i \neq \emptyset$.*

PROOF. Assume first that $k^0 \in D^i$ (the proof for $-k^0 \in D^i$ being the same). Let $y \in Y$. Because $D - k^0$ is absorbing, there exists $t > 0$ such that $ty \in D - k^0$, and so $y \in D - t^{-1}k^0 \subset D + \mathbb{R}k^0$. Hence $Y = D + \mathbb{R}k^0$. Assume now that $Y = D + \mathbb{R}k^0$. Of course, $k^0 \neq 0$ and $Y_0 := D - D$ is a linear subspace. Moreover, $Y = Y_0 + \mathbb{R}k^0$. Suppose that $Y_0 \neq Y$. Let $y \in Y_0 \subset Y$. Then $y = v + \lambda k^0$ for some $v \in D$ and $\lambda \in \mathbb{R}$, whence $\lambda k^0 = y - v \in Y_0 - D \subset Y_0$. Since $k^0 \notin Y_0$ (otherwise, $Y_0 + \mathbb{R}k^0 = Y_0$), we obtain that $\lambda = 0$, and so $Y_0 \subset D \subset Y_0$. This contradicts the fact that D is not a linear subspace. Hence $Y_0 = Y$. Let us show now that $D^i \neq \emptyset$. First, because $k^0 \in D - D$, $k^0 = v_1 - v_2$ with $v_1, v_2 \in D$. Let $\bar{v} := v_1 + v_2 \in D$. Consider $\lambda, \mu \geq 0$. Then

$$D + \lambda k^0 = D + 2\lambda v_1 - \lambda \overline{v} \subset D - \mathbb{R}_+ \overline{v}, \quad D - \mu k^0 = D + 2\mu v_2 - \mu \overline{v} \subset D - \mathbb{R}_+ \overline{v}.$$

Therefore $Y = D - \mathbb{R}_+ \overline{v}$. Consider $y \in Y$; then $y + \overline{v} \in Y = D - \mathbb{R}_+ \overline{v}$, and so there exists $\lambda \geq 0$ such that $(1 + \lambda)\overline{v} + y \in D$, whence $(1 + \lambda)^{-1} y + \overline{v} \in D$. Since D is convex, this implies that $\overline{v} \in D^i$. Hence $D^i \neq \emptyset$. Suppose by contradiction that $\{k^0, -k^0\} \cap D^i = \emptyset$. By the algebraic separation theorem, there exist $\varphi, \psi \in X' \setminus \{0\}$ (linear functionals on X) such that

$$\varphi(k^0) \leq 0 \leq \varphi(y), \quad \psi(-k^0) \leq 0 \leq \psi(y) \quad \forall y \in D. \tag{2.32}$$

If $\varphi(k^0) = 0$, then $\varphi(y + \lambda k^0) \geq 0$ for all $y \in D$ and $\lambda \in \mathbb{R}$, and so we get the contradiction $\varphi = 0$ because $D + \mathbb{R}k^0 = Y$. Hence $\varphi(k^0) < 0$. Similarly, $\psi(-k^0) < 0$. Therefore there exists $\alpha > 0$ such that $(\alpha\varphi + \psi)(k^0) = 0$. From (2.32) we obtain that $(\alpha\varphi + \psi)(y) \geq 0$ for every $y \in D$. As above, it follows that $\alpha\varphi + \psi = 0$. Hence $\varphi(y - y') = \varphi(y) + \alpha^{-1}\psi(y') \geq 0$ for all $y, y' \in D$. Since $D - D = Y$, we obtain the contradiction $\varphi = 0$. Hence $\{k^0, -k^0\} \cap D^i \neq \emptyset$. $\square$

Remark 2.3.8. By a similar proof, if $D \subset Y$ is a convex cone and $k^0 \in Y$, then $D - \mathbb{R}_+ k^0 = Y \Leftrightarrow k^0 \in D^i$.

The function φ_{D,k^0} defined by (2.23) was introduced by Gerstewitz (Tammer) and Iwanow [129]. The most part of the properties of this function established in Theorem 2.3.1 and its corollaries were stated by Zălinescu [388], Gerth (Tammer), and Weidner [131], Tammer [337] and Göpfert, Tammer, and Zălinescu [144]. Theorem 2.3.6 is stated by Gerth (Tammer), and Weidner [131]; here one can find versions of nonconvex separation theorems without interiority conditions. Proposition 2.3.7 can be found in Zălinescu [388].

2.4 Convexity Notions for Sets and Multifunctions

Let X be a topological vector space over the reals.

Definition 2.4.1. *Let $A \subset X$ be a nonempty set. We say that A is α-convex, where $\alpha \in {]0,1[}$, if $\alpha x + (1 - \alpha)y \in A$ for all $x, y \in A$. The set A is **mid-convex** if A is $\frac{1}{2}$-convex. The set A is **nearly convex** if A is α-convex for some $\alpha \in {]0,1[}$. The empty set is α-convex for all $\alpha \in {]0,1[}$ (and so nearly convex).*

Of course, A is convex if and only if A is α-convex for every $\alpha \in {]0,1[}$.
Let $\alpha \in {]0,1[}$ and $\Lambda^\alpha := \bigcup_{n \geq 0} \Lambda_n^\alpha$, where

$$\Lambda_0^\alpha := \{0,1\}, \quad \Lambda_{n+1}^\alpha := \{\alpha t + (1 - \alpha)s \mid t, s \in \Lambda_n^\alpha\}, \ \forall n \geq 0.$$

It is obvious that $\Lambda_n^\alpha \subset \Lambda_{n+1}^\alpha$ for every $n \geq 0$ and Λ^α is α-convex. Moreover, if A is α-convex, then $\lambda x + (1 - \lambda)y \in A$ for all $x, y \in A$ and $\lambda \in \Lambda^\alpha$. Indeed, fixing $x, y \in A$ and taking $\Lambda_{x,y} := \{\lambda \in [0,1] \mid \lambda x + (1 - \lambda)y \in A\}$, one obtains easily, by induction, that $\Lambda_n^\alpha \subset \Lambda_{x,y}$ for every $n \geq 0$.

Lemma 2.4.2. *Let $\Lambda \subset [0,1]$ be such that $0,1 \in \Lambda$ and*

$$\exists \bar{\delta} \in {]0, 1/2]}, \ \forall t,s \in \Lambda, \ \exists \lambda \in [\bar{\delta}, 1-\bar{\delta}] \ : \ \lambda t + (1-\lambda)s \in \Lambda.$$

Then $\operatorname{cl}\Lambda = [0,1]$. *In particular,* $\operatorname{cl}\Lambda^\alpha = [0,1]$.

PROOF. Suppose that $\operatorname{cl}\Lambda \neq [0,1]$. Then there exists $\bar{t} \in [0,1] \setminus \operatorname{cl}\Lambda = {]0,1[} \setminus \operatorname{cl}\Lambda$. Since the last set is open in $\mathbb{R}$, it follows that

$$\bar{\alpha} := \sup\{t \in [0,\bar{t}] \mid t \in \operatorname{cl}\Lambda\} < \bar{t}, \quad \bar{\beta} := \inf\{t \in [\bar{t},1] \mid t \in \operatorname{cl}\Lambda\} > \bar{t}$$

and $\bar{\alpha}, \bar{\beta} \in \operatorname{cl}\Lambda$. Let $\delta := \bar{\delta}(\bar{\beta} - \bar{\alpha}) > 0$. From the definitions of $\bar{\alpha}$ and $\bar{\beta}$, there exist $\alpha, \beta \in \Lambda$ such that $\bar{\alpha} - \delta < \alpha \leq \bar{\alpha}$, $\bar{\beta} \leq \beta < \bar{\beta} + \delta$. Since $\alpha, \beta \in \Lambda$, there exists $\lambda \in [\delta_0, 1 - \delta_0]$ with $\lambda\alpha + (1-\lambda)\beta \in \Lambda$. But

$$\lambda\alpha + (1-\lambda)\beta \leq \lambda\bar{\alpha} + (1-\lambda)(\bar{\beta} + \delta) = \bar{\beta} + \delta - \lambda(\bar{\beta} - \bar{\alpha} + \delta)$$
$$\leq \bar{\beta} + \delta - \bar{\delta}(\bar{\beta} - \bar{\alpha} + \delta) = \bar{\beta} - \delta\bar{\delta} < \bar{\beta},$$
$$\lambda\alpha + (1-\lambda)\beta \geq \lambda(\bar{\alpha} - \delta) + (1-\lambda)\bar{\beta} = \bar{\alpha} - \delta + (1-\lambda)(\bar{\beta} - \bar{\alpha} + \delta)$$
$$\geq \bar{\alpha} - \delta + \bar{\delta}(\bar{\beta} - \bar{\alpha} + \delta) = \bar{\alpha} + \delta\bar{\delta} > \bar{\alpha},$$

contradicting the fact that $\Lambda \cap {]\bar{\alpha}, \bar{\beta}[} = \emptyset$. Therefore $\operatorname{cl}\Lambda = [0,1]$.

It is obvious that $\Lambda = \Lambda^\alpha$ satisfies the hypothesis with $\bar{\delta} = \min\{\alpha, 1 - \alpha\}$. $\square$

The first statement of the next result can be found in Zălinescu [390, Prop. 2.4].

Proposition 2.4.3. *Let X be a topological vector space and $A \subset X$ a nonempty nearly convex set. Then*

(i) $\operatorname{cl}A$ *is convex.*

(ii) *If $x \in \operatorname{aint}A$ and $y \in A$, then $[x,y] \subset A$. Moreover, if $x \in \operatorname{int}A$ and $y \in A$, then $[x,y[\subset \operatorname{int}A$.*

(iii) *If $\operatorname{int}A \neq \emptyset$, then $\operatorname{int}A$ is convex and $\operatorname{aint}A = \operatorname{int}A$.*

PROOF. By hypothesis there exists $\alpha \in {]0,1[}$ such that A is α-convex. Note first that for all $\gamma \in \Lambda^\alpha$ and $x,y \in A$ we have that $\gamma x + (1-\gamma)y \in A$.

(i) Consider $x,y \in \operatorname{cl}A$ and $\lambda \in {]0,1[}$. Since $\lambda \in \operatorname{cl}\Lambda^\alpha$, there exist the nets $(x_i), (y_i) \subset A$, and $(\lambda_i) \subset \Lambda^\alpha$ converging to x, y, and λ, respectively. Since $\lambda_i x_i + (1-\lambda_i)y_i \in A$ for every i, we obtain that $\lambda x + (1-\lambda)y \in \operatorname{cl}A$. Therefore $\operatorname{cl}A$ is convex.

(ii) Let $x \in \operatorname{aint}A$ and $y \in A$. Suppose that there exists $\lambda_0 \in {]0,1[}$ such that $\lambda_0 x + (1 - \lambda_0)y \notin A$; of course, $x \neq y$. Since $x \in \operatorname{aint}A$, there exists $\bar{\delta} \in {]0,1[}$ such that $(1-\lambda)x + \lambda y \in A$ for every $\lambda \in [-\bar{\delta}, \bar{\delta}]$. Let

$$\bar{\gamma} := \sup\{\gamma \in {]0,1[} \mid (1-\lambda)x + \lambda y \in A \ \forall \lambda \in [0,\gamma[\}.$$

Of course, $\bar{\delta} \leq \bar{\gamma} \leq \lambda_0$ and $(1-\lambda)x + \lambda y \in A$ for every $\lambda \in [0, \bar{\gamma}[$. Let $\gamma \in \Lambda^\alpha \cap [0, \bar{\gamma}[$. It follows that for all $\lambda \in [0, \bar{\gamma}[$ we have

$$(1 - \gamma)\,((1 - \lambda)x + \lambda y) + \gamma y = (1 - (\lambda + \gamma - \lambda\gamma))\,x + (\lambda + \gamma - \lambda\gamma)y \in A.$$

But $\overline{\gamma} \in [\gamma, \overline{\gamma} + \gamma(1 - \overline{\gamma})[= \{\lambda + \gamma - \lambda\gamma \mid \lambda \in [0, \overline{\gamma}]\}$. From the above relation it follows that $(1 - \lambda)x + \lambda y \in A$ for every $\lambda \in [0, \overline{\gamma} + \gamma(1 - \overline{\gamma})[$, contradicting the choice of $\overline{\gamma}$.

Now let $x \in \operatorname{int} A$, $y \in A$, and $\lambda \in\,]0, 1[$. There exists a neighborhood V of the origin such that $x + V \subset \operatorname{int} A \subset \operatorname{aint} A$. From the first part we have $\lambda x + (1 - \lambda)y + \lambda V \subset A$, which shows that $\lambda x + (1 - \lambda)y \in \operatorname{int} A$.

(iii) Suppose that $\operatorname{int} A \neq \emptyset$ and fix $x_0 \in \operatorname{int} A$. Consider $x \in \operatorname{aint} A$. There exist $y \in A$ and $\lambda \in\,]0, 1[$ such that $x = (1 - \lambda)x_0 + \lambda y$. From the second part of (ii) we get that $x \in \operatorname{int} A$. □

An immediate consequence of the preceding proposition is the next corollary.

Corollary 2.4.4. *Let X be a topological vector space and $A \subset X$ a nonempty nearly convex set. If A is open or closed, then A is convex.* □

Remark 2.4.5. If the nonempty subset A of the t.v.s. X satisfies the condition

$$\exists\, \overline{\delta} \in\,]0, 1/2], \ \forall\, x, y \in A, \ \exists \lambda \in [\overline{\delta}, 1 - \overline{\delta}] \ : \ \lambda x + (1 - \lambda)y \in A,$$

then $\operatorname{cl} A$ is convex, but $\operatorname{int} A$ is not necessarily convex.

Indeed, the proof of the fact that $\operatorname{cl} A$ is convex is similar to that of Proposition 2.4.3, taking into account Lemma 2.4.2. The set $A = \mathbb{R} \setminus \{0\}$ satisfies the above condition, is open, but is not convex.

Note that the intersection of an arbitrary family of α-convex sets is also α-convex, but the intersection of two nearly convex sets may not be nearly convex; indeed, $\Lambda^{\frac{1}{2}} \cap \Lambda^{\frac{1}{3}} = \{0, 1\}$ is not nearly convex.

If X, Y are real vector spaces, $T : X \to Y$ is a linear operator and $A \subset X$, $B \subset Y$ are α-convex, then $T(A)$ and $T^{-1}(B)$ are α-convex, too.

Let X, Y be arbitrary nonempty sets. A function $\Gamma : X \to 2^Y$ is called a **multifunction**, and is denoted by $\Gamma : X \rightrightarrows Y$. So, if $\Gamma : X \rightrightarrows Y$ is a multifunction, the image of Γ at $x \in X$ is a (possibly empty) subset $\Gamma(x)$ of Y. The **domain** of Γ is $\operatorname{dom} \Gamma := \{x \in X \mid \Gamma(x) \neq \emptyset\}$, while its **image** is $\operatorname{Im} \Gamma := \{y \in Y \mid \exists x \in X \ : \ y \in \Gamma(x)\}$. The multifunction $\Gamma : X \rightrightarrows Y$ is usually identified with its **graph**, $\operatorname{gr} \Gamma := \{(x, y) \in X \times Y \mid y \in \Gamma(x)\}$. In this way, with each multifunction one associates a relation and vice versa. It is obvious that $\operatorname{dom} \Gamma = \operatorname{proj}_X(\operatorname{gr} \Gamma)$ and $\operatorname{Im} \Gamma = \operatorname{proj}_Y(\operatorname{gr} \Gamma)$. The image by Γ of the set $A \subset X$ is $\Gamma(A) := \bigcup_{x \in A} \Gamma(x)$; so, $\operatorname{Im} \Gamma = \Gamma(X)$. The inverse image by Γ of the set $B \subset Y$ is $\Gamma^{-1}(B) := \{x \in X \mid \Gamma(x) \cap B \neq \emptyset\}$. In fact, $\Gamma^{-1}(B)$ is the image of B by the inverse multifunction $\Gamma^{-1} : Y \rightrightarrows X$ defined by $\Gamma^{-1}(y) := \{x \in X \mid y \in \Gamma(x)\}$; so $\Gamma^{-1}(y) = \Gamma^{-1}(\{y\})$. Of course, $\operatorname{dom} \Gamma^{-1} = \operatorname{Im} \Gamma$ and $\operatorname{Im} \Gamma^{-1} = \operatorname{dom} \Gamma$. We shall use in the sequel also another type of inverse image: $\Gamma^{+1}(B) := \{x \in X \mid \Gamma(x) \subset B\}$. When $\Delta : Y \rightrightarrows Z$ is another multifunction, the composition of Δ and Γ is the multifunction $\Delta \circ \Gamma : X \rightrightarrows Z$ defined by $\Delta \circ \Gamma(x) := \{z \in Z \mid \exists y \in \Gamma(x)$ with $z \in \Delta(y)\}$; note that $\mathrm{gr}(\Delta \circ \Gamma) = \mathrm{Pr}_{X \times Z}\left(\mathrm{gr}\,\Gamma \times Z \cap X \times \mathrm{gr}\,\Delta\right)$. When Y is a linear space and $\Gamma, \Gamma_1, \Gamma_2 : X \rightrightarrows Y$, we define the sum $\Gamma_1 + \Gamma_2$ and multiplication by a scalar $\gamma\Gamma$ as the multifunctions $\Gamma_1 + \Gamma_2, \gamma\Gamma : X \rightrightarrows Y$ defined by $(\Gamma_1 + \Gamma_2)(x) := \Gamma_1(x) + \Gamma_2(x)$, $(\gamma\Gamma)(x) := \gamma \cdot \Gamma(x)$ with the usual conventions that $A + \emptyset := \emptyset + A := \emptyset$, $\gamma \cdot \emptyset = \emptyset$ for $A \subset Y$.

Suppose now that X, Y are real vector spaces and $\Gamma : X \rightrightarrows Y$. We say that Γ is α-**convex (mid-convex, nearly convex, convex)** if $\mathrm{gr}\,\Gamma$ is α-convex (mid-convex, nearly convex, convex). It is obvious that if Γ is α-convex (mid-convex, nearly convex, convex), so are $\mathrm{dom}\,\Gamma$ and $\mathrm{Im}\,\Gamma$. It is easy to see that Γ is α-convex if and only if

$$\alpha\Gamma(x) + (1 - \alpha)\Gamma(y) \subset \Gamma\left(\alpha x + (1 - \alpha)y\right) \quad \forall x, y \in \mathrm{dom}\,\Gamma.$$

Let $C \subset Y$ be a convex cone. We say that Γ is C-α-**convex (C-mid-convex, C-nearly convex, C-convex)** if the multifunction

$$\Gamma_C : X \rightrightarrows Y, \quad \Gamma_C(x) := \Gamma(x) + C,$$

is α-convex (mid-convex, nearly convex, convex). Of course, Γ is C-α-convex if and only if

$$\alpha\Gamma(x) + (1 - \alpha)\Gamma(y) \subset \Gamma\left(\alpha x + (1 - \alpha)y\right) + C \quad \forall x, y \in \mathrm{dom}\,\Gamma.$$

By inverting the inclusion in these definitions we get concavity notions for multifunctions; so Γ is α-**concave** or C-α-**concave** if

$$\alpha\Gamma(x) + (1 - \alpha)\Gamma(y) \supset \Gamma\left(\alpha x + (1 - \alpha)y\right) \quad \forall x, y \in \mathrm{dom}\,\Gamma$$

or

$$\alpha\Gamma(x) + (1 - \alpha)\Gamma(y) + C \supset \Gamma\left(\alpha x + (1 - \alpha)y\right) \quad \forall x, y \in \mathrm{dom}\,\Gamma,$$

respectively.

Note that sometimes $\mathrm{gr}\,\Gamma_C$ is denoted by $\mathrm{epi}_C\,\Gamma$, or simply $\mathrm{epi}\,\Gamma$, and is called the **epigraph** of Γ. The **sublevel set of Γ of height y** (w.r.t. C) is the set

$$\mathrm{lev}_\Gamma(y) := \{x \in X \mid \Gamma(x) \cap (y - C) \neq \emptyset\};$$

when Y is a topological vector space and $\mathrm{int}\,C \neq \emptyset$ we also consider the **strict sublevel set of Γ of height y** (w.r.t. C) defined by

$$\mathrm{lev}_\Gamma^<(y) := \{x \in X \mid \Gamma(x) \cap (y - \mathrm{int}\,C) \neq \emptyset\}.$$

In this way we get the **sublevel** and **strict sublevel multifunctions** lev_Γ, $\mathrm{lev}_\Gamma^< : Y \rightrightarrows X$.

We say that Γ is C-α-**quasiconvex (C-mid-quasiconvex, C-nearly quasiconvex, C-quasiconvex)** if for every $z \in Y$ the sublevel set $\mathrm{lev}_\Gamma(z)$

is α-convex (mid-convex, nearly convex, convex). An equivalent definition of C-α-quasiconvexity is that

$$(\Gamma(x) + C) \cap (\Gamma(y) + C) \subset \Gamma(\alpha x + (1 - \alpha)y) + C \quad \forall\, x, y \in \operatorname{dom} \Gamma.$$

Notice that Γ is C-α-quasiconvex if

$$\Gamma(x) \subset \Gamma(\alpha x + (1 - \alpha)y) + C \text{ or } \Gamma(y) \subset \Gamma(\alpha x + (1 - \alpha)y) + C$$

for all $x, y \in \operatorname{dom} \Gamma$. Note also that Γ is C-α-quasiconvex (C-mid-quasiconvex, C-nearly quasiconvex, C-quasiconvex) whenever Γ is C-α-convex (C-mid-convex, C-nearly convex, C-convex).

In order to characterize the C-α-quasiconvexity of Γ we use the scalarization function φ introduced in relation (2.23) (for D replaced by C).

Proposition 2.4.6. ([246, Prop. 2.3], [69, Lemma 2.3]) *Let X, Y be topological vector spaces, $\Gamma : X \rightrightarrows Y$ have nonempty domain, $C \subset Y$ be a convex cone, and $k^0 \in \operatorname{int} C$ be fixed. Then Γ is C-α-quasiconvex if and only if $\varphi_z \circ \Gamma : X \rightrightarrows \mathbb{R}$ is $\mathbb{R}_+$-α-quasiconvex for every $z \in Y$, where $\varphi_z(y) := \varphi(y - z)$ and $\varphi_z \circ \Gamma(x) := \{\varphi_z(y) \mid y \in \Gamma(x)\}$.*

PROOF. For $t \in \mathbb{R}$ and $z \in Y$ we have that

$$
\begin{aligned}
\operatorname{lev}_{\varphi_z \circ \Gamma}(t) &= \{x \in X \mid \varphi_z \circ \Gamma(x) \cap (t - \mathbb{R}_+) \neq \emptyset\} \\
&= \{x \in M \mid \exists\, y \in \Gamma(x) : \varphi_z(y) \leq t\} \\
&= \{x \in M \mid \exists\, y \in \Gamma(x) : \varphi(y - z) \leq t\} \\
&= \{x \in M \mid \exists\, y \in \Gamma(x) : y - z \in tk^0 - C\} \quad \text{(see Corollary 2.3.5)} \\
&= \{x \in M \mid \Gamma(x) \cap (z + tk^0 - C) \neq \emptyset\} \\
&= \operatorname{lev}_\Gamma(z + tk^0).
\end{aligned}
$$

The conclusion follows. $\qquad\square$

We introduce some useful notions and notation related to vector-valued functions. To Y we adjoin a greatest element ∞ ($\notin Y$), thereby obtaining $Y^\bullet := Y \cup \{\infty\}$ and $C^\bullet := C \cup \{\infty\}$. We consider that $y + \infty = \infty$, $\lambda \cdot \infty = \infty$, and $y \leq_C \infty$ for all $y \in Y^\bullet$ and $\lambda \in \mathbb{R}_+$. Of course, if $f : X \to Y^\bullet$, the **domain** of f is $\operatorname{dom} f := \{x \in X \mid f(x) \in Y\}$, the **sublevel** and **strict sublevel sets of f of height** y are $\operatorname{lev}_f(y) := \{x \in X \mid f(x) \leq_C y\}$ and $\operatorname{lev}_f^<(y) := \{x \in X \mid y - f(x) \in \operatorname{int} C\}$, and the **epigraph** of f is $\operatorname{epi} f := \{(x, y) \in X \times Y \mid f(x) \leq_C y\}$. With such an f we associate the multifunction $\Gamma_{f,C} : X \rightrightarrows Y$ whose graph is $\operatorname{epi} f$. So, $\Gamma_{f,C}(x) = f(x) + C$ for every $x \in \operatorname{dom} \Gamma_{f,C} = \operatorname{dom} f$, $\operatorname{epi} f = \operatorname{epi} \Gamma_{f,C}$, $\operatorname{lev}_f(y) = \operatorname{lev}_{\Gamma_{f,C}}(y)$, and $\operatorname{lev}_f^<(y) = \operatorname{lev}_{\Gamma_{f,C}}^<(y)$ for every $y \in Y$; in particular, $\Gamma_{f,C} = (\Gamma_{f,C})_C$. We say that $f : X \to Y^\bullet$ is C-α-**convex** (C-**mid-convex**, C-**nearly convex**, C-**convex**, C-α-**quasiconvex**, C-**mid-quasiconvex**, C-**nearly quasiconvex**, C-**quasiconvex**) if the multifunction $\Gamma_{f,C}$ is C-α-convex (C-mid-convex, C-nearly convex, C-convex, C-α-quasiconvex,

C-mid-quasiconvex, C-nearly quasiconvex, C-quasiconvex); in particular, f is C-convex if and only if

$$f\left(\alpha x_1 + (1 - \alpha)x_2\right) \leq_C \alpha f(x_1) + (1 - \alpha)f(x_2) \quad \forall x_1, x_2 \in X, \ \forall \alpha \in [0, 1].$$

If f is C-α-convex (C-mid-convex, C-nearly convex, C-convex), then dom f is so, and f is C-α-quasiconvex (C-mid-quasiconvex, C-nearly quasiconvex, C-quasiconvex).

We also remark that f is C-α-quasiconvex, provided that

$$\{f(x), f(y)\} \cap \left(f\left(\alpha x + (1 - \alpha)y\right) + C\right) \neq \emptyset \quad \forall x, y \in \mathrm{dom}\, f.$$

For the converse we need the condition $C \cup (-C) = Y$. Indeed, let $x, y \in \mathrm{dom}\, f$; suppose that $f(x) \notin f\left(\alpha x + (1 - \alpha)y\right) + C$, that is, $\alpha x + (1 - \alpha)y \notin \mathrm{lev}_f(f(x))$. Since $\mathrm{lev}_f(f(x))$ is α-convex, we deduce that $y \notin \mathrm{lev}_f(f(x))$, which means that $f(x) - f(y) \notin C$. This fact, together with $C \cup (-C) = Y$, implies that $f(y) - f(x) \in C$, that is, $x \in \mathrm{lef}_f(f(y))$. Since $\mathrm{lef}_f(f(y))$ is α-convex, we obtain that $f(y) \in f\left(\alpha x + (1 - \alpha)y\right) + C$, which completes the proof.

This remark shows that $\mathbb{R}_+$-$(\alpha$-$)$quasiconvexity of real-valued functions reduces to usual $(\alpha$-$)$quasiconvexity. So, in Proposition 2.4.6 we may replace the $\mathbb{R}_+$-α-quasiconvexity of $\varphi_z \circ f$ with the usual α-quasiconvexity.

Let Y be ordered by the proper convex cone C and $f : X \to Y^\bullet$.

Definition 2.4.7. *We call the **subdifferential** of f at $x_0 \in \mathrm{dom}\, f$ the set*

$$\partial^{\leq_C} f(x_0) := \{T \in L(X, Y) \mid T(x - x_0) \leq_C f(x) - f(x_0) \ \forall x \in X\},$$

where $L(X, Y)$ denotes the class of all continuous linear operators from X into Y.

It is obvious that $\partial^{\leq_C} f(x_0)$ is a convex subset of $L(X, Y)$. When $f : X \to Y^\bullet$ is a **sublinear operator**; i.e., $f(x_1 + x_2) \leq_C f(x_1) + f(x_2)$, $f(0) = 0$, and $f(\alpha x) = \alpha f(x)$ for all $x, x_1, x_2 \in X$ and $\alpha \in (0, \infty)$, and C is pointed, the formula

$$\partial^{\leq_C} f(x_0) = \{T \in \partial^{\leq_C} f(0) \mid T(x_0) = f(x_0)\}$$

holds for all $x_0 \in \mathrm{dom}\, f$. The following important result was stated by Valadier [360].

Theorem 2.4.8. *Let $(X, \|\cdot\|)$ and $(Y, \|\cdot\|)$ be real reflexive Banach spaces and $C \subset Y$ a proper convex cone with a weak compact base. If $f : X \to Y^\bullet$ is a C-convex operator, continuous at some point of its domain, then*

$$y^* \circ \partial^{\leq_C} f(x) = \partial(y^* \circ f)(x) \quad \forall x \in \mathrm{int}(\mathrm{dom}\, f), \ \forall y^* \in C^+,$$

where we use the convention that $y^*(\infty) = \infty$ for $y^* \in C^+$.

2.5 Continuity Notions for Multifunctions

In this section X and Y are separated (in the sense of Hausdorff) topological spaces and $\Gamma : X \rightrightarrows Y$ a multifunction. When mentioned explicitly, Y is a separated topological vector space (s.t.v.s.).

Definition 2.5.1. *Let $x_0 \in X$. We say that*

(a) *Γ is **upper continuous** (u.c.) at x_0 if*

$$\forall D \subset Y, \ D \text{ open}, \ \Gamma(x_0) \subset D, \ \exists U \in \mathcal{V}_X(x_0) \ \forall x \in U \ : \ \Gamma(x) \subset D, \tag{2.33}$$

i.e., $\Gamma^{+1}(D)$ is a neighborhood of x_0 for each open set $D \subset Y$ such that $\Gamma(x_0) \subset D$;

(b) *Γ is **lower continuous** (l.c.) at x_0 if*

$$\forall D \subset Y, \ D \text{ open}, \ \Gamma(x_0) \cap D \neq \emptyset, \ \exists U \in \mathcal{V}_X(x_0), \ \forall x \in U \ : \ \Gamma(x) \cap D \neq \emptyset, \tag{2.34}$$

i.e., $\Gamma^{-1}(D)$ is a neighborhood of x_0 for each open set $D \subset Y$ such that $\Gamma(x_0) \cap D \neq \emptyset$.

(c) *Γ is **continuous** at x_0 if Γ is u.c. and l.c. at x_0.*

(d) *Γ is **upper continuous** (**lower continuous, continuous**) if Γ is so at every $x \in X$;*

(e) *Γ is **lower continuous** at $(x_0, y_0) \in X \times Y$ if*

$$\forall V \in \mathcal{V}_Y(y_0), \ \exists U \in \mathcal{V}_X(x_0), \ \forall x \in U \ : \ \Gamma(x) \cap V \neq \emptyset.$$

It follows from the definition that $x_0 \in \mathrm{int}(\mathrm{dom}\,\Gamma)$ and $y_0 \in \mathrm{cl}\,(\Gamma(x_0))$ if Γ is l.c. at (x_0, y_0) and Γ is l.c. at $x_0 \in \mathrm{dom}\,\Gamma$ if and only if Γ is l.c. at every (x_0, y) with $y \in \Gamma(x_0)$; moreover, Γ is l.c. at every $x_0 \in X \setminus \mathrm{dom}\,\Gamma$. If $x_0 \in X \setminus \mathrm{dom}\,\Gamma$, then Γ is u.c. at x_0 if and only if $x_0 \in \mathrm{int}(X \setminus \mathrm{dom}\,\Gamma)$. So, if Γ is u.c., then $\mathrm{dom}\,\Gamma$ is closed, while if Γ is l.c., then $\mathrm{dom}\,\Gamma$ is open.

The next result follows immediately from the definitions.

Proposition 2.5.2. (i) *Γ is upper continuous if and only if $\Gamma^{+1}(D)$ is open for every open set $D \subset Y$;*

(ii) *Γ is lower continuous if and only if $\Gamma^{-1}(D)$ is open for every open set $D \subset Y$.*

The **limit inferior** of Γ at $x_0 \in X$ is defined by

$$\liminf_{x \to x_0} \Gamma(x) := \{ y \in Y \mid \forall V \in \mathcal{V}_Y(y), \ \exists U \in \mathcal{V}_X(x_0),$$

$$\forall x \in U^\bullet \ : \ \Gamma(x) \cap V \neq \emptyset \},$$

while the **limit superior** of Γ at $x_0 \in X$ is defined by

$$\limsup_{x \to x_0} \Gamma(x) := \{y \in Y \mid \forall V \in \mathcal{V}_Y(y), \ \forall U \in \mathcal{V}_X(x_0),$$

$$\exists x \in U^\bullet \ : \ \Gamma(x) \cap V \neq \emptyset\}$$

$$= \bigcap_{U \in \mathcal{V}_X(x_0)} \mathrm{cl}\,(\Gamma(U^\bullet)),$$

where for $U \in \mathcal{V}_X(x_0)$, $U^\bullet := U \setminus \{x_0\}$.

Note that if $\limsup_{x \to x_0} \Gamma(x) \neq \emptyset$, then x_0 is an accumulation point of dom Γ, while if $\liminf_{x \to x_0} \Gamma(x) \neq \emptyset$, then $x_0 \in \mathrm{int}\,(\mathrm{dom}\,\Gamma \cup \{x_0\})$; if x_0 is an isolated point of X (i.e., $\{x_0\} \in \mathcal{V}_X(x_0)$), then $\liminf_{x \to x_0} \Gamma(x) = Y$. Of course, if x_0 is an accumulation point of dom Γ, then

$$\liminf_{x \to x_0} \Gamma(x) \subset \limsup_{x \to x_0} \Gamma(x),$$

both of them being closed sets.

Sometimes in the definitions of $\liminf_{x \to x_0} \Gamma(x)$ and $\limsup_{x \to x_0} \Gamma(x)$ one takes $x \in \mathrm{dom}\,\Gamma$. Note that this situation reduces to the preceding one by considering the restriction of Γ at $\{x_0\} \cup \mathrm{dom}\,\Gamma$.

Recall first that if (X, d) is a metric space and $A, B \subset X$, then the **excess of A over B** is

$$e(A, B) = \sup_{x \in A} \mathrm{dist}(x, B) \ \text{if } A, B \neq \emptyset, \quad e(\emptyset, B) = 0, \quad e(A, \emptyset) = \infty \ \text{if } A \neq \emptyset,$$

where $\mathrm{dist}(x, A) := \inf_{a \in A} d(x, a)$; in particular, $\mathrm{dist}(x, \emptyset) = \infty$. It is simple to show that for a nonempty and compact set $A \subset X$ and an open set $D \subset X$, if $A \subset D$, then there exists $\varepsilon > 0$ such that $A_\varepsilon := \{x \in X \mid \mathrm{dist}(x, A) < \varepsilon\} = \{x \mid A \cap B(x, \varepsilon) \neq \emptyset\} \subset D$, where $B(x, \varepsilon) := \{x' \in X \mid d(x, x') < \varepsilon\}$.

In particular, cases for X or/and Y one has useful characterizations for the elements of $\liminf_{x \to x_0} \Gamma(x)$ and $\limsup_{x \to x_0} \Gamma(x)$.

Proposition 2.5.3. *Let $x_0 \in X$ and $y \in Y$.*

(i) $y \in \liminf_{x \to x_0} \Gamma(x)$ *if and only if*

$$\forall X \setminus \{x_0\} \supset (x_i)_{i \in I} \to x_0, \ \exists (x_{\varphi(j)})_{j \in J}, \ \exists Y \supset (y_j)_{j \in J} \to y,$$
$$\forall j \in J \ : \ y_j \in \Gamma(x_{\varphi(j)}), \tag{2.35}$$

and $y \in \limsup_{x \to x_0} \Gamma(x)$ if and only if

$$\exists X \setminus \{x_0\} \supset (x_i)_{i \in I} \to x_0, \ \exists Y \supset (y_i)_{i \in I} \to y, \ \forall i \in I \ : \ y_i \in \Gamma(x_i). \tag{2.36}$$

(ii) *If X and Y are first countable, then $y \in \liminf_{x \to x_0} \Gamma(x)$ if and only if*

$$\forall X \setminus \{x_0\} \supset (x_n) \to x_0, \ \exists Y \supset (y_n) \to y, \ \exists n_0 \in \mathbb{N}, \ \forall n \geq n_0 \ :$$
$$y_n \in \Gamma(x_n), \tag{2.37}$$

and $y \in \limsup_{x \to x_0} \Gamma(x)$ if and only if

$$\exists X \setminus \{x_0\} \supset (x_n) \to x_0, \ \exists Y \supset (y_n) \to y, \ \forall n \in \mathbb{N} \ : \ y_n \in \Gamma(x_n). \tag{2.38}$$

(iii) *If (Y, ρ) is a metric space, then*

$$y \in \liminf_{x \to x_0} \Gamma(x) \Leftrightarrow \lim_{x \to x_0} \operatorname{dist}(y, \Gamma(x)) = 0, \tag{2.39}$$

$$y \in \limsup_{x \to x_0} \Gamma(x) \Leftrightarrow \liminf_{x \to x_0} \operatorname{dist}(y, \Gamma(x)) = 0. \tag{2.40}$$

PROOF. (i) Suppose that $y \in \liminf_{x \to x_0} \Gamma(x)$ and take $X \setminus \{x_0\} \supset (x_i)_{i \in I} \to x_0$. Then for every $V \in \mathcal{V}_Y(y)$ there exists $i_V \in I$ such that $\Gamma(x_i) \cap V \neq \emptyset$ for $i \succeq i_V$. Consider $J := I \times \mathcal{V}_Y(y)$ ordered by $(i, V) \succeq (i', V')$ if and only if $i \succeq i'$ and $V \subset V'$. Consider $\varphi : J \to I$ such that $\varphi(i, V) \succeq i, i_V$ and take $y_j \in \Gamma(x_{\varphi(j)}) \cap V$ for $j = (i, V)$. It is obvious that $(x_{\varphi(j)})_{j \in J}$ is a subnet of $(x_i)_{i \in I}$. Moreover, $(y_j)_{j \in J} \to y$. Therefore (2.35) holds.

Suppose now that (2.35) holds, but $y \notin \liminf_{x \to x_0} \Gamma(x)$. Then there exists $V_0 \in \mathcal{V}_Y(y)$ such that for every $U \in \mathcal{V}_X(x_0)$ there exists $x_U \in U^{\bullet}$ with $\Gamma(x_U) \cap V = \emptyset$. Of course, $(x_U) \to x_0$. Therefore there exists a subnet $(x_{\varphi(j)})_{j \in J}$ and a net $(y_j)_{j \in J} \to y$ such that $y_j \in \Gamma(x_{\varphi(j)})$ for $j \in J$. Since $V_0 \in \mathcal{V}_Y(y)$, there exists $j_0 \in J$ with $y_j \in V_0$ for $j \succeq j_0$. It follows that $y_{j_0} \in \Gamma(x_{\varphi(j_0)}) \cap V_0 = \emptyset$, a contradiction.

Suppose now that $y \in \limsup_{x \to x_0} \Gamma(x)$. Then for every $(U, V) \in \mathcal{V}_X(x_0) \times \mathcal{V}_Y(y) =: I$ there exist $x_{U,V} \in U^{\bullet}$ and $y_{U,V} \in \Gamma(x_{U,V}) \cap V$. Defining, as usual, $(U, V) \succeq (U', V')$ iff $U \subset U'$ and $V \subset V'$, it is obvious that (2.36) holds. The converse implication is obtained easily by contradiction.

(ii) Consider $(U_n)_{n \in \mathbb{N}}$ and $(V_n)_{n \in \mathbb{N}}$ decreasing bases of neighborhoods for x_0 and y, respectively.

Suppose first that $y \in \liminf_{x \to x_0} \Gamma(x)$. Then for $k \in \mathbb{N}$ there exists $m_k \in \mathbb{N}$ such that $\Gamma(x) \cap V_k \neq \emptyset$ for $x \in U_{m_k}^{\bullet}$; without loss of generality we may suppose that (m_k) is increasing. Take $X \setminus \{x_0\} \supset (x_n) \to x_0$. For every $k \in \mathbb{N}$ there exists $n_k \in \mathbb{N}$ such that $x_n \in U_{m_k}$ for $n \geq n_k$; once again, we may suppose that $n_{k+1} > n_k$ for every k. For $k \in \mathbb{N}$ and $n_k \leq n < n_{k+1}$ take $y_n \in \Gamma(x_n) \cap V_k$. Of course, $(y_n) \to y$ and $y_n \in \Gamma(x_n)$ for $n \geq n_0$.

Conversely, suppose that the right-hand side of (2.37) holds, but $y \notin \liminf_{x \to x_0} \Gamma(x)$. Then there exists $W_0 \in \mathcal{V}_Y(y)$ such that for every $U \in \mathcal{V}_X(x_0)$ there exists $x_U \in U^{\bullet}$ with $\Gamma(x_U) \cap W_0 = \emptyset$. Therefore, for every $n \in \mathbb{N}$ there exists $x_n \in U_n$ such that $\Gamma(x_n) \cap W_0 = \emptyset$. By hypothesis, there exists $Y \supset (y_n) \to y$ and $n_0 \in \mathbb{N}$ such that $y_n \in \Gamma(x_n)$ for $n \geq n_0$. Since $(y_n) \to y$, $y_n \in W_0$ for every $n \geq n_1$. Taking $\bar{n} = \max\{n_0, n_1\}$, we obtain that $\Gamma(x_{\bar{n}}) \cap W_0 \neq \emptyset$, a contradiction.

The proof of (2.38) is similar.

(iii) Let Y be a metric space. Recall that for $f : X \to \overline{\mathbb{R}}$ and $x_0 \in X$,

$$\liminf_{x \to x_0} f(x) := \sup_{U \in \mathcal{V}_X(x_0)} \inf_{x \in U^{\bullet}} f(x).$$

So,

$$y \in \limsup_{x \to x_0} \Gamma(x) \Leftrightarrow \forall \varepsilon > 0, \ \forall U \in \mathcal{V}_X(x_0), \ \exists x \in U^\bullet : \ \Gamma(x) \cap B(y, \varepsilon) \neq \emptyset$$

$$\Leftrightarrow \forall \varepsilon > 0, \ \forall U \in \mathcal{V}_X(x_0), \ \exists x \in U^\bullet : \ \mathrm{dist}(y, \Gamma(x)) < \varepsilon$$

$$\Leftrightarrow \forall U \in \mathcal{V}_X(x_0), \ \forall \varepsilon > 0 : \ \inf_{x \in U^\bullet} \mathrm{dist}(y, \Gamma(x)) < \varepsilon$$

$$\Leftrightarrow \liminf_{x \to x_0} \mathrm{dist}(y, \Gamma(x)) = 0,$$

i.e., (2.40) holds. The proof of (2.39) is similar.   □

Using the above remark, we have the following characterization of upper continuity in a special case.

Proposition 2.5.4. *Suppose that* (Y, ρ) *is a metric space and* $\Gamma(x_0)$ *is compact. Then* Γ *is u.c. at* x_0 *if and only if* $\lim_{x \to x_0} e(\Gamma(x), \Gamma(x_0)) = 0$.

PROOF. The (easy) proof is left to the reader.   □

In the next results we give (other) characterizations for upper and lower continuity at a point.

Proposition 2.5.5. *Let* X, Y *be first countable (in particular, let* X, Y *be metric spaces) and* $x \in \mathrm{dom}\,\Gamma$. *The following statements are equivalent:*

(i) *the multifunction* Γ *is u.c. at* x;
(ii) *for every closed set* $F \subset Y$ *with* $\Gamma(x) \cap F = \emptyset$, *there exists* $U \in \mathcal{V}_X(x)$ *such that* $\Gamma(U) \cap F = \emptyset$;
(iii) *for every closed set* $F \subset Y$ *and every sequence* $X \supset (x_n) \to x$ *with* $\Gamma(x_n) \cap F \neq \emptyset$ *for* $n \in \mathbb{N}$, *we have* $\Gamma(x) \cap F \neq \emptyset$;
(iv) *for every open set* $D \subset Y$ *with* $\Gamma(x) \subset D$, *and every sequence* $X \supset (x_n) \to x$ *there exists* $n_D \in \mathbb{N}$ *such that* $\Gamma(x_n) \subset D$ *for* $n \geq n_D$;
(v) *for all sequences* $(x_n) \subset X$, $(y_n) \subset Y$ *with* $(x_n) \to x$ *and* $y_n \in \Gamma(x_n) \setminus \Gamma(x)$ *for* $n \in \mathbb{N}$, *there exists a subsequence* $(y_{n_k}) \to y \in \Gamma(x)$.

If X, Y *are not first countable, the conditions* (i)–(iv) *remain equivalent by replacing sequences by nets.*

PROOF. The equivalence of conditions (i)–(iv) is quite simple (and known), so it is left to the reader.

(iii) $\Rightarrow$ (v) Consider $(V_n)_{n \in \mathbb{N}}$ a countable basis of neighborhoods for $y \in Y$; we suppose (without loss of generality) that $V_{n+1} \subset V_n$ for every $n \in \mathbb{N}$.

Let $(x_n) \subset X$, $(y_n) \subset Y$ with $(x_n) \to x$ and $y_n \in \Gamma(x_n) \setminus \Gamma(x)$ for $n \in \mathbb{N}$, and consider $F := \mathrm{cl}\{y_n \mid n \in \mathbb{N}\}$. Since F is closed and $\Gamma(x_n) \cap F \neq \emptyset$ for every $n \in \mathbb{N}$, there exists $y \in \Gamma(x) \cap F$. Of course, $y_n \neq y$ for every $n \in \mathbb{N}$. Since $y \in F$, there exists $y_{n_1} \in V_1$. Since Y is separated, there exists $m_1 \in \mathbb{N}$ such that $V_{m_1} \cap \{y_k \mid 0 \leq k \leq n_1\} = \emptyset$. There exists $n_2 \in \mathbb{N}$ such that $y_{n_2} \in V_{m_1}$. Similarly, there exists $m_2 \in \mathbb{N}$ such that $V_{m_2} \cap \{y_k \mid 0 \leq k \leq n_2\} = \emptyset$. There exists $n_3 \in \mathbb{N}$ such that $y_{n_3} \in V_{m_2}$. The choice of m_2 shows that $m_2 > m_1$ and $n_3 > n_2$. Continuing in this way we find increasing sequences $(m_k), (n_k) \subset \mathbb{N}$ such that $y_{n_k} \in V_{m_k}$ for every $k \in \mathbb{N}$. It is obvious that $(y_{n_k}) \to y$.

(v) $\Rightarrow$ (iv) In the contrary case, there exist an open set $D \subset Y$ with $\Gamma(x) \subset D$ and a sequence $X \supset (x_n) \to x$ such that $P := \{n \in \mathbb{N} \mid \Gamma(x_n) \not\subset D\}$ is infinite. For each $n \in P$ take $y_n \in \Gamma(x_n) \setminus D \subset \Gamma(x_n) \setminus \Gamma(x)$. Letting $P = \{n_0, n_1, \ldots, n_k, \ldots\}$ with $n_0 < n_1 < \cdots < n_k < \cdots$, by (v) there exists the subsequence $(y_{n_{k_p}}) \to y \in \Gamma(x) \setminus D$, a contradiction. □

Proposition 2.5.6. *Let $(x_0, y_0) \in X \times Y$. The following statements are equivalent:*

(i) *the multifunction Γ is l.c. at (x_0, y_0);*
(ii) *for every net $(x_i)_{i \in I} \to x_0$ there exist a subnet $(x_{\varphi(j)})_{j \in J}$ of (x_i) and a net $(y_j)_{j \in J} \to y_0$ such that $y_j \in \Gamma(x_{\varphi(j)})$ for $j \in J$.*
(iii) $y_0 \in \liminf_{x \to x_0} \Gamma(x) \cap \operatorname{cl} \Gamma(x_0)$.

Suppose now that X, Y are first countable. Then (i) *is equivalent to*

(iv) *for every sequence $X \supset (x_n) \to x_0$ there exist a sequence $Y \supset (y_n) \to y_0$ and $n_0 \in \mathbb{N}$ such that $y_n \in \Gamma(x_n)$ for every $n \geq n_0$.*

PROOF. The proofs of the equivalences (i) $\Leftrightarrow$ (ii) and (i) $\Leftrightarrow$ (iv) are, mainly, the same as those of the first parts of Proposition 2.5.3 (i) and (ii), respectively. The equivalence of (i) and (iii) is immediate. □

From the preceding result we obtain that

$$\Gamma \text{ is l.c. at } x_0 \Leftrightarrow \Gamma(x_0) \subset \liminf_{x \to x_0} \Gamma(x).$$

Definition 2.5.7. *We say that*

(i) Γ *is **closed** if* $\operatorname{gr} \Gamma$ *is a closed subset of $X \times Y$;*
(ii) Γ *is **closed** at $x \in X$ if for every net $((x_i, y_i))_{i \in I} \subset \operatorname{gr} \Gamma$ converging to (x, y) we have that $y \in \Gamma(x)$;*
(iii) Γ *is **closed-valued** if $\Gamma(x)$ is closed for every $x \in X$;*
(iv) Γ *is **compact** at $x \in X$ if for every net $((x_i, y_i))_{i \in I} \subset \operatorname{gr} \Gamma$ with $(x_i) \to x$, there exists a subnet $(y_{\varphi(j)})_{j \in J}$ converging to some $y \in \Gamma(x)$.*

Of course, Γ is closed if and only if Γ is closed at every $x \in X$; moreover, if Γ is closed, then Γ is closed-valued. If Γ is compact at x, then Γ is closed at x; moreover, if Γ is closed (compact) at x, then $\Gamma(x)$ is closed (compact). Note also that Γ is compact at $x \in X \setminus \operatorname{dom} \Gamma$ if and only if $x \in \operatorname{int}(X \setminus \operatorname{dom} \Gamma)$, but Γ may be closed at some $x \in \operatorname{cl}(\operatorname{dom} \Gamma) \setminus \operatorname{dom} \Gamma$. Indeed, consider $\Gamma : \mathbb{R} \rightrightarrows \mathbb{R}$ with $\operatorname{dom} \Gamma =]0, 1[$ and $\Gamma(x) = \{x^{-1}\}$ for $x \in \operatorname{dom} \Gamma$; Γ is closed at 0 (but not compact).

Proposition 2.5.8. *Assume that $\emptyset \neq A \subset \operatorname{dom} \Gamma$ and Γ is u.c. at every $x \in A$.*

(i) *If A is compact and $\Gamma(x)$ is compact for every $x \in A$, then $\Gamma(A)$ is compact.*

(ii) *If A is connected and $\Gamma(x)$ is connected for every $x \in A$, then $\Gamma(A)$ is connected.*

Recall that the nonempty subset A of the topological space (X, τ) is **connected** if $A \cap D_1 \cap D_2 \neq \emptyset$ whenever D_1, D_2 are open subsets of X such that $A \cap D_1 \neq \emptyset$, $A \cap D_2 \neq \emptyset$ and $A \subset D_1 \cup D_2$. Recall that the connected subsets of $\mathbb{R}$ are the intervals, while in a topological vector space every convex set is connected.

PROOF. Replacing Γ by $\Gamma_A : A \rightrightarrows Y$ defined by $\Gamma_A(x) := \Gamma(x)$ for $x \in A$, we may assume that $A = X = \operatorname{dom}\Gamma$.

(i) Let $(D_i)_{i \in I}$ be an open cover of $\Gamma(X)$. Then for every $x \in X$ we have that $\Gamma(x) \subset \bigcup_{i \in I} D_i$. Since $\Gamma(x)$ is compact, there exists a finite set $I_x \subset I$ such that $\Gamma(x) \subset \bigcup_{i \in I_x} D_i =: D_x$; of course, $x \in \Gamma^{+1}(D_x)$. Hence $X = \bigcup_{x \in X} \Gamma^{+1}(D_x)$. Since D_x is open, $\Gamma^{+1}(D_x)$ is open, and so $\left(\Gamma^{+1}(D_x)\right)_{x \in X}$ is an open cover of the compact set X. Therefore there exists a finite set $X_0 \subset X$ such that $X = \bigcup_{x \in X_0} \Gamma^{+1}(D_x)$. Let $I_0 := \bigcup_{x \in X_0} I_x$. Of course, I_0 is finite. Let $y \in \Gamma(X)$; hence $y \in \Gamma(x)$ for some $x \in X = \bigcup_{x' \in X_0} \Gamma^{+1}(D_{x'})$. Therefore there exists $x_0 \in X_0$ such that $x \in \Gamma^{+1}(D_{x_0})$. It follows that $y \in \Gamma(x) \subset D_{x_0} = \bigcup_{i \in I_{x_0}} D_i \subset \bigcup_{i \in I_0} D_i$. Hence $\Gamma(X)$ is compact.

(ii) Now let $D_1, D_2 \subset Y$ be open sets such that $\Gamma(X) \cap D_1 \neq \emptyset$, $\Gamma(X) \cap D_2 \neq \emptyset$ and $\Gamma(X) \subset D_1 \cup D_2$. We have to show that $\Gamma(X) \cap D_1 \cap D_2 \neq \emptyset$. Assume that $\Gamma(x) \cap D_1 \cap D_2 = \emptyset$ for every $x \in X$. Because $\Gamma(x)$ is connected and $\Gamma(x) \subset D_1 \cup D_2$ we have that $\Gamma(x) \cap D_1 = \emptyset$ or $\Gamma(x) \cap D_2 = \emptyset$; hence $\Gamma(x) \subset D_2$ or $\Gamma(x) \subset D_1$. It follows that $X = \Gamma^{+1}(D_1) \cup \Gamma^{+1}(D_2)$. From our hypotheses we have that $\Gamma^{+1}(D_1)$ and $\Gamma^{+1}(D_2)$ are nonempty open sets. Because $X = A$ is connected we obtain that there is some $x_0 \in \Gamma^{+1}(D_1) \cap \Gamma^{+1}(D_2)$. Then $\Gamma(x_0) \subset D_1 \cap D_2$, whence the contradiction $\Gamma(x_0) \cap D_1 \cap D_2 \neq \emptyset$. Therefore $\Gamma(x) \cap D_1 \cap D_2 \neq \emptyset$ for some $x \in X$, and so $\Gamma(X) \cap D_1 \cap D_2 \neq \emptyset$. Hence $\Gamma(X)$ is connected. $\square$

When Y is a metric space or a topological vector space, Proposition 2.5.8(i) follows also from the next result.

Proposition 2.5.9. *Let $x \in X$. The following assertions hold:*

(i) *Γ is compact at x if and only if $\Gamma(x)$ is compact and Γ is u.c. at x.*

(ii) *If (Y, ρ) is a metric space, then Γ is compact at x if and only if $\Gamma(x)$ is compact and $\lim_{x' \to x} e(\Gamma(x'), \Gamma(x)) = 0$.*

(iii) *Let Y be first countable. If Γ is compact at x, then for every sequence $((x_n, y_n))_{n \in \mathbb{N}} \subset \operatorname{gr}\Gamma$ with $(x_n) \to x$, there exists a subsequence $(y_{n_k}) \to y \in \Gamma(x)$. If X is first countable and Y is a metric space, the converse is true.*

PROOF. If $x \notin \operatorname{dom}\Gamma$, the hypotheses of every assertion of the proposition imply that $x \in \operatorname{int}(X \setminus \operatorname{dom}\Gamma)$; so the conclusion is obvious. So we suppose, during the proof, that $x \in \operatorname{dom}\Gamma$.

(i) We noted already that $\Gamma(x)$ is compact. Suppose that Γ is not u.c. at x. Then there exists an open set $D \subset Y$ such that for every $U \in \mathcal{V}_X(x)$ there exists $x_U \in U$ and $y_U \in \Gamma(x_U) \setminus D$. Since Γ is compact at x, there exists a subnet $(y_{\varphi(j)})_{j \in J}$ converging to $y \in \Gamma(x)$. Since $(y_{\varphi(j)})_{j \in J} \subset Y \setminus D$ and D is open, it follows that $y \in Y \setminus D$, contradicting $\Gamma(x) \subset D$.

(ii) Let $((x_i, y_i))_{i \in I} \subset \operatorname{gr} \Gamma$ with $(x_i) \to x$. For every $j \in I$ consider the set $F_j := \operatorname{cl}\{y_i \mid i \succeq j\}$. Suppose that for some $j_0 \in I$, $F_{j_0} \cap \Gamma(x) = \emptyset$. Then $\Gamma(x) \subset Y \setminus F_{j_0}$. Since $Y \setminus F_{j_0}$ is open and Γ is u.c. at x, there exists $i_0 \in I$ such that $\Gamma(x_i) \subset Y \setminus F_{j_0}$ for every $i \succeq i_0$. Taking $i \in I$ such that $i \succeq i_0$ and $i \succeq j_0$ we get the contradiction $y_i \in F_{j_0} \cap (Y \setminus F_{j_0})$. Therefore $F_i \cap \Gamma(x) \neq \emptyset$ for every $i \in I$. Since $\Gamma(x)$ is compact and the family $(F_i \cap \Gamma(x))_{i \in I}$ has the finite intersection property, there exists $y \in \bigcap_{i \in I} (F_i \cap \Gamma(x)) = \Gamma(x) \cap \bigcap_{i \in I} F_i$. It follows that for every $i \in I$ and every $V \in \mathcal{V}_Y(y)$ there exists $\varphi(i, V) \in I$ such that $y_{\varphi(i,V)} \in V$ and $\varphi(i, V) \succeq i$. Let $J := I \times \mathcal{V}_Y(y)$ in which $(i, V) \succeq (i', V')$ iff $i \succeq i'$ and $V \subset V'$. Then $(y_{\varphi(j)})_{j \in J}$ is a subnet of $(y_i)_{i \in I}$ converging to y. The conclusion follows.

(iii) Suppose that Γ is compact at x, but $\lim_{x' \to x} e(\Gamma(x'), \Gamma(x))$ does not exist or is different from 0. Then there exists $\varepsilon_0 > 0$ such that for every $U \in \mathcal{V}_X(x)$ there exist $x_U \in U$ and $y_U \in \Gamma(x_U)$ such that $d(y_U, \Gamma(x)) := \inf_{y \in \Gamma(x)} \rho(y_U, y) > \varepsilon_0$. By hypothesis, there exists a subnet $(y_{\varphi(j)})_{j \in J} \to \overline{y} \in \Gamma(x)$. Since $d(\cdot, \Gamma(x))$ is continuous, we obtain the contradiction $0 = d(\overline{y}, \Gamma(x)) \geq \varepsilon_0$.

Conversely, suppose that $\Gamma(x)$ is compact and $\lim_{x' \to x} e(\Gamma(x'), \Gamma(x)) = 0$, and take the net $((x_i, y_i))_{i \in I} \subset \operatorname{gr} \Gamma$ with $(x_i) \to x$. It follows that $\lim_{i \in I} d(y_i, \Gamma(x)) = 0$. Therefore, for every $\varepsilon > 0$ there exists $i_\varepsilon \in I$ such that $d(y_i, \Gamma(x)) < \varepsilon$ for $i \succeq i_\varepsilon$. Let $J := I \times \,]0, \infty[$ ordered by $(i, \varepsilon) \succeq (i', \varepsilon')$ iff $i \succeq i'$ and $\varepsilon \leq \varepsilon'$. For every $j = (i, \varepsilon) \in J$ consider $\varphi(j) \in I$ such that $\varphi(j) \succeq i, i_\varepsilon$. Consider also $\overline{y}_{\varphi(j)} \in \Gamma(x)$ such that $\rho(y_{\varphi(j)}, \overline{y}_{\varphi(j)}) < \varepsilon$. Since $\Gamma(x)$ is compact, $\left(\overline{y}_{\varphi(j)} \right)$ contains a subnet $\left(\overline{y}_{\varphi \circ \psi(k)} \right)_{k \in K}$ converging to $\overline{y} \in \Gamma(x)$. It follows that $\left(y_{\varphi \circ \psi(k)} \right)_{k \in K} \to \overline{y}$.

(iv) Let Y be first countable and Γ compact at x. Consider the sequence $((x_n, y_n))_{n \in \mathbb{N}} \subset \operatorname{gr} \Gamma$ such that $(x_n) \to x$. By hypothesis there exists a subnet $(y_{\varphi(j)})_{j \in J} \to y \in \Gamma(x)$. There exists a base $(V_k)_{k \in \mathbb{N}}$ of neighborhoods of y. For every $k \in \mathbb{N}$ there exists $j_k \in J$ such that $y_{\varphi(j)} \in V_k$ for $j \succeq j_k$. Of course, taking $n_0 = \varphi(j_0)$, for every $k \in \mathbb{N}$ there exists $n_{k+1} \in \mathbb{N}$ such that $n_{k+1} \geq \max\{\varphi(j_{k+1}), n_k + 1\}$. It is obvious that $(y_{n_k}) \to y$.

For the converse implication we must show that $\lim_{x' \to x} e(\Gamma(x'), \Gamma(x)) = 0$. Otherwise, since X is first countable, there exist $\varepsilon_0 > 0$ and $(x_n) \to x$ such that $e(\Gamma(x_n), \Gamma(x)) > \varepsilon_0$ for every $n \in \mathbb{N}$. Proceeding as in the proof of the first part of (iii), we get a contradiction. $\square$

Related to closedness of multifunctions we have the following result.

Proposition 2.5.10. *Let $x \in X$. The following assertions hold:*

(i) *Γ is closed at x if and only if for every $y \in Y \setminus \Gamma(x)$ there exist $U \in \mathcal{V}_X(x)$ and $V \in \mathcal{V}_Y(y)$ such that $\Gamma(U) \cap V = \emptyset$.*

(ii) *Γ is closed at x if and only if $\limsup_{x' \to x} \Gamma(x') \subset \Gamma(x) = \operatorname{cl} \Gamma(x)$.*

(iii) *Suppose that X, Y are first countable. Then Γ is closed at x if and only if for every sequence $((x_n, y_n))_{n \in \mathbb{N}} \to (x, y)$, $y \in \Gamma(x)$.*

(iv) *If Γ is u.c. at x and $\Gamma(x)$ is compact, then Γ is closed at x.*

(v) *If Y is regular (in particular, if Y is a metric space or a topological vector space), Γ is u.c. at x, and $\Gamma(x)$ is closed, then Γ is closed at x.*

PROOF. The proof, which is not difficult, is left to the reader. □

The next result gives sufficient conditions for the upper continuity of the intersection of two multifunctions.

Proposition 2.5.11. *Consider $\Gamma_1, \Gamma_2 : X \rightrightarrows Y$ and $\Gamma(x) = \Gamma_1(x) \cap \Gamma_2(x)$ for every $x \in X$. Then Γ is u.c. at $x_0 \in X$ if one of the following conditions holds:*

(i) *Γ_1 is closed at x_0, Γ_2 is u.c. at x_0, and $\Gamma_2(x_0)$ is compact;*

(ii) *Y is normal (in particular, Y is a metric space), Γ_1, Γ_2 are u.c. at x_0, and $\Gamma_1(x_0)$, $\Gamma_2(x_0)$ are closed.*

PROOF. (i) Let $D \subset Y$ be an open set such that $\Gamma(x_0) \subset D$. Since Γ_1 is closed at x_0, for every $y \in \Gamma_2(x_0) \setminus \Gamma_1(x_0)$ there exist $U_y \in \mathcal{V}_X(x_0)$ and $V_y \in \mathcal{V}_Y(y)$ such that $\Gamma_1(U_y) \cap V_y = \emptyset$. It follows that $\Gamma_2(x_0) \subset D \cup \bigcup_{y \in \Gamma_2(x_0) \setminus \Gamma_1(x_0)} \text{int } V_y$. Since $\Gamma_2(x_0)$ is compact, there exist $y_1, \ldots, y_n \in \Gamma_2(x_0) \setminus \Gamma_1(x_0)$ such that $\Gamma_2(x_0) \subset D \cup \bigcup_{i=1}^n \text{int } V_{y_i}$. Since Γ_2 is u.c. at x_0, there exists $U_0 \in \mathcal{V}_X(x_0)$ such that $\Gamma_2(U_0) \subset D \cup \bigcup_{i=1}^n \text{int } V_{y_i}$. Let $x \in U := U_0 \cap \bigcap_{i=1}^n U_{y_i}$ and $y \in \Gamma(x)$. Since $x \in \Gamma_1(U_{y_i})$, $y \notin V_{y_i}$ for every i, $1 \leq i \leq n$. It follows that, necessarily, $y \in D$. Therefore Γ is u.c. at x_0.

(ii) Let $D \subset Y$ be an open set such that $\Gamma(x_0) \subset D$. Then $(\Gamma_1(x_0) \setminus D) \cap (\Gamma_2(x) \setminus D) = \emptyset$. Since the sets $\Gamma_1(x_0) \setminus D$ and $\Gamma_2(x_0) \setminus D$ are closed and Y is normal, there exist the disjoint open sets D_1, D_2 such that $\Gamma_i(x_0) \setminus D \subset D_1$ for $i = 1, 2$. Since Γ_i is u.c. at x_0 and $\Gamma_i(x_0) \subset D \cup D_i$, there exists $U_i \in \mathcal{V}_X(x_0)$ such that $\Gamma_i(U_i) \subset D \cup D_i$ for $i = 1, 2$. Let $x \in U_1 \cap U_2$ and $y \in \Gamma(x)$. It follows that $y \in (D \cup D_1) \cap (D \cup D_2) = D$. Therefore Γ is u.c. at x_0. □

When Y is a topological vector space, $\mathcal{V}_Y$ denotes the class of balanced neighborhoods of $0 \in Y$.

Definition 2.5.12. *Let Y be a topological vector space and $x_0 \in X$. We say that*

(a) *Γ is **Hausdorff upper continuous (H-u.c.) at** x_0 if*

$$\forall V \in \mathcal{V}_Y, \ \exists U \in \mathcal{V}_X(x_0), \ \forall x \in U \ : \ \Gamma(x) \subset \Gamma(x_0) + V. \qquad (2.41)$$

(b) *Γ is **Hausdorff lower continuous (H-l.c.) at** x_0 if*

$$\forall V \in \mathcal{V}_Y, \ \exists U \in \mathcal{V}_X(x_0), \ \forall x \in U \ : \ \Gamma(x_0) \subset \Gamma(x) + V. \qquad (2.42)$$

(c) *Γ is **Hausdorff continuous at** x_0 if Γ is H-u.c. and H-l.c. at x_0.*

(d) Γ is **Hausdorff upper continuous (Hausdorff lower continuous, Hausdorff continuous)** if Γ is so at every $x \in X$.

The above definition can be given when Y is a metric space, too; just replace $V \in \mathcal{V}_Y$ by $\varepsilon > 0$ and $\Gamma(x_0) + V$ by $(\Gamma(x_0))_\varepsilon$. Of course, if Y is a metric space, then Γ is H-u.c. at x_0 if and only if $\lim_{x \to x_0} e\left(\Gamma(x), \Gamma(x_0)\right) = 0$, and Γ is H-l.c. at x_0 if and only if $\lim_{x \to x_0} e\left(\Gamma(x_0), \Gamma(x)\right) = 0$.

Concerning the continuity of the sum of multifunctions and of the multiplication by scalars, we have the following result.

Proposition 2.5.13. *Let Y be a topological vector space, $\Gamma, \Gamma_1, \Gamma_2 : X \rightrightarrows Y$, $x_0 \in X$, and $\alpha \in \mathbb{R}$.*

(i) *If Γ is u.c. (resp. l.c., H-u.c., H-l.c.) at x_0, then $\alpha \Gamma$ is u.c. (resp. l.c., H-u.c., H-l.c.) at x_0.*

(ii) *If Γ_1 and Γ_2 are l.c. (resp. H-u.c., H-l.c.) at x_0, then $\Gamma_1 + \Gamma_2$ is l.c. (resp. H-u.c., H-l.c.) at x_0.*

PROOF. The proof of (i) is immediate.

(ii) Assume that Γ_1 and Γ_2 are l.c. at x_0 and let $D \subset Y$ be an open set such that $\left(\Gamma_1(x_0) + \Gamma_2(x_0)\right) \cap D \neq \emptyset$. Let $y_1 \in \Gamma_1(x_0)$ and $y_2 \in \Gamma_2(x_0)$ be such that $y_1 + y_2 \in D$. Let $V \in \mathcal{V}_Y$ be such that $y_1 + y_2 + V \subset D$. There exists $V_0 \in \mathcal{V}_Y$ with $V_0 + V_0 \subset V$. Because Γ_1 and Γ_2 are l.c. at x_0, there exist $U_1, U_2 \in \mathcal{V}_X(x_0)$ such that $\Gamma_i(x) \cap (y_i + V_0) \neq \emptyset$ for $i \in \{1, 2\}$ and all $x \in U_i$. It follows that $\left(\Gamma_1(x) + \Gamma_2(x)\right) \cap (y_1 + V_0 + y_2 + V_0) \neq \emptyset$ for $x \in U := U_1 \cap U_2 \in \mathcal{V}_X(x_0)$. Hence $(\Gamma_1 + \Gamma_2)(x) \cap D \neq \emptyset$ for $x \in U$, and so $\Gamma_1 + \Gamma_2$ is l.c. at x_0.

Assume now that Γ_1 and Γ_2 are H-u.c. at x_0 and take $V \in \mathcal{V}_Y$. Consider $V_0 \in \mathcal{V}_Y$ with $V_0 + V_0 \subset V$. Then there exist $U_1, U_2 \in \mathcal{V}_X(x_0)$ such that $\Gamma_i(x) \subset \Gamma_i(x_0) + V_0$ for $i \in \{1, 2\}$ and $x \in U_i$. It follows that $(\Gamma_1 + \Gamma_2)(x) \subset \Gamma_1(x_0) + V_0 + \Gamma_2(x_0) + V_0 \subset (\Gamma_1 + \Gamma_2)(x_0) + V$ for $x \in U := U_1 \cap U_2 \in \mathcal{V}_X(x_0)$, and so $\Gamma_1 + \Gamma_2$ is H-u.c. at x_0. The proof for H-lower continuity is similar. $\square$

Note that we have not a similar result to Proposition 2.5.13(ii) for upper continuity. To see this, consider the multifunctions $\Gamma_1, \Gamma_2 : \mathbb{R} \rightrightarrows \mathbb{R}^2$ defined by $\Gamma_1(x) := \mathbb{R} \times \{0\}$ and $\Gamma_2(x) := \{(0, x)\}$. It is obvious that Γ_1 and Γ_2 are u.c. at every $x \in \mathbb{R}$, but $\Gamma_1 + \Gamma_2$ is not u.c. at every $x \in \mathbb{R}$. Indeed, $(\Gamma_1 + \Gamma_2)(0) = \mathbb{R} \times \{0\} \subset D := \{(u, v) \in \mathbb{R}^2 \mid |v| < \exp(-u)\}$, but $(\Gamma_1 + \Gamma_2)(x) = \mathbb{R} \times \{x\} \not\subset D$ for every $x \in \mathbb{R} \setminus \{0\}$.

Note that if Γ is u.c. at x_0, then Γ is H-u.c. at x_0; the converse implication is true when $\Gamma(x_0)$ is compact. On the other hand, if Γ is H-l.c. at x_0, then Γ is l.c. at x_0, the converse being true if $\Gamma(x_0)$ is compact.

We can characterize Hausdorff upper and lower continuities by using nets, and even sequences when X and Y are first countable.

Proposition 2.5.14. *Suppose that Y is a topological vector space and $x \in X$. Then:*

(i) Γ is H-l.c. at x if and only if for all nets $(x_i)_{i \in I} \subset X$ with $(x_i) \to x$ and $(\overline{y}_i)_{i \in I} \subset \Gamma(x)$ there exist a subnet $(x_{\varphi(j)})$ and a net $(y_j)_{j \in J}$ such that $y_j - \overline{y}_{\varphi(j)} \to 0$ and $y_j \in \Gamma(x_{\varphi(j)})$ for all $j \in J$;

(ii) Γ is H-u.c. at x if and only if for every net $((x_i, y_i))_{i \in I} \subset \mathrm{gr}\,\Gamma$ with $(x_i) \to x$ there exists a subnet $(y_{\varphi(j)})_{j \in J}$ and a net $(\overline{y}_j)_{j \in J} \subset \Gamma(x)$ such that $y_{\varphi(j)} - \overline{y}_j \to 0$.

Suppose now that X and Y are first countable.

(iii) Γ is H-l.c. at x if and only if for every sequence $(x_n)_{n \in \mathbb{N}} \subset X$ with $(x_n) \to x$ and every sequence $(\overline{y}_n) \subset \Gamma(x)$ there exists a sequence (y_n) such that $y_n - \overline{y}_n \to 0$ and $y_n \in \Gamma(x_n)$ for all $n \geq n_0$;

(iv) Γ is H-u.c. at x if and only if for every sequence $((x_n, y_n))_{n \in \mathbb{N}} \subset \mathrm{gr}\,\Gamma$ with $(x_n) \to x$ there exists a sequence $(\overline{y}_n) \subset \Gamma(x)$ such that $y_n - \overline{y}_n \to 0$.

PROOF. We prove (i) and (iv), the proofs of (ii) and (iii) being similar.

(i) Suppose that Γ is H-l.c. at x and consider the nets $(x_i) \to x$ and $(\overline{y}_i)_{i \in I} \subset \Gamma(x)$. For every $V \in \mathcal{V}_Y$ there exists $i_V \in I$ such that $\Gamma(x) \subset \Gamma(x_i) + V$ for all $i \succeq i_V$. Ordering $J := I \times \mathcal{V}_Y$ in the usual way, consider $\varphi : J \to I$ with $\varphi(i, V) \succeq i, i_V$ and take $y_j \in \Gamma(x_{\varphi(j)}) \cap (\overline{y}_{\varphi(j)} + V)$ for $j = (i, V)$. It is obvious that $(y_j)_{j \in J}$ satisfies the needed conditions.

We prove by contradiction the converse implication. So, assume that Γ is not H-l.c. at (x, y). Then there exists $V_0 \in \mathcal{V}_Y$ such that for every $U \in \mathcal{V}_X(x)$ there exist $x_U \in U$ and $\overline{y}_U \in \Gamma(x)$ with $\overline{y}_U \notin \Gamma(x_U) + V_0$. By hypothesis, there exist a subnet $(x_{\varphi(j)})$ and a net $(y_j)_{j \in J}$ such that $y_j - \overline{y}_{\varphi(j)} \to 0$ and $y_j \in \Gamma(x_{\varphi(j)})$ for all $j \in J$. Since $V_0 \in \mathcal{V}_Y$, there exists $j_0 \in J$ such that $y_j - \overline{y}_{\varphi(j)} \in V_0$ for $j \succeq j_0$. It follows that $\overline{y}_{\varphi(j_0)} \in y_{j_0} + V_0 \subset \Gamma(x_{\varphi(j_0)}) + V_0$, a contradiction.

(iv) Suppose that Γ is H-u.c. at x and consider $((x_n, y_n))_{n \in \mathbb{N}} \subset \mathrm{gr}\,\Gamma$ with $(x_n) \to x$. Let $(V_k)_{k \in \mathbb{N}}$ be a base of neighborhoods of $0 \in Y$. For every $k \in \mathbb{N}$ there exists $U_k \in \mathcal{V}_X(x_0)$ such that $\Gamma(x') \subset \Gamma(x) + V_k$ for $x' \in U_k$. Since $(x_n) \to x$, there exists $n_k \in \mathbb{N}$ such that $x_n \in U_k$ for every $n \geq n_k$. Without loss of generality we may suppose that (n_k) is increasing. For every n such that $n_k \leq n < n_{k+1}$ we take $\overline{y}_n \in \Gamma(x)$ such that $y_n - \overline{y}_n \in V_k$. The conclusion follows. The converse part follows immediately by contradiction. $\square$

Proposition 2.5.15. *Suppose that Γ is upper continuous and closed-valued. Then $\mathrm{gr}\,\Gamma$ is closed. The same conclusion holds if Y is a topological vector space and Γ is H-u.c. instead of being u.c.*

PROOF. We give the proof for the second case. So, let Γ be H-u.c. and closed-valued. Consider $(x, y) \in X \times Y \setminus \mathrm{gr}\,\Gamma$. If $x \notin \mathrm{dom}\,\Gamma$, since Γ is H-u.c., there exists $U \in \mathcal{V}_X(x)$ such that $U \subset \mathrm{dom}\,\Gamma$. Hence $U \times Y \cap \mathrm{gr}\,\Gamma = \emptyset$. It follows that $(x, y) \notin \mathrm{cl}(\mathrm{gr}\,\Gamma)$. Suppose now that $x \in \mathrm{dom}\,\Gamma$, and so $y \notin \Gamma(x)$. Since $\Gamma(x)$ is closed, there exists $V \in \mathcal{V}_Y$ such that $(y + V) \cap \Gamma(x) = \emptyset$. Let $W \in \mathcal{V}_Y$ with $W + W \subset V$. Since Γ is H-u.c. at x, there exists $U \in \mathcal{V}_X(x)$

such that $\Gamma(U) \subset \Gamma(x) + W$. It follows that $(U \times (y + W)) \cap \operatorname{gr} \Gamma = \emptyset$, and so $(x, y) \notin \operatorname{cl}(\operatorname{gr} \Gamma)$. Therefore $\operatorname{gr} \Gamma$ is closed in $X \times Y$. $\qquad\square$

Note that requiring the upper continuity of Γ only on $\operatorname{dom} \Gamma$ is not sufficient for the conclusion of the preceding result. Taking for example $X = Y = \mathbb{R}$ and $\Gamma(x) = \mathbb{R}_+$ for $x \in]0, 1[$ and $\Gamma(x) = \emptyset$ otherwise, we have that Γ is u.c. (and so H-u.c.) at every $x \in \operatorname{dom} \Gamma$, is closed-valued, but $\operatorname{gr} \Gamma =]0, 1[\times \mathbb{R}_+$ is not closed in $X \times Y$.

Let $(A_n)_{n \in \mathbb{N}} \subset \mathcal{P}(Y)$. Recall that

$$\liminf_{n \to \infty} A_n := \{y \in Y \mid \exists (y_n) \to y \text{ such that } y_n \in A_n \text{ for } n \geq n_0\},$$

$$\limsup_{n \to \infty} A_n := \{y \subset Y \mid \exists (y_{n_k}) \to y \text{ such that } y_{n_k} \in A_{n_k} \text{ for } k \in \mathbb{N}\};$$

of course, $\liminf_{n \to \infty} A_n \subset \limsup_{n \to \infty} A_n$. We say that (A_n) **converges in the sense of Kuratowski–Painlevé** to $A \subset Y$ if $\limsup_{n \to \infty} A_n \subset A \subset \liminf_{n \to \infty} A_n$.

Having $A, A_n \subset Y$, $n \in \mathbb{N}$, and taking $X := \mathbb{N} \cup \{\infty\} \subset \overline{\mathbb{R}}$ (X endowed with the topology induced by that of $\overline{\mathbb{R}}$) we may consider the multifunction $\Gamma : X \rightrightarrows Y$ defined by $\Gamma(n) = A_n$ for $n \in \mathbb{N}$ and $\Gamma(\infty) = A$. When Y is first countable, $\liminf_{n \to \infty} A_n$ is exactly $\liminf_{n \to \infty} \Gamma(n)$, and $\limsup_{n \to \infty} A_n$ is $\limsup_{n \to \infty} \Gamma(n)$.

Definition 2.5.16. *Suppose that Y is a separated topological vector space and $C \subset Y$ is a convex cone. We say that Γ is C-u.c., C-l.c., H-C-u.c., or H-C-l.c. at $x_0 \in X$ if relation (2.33), (2.34), (2.41) or (2.42) holds with $\Gamma(x) \subset D + C$, $\Gamma(x) \cap (D - C) \neq \emptyset$, $\Gamma(x) \subset \Gamma(x_0) + V + C$, $\Gamma(x_0) \subset \Gamma(x) + V + C$ instead of $\Gamma(x) \subset D$, $\Gamma(x) \cap D \neq \emptyset$, $\Gamma(x) \subset \Gamma(x_0) + V$, $\Gamma(x_0) \subset \Gamma(x) + V$, respectively. Γ is C-continuous (C-Hausdorff continuous) at x_0 if Γ is C-u.c. and C-l.c. (H-C-u.c. and H-C-l.c.) at x_0.*

Remark 2.5.17. As in Proposition 2.5.13 one can prove that having the multifunctions $\Gamma, \Gamma_1, \Gamma_2 : X \rightrightarrows Y$ that are C-l.c. (resp. C-H-u.c., C-H-l.c.) at $x_0 \in X$, Y being a t.v.s., $C \subset Y$ a convex cone, and $\alpha \in \mathbb{R}_+$, then $\Gamma_1 + \Gamma_2$ and $\alpha\Gamma$ are C-l.c. (resp. C-H-u.c., C-H-l.c.) at x_0.

When $P \subset C$ is another convex cone, if Γ is P-l.c. (P-u.c., H-P-l.c., or H-P-u.c.) at x_0, then Γ is C-l.c. (C-u.c., H-C-l.c., or H-C-u.c.) at x_0. Note also that Γ is C-l.c., H-C-u.c., or H-C-l.c. if and only if Γ_C is l.c., H-u.c., or H-l.c., respectively, but such an equivalence is not true for upper continuity. In fact, we have that Γ is C-u.c. at x_0 if Γ_C is u.c. at x_0, but the converse is not true even if $\Gamma = \Gamma_C$, as shown by the following example: $\Gamma : [0, \infty[\rightrightarrows \mathbb{R}^2$, $\Gamma(x) = [0, \infty[\times [0, x]$, and $C = [0, \infty[\times \{0\}$.

Definition 2.5.18. *When Y is a topological vector space, we say that Γ is uniformly C-l.c. at x_0 on A if $A \subset \Gamma(x_0)$ and*

$$\forall W \in \mathcal{V}_Y, \ \exists U \in \mathcal{V}_X(x_0), \ \forall x \in U \ : \ A \subset \Gamma(x) + W + C.$$

Remark 2.5.19. If Γ is C-l.c. at (x_0, y) for every $y \in A \subset \Gamma(x_0)$ and A is compact, then Γ is uniformly C-l.c. at x_0 on A.

Indeed, let $W \in \mathcal{V}_Y$ and consider $W_1 \in \mathcal{V}_Y$ open with $W_1 + W_1 \subset W$. For $y \in A$, since Γ is C-l.c. at (x_0, y), there exists $U_y \in \mathcal{V}_X(x_0)$ such that $y \in \Gamma(x) + W_1 + C$ for every $x \in U_y$. Since A is compact and $A \subset \bigcup_{y \in A}(y + W_1)$, there exist $y_1, \dots, y_n \in A$ such that $A \subset \bigcup_{i=1}^{n}(y_i + W_1)$. Let $x \in U := \bigcap_{i=1}^{n} U_{y_i}$ and $y \in A$. There exists i, $1 \le i \le n$, such that $y \in y_i + W_1$. Since $x \in U_{y_i}$, $y_i \in \Gamma(x) + W_1 + C$, whence $y \in \Gamma(x) + W + C$. Therefore $A \subset \Gamma(x) + W + C$ for every $x \in U$.

With any multifunction $\Gamma : X \rightrightarrows Y$ we associate its closure $\overline{\Gamma} : X \rightrightarrows Y$ defined by $\overline{\Gamma}(x) := \mathrm{cl}\,(\Gamma(x))$. Of course, $\mathrm{dom}\,\Gamma = \mathrm{dom}\,\overline{\Gamma}$. Continuity properties of Γ and $\overline{\Gamma}$ are deeply related.

Proposition 2.5.20. *Let $x \in X$ and $y \in Y$. The following assertions hold:*

(i) Γ *is l.c. at* (x, y) *if and only if* $\overline{\Gamma}$ *is l.c. at* (x, y).

(ii) Γ *is l.c. at* x *if and only if* $\overline{\Gamma}$ *is l.c. at* x.

(iii) *If* Y *is a topological vector space and* $C \subset Y$ *is a convex cone, then* Γ *is H-C-u.c. (H-C-l.c.) at* x *if and only if* $\overline{\Gamma}$ *is H-C-u.c. (H-C-l.c.) at* x.

(iv) *If* Γ *is u.c. at* x *and either* Y *is normal or* Y *is regular and* $\mathrm{cl}\,\Gamma(x)$ *is compact, then* $\overline{\Gamma}$ *is u.c. at* x.

(v) *If* Γ *is u.c. at* x *and either* Y *is regular or* $\mathrm{cl}\,\Gamma(x)$ *is compact, then* $\overline{\Gamma}$ *is closed at* x.

(vi) Γ *is closed at* x *if and only if* $\Gamma(x)$ *is closed and* $\overline{\Gamma}$ *is closed at* x.

(vii) *If* Y *is a metric space, then* Γ *is compact at* x *if and only if* $\Gamma(x)$ *is closed and* $\overline{\Gamma}$ *is compact at* x.

PROOF. The proofs of assertions (i)–(iii) are not complicated. One uses that for $A, D \subset Y$ with D open, $D \cap \mathrm{cl}\,A \ne \emptyset \Rightarrow D \cap A \ne \emptyset$, and, when Y is a topological vector space, $\mathrm{cl}\,A = \bigcap_{W \in \mathcal{V}_Y}(A + W)$. For (vi) take into consideration Proposition 2.5.10 (i), while for (vii) take into account Proposition 2.5.9 (ii).

(iv) Let $D \subset Y$ be an open set such that $\overline{\Gamma}(x) \subset D$. If Y is normal, there exists $D_0 \subset Y$ open such that $\mathrm{cl}\,\Gamma(x) \subset D_0 \subset \mathrm{cl}\,D_0 \subset D$. If Y is regular and $\mathrm{cl}\,\Gamma(x)$ is compact, for every $y \in \mathrm{cl}\,\Gamma(x)$ there exist disjoint open sets $V_y, D_y \subset Y$ such that $y \in V_y$ and $Y \setminus D \subset D_y$. Since $\{V_y \mid y \in \mathrm{cl}\,\Gamma(x)\}$ is an open cover of $\mathrm{cl}\,\Gamma(x)$, there exist $y_1, \dots, y_n \in \mathrm{cl}\,\Gamma(x)$ such that $\mathrm{cl}\,\Gamma(x) \subset D_0 := V_{y_1} \cup \cdots \cup V_{y_n}$. Of course, D_0 and $D_1 := D_{y_1} \cap \cdots \cap D_{y_n} (\supset Y \setminus D)$ are disjoint open sets. Therefore $D_0 \subset \mathrm{cl}\,D_0 \subset Y \setminus D_1 \subset D$. Since Γ is u.c. at x, there exists $U \in \mathcal{V}_X(x)$ such that $\Gamma(x') \subset D_0$, whence $\overline{\Gamma}(x') \subset \mathrm{cl}\,D_0 \subset D$, for every $x' \in U$.

(v) Let $y \in Y \setminus \overline{\Gamma}(x)$. If Y is regular, there exist $V, D \subset Y$ disjoint open sets such that $y \in V$ and $\overline{\Gamma}(x) \subset D$. When $\overline{\Gamma}(x)$ is compact, for every $z \in \overline{\Gamma}(x)$ there exist disjoint open sets $V_z, D_z \subset Y$ such that $y \in V_z$ and $z \in D_z$. Since $\overline{\Gamma}(x)$ is compact, there exist $z_1, \dots, z_n \in \overline{\Gamma}(x)$ such that $D := D_{z_1} \cup \cdots \cup D_{z_n}$. Taking $V := V_{z_1} \cap \cdots \cap V_{z_n}$, we have again that V, D are open and disjoint,

$y \in V$, and $\overline{\Gamma}(x) \subset D$. Since Γ is u.c. at x and $\Gamma(x) \subset D$, there exists $U \in \mathcal{V}_X(x)$ with $\Gamma(U) \subset D$. Because V is open, we have even that $\overline{\Gamma}(U) \cap V = \emptyset$. Therefore $\overline{\Gamma}$ is closed at x. $\square$

Note that if $\overline{\Gamma}$ is u.c. at x, it does not follow that Γ is u.c. at x (but the implication is true if $\Gamma(x)$ is closed). Take $\Gamma : \mathbb{R} \rightrightarrows \mathbb{R}$, $\Gamma(x) =]x, x+1[$; Γ is not u.c. at 0, but $\overline{\Gamma}$ is.

From now on (in this section), Y is a separated topological vector space and $C \subset Y$ is a convex cone. Recall that the convex cone C determines a preorder $\leq_C$ defined by $y_1 \leq_C y_2 \Leftrightarrow y_2 - y_1 \in C$.

Definition 2.5.21. *We say that*

(i) Γ *is C-**lower semicontinuous** (C-l.s.c. for short) if* $\mathrm{lev}_\Gamma(y) := \{x \in X \mid \Gamma(x) \cap (y - C) \neq \emptyset\}$ *is closed for every* $y \in Y$;
(ii) Γ *is C-**upper semicontinuous** (C-u.s.c. for short) if* $\mathrm{lev}_\Gamma^<(y) := \{x \in X \mid \Gamma(x) \cap (y - \mathrm{int}\, C) \neq \emptyset\}$ *is open in* X *for every* $y \in Y$.

Note that $\mathrm{lev}_\Gamma(y) = (\Gamma_C)^{-1}(y)$ and $\mathrm{lev}_\Gamma^<(y) = (\Gamma_{\mathrm{int}\, C})^{-1}(y)$ for every $y \in Y$, so that Γ is C-l.s.c. iff $(\Gamma_C)^{-1}$ is closed-valued, and Γ is C-u.s.c. iff $(\Gamma_{\mathrm{int}\, C})^{-1}$ is open valued. Moreover, if $y_1 \leq_C y_2$, then $\mathrm{lev}_\Gamma(y_1) \subset \mathrm{lev}_\Gamma(y_2)$ and $\mathrm{lev}_\Gamma^<(y_1) \subset \mathrm{lev}_\Gamma^<(y_2)$.

When $\mathrm{int}\, C \neq \emptyset$ we have that $\mathrm{int}\big(\mathrm{dom}(\mathrm{lev}_\Gamma)\big) = \mathrm{dom}(\mathrm{lev}_\Gamma^<)$. First observe that $\mathrm{dom}(\mathrm{lev}_\Gamma^<)$ is open. Indeed, let $y \in \mathrm{dom}(\mathrm{lev}_\Gamma^<)$. Then there exist $x \in X$ and $z \in \Gamma(x) \cap (y - \mathrm{int}\, C)$. Hence $z = y - k$ for some $k \in \mathrm{int}\, C$. Taking $v \in V := \mathrm{int}\, C - k$, we have that $z = y - k = (y+v) - (v+k) \subset (y+v) - \mathrm{int}\, C$, which shows that $\Gamma(x) \cap (y - \mathrm{int}\, C) \neq \emptyset$. Therefore $y + V \subset \mathrm{dom}(\mathrm{lev}_\Gamma^<)$, and so $\mathrm{dom}(\mathrm{lev}_\Gamma^<)$ is open. From the obvious inclusion $\mathrm{dom}(\mathrm{lev}_\Gamma^<) \subset \mathrm{dom}(\mathrm{lev}_\Gamma)$ we obtain that $\mathrm{dom}(\mathrm{lev}_\Gamma^<) \subset \mathrm{int}\big(\mathrm{dom}(\mathrm{lev}_\Gamma^<)\big)$. For the converse inclusion consider $y \in \mathrm{int}(\mathrm{dom}(\mathrm{lev}_\Gamma))$. Let $k \in \mathrm{int}\, C$; there exists $t > 0$ such that $y - tk \in \mathrm{dom}(\mathrm{lev}_\Gamma)$. Therefore, there exists $x \in X$ such that $\Gamma(x) \cap (y - tk - C) \neq \emptyset$. Since $tk + C \subset \mathrm{int}\, C$, we have that $y \in \mathrm{dom}(\mathrm{lev}_\Gamma^<)$. The following result holds.

Proposition 2.5.22. *The following assertions hold:*

(i) *If* $\mathrm{epi}\, \Gamma$ *is closed, then* Γ *is C-lower semicontinuous.*
(ii) *Suppose that* Y *is a locally convex space,* C *is closed,* $\mathrm{int}\, C \neq \emptyset$, Γ *is C-lower semicontinuous, and either* $\Gamma(x)$ *is weakly compact or* $\Gamma(x) + C$ *is closed for every* $x \in X$. *Then* $\mathrm{epi}\, \Gamma$ *is closed. Moreover, if* $y_0, y_1 \in Y$ *are such that* $y_1 - y_0 \in \mathrm{int}\, C$ *and* $\mathrm{lev}_\Gamma(y_1)$ *is compact, then* $\mathrm{lev}_\Gamma(y_0)$ *is compact and* lev_Γ *is upper continuous at* y_0.

PROOF. (i) Suppose that $\mathrm{epi}\, \Gamma = \mathrm{gr}\, \Gamma_C$ is closed. Then $\mathrm{gr}(\Gamma_C)^{-1}$ is also closed, whence $(\Gamma_C)^{-1}$ is closed-valued. Hence Γ is C-l.s.c.

(ii) Suppose now that Γ is C-l.s.c., Y is a locally convex space, C is closed, $\mathrm{int}\, C \neq \emptyset$ and either $\Gamma(x)$ is weakly compact or $\Gamma(x) + C$ is closed for every $x \in X$. Let us consider the net $((x_i, y_i))_{i \in I} \subset \mathrm{epi}\, \Gamma$ converging to $(x, y) \in X \times Y$. Let $k \in \mathrm{int}\, C$ be fixed. Since $k - C \in \mathcal{V}_Y(0)$ and $(y_i) \to y$, there exists

$i_k \in I$ such that $y_i - y \in k - C$ for $i \succeq i_k$. It follows that $(x_i)_{i \succeq i_k} \subset \mathrm{lev}_\Gamma(y+k)$, whence $x \in \mathrm{lev}_\Gamma(y+k)$ for every $k \in \mathrm{int}\, C$. Fixing $k_0 \in \mathrm{int}\, C$ we obtain that $x \in \mathrm{lev}_\Gamma(y + n^{-1}k_0)$ for every $n \in \mathbb{N}^*$. Therefore, for every n there exists $z_n \in \Gamma(x) \cap (y + n^{-1}k_0 - C)$; in particular, $y + n^{-1}k_0 \in \Gamma(x) + C$. If $\Gamma(x) + C$ is closed, we get $y \in \Gamma(x) + C$, and so $(x, y) \in \mathrm{epi}\, \Gamma$. If $\Gamma(x)$ is weakly compact, the sequence $(z_n)_{n \in \mathbb{N}^*}$ contains a subnet converging weakly to some $z \in \Gamma(x)$. Since C is (weakly) closed, we obtain that $z \in y - C$, which shows that $(x, y) \in \mathrm{epi}\, \Gamma$ in this case, too.

Now let $y_0, y_1 \in Y$ be such that $y_1 - y_0 \in \mathrm{int}\, C$ and $\mathrm{lev}_\Gamma(y_1)$ is compact. Of course, $\mathrm{lev}_\Gamma(y_0) \subset \mathrm{lev}_\Gamma(y_1)$. Since $\mathrm{lev}_\Gamma(y_0)$ is closed and $\mathrm{lev}_\Gamma(y_1)$ is compact, we have that $\mathrm{lev}_\Gamma(y_0)$ is compact. Suppose that lev_Γ is not u.c. at y_0. Then there exist an open set $D \subset X$ and the nets $(y_i)_{i \in I} \subset Y$, $(x_i)_{i \in I} \subset X$ such that $\mathrm{lev}_\Gamma(y_0) \subset D$, $(y_i)_{i \in I}$ converges to y_0, and $x_i \in \mathrm{lev}_\Gamma(y_i) \setminus D$ for every $i \in I$. Since $y_0 \in y_1 - \mathrm{int}\, C$, there exists $i_1 \in I$ such that $y_i \in y_1 - C$; i.e., $y_i \leq_C y_1$, for every $i \succeq i_1$. It follows that $x_i \in \mathrm{lev}_\Gamma(y_1)$ for every $i \succeq i_1$. Since $\mathrm{lev}_\Gamma(y_1)$ is compact, there exists a subnet $(x_{\varphi(j)})_{j \in J}$ of (x_i) converging to x. Of course, $x \notin D$. Since $(x_i, y_i) \in \mathrm{epi}\, \Gamma$ for every $i \in I$ and $\mathrm{epi}\, \Gamma$ is closed in our conditions (as seen above), we obtain that $(x, y_0) \in \mathrm{epi}\, \Gamma$, whence $x \in \mathrm{lev}_\Gamma(y_0) \subset D$. This is a contradiction because $x \notin D$. $\qquad\square$

Proposition 2.5.23. *Consider the multifunction* $\Gamma : X \rightrightarrows Y$.

(i) *Suppose that C is closed and $P \subset C$ is another convex cone. If Γ is P-upper continuous, then Γ is C-lower semicontinuous.*

(ii) *Suppose that $\mathrm{int}\, C \neq \emptyset$; then Γ is C-lower continuous if and only if Γ is C-upper semicontinuous.*

PROOF. (i) Let $y \in Y$ and $x \notin \mathrm{lev}_\Gamma(y)$. It follows that $\Gamma(x) \subset Y \setminus (y - C)$. Since Γ is P-upper continuous at x, there exists $U \in \mathcal{V}_X(x)$ such that $\Gamma(x') \subset (Y \setminus (y - C)) + P$, for every $x' \in U$. Since $(Y \setminus (y - C)) + P \subset Y \setminus (y - C)$, we obtain that $\Gamma(x') \cap (y - C) = \emptyset$ for each $x' \in U$. Therefore $U \cap \mathrm{lev}_\Gamma(y) = \emptyset$, which shows that $x \notin \mathrm{cl}\,(\mathrm{lev}_\Gamma(y))$. Hence $\mathrm{lev}_\Gamma(y)$ is closed for every $y \in Y$.

(ii) Suppose first that Γ is C-upper semicontinuous and fix $x_0 \in X$. Let D be an open set in Y such that $\Gamma(x_0) \cap D \neq \emptyset$; then $D - \Gamma(x_0)$ is a neighborhood of 0 in Y. Since $0 \in \mathrm{cl}(\mathrm{int}\, C)$, we deduce $(D - \Gamma(x_0)) \cap \mathrm{int}\, C \neq \emptyset$. Let $y_0 \in D$ be such that $(y_0 - \Gamma(x_0)) \cap \mathrm{int}\, C \neq \emptyset$; then $x_0 \in U := \mathrm{lev}_\Gamma^<(y_0)$, and U is open because Γ is C-u.s.c.; hence $U \in \mathcal{V}_X(x_0)$ and $\emptyset \neq \Gamma(x) \cap (y_0 - \mathrm{int}\, C) \subset \Gamma(x) \cap (D - C)$ for each $x \in U$. It follows that Γ is C-l.c. at x_0.

Suppose now that Γ is C-l.c. and fix $y \in Y$. Let $x_0 \in \mathrm{lev}_\Gamma^<(y)$; this means that $\Gamma(x_0) \cap D \neq \emptyset$, where $D := y - \mathrm{int}\, C$ is an open set by our hypothesis. Because Γ is C-l.c. at x_0, there exists $U \in \mathcal{V}_X(x_0)$ such that

$$\emptyset \neq \Gamma(x) \cap (D - C) = \Gamma(x) \cap (y - \mathrm{int}\, C - C) = \Gamma(x) \cap (y - \mathrm{int}\, C)$$

for all $x \in U$. It follows that $U \subset \mathrm{lev}_\Gamma^<(y)$. Therefore $\mathrm{lev}_\Gamma^<(y)$ is open, and so Γ is C-u.s.c. The proof is complete. $\qquad\square$

From the preceding result we obtain that Γ is C-l.s.c. and $-C$-l.s.c. when Γ is upper continuous (and C is closed).

The following interesting result holds.

Proposition 2.5.24. *Suppose that* $\operatorname{int} C \neq \emptyset$ *and* $y \in \operatorname{dom}(\operatorname{lev}_\Gamma)$. *If* $\operatorname{lev}_\Gamma$ *is lower continuous at* y, *then* $\operatorname{lev}_\Gamma(y) \subset \operatorname{cl}\left(\operatorname{lev}_\Gamma^<(y)\right)$. *Conversely, if* X *is a topological vector space (or a metric space),* $\operatorname{lev}_\Gamma(y)$ *is totally bounded, and* $\operatorname{lev}_\Gamma(y) \subset \operatorname{cl}\left(\operatorname{lev}_\Gamma^<(y)\right)$, *then* $\operatorname{lev}_\Gamma$ *is Hausdorff lower continuous at* y. *Moreover, if* $\operatorname{lev}_\Gamma(y)$ *is compact, then* $\operatorname{lev}_\Gamma$ *is lower continuous at* y.

PROOF. Suppose first that $\operatorname{lev}_\Gamma$ is lower continuous at y and consider $x \in \operatorname{lev}_\Gamma(y)$. Fix $k \in \operatorname{int} C$ and take $y_n := y - n^{-1}k$ for $n \in \mathbb{N}^*$. By Proposition 2.5.6 there exist a subnet $(y_{\varphi(j)})_{j\in J}$ of $(y_n)_{n\in\mathbb{N}^*}$ and a net $(x_j)_{j\in J}$ convergent to x such that $x_j \in \operatorname{lev}_\Gamma(y_{\varphi(j)})$ for every $j \in J$. Since $y_{\varphi(j)} - C \subset y - \operatorname{int} C$, we obtain that $x_j \in \operatorname{lev}_\Gamma^<(y)$ for every $j \in J$. Therefore $x \in \operatorname{cl}\left(\operatorname{lev}_\Gamma^<(y)\right)$. Suppose now that X is a t.v.s., $\operatorname{lev}_\Gamma(y)$ is totally bounded, and $\operatorname{lev}_\Gamma(y) \subset \operatorname{cl}\left(\operatorname{lev}_\Gamma^<(y)\right)$. We show first that

$$\forall U \in \mathcal{V}_X, \ \exists k \in \operatorname{int} C \ : \ \operatorname{lev}_\Gamma(y) \subset \operatorname{lev}_\Gamma(y - k) + U. \qquad (2.43)$$

So, let $U \in \mathcal{V}_X(0)$; there exists a balanced neighborhood W of $0 \in X$ such that $W + W \subset U$. Since $\operatorname{lev}_\Gamma(y)$ is totally bounded, there exist $x_1, \ldots, x_p \in \operatorname{lev}_\Gamma(y)$ such that

$$\operatorname{lev}_\Gamma(y) \subset \bigcup_{l=1}^{p}(x_l + W). \qquad (2.44)$$

Since $\operatorname{lev}_\Gamma(y) \subset \operatorname{cl}\left(\operatorname{lev}_\Gamma^<(y)\right)$, for every $l \in \{1, \ldots, p\}$ there exists $x_l' \in (x_l + W) \cap \operatorname{lev}_\Gamma^<(y)$. It follows that for every l there exists $k_l' \in \operatorname{int} C$ such that $y_l := y - k_l' \in \Gamma(x_l')$. Let $t_1 := 1$ and $k_1 := t_1 k_1'$. Since $k_1 - t k_2' \to k_1 \in \operatorname{int} C$ for $t \to 0$, there exists $t_2 \in]0,1]$ such that $k_1 - t_2 k_2' \in \operatorname{int} C$. Let $k_2 := t_2 k_2'$. Similarly, there exists $t_3 \in]0,1]$ such that $k_2 - t_3 k_3' \in \operatorname{int} C$. Let $k_3 := t_3 k_3'$. Continuing in this way, for every $l \in \mathbb{N}^*$, we find $t_l \in]0,1]$ and $k_l := t_l k_l'$ such that $k_l - k_{l+1} \in \operatorname{int} C$ for $1 \le l \le p-1$. Since $y_l = y - k_l - (1-t_l)k_l' \subset y - k_l - C$, we have that $x_l' \in \operatorname{lev}_\Gamma(y - k_l)$ for every l. Moreover,

$$\operatorname{lev}_\Gamma(y - k_1) \subset \operatorname{lev}_\Gamma(y - k_2) \subset \ldots \subset \operatorname{lev}_\Gamma(y - k_p).$$

Now let $x \in \operatorname{lev}_\Gamma(y)$. From (2.44) there exists $1 \le l \le p$ such that $x \in x_l + W$. Since $x_l' \in x_l + W$, we obtain that

$$x \in x_l' + W - W \subset x_l' + U \subset \operatorname{lev}_\Gamma(y - k_l) + U \subset \operatorname{lev}_\Gamma(y - k_p) + U.$$

Therefore $\operatorname{lev}_\Gamma(y) \subset \operatorname{lev}_\Gamma(y - k) + U$ with $k := k_p \in \operatorname{int} C$; i.e., (2.43) holds. Taking $V = y + C - k$, V is a neighborhood of y and

$$\operatorname{lev}_\Gamma(y) \subset \operatorname{lev}_\Gamma(y - k) + U \subset \operatorname{lev}_\Gamma(y') + U \quad \forall y' \in V.$$

Therefore lev_Γ is H-l.c. at y. When X is a metric space instead of U one takes $\varepsilon > 0$, and instead of $x_l + W$ one takes $B(x_l, \varepsilon/2)$. When $\mathrm{lev}_\Gamma(y)$ is compact the conclusion is obvious. □

It is quite interesting that continuity properties for **vector-valued functions** and those of multifunctions are related. We give first some continuity notions for extended vector-valued functions.

Definition 2.5.25. *Let* $f : X \to Y^\bullet$ *and* $x_0 \in \mathrm{dom}\, f$.

(i) f *is* C-**lower continuous** *(C-l.c.) at* x_0 *if*

$$\forall V \in \mathcal{V}_Y, \ \exists U \in \mathcal{V}_X(x_0), \ \forall x \in U \ : \ f(x) \in f(x_0) + V - C^\bullet;$$

(ii) f *is* C-**upper continuous** *(C-u.c.) at* x_0 *if*

$$\forall V \in \mathcal{V}_Y, \ \exists U \in \mathcal{V}_X(x_0), \ \forall x \in U \ : \ f(x) \in f(x_0) + V + C;$$

(iii) f *is* C-**lower semicontinuous** *if* $\mathrm{lev}_f(y)$ *is closed for every* $y \in Y$;
(iv) f *is* C-**upper semicontinuous** *if* $\mathrm{lev}_f^<(y)$ *is open for every* $y \in Y$.

From Proposition 2.5.23(ii) we obtain that the function $f : X \to Y$ is C-upper semicontinuous if and only if f is C-lower continuous.

Consider $x_0 \in \mathrm{dom}\, f$. If f is C-u.c. at x_0, then $x_0 \in \mathrm{int}(\mathrm{dom}\, f)$. On the other hand, if $\Gamma_{f,C}$ is closed at x_0, then C is closed; if C is closed and $\Gamma_{f,C}$ is u.c. at x_0, then $\Gamma_{f,C}$ is closed at x_0. Moreover, f is $\{0\}$-u.c. at $x_0 \Leftrightarrow$ if f is continuous at $x_0 \Leftrightarrow x_0 \in \mathrm{int}(\mathrm{dom}\, f)$ and f is $\{0\}$-l.c. at x_0.

Proposition 2.5.26. *Let* $f : X \to Y^\bullet$, $\Gamma_{f,C} : X \rightrightarrows Y$ *its associated multifunction, and* $x_0 \in \mathrm{dom}\, f$. *Then:*

(i) f *is* C-l.c. *at* $x_0 \Leftrightarrow \Gamma_{f,C}$ *is* C-u.c. *at* $x_0 \Leftrightarrow \Gamma_{f,C}$ *is* H-u.c. *at* x_0.
(ii) f *is* C-u.c. *at* $x_0 \Leftrightarrow \Gamma_{f,C}$ *is* l.c. *at* $x_0 \Leftrightarrow \Gamma_{f,C}$ *is* H-l.c. *at* x_0.

PROOF. (i) Suppose that f is C-l.c. at x_0 and consider $D \subset Y$ an open set such that $\Gamma_{f,C}(x_0) = f(x_0) + C \subset D$. Since $f(x_0) \in D$, there exists $U \in \mathcal{V}_X(x_0)$ such that $f(x) \in D + C^\bullet$ for every $x \in U$. Hence $\Gamma_{f,C}(x) \subset D + C$ for every $x \in U$, and so $\Gamma_{f,C}$ is C-u.c. at x_0.

Suppose now that $\Gamma_{f,C}$ is H-u.c. at x_0 and consider $V \in \mathcal{V}_Y$. By definition, there exists $U \in \mathcal{V}_X(x_0)$ such that $\Gamma_{f,C}(x) \subset \Gamma_{f,C}(x_0) + V = f(x_0) + V + C$ for all $x \in U$. Therefore $f(x) \in f(x_0) + V + C^\bullet$ for every $x \in U$; i.e., f is C-l.c. at x_0. Since the other implication is immediate, the proof of (i) is complete.

(ii) Suppose that f is C-u.c. at x_0 and consider $V \in \mathcal{V}_Y$; there exists $U \in \mathcal{V}_X(x_0)$ such that $f(x) \in f(x_0) + V - C$ for every $x \in U$. It follows that $f(x_0) \in \Gamma_{f,C}(x) + V$, whence $\Gamma_{f,C}(x_0) \subset \Gamma_{f,C}(x) + V$ for every $x \in U$. Therefore $\Gamma_{f,C}$ is H-l.c. at x_0.

Suppose now that $\Gamma_{f,C}$ is l.c. at x_0 and consider $V \in \mathcal{V}_Y$. Since $f(x_0) \in (f(x_0) + \mathrm{int}\, V) \cap \Gamma(x_0)$, there exists $U \in \mathcal{V}_X(x_0)$ such that $(f(x_0) + \mathrm{int}\, V) \cap \Gamma(x) \neq \emptyset$ for all $x \in U$. Therefore $f(x) \in f(x_0) + V - C$ for every $x \in U$; i.e.,

f is C-u.c. at x_0. Since Hausdorff lower continuity implies lower continuity, the proof is complete.   □

When $Y = \mathbb{R}$ and $C = \mathbb{R}_+$ we have more refined statements; $\Gamma_f := \Gamma_{f,\mathbb{R}_+}$ in this case.

Proposition 2.5.27. *Let* $f : X \to \overline{\mathbb{R}} := \mathbb{R} \cup \{-\infty, \infty\}$ *with the associated multifunction* $\Gamma_f : X \rightrightarrows \mathbb{R}$ *whose graph is* epi f.

(i) *Suppose that* $x_0 \in \mathrm{dom}\, f \; (= \mathrm{dom}\, \Gamma_f)$. *Then* f *is l.c. at* $x_0 \Leftrightarrow \Gamma_f$ *is u.c. at* $x_0 \Leftrightarrow \Gamma_f$ *is H-u.c. at* x_0.
(ii) *Suppose that* $x_0 \in X$. *Then* Γ_f *is H-l.c. at* $x_0 \Rightarrow f$ *is u.c. at* $x_0 \Leftrightarrow \Gamma_f$ *is l.c. at* x_0; *moreover, if either* $f(x_0) > -\infty$ *or* X *is a topological vector space and* f *is convex, then* f *is u.c. at* $x_0 \Leftrightarrow \Gamma_f$ *is H-l.c. at* x_0.

PROOF. When $f(x_0) \in \mathbb{R}$ the conclusion follows from the preceding proposition; one must only mention that in this case upper continuity and $C = \mathbb{R}_+$ upper continuity for Γ_f coincide.

If $f(x_0) = \infty$, it is obvious that f is u.c. at x_0 and Γ_f is l.c. and H-l.c. at x_0 (since $\Gamma_f(x_0) = \emptyset$); therefore (ii) holds in this case.

Suppose now that $f(x_0) = -\infty$; in this case $\Gamma_f(x_0) = \mathbb{R}$.

Obviously, f is l.c. at x_0 and Γ_f is u.c. and H-u.c. at x_0. Therefore (i) holds.

Assume that Γ_f is H-l.c. at x_0. Since $\Gamma_f(x_0) = \mathbb{R}$, taking $V = [-1, 1] \in \mathcal{V}_{\mathbb{R}}(0)$, there exists $U \in \mathcal{V}_X(x_0)$ such that $\Gamma_f(x) = \mathbb{R}$ for every $x \in U$. Therefore $f(x) = -\infty$ for every $x \in U$, whence f is u.c. at x_0.

Assume that f is u.c. at x_0 and consider $D \subset \mathbb{R}$ an open set; take $y_0 \in \Gamma_f(x_0) \cap D = D$. There exists $\varepsilon > 0$ such that $]y_0 - \varepsilon, y_0 + \varepsilon[\subset D$. Since f is u.c. at x_0, there exists $U \in \mathcal{V}_X(x_0)$ such that $f(x) < y_0$ for every $x \in U$. It follows that $y_0 \in \Gamma_f(x) \cap D$ for every $x \in U$. Therefore Γ_f is l.c. at x_0.

Assume that Γ_f is l.c. at x_0 and take $\lambda \in \mathbb{R}$. Since $\Gamma_f(x_0) \cap]-\infty, \lambda[\neq \emptyset$, there exists $U \in \mathcal{V}_X(x_0)$ such that $\Gamma_f(x) \cap]-\infty, \lambda[\neq \emptyset$ for every $x \in U$. It follows that $f(x) < \lambda$ for every $x \in U$. Therefore f is u.c. at x_0.

Of course, if f takes the value $-\infty$ on a neighborhood of x_0, then Γ_f is constant on that neighborhood, and so it is H-l.c. at x_0. This is the case in which f is convex, $f(x_0) = -\infty$, and f is u.c. at x_0. The proof is complete. □

We are now interested in continuity properties of the composition of two multifunctions or of a function with a multifunction. So consider another topological space $\mathcal{U}$, the functions $f : X \times \mathcal{U} \to Y$, $g : X \to Y$, and the multifunctions $\Lambda : \mathcal{U} \rightrightarrows X$, $F : X \times \mathcal{U} \rightrightarrows Y$ and $G : X \rightrightarrows Y$; we associate the multifunctions $f\Lambda, g\Lambda, F\Lambda, G\Lambda : \mathcal{U} \rightrightarrows Y$ defined by $(f\Lambda)(u) := f(\Lambda(u) \times \{u\})$, $(g\Lambda)(u) := g(\Lambda(u))$, $(F\Lambda)(u) = F(\Lambda(u) \times \{u\})$, and $(G\Lambda)(u) := G(\Lambda(u))$.

Proposition 2.5.28. *Let* f, g *and* Λ, F, G *be as above and* $u_0 \in \mathrm{dom}\, \Lambda$.

(i) *If* Λ *is u.c. at* u_0 *and* g *is* C-l.c. on $\Lambda(u_0)$, *then* $g\Lambda$ *is* C-u.c. *at* u_0. *Moreover, if* $\Lambda(u_0)$ *is compact, then* $g\Lambda$ *is H-C-u.c. at* u_0.

(ii) *If Λ is u.c. at u_0, F is C-u.c. at (x, u_0) for all $x \in \Lambda(u_0)$, and $\Lambda(u_0)$ is compact, then $F\Lambda$ is C-u.c. at u_0. If f is C-l.c. on $\Lambda(u_0)$, then $f\Lambda$ is H-C-u.c. at u_0.*

(iii) *If Λ is l.c. at $(x_0, u_0) \in \mathrm{gr}\,\Lambda$ and F is C-l.c. at $(x_0, u_0; y_0)$, then $F\Lambda$ is C-l.c. at (u_0, y_0). In particular, if f is C-u.c. at (x_0, u_0), then $f\Lambda$ is C-l.c. at (u_0, y_0), where $y_0 = f(u_0, x_0)$.*

(iv) *If Λ is compact at u_0 and F is closed at (x, u_0) for all $x \in \Lambda(u_0)$, then $F\Lambda$ is closed at u_0.*

(v) *If Λ is compact at every $u \in \mathcal{U}$ and G is C-l.s.c., then $G\Lambda$ is C-l.s.c., too.*

Suppose now that X is a topological vector space, too.

(vi) *If Λ is H-u.c. at u_0 and f is equi-C-l.c. on $\Lambda(u_0) \times \{u_0\}$, then $f\Lambda$ is H-C-u.c. at u_0.*

(vii) *If Λ is H-l.c. at u_0 and f is equi-C-u.c. on $\Lambda(u_0) \times \{u_0\}$, then $f\Lambda$ is H-C-l.c. at u_0.*

PROOF. (i) Let $D \subset Y$ be an open set such that $g(\Lambda(u_0)) \subset D$. For every $x \in \Lambda(u_0)$, since g is C-l.c. at x and $D \in \mathcal{V}_Y(g(x))$, there exists an open neighborhood V_x of x in X such that $g(V_x) \subset D + C$. The set $D' := \bigcup_{x \in \Lambda(u_0)} V_x$ is open and contains $\Lambda(u_0)$. Therefore there exists $U_0 \in \mathcal{V}_{\mathcal{U}}(u_0)$ such that $\Lambda(u) \subset D'$ for every $u \in U_0$. It follows that $(g\Lambda)(u) = g(\Lambda(u)) \subset g(D') \subset D + C$ for $u \in U_0$. Hence $g\Lambda$ is C-u.c. at u_0. If $\Lambda(u_0)$ is compact, then $(g\Lambda)(u_0)$ is also compact, and so $g\Lambda$ is H-C-u.c. at u_0.

(ii) Let $D \subset Y$ be an open set such that $F(\Lambda(u_0) \times \{u_0\}) \subset D$. Then for every $x \in \Lambda(u_0)$, $F(x, u_0) \subset D$. Since F is C-u.c. at (x, u_0), there exists an open neighborhood V_x of x in X and $U_x \in \mathcal{V}_{\mathcal{U}}(u_0)$ such that $F(V_x \times U_x) \subset D + C$. Since $\{V_x \mid x \in \Lambda(u_0)\}$ is an open cover for the compact set $\Lambda(u_0)$, there exist $x_1, \dots, x_n \in \Lambda(u_0)$ such that $\Lambda(u_0) \subset V_{x_1} \cup \cdots \cup V_{x_n} =: V$; of course, V is an open set. Since Λ is u.c. at u_0, there exists $U_0 \in \mathcal{V}_{\mathcal{U}}(u_0)$ such that $\Lambda(U_0) \subset V$. Let $U := U_0 \cap U_{x_1} \cap \cdots \cap U_{x_n}$, $u \in U$, and $x \in \Lambda(u)$ $(\subset V)$. It follows that $x \in V_{x_i}$ for some $1 \le i \le n$. Since $u \in U_{x_i}$, $F(x, u) \subset D + C$. Hence $F\Lambda(u) \subset D + C$. Therefore $F\Lambda$ is C-u.c.

If f is C-l.c. on $\Lambda(u_0) \times \{u_0\}$, by applying what precedes for the multifunction $F := \mathrm{gr}\, f$, one obtains that $f\Lambda$ is C-u.c. at u_0. To obtain the stronger result note that the multifunction $\widetilde{\Lambda} : \mathcal{U} \rightrightarrows X \times \mathcal{U}$ defined by $\widetilde{\Lambda}(u) := \Lambda(u) \times \{u\}$ is l.c. at u_0; by applying then (i), the conclusion follows. The fact that $\widetilde{\Lambda}$ is l.c. at u_0 is obtained by using a similar argument to that used above, so that the proof is left to the reader.

(iii) Let $W \in \mathcal{V}_Y(y_0)$. Since F is C-u.c. at $(x_0, u_0; y_0)$, there exist $V \in \mathcal{V}_X(x_0)$ and $U_1 \in \mathcal{V}_{\mathcal{U}}(u_0)$ such that $F(x, u) \cap (W - C) \ne \emptyset$ for every $(x, u) \in V \times U_1$. Since Λ is l.c. at (u_0, x_0), there exists $U_2 \in \mathcal{V}_{\mathcal{U}}(u_0)$ such that $\Lambda(u) \cap V \ne \emptyset$ for all $u \in U_2$. Let $u \in U_1 \cap U_2$ and take $x \in \Lambda(u) \cap V$. It follows that there exists $y \in F(x, u) \cap (W - C)$. Therefore $y \in F\Lambda(u) \cap (W - C)$. Hence $F\Lambda$ is C-l.c. at (u_0, y_0).

When f is C-u.c. at (x_0, u_0), by applying what precedes to gr f, we get the desired conclusion.

(iv) Let gr $F\Lambda \supset ((u_i, y_i))_{i \in I} \to (u_0, y_0)$. There exists $X \supset (x_i)_{i \in I}$ such that $((u_i, x_i))_{i \in I} \subset$ gr Λ and $((x_i, u_i; y_i))_{i \in I} \subset$ gr F. Since Λ is compact at u_0, there exists a subnet $(x_{\varphi(j)})_{j \in J} \to x_0 \in \Lambda(u_0)$. Since F is closed at (x_0, y_0), $(x_0, u_0; y_0) \in$ gr F. It follows that $(u_0, y_0) \in$ gr $F\Lambda$. Therefore $F\Lambda$ is closed at u_0.

(v) Let $y \in Y$ and let $(u_i)_{i \in I}$ be a net in $\mathrm{lev}_{G\Lambda}(y)$ converging to $u \in \mathcal{U}$. There exists $(x_i)_{i \in I} \subset \mathrm{lev}_G(y)$ such that $x_i \in \Lambda(u_i)$ for every $i \in I$. Since Λ is compact at u, there exists the subnet $(x_{\varphi(j)})_{j \in J}$ of $(x_i)_{i \in I}$ convergent to $x \in \Lambda(u)$. Since G is C-l.s.c., $x \in \mathrm{lev}_G(y)$, whence $x \in \mathrm{lev}_{G\Lambda}(y)$. Consider now that X is a topological vector space.

(vi) Let $W \in \mathcal{V}_Y$. Since f is equi-C-l.c. on $\Lambda(u_0) \times \{u_0\}$, there exist $V \in \mathcal{V}_X$ and $U_1 \in \mathcal{V}_{\mathcal{U}}(u_0)$ such that $f((x + V) \times U_1) \subset f(x, u_0) + W + C$ for every $x \in \Lambda(u_0)$. Since Λ is H-u.c. at u_0, there exists $U_2 \in \mathcal{V}_{\mathcal{U}}(u_0)$ such that $\Lambda(U_2) \subset \Lambda(u_0) + V$. Consider $u \in U_1 \cap U_2$ and $y \in (f\Lambda)(u)$. Then $y = f(x, u)$ for some $x \in \Lambda(u)$. It follows that $x = x_0 + v$ with $x_0 \in \Lambda(u_0)$ and $v \in V$. So

$$y = f(x_0 + v, u) \in f(x_0, u_0) + W + C \subset (f\Lambda)(u_0) + W + C,$$

whence $(f\Lambda)(U_1 \cap U_2) \subset (f\Lambda)(u_0) + W + C$. Thus $f\Lambda$ is H-C-u.c. at u_0.

(vii) Let $W \in \mathcal{V}_Y$. Since f is equi-C-u.c. on $\Lambda(u_0) \times \{u_0\}$, there exist $V \in \mathcal{V}_X$ and $U_1 \in \mathcal{V}_{\mathcal{U}}(u_0)$ such that $f(x', u) \in f(x, u_0) + W - C$ for every $x \in \Lambda(u_0)$, $x' \in x + V$, and $u \in U_1$, or equivalently, $f(x, u_0) \in f(x', u) + W + C$ for every $x \in \Lambda(u_0)$, $x' \in x + V$, and $u \in U_1$. Since Λ is H-l.c. at u_0, there exists a neighborhood U_2 of u_0 such that $\Lambda(u_0) \subset \Lambda(u) + V$ for every $u \in U_2$. Let $u \in U_1 \cap U_2$ and $y \in (f\Lambda)(u_0)$. Then $y = f(x, u_0)$ for some $x \in \Lambda(u_0) \subset \Lambda(u) + V$. Hence there exists some $x' \in \Lambda(u) \cap (x + V)$. It follows that $y \in f(x', u) + W + C \subset f(\Lambda(u)) + W + C$, whence $(f\Lambda)(u_0) \subset (f\Lambda)(u) + W + C$ for every $u \in U_1 \cap U_2$. The proof is complete. $\square$

The implication "$\Rightarrow$" of assertion (i) of Proposition 2.5.9 was obtained by Smithson [324, Th. 3.1], while the implication "$\Rightarrow$" was established by Ferro [115, Lemma 2.2]. The assertion (iv) of Proposition 2.5.10 is stated in Ferro [115, Lemma 2.1], while assertion (v) is stated in Smithson [324, Prop. 1.1.2]. The assertion (i) of Proposition 2.5.11 can be found in [26], the assertion (ii) in [224], while assertions (iv) and (v) of Proposition 2.5.20 in [291], where one can also find the statement (i) of Proposition 2.5.10. The continuity notions for vector-valued functions are in accordance with those of Penot and Théra [293]. Remark 2.5.8 and Proposition 2.5.28 (ii)–(iv) may be found in [244] for $C = \{0\}$. Note that the statements (ii), (iii), (vi), and (vii), for g instead of f, of Proposition 2.5.28 can be found in [17], [18], and [19]. Other results on continuity of multifunctions can be found in the books by Aubin and Frankowska [13] and Cârjă [52].

2.6 Continuity Properties of Multifunctions Under Convexity Assumptions

Throughout this section X and Y are topological vector spaces, and $C \subset Y$ is a convex cone and $\Gamma : X \rightrightarrows Y$ is a multifunction. The next result is the analogue of a known one for convex functions.

Theorem 2.6.1. *Let Γ be C-nearly convex. Suppose that Γ is C-l.c. at $(x_0, y_0) \in \operatorname{gr} \Gamma$. Then Γ is C-l.c. at every $x \in \operatorname{int}(\operatorname{dom} \Gamma) = \operatorname{aint}(\operatorname{dom} \Gamma)$. Moreover, Γ is H-C-l.c. at every $x \in \operatorname{int}(\operatorname{dom} \Gamma)$ for which there exists a bounded set $B \subset Y$ such that $\Gamma(x) \subset B + C$.*

PROOF. Let $\alpha \in \,]0,1[$ be such that Γ is C-α-convex. Consider $V \in \mathcal{V}_Y$; there exists $W \in \mathcal{V}_Y$ with $W + W \subset V$. By hypothesis, there exists $U_0 \in \mathcal{V}_X$ such that $\Gamma(x_0 + u) \cap (y_0 + V) \neq \emptyset$, or, equivalently,

$$y_0 \in \Gamma(x_0 + u) + W + C \quad \forall u \in U_0. \tag{2.45}$$

Let $(x, y) \in \operatorname{gr} \Gamma$ with $x \in \operatorname{int}(\operatorname{dom} \Gamma)$. Of course, $\operatorname{dom} \Gamma$ is α-convex, whence, by Proposition 2.4.3 (iii), $\operatorname{int}(\operatorname{dom} \Gamma)$ is a convex set. Then there exist $\lambda \in \Lambda^{\alpha} \setminus \{0,1\}$ and $(x_1, y_1) \in \operatorname{gr} \Gamma$, with $x_1 \in \operatorname{int}(\operatorname{dom} \Gamma)$, such that $x = \lambda x_0 + (1-\lambda) x_1$; take $\overline{y} := \lambda y_0 + (1 - \lambda) y_1$. It follows that

$$\begin{aligned}
\overline{y} &\in \lambda \left(\Gamma(x_0 + u) + W + C \right) + (1 - \lambda) \Gamma(x_1) \\
&\subset \Gamma(x + \lambda u) + \lambda W + C \quad \forall u \in U_0.
\end{aligned} \tag{2.46}$$

Consider $\mu \in \Lambda^{\alpha} \setminus \{0,1\}$ such that $\mu(y - \overline{y}) \in W$. Then, by (2.46), the C-α-convexity of Γ and the choice of μ,

$$\begin{aligned}
y &\in (1 - \mu)\Gamma(x) + \mu \overline{y} + \mu(y - \overline{y}) \\
&\subset (1 - \mu)\Gamma(x) + \mu \Gamma(x + \lambda u) + \lambda \mu W + C + W \\
&\subset \Gamma(x + \lambda \mu u) + V + C \quad \forall u \in U_0.
\end{aligned} \tag{2.47}$$

Taking $U := \lambda \mu U_0$, we see that Γ is l.c. at (x, y).

Consider now $x \in \operatorname{int}(\operatorname{dom} \Gamma)$ and a bounded set $B \subset Y$ such that $\Gamma(x) \subset B + C$. For $V \in \mathcal{V}_Y$, with the same construction as above, we obtain W and U_0 for which (2.45) holds. Then we take (x_1, y_1), λ and $\overline{y}$ as above; so relation (2.46) holds, too. Taking now $\mu \in \Lambda^{\alpha} \setminus \{0,1\}$ such that $\mu(B - \overline{y}) \subset W$, we find that (2.47) holds for every $y \in \Gamma(x)$. Taking again $U := \lambda \mu U_0$, we obtain that Γ is H-C-l.c. at x. $\qquad \square$

The next result is related to the boundedness condition of the preceding theorem.

Proposition 2.6.2. *Let Γ be C-nearly convex. Suppose that $x_0 \in \operatorname{int}(\operatorname{dom} \Gamma)$ and $\Gamma(x_0) \subset B + C$ for some bounded set $B \subset Y$. Then for every $x \in \operatorname{dom} \Gamma$ there exists a bounded set $B_x \subset Y$ such that $\Gamma(x) \subset B_x + C$.*

PROOF. By hypothesis there exists $\alpha \in]0,1[$ such that Γ is C-α-convex. Let $x \in \operatorname{dom} \Gamma$; there exist $x_1 \in \operatorname{dom} \Gamma$ and $\lambda \in \Lambda^\alpha \setminus \{0,1\}$ such that $x_0 = \lambda x + (1-\lambda)x_1$. Since Γ is α-convex,

$$\lambda \Gamma(x) + (1-\lambda)\Gamma(x_1) \subset \Gamma(x_0) + C \subset B + C.$$

Taking $y_1 \in \Gamma(x_1)$, we obtain that $\Gamma(x) \subset \lambda^{-1}(B - (1-\lambda)y_1) + C = B_x + C$, where $B_x := \lambda^{-1}(B - (1-\lambda)y_1)$.   □

As an application of the preceding result we obtain that if $f : X \to \overline{\mathbb{R}}$ is nearly convex and finite-valued at $x_0 \in \operatorname{int}(\operatorname{dom} f)$, then f is finite-valued on $\operatorname{dom} f$.

Proposition 2.6.3. *Assume that Γ is C-l.c. at $(x_0, y_0) \in \operatorname{gr} \Gamma$ and C-nearly convex. If $\Gamma(x_0) \subset B + C$ for some bounded set $B \subset Y$, then Γ is H-C-u.c. at x_0.*

PROOF. Let $\alpha \in]0,1[$ be such that Γ is α-convex. Without loss of generality we may suppose that $x_0 = 0$. Let $y_0 \in \Gamma(0)$ be fixed. Consider $V \in \mathcal{V}_Y$ and take $W \in \mathcal{V}_Y$ such that $W + W + W \subset V$. Since Γ is H-C-l.c. at 0, there exists $U \in \mathcal{V}_X$ such that $y_0 \in \Gamma(u) + W + C$ for all $u \in U$. Since B is bounded, there exists $\mu \in]1/2,1[\cap \Lambda^\alpha$ with $(\mu^{-1} - 1)(B \cup \{y_0\}) \subset W$. Let $u \in (1-\mu^{-1})U \in \mathcal{V}_X$. There exists $u' \in U$ such that $0 = \mu u + (1-\mu)u'$. It follows that $y_0 = y' + w + k$ with $y' \in \Gamma(u')$, $w \in W$ and $k \in C$. Therefore $\mu \Gamma(u) + (1-\mu)y' \subset \Gamma(0)$, whence $\mu \Gamma(u) \subset \Gamma(0) + (1-\mu)(w + k - y_0)$. Hence

$$\begin{aligned}
\Gamma(u) &\subset \Gamma(0) + (\mu^{-1} - 1)\Gamma(0) + (\mu^{-1} - 1)W + C - (\mu^{-1} - 1)y_0 \\
&\subset \Gamma(0) + (\mu^{-1} - 1)B + C + W + C + W \\
&\subset \Gamma(0) + W + W + W + C \subset \Gamma(0) + V + C.
\end{aligned}$$

Therefore Γ is H-C-u.c. at x_0.   □

The preceding results lead to the following important theorem.

Theorem 2.6.4. *Let Γ be a C-nearly convex. Suppose that Γ is C-l.c. at $(x_0, y_0) \in \operatorname{gr} \Gamma$ and $\Gamma(x_0) \subset B + C$ for some bounded set $B \subset Y$. Then Γ is C-H-continuous on $\operatorname{int}(\operatorname{dom} \Gamma)$.*

PROOF. Of course, $x_0 \in \operatorname{int}(\operatorname{dom} \Gamma)$ in our conditions. From Proposition 2.6.2, $\Gamma(x) \subset B_x + C$, with B_x bounded, for every $x \in \operatorname{dom} \Gamma$. Using Theorem 2.6.1 we obtain that Γ is H-C-l.c. at every $x \in \operatorname{int}(\operatorname{dom} \Gamma)$. Then using Proposition 2.6.3 we obtain that Γ is H-C-u.c. at every $x \in \operatorname{int}(\operatorname{dom} \Gamma)$. Therefore Γ is C-H-continuous on $\operatorname{int}(\operatorname{dom} \Gamma)$.   □

From the preceding results we have the following characterizations of the continuity of nearly convex functions.

Corollary 2.6.5. *Let $f : X \to Y^\bullet$ be a nearly convex function and $x_0 \in \operatorname{int}(\operatorname{dom} f)$. Suppose that C is normal. Then f is C-u.c. at $x_0 \Leftrightarrow \Gamma_f$ is l.c. at $x_0 \Leftrightarrow \Gamma_f$ is H-continuous at every $x \in \operatorname{int}(\operatorname{dom} f) \Leftrightarrow f$ is continuous on $\operatorname{int}(\operatorname{dom} f) \Leftrightarrow f$ is continuous at x_0.*

PROOF. We have that f is C-u.c. at $x_0 \Rightarrow \Gamma_f$ is l.c. at x_0 by Proposition 2.5.26 (ii) and Γ_f is l.c. at $x_0 \Rightarrow \Gamma_f$ is H-continuous at every $x \in \operatorname{int}(\operatorname{dom} f)$ by Theorem 2.6.1 and Proposition 2.6.3. If Γ_f is H-continuous at (any) $x \in \operatorname{int}(\operatorname{dom} f)$, by Proposition 2.5.26 (i) and (ii) we obtain that f is C-l.c. and C-u.c. at x; the continuity of f at x follows from the normality of C. Of course, if f is continuous at every $x \in \operatorname{int}(\operatorname{dom} f)$, then f is continuous at x_0, whence f is C-u.c. at x_0. ∎

The next result corresponds to another well-known result of convex analysis.

Theorem 2.6.6. *Let Γ be a C-nearly convex. Suppose that there exist $x_0 \in \operatorname{int}(\operatorname{dom}\Gamma)$, $B \subset Y$ a bounded set, and $U \in \mathcal{V}_X$ such that $\Gamma(x_0) \subset B + C$ and $\Gamma(x_0 + u) \cap (B - C) \neq \emptyset$ for every $u \in U$. Then Γ is C-H-continuous on $\operatorname{int}(\operatorname{dom}\Gamma)$.*

PROOF. Let $\alpha \in {]0,1[}$ be such that Γ is α-convex. Taking into account Theorem 2.6.1 and Propositions 2.6.2, 2.6.3, it is sufficient to show that Γ is H-C-l.c. at x_0. Consider $V \in \mathcal{V}_Y$ and take $W \in \mathcal{V}_Y$ such that $W + W \subset V$. There exists $\lambda \in \Lambda^\alpha \setminus \{0,1\}$ such that $\lambda B \subset W$. It follows that

$$\Gamma(x_0) \subset \lambda\Gamma(x_0) + (1-\lambda)\Gamma(x_0) \subset \lambda(B + C) + (1-\lambda)\Gamma(x_0)$$
$$\subset (1-\lambda)\Gamma(x_0) + W + C.$$

Let $u \in U$ and $y \in \Gamma(x_0 + u) \cap (B - C)$; hence $y = b - k$ with $b \in B$ and $k \in C$. Then

$$\lambda(b - k) + (1-\lambda)\Gamma(x_0) \subset \lambda\Gamma(x_0 + u) + (1-\lambda)\Gamma(x_0) \subset \Gamma(x_0 + \lambda u) + C,$$

whence

$$(1-\lambda)\Gamma(x_0) \subset \Gamma(x_0 + \lambda u) + C - \lambda b + \lambda k \subset \Gamma(x_0 + \lambda u) + W + C.$$

Therefore

$$\Gamma(x_0) \subset \Gamma(x_0 + \lambda u) + W + C + W + C \subset \Gamma(x_0 + \lambda u) + V + C.$$

Hence $\Gamma(x_0) \subset \Gamma(x) + V + C$ for every $x \in x_0 + \lambda U \in \mathcal{V}_X(x_0)$. It follows that Γ is H-C-l.c. at x_0. ∎

From Theorem 2.6.6, also taking into account Proposition 2.5.27, we obtain that if $f : X \to \overline{\mathbb{R}}$ is nearly convex, $f(x_0) \in \mathbb{R}$, and f is bounded above on a neighborhood of x_0, then f is (finite-valued and) continuous on $\operatorname{int}(\operatorname{dom} f)$. In fact, semicontinuous nearly convex multifunctions are very close to convex multifunctions.

Proposition 2.6.7. *Suppose that $\Gamma : X \rightrightarrows Y$ is closed-valued and nearly convex. If $\operatorname{dom}\Gamma$ is convex (for example, if $\operatorname{dom}\Gamma$ is closed or open) and Γ is H-u.c. at every point of $\operatorname{dom}\Gamma$, then Γ is convex.*

PROOF. Since Γ is nearly convex, so is dom Γ; so, if dom Γ is closed or open, from Corollary 2.4.4 we obtain that dom Γ is convex. Let $\alpha \in {]}0,1{[}$ be such that Γ is α-convex. Consider $x, y \in$ dom Γ and $\lambda \in {]}0,1{[}$. There exists a sequence $(\lambda_n)_{n \in \mathbb{N}} \subset \Lambda^\alpha$ converging to λ. Of course,

$$\lambda_n \Gamma(x) + (1 - \lambda_n)\Gamma(y) \subset \Gamma(\lambda_n x + (1 - \lambda_n)y) \quad \forall n \in \mathbb{N}.$$

Let $u \in \Gamma(x)$ and $v \in \Gamma(y)$ be fixed. Consider $V \in \mathcal{V}_Y$, and take $W \in \mathcal{V}_Y$ such that $W + W \subset V$. Since Γ is H-u.c. at $x_\lambda := \lambda x + (1 - \lambda)y \in$ dom Γ, there exists $n_V \in \mathbb{N}$ such that $\Gamma(\lambda_n x + (1 - \lambda_n)y) \subset \Gamma(x_\lambda) + W$ for all $n \geq n_V$. It follows that $\lambda_n u + (1 - \lambda_n)v \in \Gamma(x_\lambda) + W$ for all $n \geq n_V$, whence $\lambda u + (1 - \lambda)v \in \mathrm{cl}\,(\Gamma(x_\lambda) + W) \subset \Gamma(x_\lambda) + W + W \subset \Gamma(x_\lambda) + V$. Hence $\lambda u + (1 - \lambda)v \in \bigcap_{V \in \mathcal{V}_Y} (\Gamma(x_\lambda) + V) = \mathrm{cl}\,\Gamma(x_\lambda) = \Gamma(x_\lambda)$. Therefore $\lambda\Gamma(x) + (1 - \lambda)\Gamma(y) \subset \Gamma(\lambda x + (1 - \lambda)y)$, which shows that Γ is convex. $\square$

The first part of Theorem 2.6.1 was obtained by Borwein [37] for Γ convex and $C = \{0\}$, while the second part was obtained by Nikodem [275] for Γ mid-convex, bounded-valued, and H-C-l.c. at x_0; also, Nikodem obtained Proposition 2.6.3 and Theorem 2.6.6 for Γ mid-convex and bounded-valued in [275], and Proposition 2.6.7 for Γ mid-convex with bounded values and H-continuous at each point of its open convex domain in [276]. The equivalence "f is continuous at $x_0 \Leftrightarrow \Gamma_f$ is l.c. at x_0" of Corollary 2.6.5 is stated by Borwein [36] for f convex.

2.7 Tangent Cones and Differentiability of Multifunctions

Throughout this section X, Y are normed vector spaces. Consider $A \subset X$ a nonempty set and $a \in X$. The **Bouligand tangent cone** or **contingent cone** of A at a is defined as the set

$$T_B(A; a) = \limsup_{t \downarrow 0} t^{-1}(A - a) := \limsup_{t \to 0} \mathcal{R}_{A,a}(t),$$

where $\mathcal{R}_{A,a} : [0, \infty[\rightrightarrows X$, $\mathcal{R}_{A,a}(t) = t^{-1}(A - a)$ for $t > 0$, $\mathcal{R}_{A,a}(0) = \emptyset$. Similarly, the **Ursescu tangent cone** or **adjacent cone** of A at a is defined by

$$T_U(A; a) = \liminf_{t \downarrow 0} t^{-1}(A - a) := \liminf_{t \to 0} \mathcal{R}_{A,a}(t).$$

Of course, $T_U(A; a) \subset T_B(A; a)$ and $T_B(A; a) = \emptyset$ if $a \notin \mathrm{cl}\,A$. Taking this into account, in the sequel we consider only the case $a \in \mathrm{cl}\,A$ when considering tangent cones. It is known (and easy to see) that

$$T_B(A; a) = T_U(A; a) = \mathrm{cl}\,({]}0, \infty{[}\cdot(A - a))$$

if A is a convex set and $a \in \mathrm{cl}\,A$.

Using Proposition 2.5.3 we have the following result; the notation $(t_n) \to 0_+$ means that $(t_n)_{n \in \mathbb{N}} \subset {]}0, \infty{[}$ and $(t_n) \to 0$.

Proposition 2.7.1. *Let $A \subset X$ and $a \in \operatorname{cl} A$. Then*

$$T_B(A; a) = \left\{ u \in X \,\middle|\, \liminf_{t \downarrow 0} \operatorname{dist}\left(u, t^{-1}(A - a)\right) = 0 \right\}$$
$$= \{ u \in X \mid \exists\, (t_n) \to 0_+,\ \exists\, (u_n) \to u,\ \forall n \in \mathbb{N} \,:\, a + t_n u_n \in A \},$$

and

$$T_U(A; a) = \left\{ u \in X \,\middle|\, \lim_{t \downarrow 0} \operatorname{dist}\left(u, t^{-1}(A - a)\right) = 0 \right\}$$
$$= \{ u \in X \mid \forall\, (t_n) \to 0_+,\ \exists\, (u_n) \to u,\ \forall n \in \mathbb{N} \,:\, a + t_n u_n \in A \}.$$

Moreover, $T_B(A; a)$ and $T_U(A; a)$ are nonempty closed cones (in particular, they contain 0).

PROOF. The formulae above follow from Proposition 2.5.3. The sets $T_B(A; a)$ and $T_U(A; a)$ are closed, since they are limit superior and limit inferior of multifunctions, respectively. Since $a \in \operatorname{cl} A$, $\operatorname{dist}\left(0, t^{-1}(A - a)\right) = 0$ for every $t > 0$. So $0 \in T_U(A; a)$. The fact that they are cones follows immediately from the sequential characterizations of their elements. $\qquad\square$

Let $\Gamma : X \rightrightarrows Y$ and $(x, y) \in X \times Y$. Consider the multifunction

$$\Gamma_{x,y} : \mathbb{R} \times X \rightrightarrows Y, \quad \Gamma_{x,y}(t, u) = \begin{cases} t^{-1}\left(\Gamma(x + tu) - y\right) & \text{if } t > 0, \\ \emptyset & \text{if } t \le 0. \end{cases}$$

Of course, the domain of $\Gamma_{x,y}$ is the set $D := \{(t, u) \mid t > 0,\ u \in X,\ x + tu \in \operatorname{dom} \Gamma\}$. So $(0, u) \in \operatorname{cl} D$ if and only if $u \in T_B(\operatorname{dom} \Gamma; x)$.

Definition 2.7.2. *Let $\Gamma : X \rightrightarrows Y$ and $(x, y) \in \operatorname{gr} \Gamma$. The **Dini upper derivative** of Γ at (x, y) in the direction $u \in X$ is the set*

$$D\Gamma(x, y)(u) := \limsup_{(t, u') \to (0, u)} (\Gamma_{x,y}) \big|_{D \cup \{(0,u)\}}(t, u') = \limsup_{(t, u') \to (0, u)} \Gamma_{x,y}(t, u'),$$

*while the **Dini lower derivative** of Γ at (x, y) in the direction $u \in X$ is the set*

$$\underline{D}\Gamma(x, y)(u) := \begin{cases} \displaystyle\liminf_{(t, u') \to (0, u)} (\Gamma_{x,y}) \big|_{D \cup \{(0,u)\}}(t, u') & \text{if } u \in T_B(\operatorname{dom} \Gamma; x), \\ \emptyset & \text{if } u \notin T_B(\operatorname{dom} \Gamma; x). \end{cases}$$

Normally, the Dini upper derivative of Γ at (x, y) would be denoted by $\overline{D}\Gamma(x, y)$ (see Penot [289]); we prefer the notation $D\Gamma(x, y)$ because the Dini lower derivative will be seldom used. It is obvious that $D\Gamma(x, y)(u) = \emptyset$ if $u \notin T_B(\operatorname{dom} \Gamma; x)$, and so

$$\underline{D}\Gamma(x, y)(u) \subset D\Gamma(x, y)(u) \ \forall u \in X,$$

and these sets are closed. Moreover,

$$\operatorname{gr} D\Gamma(x,y) = T_B\left(\operatorname{gr}\Gamma;(x,y)\right).$$

These facts can be seen from the following characterizations for $v \in D\Gamma(x,y)(u)$:

$$\exists\,(t_n) \to 0_+,\ (u_n) \to u,\ (v_n) \to v,\ \forall n \in \mathbb{N}\ :\ v_n \in \Gamma_{x,y}(t_n,u_n),$$
$$\exists\,(t_n) \to 0_+,\ (u_n,v_n) \to (u,v),\ \forall n \in \mathbb{N}\ :\ (x,y) + t_n(u_n,v_n) \in \operatorname{gr}\Gamma,$$

and for $v \in \underline{D}\Gamma(x,y)(u)$ when $u \in T_B(\operatorname{dom}\Gamma;x)$,

$$\forall D \supset ((t_n,u_n)) \to (0_+,u),\ \exists\,(v_n) \to v,\ \exists n_0,$$
$$\forall n \geq n_0\ :\ (x,y) + t_n(u_n,v_n) \in \operatorname{gr}\Gamma.$$

Definition 2.7.3. *Let* $\Gamma : X \rightrightarrows Y$ *and* $(x,y) \in \operatorname{gr}\Gamma$. *We say that* Γ *is* **semidifferentiable at** (x,y) **in the direction** u *if* $\underline{D}\Gamma(x,y)(u) = D\Gamma(x,y)(u)$, *and* Γ *is* **semidifferentiable at** (x,y) *if* Γ *is semidifferentiable at every* $u \in X$.

The following theorem establishes sufficient conditions for semidifferentiability of convex multifunctions. By raint A we mean the **algebraic relative interior** of A. For the proof of the theorem we need the following lemma.

Lemma 2.7.4. *Let* $\emptyset \neq A \subset X$ *be a convex set and* $T : X \to Y$ *be a linear operator.*

(i) *If* int $A \neq \emptyset$ *and* T *is open, then* $\operatorname{int}(]0,\infty[\cdot A) =]0,\infty[\cdot \operatorname{int} A$ *and* $T(\operatorname{int} A) = \operatorname{int} T(A)$.

(ii) *If* raint $A \neq \emptyset$ *(e.g., if* X *is finite-dimensional), then* $\operatorname{raint}(]0,\infty[\cdot A) =]0,\infty[\cdot \operatorname{raint} A$ *and* $T(\operatorname{raint} A) = \operatorname{raint} T(A)$.

PROOF. (i) The inclusions $]0,\infty[\cdot \operatorname{int} A \subset \operatorname{int}(]0,\infty[\cdot A)$ and $T(\operatorname{int} A) \subset \operatorname{int} T(A)$ are obvious.

Let $a \in \operatorname{int} A$ be fixed. Consider $x \in \operatorname{int}(]0,\infty[\cdot A)$. Then there exist $x' \in]0,\infty[\cdot A$ and $\lambda \in\,]0,1[$ such that $x = (1-\lambda)x' + \lambda a$. Therefore there exist $a' \in A$ and $t' > 0$ such that $x' = t'a'$. Since $a \in \operatorname{int} A$ and A is convex,

$$\frac{1}{(1-\lambda)t' + \lambda}x = \frac{(1-\lambda)t'}{(1-\lambda)t' + \lambda}a' + \frac{\lambda}{(1-\lambda)t' + \lambda}a \in \operatorname{int} A,$$

whence $x \in\,]0,\infty[\cdot \operatorname{int} A$.

Consider now $y \in \operatorname{int} T(A)$. Then there exist $y' \in T(A)$ and $\lambda \in\,]0,1[$ such that $y = (1-\lambda)y' + \lambda T(a)$. Take $x' \in A$ such that $y' = T(x')$ and consider $x := (1-\lambda)x' + \lambda a$. Since $a \in \operatorname{int} A$ and A is convex, $x \in \operatorname{int} A$. Therefore $y = T(x) \in T(\operatorname{int} A)$.

(ii) The proof is similar; just note that $a \in \operatorname{raint} A$ if and only if for every $x \in A$ there exists $\lambda > 0$ such that $(1+\lambda)a - \lambda x \in A$ and if $a \in \operatorname{raint} A$, $x \in A$ and $\lambda \in\,]0,1[$, then $(1-\lambda)a + \lambda x \in \operatorname{raint} A$.

Lemma 2.7.5. *Let $A \subset X \times Y$ be a nonempty convex set. Suppose that either $(x, y) \in \operatorname{int} A$ or $(x, y) \in \operatorname{raint} A$ and $\dim X < \infty$. Then for every sequence $(x_n) \subset \operatorname{Pr}_X(A)$ with $(x_n) \to x$ there exists $(y_n) \to y$ such that $(x_n, y_n) \in A$ for every $n \in \mathbb{N}$.*

PROOF. Without loss of generality we may assume that $(x, y) = (0, 0)$. Suppose first that $(0, 0) \in \operatorname{int} A$. Since $(x_n, 0) \to (0, 0) \in \operatorname{int} A$, $(x_n, 0) \in A$ for $n \geq n_0$. Taking $y_0 = 0$ for $n \geq n_0$ and y_n arbitrary such that $(x_n, y_n) \in A$ for $n < n_0$, the conclusion follows.

Suppose now that $\dim X < \infty$ and $(0, 0) \in \operatorname{raint} A$. From the preceding lemma with $T = \operatorname{Pr}_X$, we have that $0 \in \operatorname{raint}(\operatorname{Pr}_X(A))$. Let $X_0 := \operatorname{span}(\operatorname{Pr}_X(A))$. Since $\dim X < \infty$ and $\operatorname{Pr}_X(A)$ is convex, there exists $r > 0$ such that $X_0 \cap \{x \in X \mid \|x\| \leq r\} \subset \operatorname{Pr}_X(A)$. For every $n \in \mathbb{N}^*$ consider $y_n \in Y$ such that $(x_n, y_n) \in A$ and $\|y_n\| \leq \inf\{\|y\| \mid (x_n, y) \in A\} + n^{-1}$. Suppose that $(y_n) \not\to 0$. Taking eventually subsequences, we may suppose that $\|y_n\| \geq r_1 > 0$ for every $n \in \mathbb{N}$ and $(x_n / \|x_n\|) \to x$. There exists a base $\{u_1, u_2, \ldots, u_p\}$ of X_0 such that $\|u_i\| = r$ for $1 \leq i \leq p$ and $x \in \operatorname{int} C$, where $C = \{\sum_{i=1}^p \lambda_i u_i \mid \lambda_i \geq 0\}$. (For example, one takes a base $\{u_1', u_2', \ldots, u_p'\}$ of X_0, a bijective linear operator $T : X_0 \to X_0$ such that $x = \frac{1}{p} T(u_1' + \cdots + u_p')$ and $u_i = T u_i'$ for $1 \leq i \leq p$.) Let $(v_i)_{i=1}^p \subset Y$ be such that $(u_i, v_i) \in A$ for every i. Since $(x_n / \|x_n\|) \to x \in \operatorname{int} C$, there exists $n_0 \in \mathbb{N}$ such that $x_n \in C$ for every $n \geq n_0$. Therefore, there exist the sequences $(\lambda_i^n) \subset [0, \infty[$, $1 \leq i \leq p$, such that $x_n = \sum_{i=1}^p \lambda_i^n u_i$ for $n \geq n_0$. Since $(x_n) \to 0$, $(\lambda_i^n) \to 0$ for every i. We may suppose that $\sum_{i=1}^p \lambda_i^n \leq 1$ for $n \geq n_0$. Since $(0, 0) \in A$, it follows that

$$\left(x_n, \sum_{i=1}^p \lambda_i^n v_i \right) = \sum_{i=1}^p \lambda_i^n (u_i, v_i) \in A,$$

whence $\|y_n\| \leq \|\sum_{i=1}^p \lambda_i^n v_i\| + n^{-1} \leq \sum_{i=1}^p \lambda_i^n \|v_i\| + n^{-1} \to 0$, a contradiction. $\square$

Theorem 2.7.6. *Let $\Gamma : X \rightrightarrows Y$ be a convex multifunction. Suppose that one of the following conditions holds:*

(i) $\operatorname{int}(\operatorname{gr} \Gamma) \neq \emptyset$, $x \in \operatorname{int}(\operatorname{dom} \Gamma)$ *and* $y \in \Gamma(x)$;
(ii) X *and* Y *are finite-dimensional,* $x \in \operatorname{raint}(\operatorname{dom} \Gamma)$, *and* $y \in \Gamma(x)$.

Then Γ is semidifferentiable at (x, y).

PROOF. We give the proof of (ii), that of (i) being similar (in this case one takes into account that Pr_X is an open linear operator).

We already know that

$$\underline{D\Gamma}(x, y)(u) \subset D\Gamma(x, y)(u) = \{v \in Y \mid (u, v) \in T_B(\operatorname{gr} \Gamma, (x, y))\} \quad \forall u \in X.$$

Let us show that $v \in \underline{D\Gamma}(x, y)(u)$ if $(u, v) \in \operatorname{raint} T_B(\operatorname{gr} \Gamma, (x, y))$, even without asking that $x \in \operatorname{raint}(\operatorname{dom} \Gamma)$.

Without loss of generality we consider that $(x, y) = (0,0)$. Take

$$(u, v) \in \operatorname{raint} T_B(\operatorname{gr} \Gamma, (0,0)) = \operatorname{raint}(]0, \infty[\cdot \operatorname{gr} \Gamma) =]0, \infty[\cdot \operatorname{raint}(\operatorname{gr} \Gamma);$$

we used the relation $\operatorname{raint} A = \operatorname{raint}(\operatorname{cl} A)$, valid if A is a convex subset of a finite-dimensional space, for the first equality and Lemma 2.7.4 for the last one. Then there exist $\lambda' > 0$ and $(u', v') \in \operatorname{raint}(\operatorname{gr} \Gamma)$ such that $(u, v) = \lambda'(u', v')$. Let $(t_n) \to 0_+$ and $(u_n) \to u$ such that $t_n u_n = x + t_n u_n \in \operatorname{dom} \Gamma$ for every $n \in \mathbb{N}$. It follows that $(u_n) \subset \operatorname{aff}(\operatorname{dom} \Gamma) = \operatorname{span}(\operatorname{dom} \Gamma)$ and $(u_n/\lambda') \to u' \in \operatorname{raint}(\operatorname{dom} \Gamma)$. So there exists $n_0 \in \mathbb{N}$ such that $u_n/\lambda' \in \operatorname{dom} \Gamma$ for $n \geq n_0$. So, using Lemma 2.7.5, there exists a sequence $(v'_n) \subset Y$ such that $(v'_n) \to v'$ and $(u_n/\lambda', v'_n) \in \operatorname{gr} \Gamma$ for $n \geq n_0$. Let $v_n := \lambda' v'_n$ for $n \geq n_0$ and arbitrary for $n < n_0$. So $(v_n) \to v$. Since $(t_n) \to 0_+$, there exists $n_1 \geq n_0$ such that $t_n \leq 1/\lambda'$ for $n \geq n_1$. Since $(0,0) \in \operatorname{gr} \Gamma$, we obtain that $(x, y) + t_n(u_n, v_n) = t_n\lambda'(u_n/\lambda', v'_n) \in \operatorname{gr} \Gamma$ for $n \geq n_1$. Therefore $v \in \underline{D}\Gamma(x, y)(u)$.

Consider now $u \in \operatorname{raint} T_B(\operatorname{dom} \Gamma; 0) =]0, \infty[\cdot \operatorname{raint}(\operatorname{dom} \Gamma)$ and $v \in \underline{D}\Gamma(x, y)(u)$. Therefore $u \in \operatorname{raint}(]0, \infty[\cdot \operatorname{dom} \Gamma)$ and $(u, v) \in \operatorname{cl}(]0, \infty[\cdot \operatorname{gr} \Gamma)$. By Lemma 2.7.5, there exists $v_0 \in Y$ such that $(u, v) \in \operatorname{raint}(]0, \infty[\cdot \operatorname{gr} \Gamma)$. It follows that for every $\lambda \in]0,1[$,

$$(u, (1-\lambda)v + \lambda v_0) = (1-\lambda)(u, v) + \lambda(u, v_0) \in \operatorname{raint}(]0, \infty[\cdot \operatorname{gr} \Gamma).$$

By what precedes, we have that $(1-\lambda)v + \lambda v_0 \in \underline{D}\Gamma(x, y)(u)$ for every $\lambda \in]0,1[$. Taking into account that $\underline{D}\Gamma(x, y)(u)$ is a closed set, we obtain that $v \in \underline{D}\Gamma(x, y)(u)$ for $\lambda \to 0$. Therefore for every $u \in \operatorname{raint} T_B(\operatorname{dom} \Gamma, x)$,

$$\underline{D}\Gamma(x, y)(u) = \{v \in Y \mid (u, v) \in T_B(\operatorname{gr} \Gamma, (x, y))\} = D\Gamma(x, y)(u);$$

i.e., Γ is semidifferentiable at every $u \in \operatorname{raint} T_B(\operatorname{dom} \Gamma, x)$. If our initial condition $x \in \operatorname{raint}(\operatorname{dom} \Gamma)$ holds, then $T_B(\operatorname{dom} \Gamma; x) = \operatorname{span}(\operatorname{dom} \Gamma - x) = \operatorname{raint} T_B(\operatorname{dom} \Gamma, x)$, and so Γ is semidifferentiable. $\qquad \square$

Remark 2.7.7. The notion of semidifferentiable multifunction is introduced by Penot in [289]. Theorem 2.7.6 (i) is established in [289, Prop. 3.5].

Consider now another normed vector space $\mathcal{U}$, the multifunction $\Lambda : \mathcal{U} \rightrightarrows X$, and the function $f : X \times \mathcal{U} \to Y$. With these we associate the multifunction $f\Lambda : \mathcal{U} \rightrightarrows Y$ defined by $f\Lambda(u) := f(\Lambda(u) \times \{u\})$. We are interested in evaluating $Df\Lambda(\overline{u}, \overline{y})$ for some $(\overline{u}, \overline{y}) \in \operatorname{gr} f\Lambda$. Recall that a multifunction $\mathcal{S} : X \rightrightarrows Y$ is **upper Lipschitz** at $\overline{x} \in \operatorname{dom} \mathcal{S}$ if there exist $L, \rho > 0$ such that $\mathcal{S}(x) \subset \mathcal{S}(\overline{x}) + L \|x - \overline{x}\| B_Y$ for every $x \in B(\overline{x}, \rho)$.

Theorem 2.7.8. *Suppose that $\overline{x} \in \Lambda(\overline{u})$, $\overline{y} = f(\overline{x}, \overline{u})$ and that f is Fréchet differentiable at $(\overline{x}, \overline{u})$. Then*

$$\nabla_x f(\overline{x}, \overline{u})(D\Lambda(\overline{u}, \overline{x})(u)) + \nabla_u f(\overline{x}, \overline{u})(u) \subset Df\Lambda(\overline{u}, \overline{y})(u) \quad \forall u \in \mathcal{U} \quad (2.48)$$

and

$$\nabla_x f(\overline{x}, \overline{u}) \left(\underline{D} \Lambda(\overline{u}, \overline{x})(u) \right) + \nabla_u f(\overline{x}, \overline{u})(u) \subset \underline{D} f \Lambda(\overline{u}, \overline{y})(u) \quad \forall u \in \mathcal{U}. \quad (2.49)$$

Consider

$$\widetilde{\Lambda} : \mathcal{U} \times Y \rightrightarrows X, \quad \widetilde{\Lambda}(u, y) := \{ x \in \Lambda(u) \mid f(x, u) = y \}.$$

Moreover, suppose that $\dim X < \infty$ *and either* (a) $\Lambda(\overline{u}) = \{\overline{x}\}$ *and* Λ *is upper Lipschitz at* $\overline{u}$ *or* (b) $\widetilde{\Lambda}(\overline{u}, \overline{y}) = \{\overline{x}\}$ *and* $\widetilde{\Lambda}$ *is upper Lipschitz at* $(\overline{u}, \overline{y})$. *Then*

$$\nabla_x f(\overline{x}, \overline{u}) (D\Lambda(\overline{u}, \overline{x})(u)) + \nabla_u f(\overline{x}, \overline{u})(u) = Df\Lambda(\overline{u}, \overline{y})(u) \quad \forall u \in \mathcal{U}. \quad (2.50)$$

PROOF. Let $x \in D\Lambda(\overline{u}, \overline{x})(u)$ and take $y = \nabla_x f(\overline{x}, \overline{u})(x) + \nabla_u f(\overline{x}, \overline{u})(u)$. Then there exist $(t_n) \to 0_+$, $((u_n, x_n)) \to (u, x)$ such that $\overline{x} + t_n x_n \in \Lambda(\overline{u} + t_n u_n)$ for every $n \in \mathbb{N}$. Since f is differentiable at $(\overline{x}, \overline{u})$, we have that for some $Y \supset (v_n) \to 0$,

$$f(\overline{x} + t_n x_n, \overline{u} + t_n u_n) = \overline{y} + t_n \left(\nabla_x f(\overline{x}, \overline{u})(x_n) + \nabla_u f(\overline{x}, \overline{u})(u_n) + v_n \right)$$

for every $n \in \mathbb{N}$. Taking $y_n := \nabla_x f(\overline{x}, \overline{u})(x_n) + \nabla_u f(\overline{x}, \overline{u})(u_n) + v_n$, since $\nabla f(\overline{x}, \overline{u})$ is continuous, we have that $(y_n) \to y$. Since $\overline{y} + t_n y_n \in f\Lambda(\overline{u} + t_n u_n)$ for every $n \in \mathbb{N}$, $y \in Df\Lambda(\overline{u}, \overline{y})(u)$. Hence (2.48) holds.

Consider $x \in \underline{D}\Lambda(\overline{u}, \overline{x})(u)$ and y as above. Let $((t_n, u_n)) \to (0_+, u)$ with $\overline{u} + t_n u_n \in \operatorname{dom} f\Lambda (= \operatorname{dom} \Lambda)$ for every n. Then there exists $(x_n) \to x$ such that $\overline{x} + t_n x_n \in \Lambda(\overline{u} + t_n u_n)$ for $n \geq n_0$. Continuing as above, we get $(y_n) \to y$ with $\overline{y} + t_n y_n \in f\Lambda(\overline{u} + t_n u_n)$ for $n \geq n_0$. Therefore $y \in \underline{D}f\Lambda(\overline{u}, \overline{y})(u)$, whence (2.49).

Suppose now that $\dim X < \infty$ and (a) or (b) holds. Consider $y \in Df\Lambda(\overline{u}, \overline{y})(u)$; then there exist $(t_n) \to 0_+$, $((u_n, y_n)) \to (u, y)$ such that $\overline{y} + t_n y_n \in f\Lambda(\overline{u} + t_n u_n)$ for every $n \in \mathbb{N}$. Let $\widetilde{x}_n \in \Lambda(\overline{u} + t_n u_n)$ be such that $\overline{y} + t_n y_n = f(\widetilde{x}_n, \overline{u} + t_n u_n)$; i.e., $\widetilde{x}_n \in \widetilde{\Lambda}(\overline{u} + t_n u_n, \overline{y} + t_n y_n)$, for every n. Since Λ is upper Lipschitz at $\overline{u}$ and $\Lambda(\overline{u}) = \{\overline{x}\}$ in case (a) and $\widetilde{\Lambda}$ is upper Lipschitz at $(\overline{u}, \overline{y})$ and $\widetilde{\Lambda}(\overline{u}, \overline{y}) = \{\overline{x}\}$ in case (b), there exists a bounded sequence $(x_n) \subset X$ such that $\widetilde{x}_n = \overline{x} + t_n x_n$ for every $n \in \mathbb{N}$. Since $\dim X < \infty$, there exists a subsequence $(x_{n_p})_{p \in \mathbb{N}}$ converging to some $x \in X$. It follows that $x \in D\Lambda(\overline{u}, \overline{x})(u)$, and so $y \in \nabla_x f(\overline{x}, \overline{u}) (D\Lambda(\overline{u}, \overline{x})(u)) + \nabla_u f(\overline{x}, \overline{u})(u)$. Taking into account (2.48), we obtain that (2.50) holds. □

In the following example one of the conditions in (a) and (b) of Theorem 2.7.8 is not satisfied, and (2.50) does not hold.

Example 2.7.9. ([354]) Let $\mathcal{U} = X = Y = \mathbb{R}$, $\Lambda, \Lambda' : \mathcal{U} \rightrightarrows X$ and $f : X \times \mathcal{U} \to Y$ be defined by

$$\Lambda(u) := [0, \max(1, 1 + u)], \quad \Lambda'(u) := [0, \max(1, 1 + u)[, \quad f(x, u) := x^2 - x,$$

$\overline{u} = 0$, $\overline{x} = 0$, $\overline{y} = f(0, 0) = 0$. Then Λ, Λ' are upper Lipschitz at $\overline{u}$, but $\Lambda(\overline{u}) \neq \{\overline{x}\} \neq \Lambda'(\overline{u})$ (so (a) does not hold for Λ and Λ') and

$$f\Lambda(u) = \begin{cases} [-\frac{1}{4}, 0] & \text{if } u \leq 0, \\ [-\frac{1}{4}, u(1+u)] & \text{if } u > 0, \end{cases} \qquad f'\Lambda(u) = \begin{cases} [-\frac{1}{4}, 0] & \text{if } u \leq 0, \\ [-\frac{1}{4}, u(1+u)[& \text{if } u > 0. \end{cases}$$

Let $\{x_1^y, x_2^y\} = \{x \in \mathbb{R} \mid x^2 - x = y\}$, with $x_1^y \leq x_2^y$, for $y \geq -\frac{1}{4}$. Then

$$\widetilde{\Lambda}(u, y) = \begin{cases} \{x_1^y, x_2^y\} & \text{if } u \in \mathbb{R}, \ y \in [-\frac{1}{4}, 0], \\ \{x_2^y\} & \text{if } u > 0, \ y \in]0, u(1+u)], \\ \emptyset & \text{otherwise}, \end{cases}$$

$$\widetilde{\Lambda}'(u, y) = \begin{cases} \{x_1^y, x_2^y\} & \text{if } u \in \mathbb{R}, \ y \in [-\frac{1}{4}, 0[, \\ \{0\} & \text{if } u \in \mathbb{R}, \ y = 0, \\ \{x_2^y\} & \text{if } u > 0, \ y \in]0, u(1+u)[, \\ \emptyset & \text{otherwise}. \end{cases}$$

Now, $\widetilde{\Lambda}$ is upper Lipschitz at $(\overline{u}, \overline{y})$ but $\widetilde{\Lambda}(\overline{u}, \overline{y}) = \{0, 1\} \neq \{\overline{x}\}$, while $\widetilde{\Lambda}'$ is not upper Lipschitz at $(\overline{u}, \overline{y})$ (so (b) does not hold for Λ and Λ'). One obtains that

$$D\Lambda(\overline{u}, \overline{x})(u) = D\Lambda'(\overline{u}, \overline{x})(u) = \mathbb{R} \times \mathbb{R}_+, \quad \forall u \in \mathbb{R},$$
$$Df\Lambda(\overline{u}, \overline{y})(u) = Df\Lambda'(\overline{u}, \overline{y})(u) = \{(u, y) \in \mathbb{R}^2 \mid y \leq \max(0, u)\}, \quad \forall u \in \mathbb{R}.$$

Since $\nabla_x f(\overline{x}, \overline{u}) = -1$ and $\nabla_u f(\overline{x}, \overline{u}) = 0$,

$$\nabla_x f(\overline{x}, \overline{u})\left(D\Lambda(\overline{u}, \overline{x})(u)\right) + \nabla_u f(\overline{x}, \overline{u})(u) = \left] -\infty, 0 \right] \neq \left] -\infty, u \right] = Df\Lambda(\overline{u}, \overline{x})(u)$$

for every $u > 0$, and similarly for Λ'.

For other types of tangent cones one can consult Aubin and Frankowska's book [13]. Formulae (2.48) and (2.50) are stated by Tanino [354] under the supplementary conditions that $\mathcal{U}$, X, and Y are finite-dimensional and f is of class $\mathcal{C}^1$ and by Klose [212, Th. 4.1].

3

Optimization in Partially Ordered Spaces

3.1 Solution Concepts

3.1.1 Approximate Minimality

In the last years several concepts for approximately efficient solutions of a vector optimization problem were published (see [110], [127], [93, 94], [241], [271], [329], [337, 338], [362]). The reason for introducing approximately efficient solutions is the fact that numerical algorithms usually generate only approximate solutions anyhow, and moreover, the set of efficient points may be empty, whereas approximately efficient points always exist under very weak assumptions.

We begin with our concept given in [127]. Let M and B be subsets of the topological vector space Y with $0 \in \mathrm{bd}\, B$, let $k^0 \in Y$ be such that $\mathrm{cl}\, B + (0, \infty)k^0 \subset \mathrm{int}\, B$, and let $\varepsilon \geq 0$.

Definition 3.1.1. *(Approximate efficiency, εk^0-efficiency) An element $y_\varepsilon \in M$ is said to be εk^0-efficient on M w.r.t. B if*

$$M \cap \left(y_\varepsilon - \varepsilon k^0 - (B \setminus \{0\})\right) = \emptyset.$$

The set of εk^0-efficient points of M w.r.t. B will be denoted by $\mathrm{Eff}(M, B_{\varepsilon k^0})$. For the case $\varepsilon = 0$ the set $\mathrm{Eff}(M, B_{\varepsilon k^0})$ coincides with the usual set $\mathrm{Eff}(M, B)$ of efficient points of M w.r.t. B (compare Remark 2.1.3). An illustration of the εk^0-efficient points of M w.r.t. B is furnished by Figure 3.1.1; see also Figure 3.1.2

Because $B + (0, \infty)k^0 \subset \mathrm{int}\, B \subset B \setminus \{0\}$, for $0 \leq \varepsilon_1 \leq \varepsilon_2$, we have that

$$\mathrm{Eff}(M, B) \subseteq \mathrm{Eff}(M, B_{\varepsilon_1 k^0}) \subset \mathrm{Eff}(M, B_{\varepsilon_2 k^0}).$$

Approximate efficiency can also be defined by scalarization.

Definition 3.1.2. *Let C be a cone and $\varphi : Y \to \mathbb{R}$ any C-monotone functional. An element $y_\varepsilon \in M$ is said to be ε-efficient w.r.t. φ if*

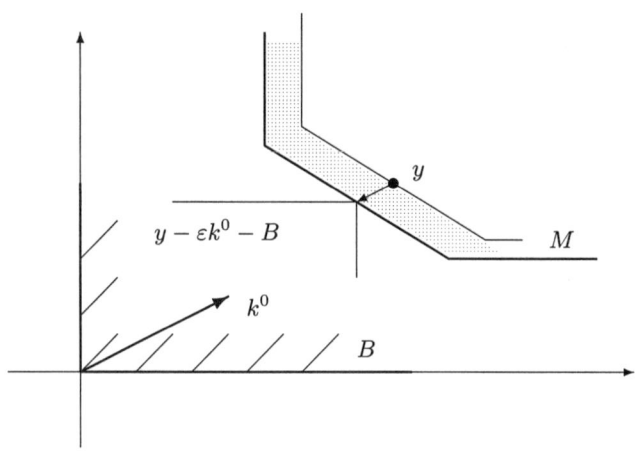

Figure 3.1.1. The set of approximately efficient elements (where the distance between it and the set of efficient elements is unbounded).

$$\forall y \in M \; : \; y \in y_\varepsilon - C \Rightarrow \varphi(y_\varepsilon) \leq \varphi(y) + \varepsilon.$$

The set of such ε-efficient points of M w.r.t. φ and C will be denoted by $\varepsilon\text{-}\mathrm{Eff}^\varphi(M,C)$.

Simple examples show that even if $k^0 \in \mathrm{int}\, C$ and $\mathrm{cl}\, B + \mathrm{int}\, C \subset B$, generally $\varepsilon\text{-}\mathrm{Eff}^\varphi(M) \neq \mathrm{Eff}(M, B_{\varepsilon k^0})$. So it seems to be very interesting to study relations between the Definitions 3.1.1 and 3.1.2. We use the functional $\varphi_{\mathrm{cl}\, B, k^0}$ introduced in Section 2.5:

$$\varphi_{\mathrm{cl}\, B, k^0}(y) := \inf\{t \in R \mid y \in tk^0 - \mathrm{cl}\, B\}.$$

Theorem 3.1.3. *Let $M, B \subset Y$ be nonempty sets such that $0 \in \mathrm{bd}\, B$, let $C \subset Y$ be a cone, $k^0 \in \mathrm{int}\, C$, and $\varepsilon \geq 0$. Assume that $\mathrm{cl}\, B + \mathrm{int}\, C \subset B$. Consider $y_\varepsilon \in \mathrm{Eff}(M, B_{\varepsilon k^0})$ and $\varphi : Y \to \mathbb{R}$, $\varphi(y) := \varphi_{\mathrm{cl}\, B, k^0}(y - y_\varepsilon)$. Then φ is a finite-valued, continuous, and strictly $(\mathrm{int}\, C)$-monotone (even strictly C-monotone if $\mathrm{cl}\, B + (C \setminus \{0\}) \subset \mathrm{int}\, B$) functional such that*

$$\forall y \in M \; : \; \varphi(y_\varepsilon) \leq \varphi(y) + \varepsilon.$$

In particular, $y_\varepsilon \in \varepsilon\text{-}\mathrm{Eff}^\varphi(M,C)$.

PROOF. By Proposition 2.3.4 we have that $\varphi_{\mathrm{cl}\, B, k^0}$ is finite-valued and continuous, while from Theorem 2.3.1(g) we obtain that $\varphi_{\mathrm{cl}\, B, k^0}$ is strictly $(\mathrm{int}\, C)$-monotone (strictly C-monotone if $\mathrm{cl}\, B + (C \setminus \{0\}) \subset \mathrm{int}\, B$). It is obvious that φ has the same properties.

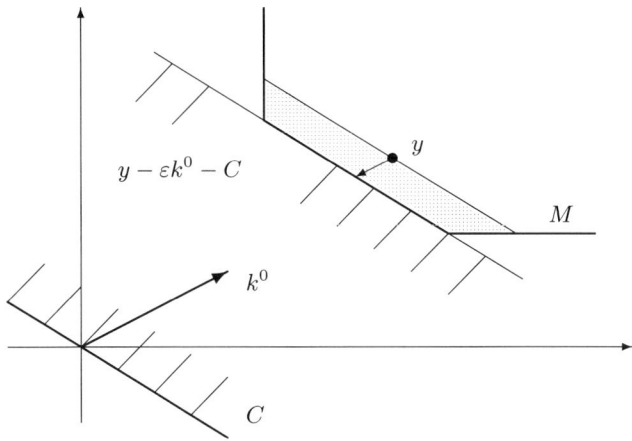

Figure 3.1.2. The set of approximately efficient elements with a bigger cone C.

Assume that there exists $y \in M$ such that $\varphi(y) + \varepsilon < \varphi(y_\varepsilon) = 0$. It follows that there exists $t < -\varepsilon$ with $y - y_\varepsilon \in tk^0 - \operatorname{cl} B$, and so

$$y \in y_\varepsilon - \varepsilon k^0 - \left(\operatorname{cl} B + (-\varepsilon - t)k^0\right) \subset y_\varepsilon - \varepsilon k^0 - \operatorname{int} B \subset y_\varepsilon - \varepsilon k^0 - (B \setminus \{0\}),$$

a contradiction with $y_\varepsilon \in \operatorname{Eff}(M, B_{\varepsilon k^0})$. □

Theorem 3.1.4. *Let $M, B \subset Y$ be proper sets, $C \subset Y$ a cone, $k^0 \in \operatorname{int} C$, and $\varepsilon \geq 0$. Assume that $\operatorname{cl} B + (C \setminus \{0\}) \subset \operatorname{int} B$. If $y_\varepsilon \in M$ is such that*

$$\forall y \in M \;:\; \varphi_{\operatorname{cl} B, k^0}(y_\varepsilon) \leq \varphi_{\operatorname{cl} B, k^0}(y) + \varepsilon, \tag{3.1}$$

then there is an open set $D \subset Y$ with $0 \in \operatorname{bd} D$ and $\operatorname{cl} D + (C \setminus \{0\}) \subset D$ such that

$$y_\varepsilon \in \operatorname{Eff}(M, D_{\varepsilon k^0}).$$

PROOF. First of all, note that $\operatorname{cl} B$ is proper (otherwise, $B = Y$) and $0 \notin \operatorname{int} C$ (otherwise, $C = Y$, and so $B = Y$). Taking into account Remark 2.3.3 we have that $\operatorname{cl} B + \operatorname{int} C \subset \operatorname{cl} B$. Using Proposition 2.3.4 we obtain that $\varphi := \varphi_{\operatorname{cl} B, k^0}$ is a finite-valued continuous function, while from Theorem 2.3.1 we have that φ is strictly C-monotone. Consider the set

$$D := \{ y \in Y \mid \varphi(y_\varepsilon - y) < \varphi(y_\varepsilon) \} = y_\varepsilon - t_\varepsilon k^0 + \operatorname{int} B,$$

where $t_\varepsilon := \varphi(y_\varepsilon)$, the last equality being a consequence of (2.29). Hence D is open. Taking again into account Remark 2.3.3, we obtain that

$$\operatorname{cl} D = y_\varepsilon - t_\varepsilon k^0 + \operatorname{cl} B, \quad \operatorname{bd} D = y_\varepsilon - t_\varepsilon k^0 + \operatorname{bd} B = \{y \in Y \mid \varphi(y_\varepsilon - y) = \varphi(y_\varepsilon)\}.$$

Therefore $0 \in \operatorname{bd} D$. From the inclusion $\operatorname{cl} B + (C \setminus \{0\}) \subset \operatorname{int} B$ and the above formulae one obtains immediately that $\operatorname{cl} D + (C \setminus \{0\}) \subset D$. Using (2.29), condition (3.1) is equivalent to $M \cap ((t_\varepsilon - \varepsilon)k^0 - \operatorname{int} B) = \emptyset$, or equivalently, $M \cap (y_\varepsilon - \varepsilon k^0 - D) = \emptyset$; i.e., $y_\varepsilon \in \operatorname{Eff}(M, D_{\varepsilon k^0})$. □

3.1.2 A General Scalarization Method

Scalarization of a given vector optimization problem (VOP) means convert-ing that problem into an optimization problem (or a family of such problems) with a (common) real-valued objective function to be minimized. If the so-lutions of the latter problems (often called **scalarized problems**) are also solutions of the given VOP, then the obtained scalarization is advantageous in solving the VOP, because methods of usual "scalar" optimization (nonlinear programming) can be used. For the finite-dimensional case see, for example, [259].

The most advantageous scalarization is obtained by using suitable mono-tone functionals; the following propositions serve as theoretical background.

Proposition 3.1.5. *Let Y be a linear space partially ordered by a nontrivial pointed convex cone C, let M be a nonempty subset of Y, and let $\varphi : Y \to \overline{\mathbb{R}}$ be a proper functional with $M \cap \operatorname{dom} f \neq \emptyset$. Assume that $y^0 \in M$ satisfies*

$$\varphi(y) \geq \varphi(y^0) \quad \forall y \in M. \tag{3.2}$$

Then $y^0 \in \operatorname{Eff}(M, C)$ if one of the following conditions is satisfied:

(i) *φ is C-monotone on M and y^0 is uniquely determined,*
(ii) *φ is strictly C-monotone on $M \cap \operatorname{dom} f$.*

PROOF. Let $y^1 \in M \cap (y^0 - C)$ and assume that $y^1 \neq y^0$. In case (i), because φ is C-monotone, we obtain that $\varphi(y^1) \leq \varphi(y^0)$, contradicting the uniqueness of y^0. In case (ii), because φ is strictly C-monotone, we obtain that $\varphi(y^1) < \varphi(y^0)$, contradicting (3.2). Therefore $y^0 \in \operatorname{Eff}(M, C)$. □

Note that condition (3.2) may be replaced by an apparently weaker one:

$$\varphi(y) \geq \varphi(y^0) \quad \forall y \in M \cap (m - C),$$

where $m \in M$ (or $m \in Y$ with $M \cap (m - C) \neq \emptyset$).

Examples of C-monotone functionals are furnished by the elements of C^+, while examples of strictly C-monotone functionals are given by the elements of $C^\#$. An easy application of the preceding result is the next corollary.

Corollary 3.1.6. *Let Y be an s.l.c.s., $C \subset Y$ a convex cone with $C^\# \neq \emptyset$, and $\emptyset \neq M \subset Y$. Then $y^0 \in \operatorname{Eff}(M, C)$ if and only if there exist $m \in M$ and $y^* \in C^\#$ such that y^0 is a solution of the following scalar minimization problem:*

$$\langle y, y^* \rangle \longrightarrow \min, \quad s.t. \ y \in M, \ y \leq_C m.$$

PROOF. If $y^0 \in \mathrm{Eff}(M, C)$, then $M \cap (y_0 - C) = \{y_0\}$, and so y^0 is a solution of the above problem for $m = y^0$ and arbitrary $y^* \in C^\#$. The converse implication follows immediately from case (ii) of the preceding proposition. □

When M is C-convex; i.e., $M + C$ is convex, the following result holds.

Proposition 3.1.7. (Jahn) *Let Y be an H.l.c.s., $C \subset Y$ a proper pointed convex cone, and $M \subset Y$ a nonempty C-convex set. If $y^0 \in \mathrm{Eff}(M, C)$ and $M + C$ has nonempty interior, then there exists $y^* \in C^+ \setminus \{0\}$ such that*

$$\langle y^0, y^* \rangle \leq \langle y, y^* \rangle \quad \forall y \in M.$$

PROOF. Since $y^0 \in \mathrm{Eff}(M, C)$, $(M + C) \cap (y^0 - C) = \{y^0\}$, and so $y^0 \notin \mathrm{int}(M + C)$. By Theorem 2.2.7 there exists $y^* \in Y^* \setminus \{0\}$ such that

$$\langle y^0, y^* \rangle \leq \langle y + c, y^* \rangle \quad \forall y \in M, \ \forall c \in C.$$

Replacing $c \in C$ by tc with $t \geq 0$ and $c \in C$, we obtain that $y^* \in C^+$. Taking $c = 0$ in the above relation, we see that the conclusion holds. □

An inspection of the proof of Proposition 3.1.5 shows that neither the convexity of C nor the fact that C is a cone was used. Also, additional properties of φ imply efficiency of y^0 w.r.t. larger sets having additional properties, as the next result shows.

Theorem 3.1.8. (Gerth–Weidner [131]) *Let B, M be nonempty subsets of the t.v.s. Y, and $\varphi : Y \to \mathbb{R}$ a strictly B-monotone function. Assume that $y^0 \in M$ is such that $\varphi(y^0) \leq \varphi(y)$ for all $y \in M$. Let $H := \{y \in Y \mid \varphi(y^0 - y) < \varphi(y^0)\}$. Then $B \setminus \{0\} \subset H$ and $y^0 \in \mathrm{Eff}(M, H)$; in particular, $y^0 \in \mathrm{Eff}(M, B)$. Moreover, (a) if φ is convex, then H is convex; (b) if φ is continuous, then H is open and $\mathrm{cl}\, H + (B \setminus \{0\}) \subset H$; (c) if φ is linear, then $H \cup \{0\}$ is a convex cone; (d) if $0 \in \mathrm{cl}(B \setminus \{0\})$, then $0 \in \mathrm{bd}\, H$.*

PROOF. From the choice of H it is clear that $M \cap (y^0 - H) = \emptyset$, and so $y^0 \in \mathrm{Eff}(M, H)$. Because φ is strictly B-monotone, it is also obvious that $B \setminus \{0\} \subset H$. Statements (a) and (c) are obvious.

(b) By the continuity of φ we have that H is open and $\mathrm{cl}\, H \subset \{y \in Y \mid \varphi(y^0 - y) \leq \varphi(y^0)\}$. Let $y \in \mathrm{cl}\, H$ and $d \in B \setminus \{0\}$. Since φ is strictly B-monotone, $\varphi(y^0 - y - d) < \varphi(y^0 - y) \leq \varphi(y^0)$, and so $y + d \in H$. (d) Of course, $0 \notin H$. Since $B \setminus \{0\} \subset H$, we have that $0 \in \mathrm{cl}(B \setminus \{0\}) \subset \mathrm{cl}\, H$, and so $0 \in \mathrm{bd}\, H$. □

Note that without asking $0 \in \mathrm{cl}(B \setminus \{0\})$ it is possible that $0 \notin \mathrm{cl}\, H$. Indeed, consider

$$Y := \mathbb{R}, \quad B := \{0\} \cup [1, \infty), \quad M := [1, 4], \quad y^0 := 1,$$

and the strictly B-monotone continuous function

$$\varphi : \mathbb{R} \to \mathbb{R}, \quad \varphi(y) := \min\left\{y + \tfrac{1}{2}, \max\{1, y\}\right\}.$$

Then $H = (\tfrac{1}{2}, \infty)$, and so $0 \notin \mathrm{cl}\, H$. Related to weakly efficient points the following result holds; we make use of the function φ_{D,k^0} considered in Section 2.3.

Theorem 3.1.9. *Let Y be an s.t.v.s., $B \subset Y$ a closed set with nonempty interior and $0 \in \mathrm{bd}\, B$, and $M \subset Y$ a nonempty set. Assume that one of the following conditions holds:*

(i) *there exists a cone $C \subset Y$ with nonempty interior such that $B + \mathrm{int}\, C \subset B$,*
(ii) *B is convex and there exists $k^0 \in Y$ such that $B + \mathbb{R}_+ k^0 \subset B$ and $B + \mathbb{R} k^0 = Y$.*

Then $y^0 \in w - \mathrm{Eff}(M, B)$ if and only if $y^0 \in M$ and there exists a continuous function $\varphi : Y \to \mathbb{R}$ such that

$$\varphi(y^0 - y) < 0 = \varphi(y^0) \le \varphi(x) \ \forall\, y \in \mathrm{int}\, B, \ \forall\, x \in M. \tag{3.3}$$

PROOF. If condition (3.3) we have that $M \cap (y^0 - \mathrm{int}\, B) = \emptyset$, and so $y^0 \in w - \mathrm{Eff}(M, B)$. For the necessity, take $k^0 \in \mathrm{int}\, C$ in case (i). In both cases consider $D := B - y^0$. The conclusion follows from the nonconvex separation theorem (Theorem 2.3.6) taking $\varphi := \varphi_{D,k^0}$. □

Remark 3.1.10. Knowing additional properties of the set B in the preceding theorem we can ask additional properties for the function φ (see Theorem 2.3.1). For example, in case (ii) of Theorem 3.1.9 we can ask that φ be convex (and surjective in all the cases).

Theorems 3.1.8 and 3.1.9 can be found in a more detailed form in Gerth and Weidner's paper [131]. Special cases of those theorems had been proved earlier; for example, scalarization results for weakly efficient points in the case $Y = \mathbb{R}^n$ and $B = y^0 - \mathbb{R}^n_+$, $k^0 \in \mathrm{int}\, \mathbb{R}^n_+$ with the functional

$$\varphi(y) = \inf\{t \in \mathbb{R} \mid y \in y^0 - \mathbb{R}^n_+ + tk^0\},$$

had been obtained by Bernau [27] and for $k^0 = (1, \dots, 1)^T$ by Brosowski and Conci [49]. Furthermore, compare Jahn [191, 192, 196, 197]. In the nineties, results of such types were useful to get general variants of Ekeland's variational principle in the case in which the performance space Y has dimension greater than 1.

3.2 Existence Results for Efficient Points

In this section we establish several existence results for maximal points with respect to transitive relations; then we apply them in topological vector spaces for preorders generated by convex cones. A comparison of existence results for efficient points is made at the end of the section.

3.2.1 Preliminary Notions and Results Concerning Transitive Relations

In the sequel X is a nonempty set and $t \subset X \times X$; i.e., t is a relation on X. If $\emptyset \neq Y \subset X$ and $t \subset X \times X$, the restriction of t to Y is denoted by t^Y; i.e., $t^Y := t \cap (Y \times Y)$.

With the relation t on X we associate the following relations:

$$t_R := t \cup \Delta_X, \qquad t_N := t \setminus t^{-1} = t \setminus (t \cap t^{-1}), \qquad t_{NR} = (t_N)_R.$$

Hazen and Morin [160] call t_N the asymmetric part of t. Some properties of these relations are given in the following lemma; the first three properties mentioned in (ii) are stated by Dolecki and Malivert in [98].

Lemma 3.2.1. *Let t be a transitive relation on X and $\emptyset \neq Y \subset X$.*

(i) t_R *is reflexive and transitive;* $t_N \cap \Delta_X = \emptyset$;
(ii) $t \circ t_N \subset t_N$, $t_N \circ t \subset t_N$, $t_N \circ t_N \subset t_N$, $(t_R)_N = t_N$, $t_R \circ t_N = t_N \circ t_R = t_N$; $(t_N)_N = t_N$;
(iii) t_{NR} *is reflexive, antisymmetric, and transitive;* $(t_N)_N = (t_{NR})_N = t_N$;
(iv) $\left(t^Y\right)_R = (t_R)^Y$, $\left(t^Y\right)_N = (t_N)^Y$.

PROOF. The proof is easy; one uses only the definitions. □

Taking into account the above lemma, we denote by t_R^Y, t_N^Y, and t_{NR}^Y the relations $\left(t^Y\right)_R$, $\left(t^Y\right)_N$, and $\left(t^Y\right)_{NR}$, respectively.

The preceding lemma shows that, with every transitive relation one can associate a partial order. It is useful to know whether they determine the same maximal and minimal points. As noted in Section 2.1, $\mathrm{Max}(Y;t) = \mathrm{Min}(Y, t^{-1})$, where $t \subset X \times X$ is a (transitive) relation and $\emptyset \neq Y \subset X$; so it is sufficient to study only the problems related to maximal points.

Corollary 3.2.2. *Let $\emptyset \neq Y \subset X$ and let t be a transitive relation on X. Then*

$$\mathrm{Max}(Y;t) = \mathrm{Max}(Y;t_R) = \mathrm{Max}(Y;t_N) = \mathrm{Max}(Y;t_{NR}).$$

PROOF. Let $\bar{y} \in Y$. Note first that $\bar{y} \in \mathrm{Max}(Y, t_N)$ iff $(\bar{y}, y) \notin t_N$ for every $y \in Y$. Using this remark we have immediately that $\mathrm{Max}(Y;t) = \mathrm{Max}(Y;t_N)$. Applying the preceding lemma, we get $t_N = (t_R)_N = (t_N)_N = (t_{NR})_N$. The other equalities are now obvious. □

The above corollary shows that the problem of existence (for example) of maximal points w.r.t. a transitive relation t reduces, theoretically, to the same problem for the partial order t_{NR}. Another way to reduce this problem to one for a partial order is given by the following known result.

Proposition 3.2.3. *Let t be a transitive relation on X and $\rho = t_R \cap (t_R)^{-1}$. Then ρ is an equivalence relation, and $\hat{t} = \{(\hat{x}, \hat{y}) \mid (x, y) \in t\}$ is a partial order on $\widehat{X} = X/\rho$, where $\hat{x}$ is the class of $x \in X$ with respect to ρ. Moreover, if $x \in X$, $x \in \mathrm{Max}(X;t)$ if and only if $\hat{x} \in \mathrm{Max}(\widehat{X};\hat{t})$.*

In the sequel we shall also use the notation $Y_t^+(x)$ and $Y_t^-(x)$ for the **upper** and **lower sections** of $Y \subset X$ with respect to t and $x \in X$. So

$$Y_t^+(x) := \{y \in Y \mid (x,y) \in t_R\}, \quad Y_t^-(x) := \{y \in Y \mid (y,x) \in t_R\},$$

respectively; the most common case is for $x \in Y$. Similarly, for $Z \subset X$, we consider

$$Y_t^+(Z) := \{y \in Y \mid \exists z \in Z : (z,y) \in t_R\} = \bigcup_{z \in Z} Y_t^+(z),$$

$$Y_t^-(Z) := \{y \in Y \mid \exists z \in Z : (y,z) \in t_R\} = \bigcup_{z \in Z} Y_t^-(z).$$

Note that $Y_t^+\left(Y_t^+(Z)\right) = Y_t^+(Z)$ (so one may suppose that $Z \subset Y$) and

$$\text{Max}(Y_t^+(Z);t) = Y_t^+(Z) \cap \text{Max}(Y;t), \quad \text{Min}(Y_t^-(Z);t) = Y_t^-(Z) \cap \text{Min}(Y;t). \tag{3.4}$$

We say that $\emptyset \neq Y \subset X$ has the **domination property** (w.r.t. t) (DP) if $\text{Max}(Y_t^+(y);t) \neq \emptyset$, for every $y \in Y$ (i.e., every element of Y is dominated by a maximal element of Y).

A quite important problem is how to extend other notions related to partially ordered sets, like chain or increasing net, to sets endowed with transitive relations. Related to this problem we have the next result.

Proposition 3.2.4. *Let t be a transitive relation on X and $\emptyset \neq Y \subset X$.*

(i) *If t_R^Y is a partial order on Y, then $t_R^Y \subset t_{NR}$.*
(ii) *If μ is a total order on Y such that $\mu \subset t_{NR}$, then $\mu = t_{NR}^Y$.*

PROOF. (i) Let $\mu = t_R^Y$ be a partial order on Y and consider $(x,y) \in \mu$. If $x = y$, we have that $(x,y) \in t_{NR}$. Suppose now that $x \neq y$. It follows that $(x,y) \in t$. Assuming that $(x,y) \notin t_{NR}$, we have that $(x,y) \notin t \setminus t^{-1}$, whence $(x,y) \in t^{-1}$. Therefore $(y,x) \in t$, and so $(y,x) \in \mu$. Since μ is antisymmetric, it follows that $x = y$, a contradiction. Hence $(x,y) \in t_{NR}$.

(ii) Suppose that (Y, μ) is totally ordered and $\mu \subset t_{NR}$. Of course, $\mu \subset t_{NR}^Y$. Consider $(x,y) \in t_{NR}^Y$ $(\subset t)$ and suppose that $(x,y) \notin \mu$. Of course, $x \neq y$. Since μ is a total order on Y, we have that $(y,x) \in \mu$, and so $(y,x) \in t_N = t \setminus t^{-1}$. It follows that $(x,y) \notin t$, a contradiction. Therefore $\mu = t_{NR}^Y$. □

Note that in (ii) we cannot replace t_{NR} by t or t_R (take $X = \mathbb{R}$ and $t = \mathbb{R} \times \mathbb{R}$).

Let X be endowed with the transitive relation t and let $(x_i)_{i \in I} \subset X$ be a net. We say that (x_i) is t-**increasing** [t-**decreasing**] if $(x_i, x_j) \in t$ [$(x_j, x_i) \in t$] for all $i, j \in I$ with $i \preceq j$ and $i \neq j$; (x_i) is **strictly t-increasing** [**strictly t-decreasing**] if (x_i) is t_N-increasing [t_N-decreasing].

In the sequel we say that $\emptyset \neq A \subset (X, t)$, with t a transitive relation on X, is a **chain** (w.r.t. to t) if t_R^A is a total order on A, while A is **well-ordered** (w.r.t. to t) if A is well-ordered by t_R^A.

The following interesting result is due to Gajek and Zagrodny (see [123]).

Proposition 3.2.5. *Let t be a transitive relation on X. Then there exists a nonempty and well-ordered subset $W \subset X$ such that for every $x \in X \setminus W$ there exists $w \in W$ with $(x, w) \in t$ or $(w, x) \notin t$.*

PROOF. Let $\mathcal{L} = \{U \subset (X, t) \mid U \text{ well-ordered}\}$. Then $\mathcal{L}$ is nonempty because $\{x\} \in \mathcal{L}$ for every $x \in X$. Consider the relation

$$\eta = \{(U, V) \in \mathcal{L} \times \mathcal{L} \mid U \subset V \text{ and } \forall v \in V, \ \forall u \in U \ : \ (v, u) \in t \Rightarrow v \in U\}.$$

It is easy to see that $(\mathcal{L}, \eta)$ is a partially ordered set. Let us show that $(\mathcal{L}, \eta)$ has maximal elements. For this let $\mathcal{U} \subset \mathcal{L}$ be a (nonempty) chain, and show that $\mathcal{U}$ is upper bounded in $\mathcal{L}$. Consider $\overline{U} = \cup \mathcal{U}$. We have that $\overline{U} \in \mathcal{L}$; for this note first that t is antisymmetric on $\overline{U}$. Indeed, let $x, y \in \overline{U}$ be such that $(x, y) \in t$ and $(y, x) \in t$. Because $\mathcal{U}$ is a chain, there exists $U \in \mathcal{U} \subset \mathcal{L}$ such that $x, y \in U$. Since t is antisymmetric on U, we obtain that $x = y$. Similarly, we obtain that $\overline{U}$ is totally ordered. Now let $\emptyset \ne A \subset \overline{U}$. There exists $U_0 \in \mathcal{U}$ such that $A_0 = A \cap U_0 \ne \emptyset$. Let $a_0 \in A_0$ be the least element of A_0 as a subset of U_0. Let us show that a_0 is the least element of A. Let $a \in A \setminus \{a_0\}$; there exists $U \in \mathcal{U}$ such that $a \in U$. If $U \eta U_0$, then $a \in A \cap U_0$, and so $(a_0, a) \in t$. In the contrary case, $U_0 \eta U$. Of course, a and a_0 are comparable. Suppose that $(a, a_0) \in t$; from the definition of η, we obtain that $a \in U_0$, whence $a \in A_0$. It follows that $(a_0, a) \in t$. Therefore a_0 is the least element of A. Hence $\overline{U} \in \mathcal{L}$. It is easy to see that $U \eta \overline{U}$ for every $U \in \mathcal{U}$. Indeed, let $\overline{u} \in \overline{U}$ and $u \in U \in \mathcal{U}$ such that $(\overline{u}, u) \in t$. There exists $U' \in \mathcal{U}$ such that $\overline{u} \in U'$. If $U' \eta U$, there is nothing to prove. If $U \eta U'$, by the very definition of η, we have that $\overline{u} \in U$. By Zorn's lemma, $\mathcal{L}$ has a maximal element W. Let $x \in X \setminus W$. Suppose that the conclusion is not true. Then for every $w \in W$ we have that $(w, x) \in t$ and $(x, w) \notin t$. Consider $W' = W \cup \{x\}$. It is obvious that W' is antisymmetric and totally ordered w.r.t. t_R. Let $\emptyset \ne A' \subset W'$; if $A' = \{x\}$, of course, x is the least element of A'. In the contrary case we have that $\emptyset \ne A = A' \setminus \{x\} \subset W$. Taking a the least element of A, it is obvious that a is also the least element of A'. Therefore $W' \in \mathcal{L}$. But $W \eta W'$, contradicting the maximality of W. $\square$

3.2.2 Existence of Maximal Elements with Respect to Transitive Relations

We begin with the following result.

Proposition 3.2.6. *Let t be a transitive relation on X and $\emptyset \ne Y \subset X$. Suppose that one of the following conditions holds:*

(i) *every nonempty set $A \subset Y$ with $A \times A \subset t \cup t^{-1} \cup \Delta_X$ is upper bounded in Y;*

(ii) *every chain in Y is upper bounded in Y;*

(iii) *every well-ordered subset of Y is upper bounded in Y.*

Then Y has the domination property. In particular, $\mathrm{Max}(Y; t) \ne \emptyset$.

PROOF. Without loss of generality we may suppose that $Y = X$.

It is obvious that (i) $\Rightarrow$ (ii) $\Rightarrow$ (iii). Assume that (iii) holds. By Gajek and Zagrodny's lemma (Proposition 3.2.5) there exists a well-ordered nonempty set $W \subset X$ such that for every $x \in X \setminus W$, there exists $w \in W$ with $(x, w) \in t$ or $(w, x) \notin t$. By hypothesis there exists $\overline{x} \in X$ such that $(w, \overline{x}) \in t_R$ for every $w \in W$. Assume that for some $u \in X$ we have $(\overline{x}, u) \in t_N$. Of course, $u \notin W$ and $(w, u) \in t_R \circ t_N \subset t_N$ for every $w \in W$. By the choice of W, there exists $w_u \in W$ such that $(u, w_u) \in t$. It follows that $(u, u) \in t \circ t_N \subset t_N$, a contradiction. Therefore $\overline{x} \in \mathrm{Max}(X; t)$. The fact that X has (DP) follows from the fact that if one of the conditions (i), (ii), (iii) holds for X, it holds also for $X_t^+(x)$ for every $x \in X$. □

Corollary 3.2.7. *Let t be a transitive relation on X and $\emptyset \neq Y \subset X$. Suppose that one of the following conditions holds:*

(i) *every t-increasing net of Y is upper bounded in Y;*
(ii) *every strictly t-increasing net of Y is upper bounded in Y;*
(iii) *every t-increasing net $(y_i)_{i \in I} \subset Y$ with I totally ordered is upper bounded in Y;*
(iv) *every strictly t-increasing net $(y_i)_{i \in I} \subset Y$ with I totally ordered is upper bounded in Y;*
(v) *every t-increasing net $(y_i)_{i \in I} \subset Y$ with I well-ordered is upper bounded in Y;*
(vi) *every strictly t-increasing net $(y_i)_{i \in I} \subset Y$ with I well-ordered is upper bounded in Y.*

Then Y has the domination property. In particular, $\mathrm{Max}(Y; t) \neq \emptyset$.

PROOF. Of course, (i) $\Rightarrow$ (ii), (iii) $\Rightarrow$ (iv), (v) $\Rightarrow$ (vi), (i) $\Rightarrow$ (iii) $\Rightarrow$ (v), and (ii) $\Rightarrow$ (iv) $\Rightarrow$ (vi). On the other hand, if (vi) holds, then condition (iii) of the preceding proposition is satisfied. Therefore the conclusions hold. □

The following result is due, essentially, to Gajek and Zagrodny [123].

Proposition 3.2.8. *Let t be a transitive relation on X and $\emptyset \neq Y \subset X$. Suppose that the following two conditions hold:*

(i) *every nonempty well-ordered subset W of Y is at most countable;*
(ii) *every strictly t-increasing sequence of Y is upper bounded in Y.*

Then Y has the domination property. In particular, $\mathrm{Max}(Y; t) \neq \emptyset$.

PROOF. Without loss of generality we may suppose that $Y = X$.

In order to apply Proposition 3.2.6 we show that every well-ordered subset W of X is upper bounded. So let $W \subset X$ be a well-ordered set. By hypothesis W is at most countable. If W has a greatest element, this is also an upper bound for W. Suppose that W does not have a greatest element. It follows that W is not finite (otherwise, since W is also totally ordered, it has a greatest element). Therefore there exists $p : \mathbb{N} \to W$ bijective. Observe first that for every $k \in \mathbb{N}$ the set $\{n \in \mathbb{N} \mid n > k, \ (p(k), p(n)) \in t\}$ is nonempty. In the contrary case we have that $(p(n), p(k)) \in t$ for every $n > k$. Taking $p(i) = \max\{p(0), \ldots, p(k)\}$, $p(i)$ is the greatest element of W, a contradiction. Let $n_0 = 0$, $n_1 = \min\{n > n_0 \mid (p(n_0), p(n)) \in t\}$, and so on; therefore $n_{k+1} = \min\{n > n_k \mid (p(n_k), p(n)) \in t\}$ for every $k \in \mathbb{N}$. Define $x_k = p(n_k) \in X$. It is clear that (x_k) is a strictly t-increasing sequence. Therefore there exists $\overline{x}$ such that $(x_k, \overline{x}) \in t_R$ for every k. We have that $(p(n), \overline{x}) \in t_R$ for every n. In the contrary case for some $\overline{n} \in \mathbb{N}$, $(p(\overline{n}), \overline{x}) \notin t_R$. It follows that $(p(\overline{n}), p(n_k)) \notin t_R$, and so, by Proposition 3.2.4, $(p(n_k), p(\overline{n})) \in t_N$, for every k; in particular, $\overline{n} \neq n_k$ for every k. Since $\overline{n} > 0 = n_0$, it follows that $\overline{n} \geq n_1$, and so $\overline{n} > n_1$. Continuing in this way we get that $\overline{n} \geq n_k$ for every k, which, of course, is not possible. It follows that $\overline{x}$ is an upper bound of W. Therefore condition (iii) of Proposition 3.2.6 is satisfied, and so the conclusions hold. $\qquad \square$

Note that condition (i) of Corollary 3.2.7 is equivalent to

$$\forall (y_i)_{i \in I} \subset Y, \ (y_i) \ t\text{-increasing} \ : \ Y \cap \left(\bigcap_{i \in I} X_t^+(y_i) \right) \neq \emptyset, \qquad (3.5)$$

while (ii) is equivalent to

$$\forall (y_i)_{i \in I} \subset Y \ \text{strictly } t\text{-increasing} \ : \ Y \cap \left(\bigcap_{i \in I} X_{t_{NR}}^+(y_i) \right) \neq \emptyset.$$

A sufficient condition for (3.5) is the following one:

$$\forall A \subset Y : Y \subset \bigcup_{a \in A} \left(X \setminus X_t^+(a) \right) \Rightarrow \exists a_1, \ldots, a_n \in A : Y \subset \bigcup_{i=1}^{n} \left(X \setminus X_t^+(a_i) \right).$$
$$(3.6)$$

If τ is a topology on X and the upper sections of X are closed (for example, if t is closed in $X \times X$), then the sets $X \setminus X_t^+(a)$ are open; in this situation condition (3.6) is a kind of compactness of Y.

The following result is related to this kind of condition.

Proposition 3.2.9. *Let t be a transitive relation on X and $\emptyset \neq Y \subset X$. Assume that there exists a relation s on X such that*

$$s \circ t_N \subset t_R \qquad (3.7)$$

and

$$\forall (y_i)_{i \in I} \subset Y \ \text{strictly } t\text{-increasing} \ : \ Y \cap \left(\bigcap_{i \in I} X_s^+(y_i) \right) \neq \emptyset. \qquad (3.8)$$

Then Y has the domination property. In particular, $\mathrm{Max}(Y; t) \neq \emptyset$.

PROOF. Once again, without loss of generality we may suppose that $Y = X$. In order to apply Proposition 3.2.6 (ii) we show that every chain C of X is upper bounded. So let $\emptyset \neq C \subset X$ be a chain. By hypothesis there exists $\overline{x} \in \bigcap_{c \in C}\{x \in X \mid (c, x) \in s\}$. If C has a greatest element, there is nothing to prove. Let the contrary case hold and $c \in C$. Then there exists $c' \in C$ such that $(c, c') \in t_N$. Of course, $(c', \overline{x}) \in s$; using (3.7) we obtain that $(c, \overline{x}) \in t_R$. Hence $\overline{x}$ is an upper bound for C. Using Proposition 3.2.6 the conclusions follow. $\qquad\square$

The above proof shows that the conclusions of Proposition 3.2.9 hold if in (3.8) I is totally ordered [even well-ordered, applying in this case Proposition 3.2.6 (ii)].

In the sequel, in this section, we suppose that (X, τ) is a topological space. In this situation one can formulate other conditions for the existence of maximal points. Having a net $(x_i)_{i \in I}$, by $(x_{\varphi(j)})_{j \in J}$, or simply $(x_{\varphi(j)})$, we denote a subnet of (x_i).

Corollary 3.2.10. *Let t be a transitive relation on the topological space (X, τ) and $\emptyset \neq Y \subset X$. Assume that one of the following conditions holds:*

$$\forall (y_i)_{i \in I} \subset Y \ t\text{-increasing}, \ \exists (y_{\varphi(j)})_{j \in J} \to y \in Y, \ \forall j \in J \ : \ (y_{\varphi(j)}, y) \in t_R, \tag{3.9}$$

$$\forall (y_i)_{i \in I} \subset Y \ \text{strictly } t\text{-increasing}, \ \exists (y_{\varphi(j)})_{j \in J} \to y \in Y,$$
$$\forall j \in J \ : \ (y_{\varphi(j)}, y) \in t_R. \tag{3.10}$$

Then Y has the domination property. In particular, $\mathrm{Max}(Y; t) \neq \emptyset$.

PROOF. If (3.9) or (3.10) holds, then condition (i) or (ii) of the preceding corollary is satisfied, respectively. Indeed, suppose that (3.9) holds and let $(y_i)_{i \in I} \subset Y$ be t-increasing. There exists a subnet $(y_{\varphi(j)})_{j \in J} \to y \in Y$ such that $(y_{\varphi(j)}, y) \in t_R$ for every $j \in J$. Fix $k \in I$. For every $i \in I$ there is some $j_i \in J$ with $i \preceq \varphi(j)$ for all $j \in J$, $j_i \preceq j$. Since (y_i) is t-increasing, $(y_k, y_{\varphi(j_k)}) \in t_R$; since $(y_{\varphi(j_k)}, y) \in t_R$, we get $(y_k, y) \in t$. Therefore y is an upper bound for (y_i) in Y. The conclusion holds. $\qquad\square$

When the upper or lower sections of X are closed, we may consider other conditions, too.

Proposition 3.2.11. *Let t be a transitive relation on the topological space (X, τ) and $\emptyset \neq Y \subset X$. Consider the following conditions:*

$$\forall (y_i)_{i \in I} \subset Y \ t\text{-increasing} \ : \ \exists (y_{\varphi(j)})_{j \in J} \to y \in Y, \tag{3.11}$$

$$\forall (y_i)_{i \in I} \subset Y \ \text{strictly } t\text{-increasing} \ : \ \exists (y_{\varphi(j)})_{j \in J} \to y \in Y, \tag{3.12}$$

$$\forall (y_i)_{i \in I} \subset Y \ t\text{-increasing}, \ \exists y \in Y \ : \ y_i \to y, \tag{3.13}$$

and

$$\forall\, (y_i)_{i\in I} \subset Y \; strictly\; t\text{-}increasing, \; \exists\, y\in Y\; :\; y_i \to y. \tag{3.14}$$

If the upper sections of X are closed, then (3.11) $\Leftrightarrow$ (3.9) *and* (3.12) $\Leftrightarrow$ (3.10), *while if t is a partial order and the upper and lower sections of X (w.r.t. t) are closed, then* (3.11) $\Leftrightarrow$ (3.13) *and* (3.12) $\Leftrightarrow$ (3.14).

PROOF. The implications (3.9) $\Rightarrow$ (3.11), (3.13) $\Rightarrow$ (3.11), (3.10) $\Rightarrow$ (3.12), and (3.14) $\Rightarrow$ (3.12) are always true. Let us show that (3.11) $\Rightarrow$ (3.9) and (3.13) $\Rightarrow$ (3.9) under the mentioned supplementary conditions, the other two being proved similarly.

Suppose that the upper sections of X are closed and (3.11) holds. Consider $(y_i)_{i\in I} \subset Y$ a t-increasing net. There exists the subnet $(y_{\varphi(j)})_{j\in J}$ converging to $y \in Y$. With the notation from the proof of the preceding proposition, we have that $(y_{\varphi(j)})_{j_i \preceq j} \subset X_t^+(y_i)$. Taking the limit, we get that $\lim y_{\varphi(j)} = y \in \mathrm{cl}\, X_t^+(y_i) = X_t^+(y_i)$ for all $i \in I$. Therefore y is an upper bound for $(y_i)_{i\in I}$ in Y, whence y is an upper bound for $(y_{\varphi(j)})$, too.

Suppose now that t is a partial order and the upper and lower sections of X are closed. Let us show that (3.11)$\Rightarrow$(3.13). So, suppose that (3.11) holds and $(y_i)_{i\in I} \subset Y$ is t-increasing. By hypothesis there exists a subnet $(y_{\varphi(j)})_{j\in J}$ converging to $y \in Y$. As in the proof of (3.11)$\Rightarrow$(3.9), we have that $y_i \in X_t^-(y)$ for every $i \in I$. Suppose that $y_i \not\to y$. It follows that there exists a neighborhood V_0 of y in X such that $I_0 := \{i \in I \mid y_i \notin V_0\}$ is cofinal in I. Of course, $(y_i)_{i\in I_0}$ is a t-increasing net of Y. Therefore, by (3.11), there exists $(y_{\psi(k)})_{k\in K}$ a subnet of $(y_i)_{i\in I_0}$ such that $y_{\psi(k)} \to y_0 \in Y$; of course, $(y_{\psi(k)})_{k\in K}$ is also a subnet of $(y_i)_{i\in I}$. The same argument as above gives $y_i \in X_t^-(y_0)$ for every $i \in I$. From $(y_{\varphi(j)})_{j\in J} \subset X_t^-(y_0)$ we get $y \in X_t^-(y_0)$, while from $(y_{\psi(k)})_{k\in K} \subset X_t^-(y_0)$ we get $y_0 \in X_t^-(y)$. Since t is antisymmetric, it follows that $y = y_0$, which contradicts the choice of I_0 (because V_0 is a neighborhood of y_0 in this case). $\square$

Corollary 3.2.12. *Let t be a transitive relation on the separated topological space (X,τ) and $\emptyset \neq Y \subset X$. Assume that the upper sections of X are closed and Y is compact. Then Y has the domination property. In particular, $\mathrm{Max}(Y;t) \neq \emptyset$.*

PROOF. In the hypotheses of the corollary condition (3.11) holds, and because the upper sections of X are closed, (3.9) holds, too. The conclusions follow from Corollary 3.2.10. $\square$

Hazen and Morin proved the result of Corollary 3.2.12 in [160, Cor. 2.8]; note that Theorem 2.2 of [160] does not follow from any of the preceding results.

One can ask naturally whether a kind of converse of Proposition 3.2.6 (Corollary 3.2.7) is true, in the sense that if $(X;t)$ has the domination property, is it true that every chain (every t-increasing net) of X is upper bounded? The answer is negative, as the following example shows.

Example 3.2.13. Let $X = \{f_n \mid n \in \mathbb{N}^*\} \cup \{g_n \mid n \in \mathbb{N}^*\}$, where $f_n, g_n :$ $[0,1] \to \mathbb{R}$, $f_n(x) = x^n$ and

$$g_n(x) = \begin{cases} x^n & \text{if } x \in [0, 2^{-1/n}], \\ \frac{2^{1/n}}{2(2^{1/n}-1)}(1-x) & \text{if } x \in (2^{-1/n}, 1]. \end{cases}$$

For $f, g \in X$ we put $f \preceq g$ iff $g(x) \le f(x)$ for every $x \in [0,1]$. Note that g_n is strictly increasing on $[0, 2^{-1/n}]$ and strictly decreasing on $[2^{-1/n}, 1]$, and so attains its greatest value, $\frac{1}{2}$, for $x = 2^{-1/n}$. It follows that for $n, m \in \mathbb{N}^*$, $n \neq m$, g_n, and g_m are not comparable. Moreover, $f_n \preceq f_{n+1} \preceq g_{n+1}$ for every $n \in \mathbb{N}^*$, but $f_n \npreceq g_m$ for $n, m \in \mathbb{N}^*$, $n > m$. It follows that the set of maximal points of X is $\{g_n \mid n \in \mathbb{N}^*\}$. So $(X, \preceq)$ has the domination property, but the chain $\{f_n \mid n \in \mathbb{N}^*\}$ is not upper bounded.

3.2.3 Existence of Efficient Points with Respect to Cones

To begin with, let X be a real vector space and $\emptyset \neq C \subset X$ a convex cone. As usual, with C we associate (see Theorem 2.1.13) the reflexive and transitive relation

$$\le_C := t := \{(x,y) \in X \times X \mid y - x \in C\}. \tag{3.15}$$

Taking

$$L := C \cap (-C),$$

the **lineality space** of C, the equivalence relation $\rho := t \cap t^{-1}$ is $\{(x,y) \in X \times X \mid y - x \in L\}$. So we get

$$t_N = \{(x,y) \in X \times X \mid y - x \in C \setminus L\}. \tag{3.16}$$

Using Lemma 3.2.1 we obtain that

$$C + (C \setminus L) = C \setminus L, \qquad (C \setminus L) + (C \setminus L) \subset C \setminus L;$$

the above formulae were obtained by Luc in [243, 244]. So,

$$t_{NR} = \{(x,y) \in X \times X \mid y - x \in (C \setminus L) \cup \{0\}\}.$$

It follows that $(C \setminus L) \cup \{0\}$ is a pointed convex cone. Note that for $x \in X$ and $Y \subset X$ one has $Y_C^+(x) := Y_{\le_C}^+(x) = Y \cap (x+C)$ and $Y_C^-(x) = Y \cap (x-C)$. Therefore the upper and lower sections of X (w.r.t. C) are closed if and only if C is closed. Similarly, for $Z \subset X$, $Y_C^+(Z) := Y \cap (Z + C)$ and $Y_C^-(Z) := Y \cap (Z - C)$.

In accordance with the notions introduced in Section 3.2.1, the net $(x_i)_{i \in I} \subset X$ is **[strictly]** C-**increasing** if $x_j - x_i \in C$ $[x_j - x_i \in (C \setminus L) \cup \{0\}]$ for all $i, j \in I$, $i \preceq j$, $i \neq j$; $(x_i)_{i \in I}$ is C-**decreasing** if $(-x_i)_{i \in I}$ is C-increasing. Of course, the set $Y \subset X$ is C-**upper (lower) bounded** if $Y \subset x_0 - C$ $(Y \subset x_0 + C)$ for some $x_0 \in X$. Moreover, the set $\text{Max}(Y, \le_C)$ will be denoted by $\mathrm{Max}(Y;C)$. An element $y \in \mathrm{Max}(Y;C)$ is called an **efficient point** of Y (w.r.t. C).

It is obvious that for $\emptyset \neq Y \subset X$ and $C = X$ or $C = \{0\}$ we have that $\mathrm{Max}(Y;C) = Y$. Taking into account this fact, in the sequel we shall suppose that C is proper, i.e., $\{0\} \neq C \neq X$.

To Proposition 3.2.6 and Corollary 3.2.7 corresponds the following result.

Proposition 3.2.14. *Let $\emptyset \neq Y \subset X$. Suppose that one of the following conditions holds:*

(i) *every nonempty set $Z \subset Y$ such that $Z - Z \subset C \cup (-C)$ is C-upper bounded in Y;*
(ii) *every chain in Y w.r.t. $\leq_C$ is C-upper bounded in Y;*
(iii) *every well-ordered subset of Y (w.r.t. $\leq_C$) is C-upper bounded in Y;*
(iv) *every C-increasing net of Y is C-upper bounded in Y.*

Then Y has the domination property. In particular, $\mathrm{Max}(Y;C) \neq \emptyset$.

In the following proposition we gather some properties of efficient sets, most of them appearing in the literature in particular cases.

Proposition 3.2.15. *Let $C, K \subset X$ be convex cones such that $C \subset K$, $x \in X$, and $\emptyset \neq Y, Z \subset X$.*

(i) $\mathrm{Max}(Y_K^+(Z);C) = Y_K^+(Z) \cap \mathrm{Max}(Y;C)$. *In particular, we have that* $\mathrm{Max}(Y_C^+(x);C) = Y_C^+(x) \cap \mathrm{Max}(Y;C)$.
(ii) *If $Y \subset Z \subset Y - K$ and*
$$K \cap (-C) \subset C, \tag{3.17}$$
then
$$\mathrm{Max}(Z;K) \subset \mathrm{Max}(Y;C) + (K \cap (-K)).$$

In particular, if $\mathrm{Max}(Z;K) \neq \emptyset$, then $\mathrm{Max}(Y;C) \neq \emptyset$.
(iii) *If (3.17) holds, then $\mathrm{Max}(Y;K) \subset \mathrm{Max}(Y;C)$.*
(iv) *If $Y \subset Z \subset Y - C$, then $\mathrm{Max}(Y;K) \subset \mathrm{Max}(Z;K)$; moreover, if $K \cap (-C) = \{0\}$, then $\mathrm{Max}(Y;K) = \mathrm{Max}(Z;K)$.*
(v) *Suppose that (3.17) and*
$$K + (C \setminus L) \subset C \tag{3.18}$$
hold. If $Y \subset \mathrm{Max}(Y;K) - K$, then $Y \subset \mathrm{Max}(Y;C) - C$ (i.e., if Y has (DP) w.r.t. K, then Y has (DP) with respect to C).
(vi) *Suppose that $K \cap (-C) = \{0\}$ and $Y \subset \mathrm{Max}(Y;K) - C$. Then $\mathrm{Max}(Y;C) = \mathrm{Max}(Y;K)$.*

PROOF. (i) Let $\bar{y} \in \mathrm{Max}\left(Y_K^+(Z);C\right)$. Then $\bar{y} \in Y_K^+(Z)$ and $Y_K^+(Z) \cap (\bar{y}+C) \subset \bar{y} - C$. Let $y' \in Y \cap (\bar{y}+C)$. It follows that $y' \in \bar{y} + C \subset z + K + C \subset z + K$ for some $z \in Z$, and so $y' \in Y_K^+(Z)$. Therefore $y' \in \bar{y} - C$, which shows that $\bar{y} \in \mathrm{Max}(Y;C)$. Hence $\mathrm{Max}\left(Y_K^+(Z);C\right) \subset Y_K^+(Z) \cap \mathrm{Max}(Y;C)$; since the converse inclusion is immediate, we have the first relation. The other one follows immediately taking $K = C$ and $Z = \{x\}$.

(ii) Let $\bar{z} \in \mathrm{Max}(Z; K) \subset Z \subset Y - K$. Then $\bar{z} = \bar{y} - \bar{k}$ with $\bar{y} \in Y$ and $\bar{k} \in K$. Since $\bar{y} \in Z$, we obtain that $\bar{k} = \bar{y} - \bar{z} \in (Z - \bar{z}) \cap K \subset -K$, and so $\bar{k} \in K \cap (-K)$. Let us show that $\bar{y} \in \mathrm{Max}(Y; C)$. For this let $c \in (Y - \bar{y}) \cap C$. It follows that $\bar{y} + c \in Y \subset Z$, whence $\bar{y} + c - \bar{z} = \bar{y} + c - \bar{y} + \bar{k} = c + \bar{k} \subset C + K \subset K$. Therefore $c + \bar{k} \in -K$, and so $-c \in K + \bar{k} \subset K$. Using (3.17) we get $c \in -C$. Hence $\bar{y} \in \mathrm{Max}(Y; C)$.

(iii) Although (iii) is not a particular case of (ii) (considering $Z = Y$), the proof is the same taking $\bar{k} = 0$.

(iv) Let $y \in \mathrm{Max}(Y; K) \subset Y \subset Z$. If $k \in (Z - y) \cap K$, then $y + k \in Z$, whence $y + k = y' - c'$ with $y' \in Y$ and $c' \in C$; so $y' - y = k + c' \in (Y - y) \cap K \subset -K$. It follows that $k \in -c' - K \subset -K$. Therefore $y \in \mathrm{Max}(Z; K)$.

Suppose that $C \cap (-K) = \{0\}$ and consider $z \in \mathrm{Max}(Z; K)$. It follows that $z = y - c$ for some $y \in Y$ and $c \in C$. Hence $y \in Z \cap (z + K) \subset z - K$. Therefore $y = z - k$ for some $k \in K$. We obtain that $c + k = 0$, whence $c = -k \in C \cap (-K) = \{0\}$. Thus $z = y \in Y$ and $Y \cap (z - K) \subset Z \cap (z + K) \subset z - K$, which shows that $z \in \mathrm{Max}(Y; K)$. Therefore $\mathrm{Max}(Y; K) = \mathrm{Max}(Z; K)$.

(v) Suppose that (3.17), (3.18) hold and $Y \subset \mathrm{Max}(Y; K) - K$. Let $y \in Y$. If $y \in \mathrm{Max}(Y; C)$, there is nothing to prove. Assume that $y \notin \mathrm{Max}(Y; C)$. It follows that $y = y' - c'$ with $y' \in Y$ and $c' \in C \setminus L$. Since Y has (DP) w.r.t. K, $y' = \bar{y} - \bar{k}$ with $\bar{y} \in \mathrm{Max}(Y; K)$ and $\bar{k} \in K$. By (iii) we have that $\bar{y} \in \mathrm{Max}(Y; C)$. Moreover,

$$y = \bar{y} - \bar{k} - c' = \bar{y} - (\bar{k} + c') \in \mathrm{Max}(Y; C) - (K + (C \setminus L)) \subset \mathrm{Max}(Y; C) - C.$$

Therefore $Y \subset \mathrm{Max}(Y; C) - C$.

(vi) Since $K \cap (-C) \subset C$, by (iii), one has that $\mathrm{Max}(Y; K) \subset \mathrm{Max}(Y; C)$. Since $\mathrm{Max}(Y; K) \subset Y \subset \mathrm{Max}(Y; K) - C$ and $C \cap (-C) = \{0\}$, by (iv) one has $\mathrm{Max}(Y; C) = \mathrm{Max}(\mathrm{Max}(Y; K); C) \subset \mathrm{Max}(Y; K)$. The proof is complete. □

In the sequel we suppose that X is a real Hausdorff topological vector space (H.t.v.s. for short), its topology being denoted by τ.

Applying the results from Section 3.2.2 we obtain several existence theorems for efficient points w.r.t. cones. Before stating them, let us recall or introduce some notions or, more exactly, some possible properties of the cone C.

So, C is **(sequentially Daniell) Daniell** if every C-upper bounded and C-increasing (sequence) net in X has a least upper bound (or supremum) and converges to it. Because we suppose that X is Hausdorff, every (sequentially) Daniel cone is pointed.

Other similar conditions are:

(P1) Every C-increasing and C-upper bounded net in C is convergent to an element of C.

(P2) Every C-increasing and C-upper bounded net in X is fundamental.

(P3) Every C-increasing and τ-bounded net in C is convergent to an element of C.

(P4) Every C-increasing and τ-bounded net in X is fundamental.

(P5) Every C-increasing and τ-bounded net in X that is contained in a complete set is convergent.

The sequential variants of (P1)–(P4) are:

(SP1) Every C-increasing and C-upper bounded sequence in C is convergent to an element of C.

(SP2) Every C-increasing and C-upper bounded sequence in X is fundamental.

(SP3) Every C-increasing and τ-bounded sequence in C is convergent to an element of C.

(SP4) Every C-increasing and τ-bounded sequence in X is fundamental.

If one of the above conditions holds, then, necessarily, C is pointed.

Note that an equivalent formulation for (P1) [and similarly for (P3), (SP1), and (SP3)] is:

(P1′) Every C-increasing and C-upper bounded net in X is convergent to an element of X that is a C-upper bound for the net.

Of course, (P1′)⇒(P1). For the converse implication consider a C-increasing and C-upper bounded net $(x_i)_{i \in I} \subset X$. Fix $k \in I$. Then the net $(y_i)_{k \preceq i}$, with $y_i := x_i - x_k \in C$, is C-increasing and C-upper bounded. By (P1) there exists $y \in C$ such that $y_i \to y$. It follows that $x_i \to x = x_k + y$, and so $x_k \leq_C x$. Since $k \in I$ is arbitrary, (P1′) holds.

Note that when X is a normed space, Krasnosel'skij [217] says that a convex cone C satisfying (SP2) is *regular*, while a convex cone C satisfying (SP4) is *completely regular*. Isac [175] says that C is *completely regular* if every C-increasing and τ-bounded net of C is convergent. Of course, if X is **quasi-complete** (i.e. every closed and bounded subset of X is complete), then C is completely regular in Isac's sense if and only if C has property (P4). Nemeth [270] says that a convex cone C satisfying (P1) [(SP1)] is *regular [sequentially regular]*; moreover, he says that C is *fully regular* if C satisfies (P3). Condition (P5) was introduced by Ha [148] under the name of property (**).

We have the following scheme of implications, where (D) means Daniell, while (sD) means sequentially Daniell.

$$
\begin{array}{ccccccc}
\text{(D)} \overset{\leftarrow}{\Rightarrow} & \text{(P1)} \overset{\leftarrow}{\Rightarrow} & \text{(P2)} & \text{(P3)} \overset{\leftarrow}{\Rightarrow} & \text{(P4)} & \Rightarrow & \text{(P5)} \\
\Downarrow\uparrow & \Downarrow\uparrow & \Updownarrow & \Downarrow\uparrow & \Updownarrow & & \downarrow \\
\text{(sD)} \overset{\leftarrow}{\Rightarrow} & \text{(SP1)} \overset{\leftarrow}{\Rightarrow} & \text{(SP2)} & \text{(SP3)} \overset{\leftarrow}{\Rightarrow} & \text{(SP4)} & & \text{(P3)}.
\end{array}
$$

The implications marked by $\Rightarrow$, $\Downarrow$, and $\Updownarrow$ always hold; those marked by $\uparrow$ are valid when C is complete or when $0 \in X$ has a countable basis of neighborhoods; the implications $\downarrow$ and $\leftarrow$ are valid if C is complete; those marked $\twoheadleftarrow$ are valid for C closed. The nontrivial implications are stated in the next proposition.

Moreover, the implications $\rightarrow$ below hold if C is normal, while the implications $\leftarrow$ hold if C is *boundedly order complete*, i.e., every C-increasing and τ-bounded net of C has a supremum:

$$(P3) \leftrightarrows (P1), \ (P4) \leftrightarrows (P2), \ (SP3) \leftrightarrows (SP1) \text{ and } (SP4) \leftrightarrows (SP2). \qquad (3.19)$$

The proof of the following result uses an idea from [270]. Note that Borwein (see Proposition 2.7 (ii) of [40]) showed that C is Daniell iff C is sequentially Daniell when "C admits a strict monotone functional."

Proposition 3.2.16. *Let $C \subset X$ be a convex cone. Then*

(i) $(P2) \Leftrightarrow (SP2)$ *and* $(P4) \Leftrightarrow (SP4)$.
(ii) *If C is complete, or $0 \in X$ has a countable basis of neighborhoods, then* $(P1) \Leftrightarrow (SP1)$, $(P3) \Leftrightarrow (SP3)$, *and C is Daniell if and only if C is sequentially Daniell.*

PROOF. We need to show only the implications "$\Leftarrow$" because the others are trivial.

(i) Let us show that $(SP2) \Rightarrow (P2)$. Consider $(x_i)_{i \in I} \subset C$ a C-increasing and C-upper bounded net, and suppose that (x_i) is not fundamental. Therefore there exists V_0 a neighborhood of $0 \in X$ such that for every $i \in I$ there exist $j, k \in I$ with $i \preceq j, k$ and $x_j - x_k \notin V_0$. There exists a symmetric neighborhood U_0 of 0 such that $U_0 + U_0 \subset V_0$. For j, k found above, there exists $l \in I$ such that $j, k \preceq l$. We have that $x_j - x_l \notin U_0$ or $x_k - x_l \notin U_0$ (otherwise, $x_j - x_k \in U_0 - U_0 \subset V_0$). Taking this into account, we have that

$$\forall i \in I, \ \exists j, k \in I \ : \ i \preceq j \preceq k, \ x_j - x_k \notin U_0. \qquad (3.20)$$

Let $i_0 \in I$ be fixed. Then taking $i = i_0$ in (3.20), there exist $i_1, i_2 \in I$ such that $i_0 \preceq i_1 \preceq i_2$ and $x_{i_1} - x_{i_2} \notin U_0$. Taking $i = i_2$ in (3.20), there exist $i_3, i_4 \in I$ such that $i_2 \preceq i_3 \preceq i_4$ and $x_{i_3} - x_{i_4} \notin U_0$. Continuing in this way we obtain an increasing sequence $(i_n)_{n \in \mathbb{N}} \subset I$ such that $x_{i_{2n+1}} - x_{i_{2n+2}} \notin U_0$ for every $n \in \mathbb{N}$. Of course, the sequence $(x_{i_n})_{n \in \mathbb{N}} \subset C$ is C-increasing and C-upper bounded, and so $(x_{i_n})_{n \in \mathbb{N}}$ is fundamental by (SP2). This contradicts the choice of this sequence.

The proof of $(SP4) \Rightarrow (P4)$ is practically the same.

(ii) I. Suppose first that C is complete. The equivalence of (P1) and (SP1) is obtained from the statement (3.1) of [270] taking $K_0 = K = C$, while for the other equivalences the proofs are similar.

II. Suppose now that 0 has a countable basis of neighborhoods, say $(V_n)_{n \in \mathbb{N}}$. Without loss of generality we may suppose that $V_n \supset V_{n+1} + V_{n+1}$ for every $n \in \mathbb{N}$.

Let us show that (SP1) $\Rightarrow$ (P1). So, suppose that (SP1) holds. Since (SP1) $\Rightarrow$ (SP2), by (i) we have that (P2) holds. Consider $(x_i)_{i \in I} \subset C$ to be C-increasing and C-upper bounded. It follows that (x_i) is Cauchy. Hence for every $n \in \mathbb{N}$ there exists $i_n \in I$ such that $x_j - x_k \in V_n$ for all $j, k \in I$, $i_n \preceq j, k$. Taking into account that I is upward directed, we may consider that $(i_n)_{n \in \mathbb{N}}$ is increasing. Therefore the sequence $(x_{i_n})_{n \in \mathbb{N}} \subset C$ is C-increasing and C-upper bounded. By hypothesis $x_{i_n} \to x \in C$. For $p \in \mathbb{N}$ there exists $n_p \in \mathbb{N}$, $n_p \geq p + 1$, such that $x_{i_n} - x \in V_{p+1}$ for every $n \geq n_p$. But for $i \in I$, $(i_{p+1} \preceq) i_{n_p} \preceq i$, we have that $x_i - x_{i_{n_p}} \in V_{p+1}$. It follows that

$$x_i - x = x_i - x_{i_{n_p}} + x_{i_{n_p}} - x \in V_{p+1} + V_{p+1} \subset V_p$$

for every $i \succeq i_{n_p}$, and so $x_i \to x$. Therefore (P1) holds.

The proof for (SP3)$\Rightarrow$(P3) is the same.

Suppose now that C is sequentially Daniell and show that C is Daniell. As in the proof of (SP1)$\Rightarrow$(P1), consider the C-increasing and C-upper bounded net $(x_i)_{i \in I}$. Without loss of generality we may suppose that $(x_i)_{i \in I} \subset C$. With the notation of that proof, there exists $x \in X$ such that $x_{i_n} \to x$ and x is the least upper bound of $(x_{i_n})_{n \in \mathbb{N}}$. With the same proof, we have that $x_i \to x$. Using an argument similar to that in the proof of (P1)$\Rightarrow$(P1$'$) above, we have that x is a C-upper bound for (x_i). The proof is complete. $\square$

Before stating the next result we introduce other two notions. We say that the subset Y of X is **C-complete** (resp. **sequentially C-complete**) if every Cauchy C-increasing net (resp. Cauchy C-increasing sequence) is convergent to an element of C.

Proposition 3.2.17. *Let $Y \subset X$ be a nonempty set. Assume that one of the following conditions holds:*

(i) *C satisfies (P1), while Y is closed and C-upper bounded;*
(ii) *C is closed and satisfies (SP2), while Y is C-complete and C-upper bounded;*
(iii) *C satisfies (P3), while Y is closed and τ-bounded;*
(iv) *C is closed and satisfies (SP4), while Y is C-complete and τ-bounded.*

Then Y has the domination property. In particular, $\mathrm{Max}(Y; C) \neq \emptyset$.

PROOF. Let $(y_i)_{i \in I} \subset Y$ be a C-increasing net.

In case (i) or (iii), since C satisfies (P1) or (P3), and Y is C-bounded from above or τ-bounded, respectively, (y_i) is convergent to some $y \in X$ that is a C-upper bound for (y_i). Since Y is closed, $y \in Y$. Therefore the condition (3.9) holds.

In case (ii) and (iv), since C satisfies (SP2) or (SP4), by the preceding proposition, C satisfies (P2) or (P4), respectively. Since Y is bounded from above or τ-bounded, respectively, (y_i) is Cauchy. Since Y is C-complete, $y_i \to y \in Y$. Therefore in both cases condition (3.13) holds. Since C is closed, by

Proposition 3.2.11, the condition (3.9) holds, too. Applying Corollary 3.2.10 the conclusions follow.   □

When every well-ordered subset W of X w.r.t. $\leq_C$ (see page 88) is at most countable (i.e., $(X, \leq_C)$ is countable orderable in the sense of Gajek and Zagrodny [123]) the closedness and completeness in the preceding result can be taken in the weaker sequential sense.

Proposition 3.2.18. *Let $Y \subset X$ be a nonempty set. Assume that every well-ordered subset W of X w.r.t. $\leq_C$ is at most countable and one of the following conditions holds:*

(i) *C satisfies (SP1), while Y is sequentially closed and C-upper bounded;*
(ii) *C is closed and satisfies (SP2), while Y is sequentially C-complete and C-upper bounded;*
(iii) *C satisfies (SP3), while Y is sequentially closed and τ-bounded;*
(iv) *C is closed and satisfies (SP4), while Y is sequentially C-complete and τ-bounded.*

Then Y has the domination property. In particular, $\mathrm{Max}(Y; C) \neq \emptyset$.

PROOF. Let $W \subset Y$ be well-ordered with respect to $\leq_C$. Therefore, by our hypothesis, W is at most countable. If W is finite, then W is bounded from above by its greatest element (which is in Y). In the contrary case $W = \{y_n \mid n \in \mathbb{N}\}$. Consider $z_n := \max\{y_k \mid k \leq n\}$. It is obvious that $(z_n) \subset Y$ is a C-increasing sequence. As in the proof of the preceding proposition we obtain that (z_n) converges to some $z \in Y$ with the property that $z_n \leq_C z$ for every n. Of course, $y_n \leq_C z$ for every n; i.e., W is C-upper bounded in Y. The conclusion follows by applying Proposition 3.2.15 (iii).   □

Note that using Proposition 3.2.15, in the above two propositions one can only ask that the hypotheses on Y be satisfied by a nonempty upper section $Y_C^+(x)$, or even a nonempty set of the form $Y_C^+(Z)$, to have $\mathrm{Max}(Y; C) \neq \emptyset$. The next result is due to Ha [149].

Proposition 3.2.19. *Assume that $\mathrm{cl}\, C$ satisfies (P5) and $Y \subset X$ is nonempty, complete, and τ-bounded. Then $\mathrm{Max}(Y; C) \neq \emptyset$.*

PROOF. The proof is the same as that of assertion (iii) of the preceding proposition (but obtaining directly that (y_i) is convergent).   □

Proposition 3.2.20. *Assume that C is closed and $Y \subset X$ is nonempty and compact. Then Y has the domination property. In particular, $\mathrm{Max}(Y; C) \neq \emptyset$.*

PROOF. The relation $t = \leq_C$ and Y satisfy the conditions of Corollary 3.2.12. The conclusions follow.   □

Corollary 3.2.21. *Let $\emptyset \neq Y \subset X$ and assume that*

$$\operatorname{cl} C \cap (-C) \subset C. \tag{3.21}$$

If there exists $Z \subset X$ such that $Y_C^+(Z)$ or $Y_{\operatorname{cl} C}^+(Z)$ is nonempty and compact (even weakly compact if X is locally convex), then $\operatorname{Max}(Y;C) \neq \emptyset$.

PROOF. Suppose first that $Y_C^+(Z)$ is nonempty and compact (or weakly compact). By the preceding proposition, $\operatorname{Max}(Y_C^+(Z); \operatorname{cl} C) \neq \emptyset$. Using Proposition 3.2.15 (iii) and (i) for $K = \operatorname{cl} C$, we have that

$$\operatorname{Max}(Y_C^+(Z); \operatorname{cl} C) \subset \operatorname{Max}(Y_C^+(Z); C) \subset \operatorname{Max}(Y;C),$$

whence $\operatorname{Max}(Y;C) \neq \emptyset$. Similarly, if $Y_{\operatorname{cl} C}^+(Z)$ is nonempty and compact (or weakly compact), then

$$\emptyset \neq \operatorname{Max}(Y_{\operatorname{cl} C}^+(Z); \operatorname{cl} C) \subset \operatorname{Max}(Y; \operatorname{cl} C) \subset \operatorname{Max}(Y;C).$$

Therefore the conclusion holds in both cases.   □

Proposition 3.2.22. *Let $C, K \subset X$ be convex cones and $\emptyset \neq Y \subset X$. Assume that (3.18) and*

$$\forall (y_i)_{i \in I} \subset Y \text{ strictly } C\text{-increasing} \; : \; Y \cap \left(\bigcap_{i \in I} (y_i + K) \right) \neq \emptyset \tag{3.22}$$

hold. Then Y has the domination property. In particular, $\operatorname{Max}(Y;C) \neq \emptyset$.

PROOF. Consider t defined by (3.15) and $s := \{(x, y) \in X \times X \mid y - x \in K\}$. Taking into account (3.15) and (3.16), (3.18) is equivalent to (3.7), while from (3.22) we obtain that (3.8) holds. The conclusions follow by applying Proposition 3.2.9.   □

Note that in locally convex spaces, every result stated above has a weak version when the topology τ is replaced by the weak topology on X.

3.2.4 Types of Convex Cones and Compactness with Respect to Cones

In Sections 2.1 and 2.2 we introduced several types of cones, some of them being useful for existence of efficient points.

The following result is stated in Jameson [199, Th. 3.8.7] (there the set B being also a base for C in the sense of Definition 2.1.14).

Theorem 3.2.23. *Let X be an H.t.v.s. and C be a nontrivial well-based convex cone. Then condition (P4) holds.*

PROOF. By Proposition 3.2.16 (i) it is sufficient to show that condition (SP4) is satisfied. Let $C = \mathbb{R}_+ B$, where B is a bounded convex set with $0 \notin \operatorname{cl} B$, and $(x_n)_{n \in \mathbb{N}}$ a τ-bounded C-increasing sequence. Without loss of generality we assume that $x_0 = 0$. Let $y_n := x_n - x_{n-1} \in C$ for $n > 0$; then $y_n = t_n b_n$ with $t_n \geq 0$ and $b_n \in B$ for $n \geq 1$. Since $x_n = y_1 + \cdots + y_n = t_1 b_1 + \cdots + t_n b_n$, from the convexity of B we have that $x_n = (t_1 + \cdots + t_n) b'_n = t'_n b'_n$ with $t'_n := t_1 + \cdots + t_n$ and $b'_n \in B$ for every $n \geq 1$. Since $0 \notin \operatorname{cl} B$ and $(x_n)_{n \geq 1}$ is bounded we have that $(t'_n)_{n \geq 1}$ is bounded, and so the series $\sum_{n \geq 1} t_n$ is convergent. Fix $V \in \mathcal{N}_X$. Since B is bounded, there exists $\varepsilon > 0$ such that $[0, \varepsilon] B \subset V$. There exists $n_0 \geq 0$ such that $\sum_{k=m+1}^{n} t_k < \varepsilon$ for $n > m \geq n_0$. Then $x_n - x_m = y_n + \cdots + y_{m+1} \in (\sum_{k=m+1}^{n} t_k) B \subset [0, \varepsilon] B \subset V$ for $n > m \geq n_0$. This shows that the sequence (x_n) is Cauchy. $\qquad\square$

When X is an l.c.s. the preceding result may be reinforced (take into account Proposition 2.2.15).

Proposition 3.2.24. *Let $(X, \mathcal{P})$ be an H.l.c.s. If C is supernormal, then C satisfies condition (P4), and therefore (P2).*

PROOF. Suppose that C is supernormal and consider $(x_i)_{i \in I} \subset X$ a C-increasing and $\tau_{\mathcal{P}}$-bounded net. Consider $p \in \mathcal{P}$. Because C is supernormal, there exists $x^* \in C^+$ such that $p(x) \leq \langle x, x^* \rangle$ for every $x \in C$. Because $(x_i)_{i \in I}$ is C-increasing and $\tau_{\mathcal{P}}$-bounded, the net $(\langle x_i, x^* \rangle)_{i \in I} \subset \mathbb{R}$ is bounded and nondecreasing. It follows that $(\langle x_i, x^* \rangle)_{i \in I}$ is convergent, and so fundamental. Since $p(x_j - x_i) \leq \langle x_j - x_i, x^* \rangle$ for all $i, j \in I$, $i \preceq j$, it follows immediately that $(x_i)_{i \in I}$ is fundamental. Therefore (P4) holds. Now (P2) follows from the normality of C. $\qquad\square$

The next result is useful in the context of Proposition 3.2.18.

Proposition 3.2.25. *Assume that either* (i) *C has a base in the sense of Definition 2.1.14 or* (ii) *X is an H.t.v.s. and C is based. Then every well-ordered subset of X w.r.t. $\leq_C$ is at most countable.*

PROOF. Let $W \subset X$ be well-ordered with respect to $\leq_C$. Replacing W by $W + c_0 - w_0$ with $c_0 \in C_0 := C \setminus \{0\}$ and w_0 the least element of W, we may assume that $W \subset C_0$.

(i) Assume that C has a base. By Theorem 2.1.15, there exists a linear functional $\varphi : X \to \mathbb{R}$ such that $\varphi(x) > 0$ for $x \in C \setminus \{0\}$. Since $\varphi(x) < \varphi(x + x')$, provided that $x, x' \in C_0$, we obtain that $\varphi(W)$ is a well-ordered subset of $\mathbb{R}$ endowed with the usual order, and so $\varphi(W)$ is at most countable (see Remark 2.1.6). Hence W is at most countable.

(ii) By Definition 2.2.14, there exists a convex set B such that $0 \notin \operatorname{cl} B$ and $C = \mathbb{R}_+ B$. Consider $C_0 := C \setminus \{0\}$ and $\nu_B : C_0 \to \mathbb{R}$ defined by $\nu_B(x) := \sup\{t > 0 \mid x \in tB\}$. Of course, $\nu_B(x) > 0$ for every $x \in C_0$; moreover, $\nu_B(x) < \infty$, because otherwise, $0 \in \operatorname{cl} B$. If $x \in tB$ and $x' \in t'B$ with $t, t' > 0$, then $x + x' \in (t + t')B$, whence $\nu_B(x + x') \geq \nu_B(x) + \nu_B(x')$. So, if $x, x' \in C_0$, then $\nu_B(x) < \nu_B(x + x')$. As in (i) we obtain that W is at most countable. $\qquad\square$

Another type of cone is the (π)-cone. Suppose that X is an l.c.s.; the cone C is a (π)-**cone** if there exists $x^* \in C^+$ such that $\{x \in C \mid \langle x, x^* \rangle \leq 1\}$ is relatively weakly compact (see [332, Def. 2.1] and [60, Def. 4.1]). Sterna-Karwat showed in [332, Prop. 2.1] that

$$C \text{ is } (\pi)\text{-cone} \iff \mathrm{cl}\, C \text{ has a weakly compact base.}$$

As seen in Proposition 2.2.15, if the cone C has a bounded base B, then C is supernormal; furthermore, if B is complete, it follows easily that C is complete, too. This discussion shows that a cone with compact base is Daniell, supernormal, and complete. In particular, a closed (π)-cone is Daniell and complete w.r.t. the weak topology.

Luc in [243, Def. 2.3] and [244, Def. I.1.1] says that C is **correct** if (3.18) holds for $K = \mathrm{cl}\, C$. Every domination cone in the sense of Henig [162, p. 112] is correct.

Concerning cone compactness, recall the following notions.

Hartley [158, p. 214] says that Y is C-**compact** if $Y^+_{\mathrm{cl}\, C}(y)$ is compact for every $y \in Y$.

Corley in Definition 2.5 of [82] says that $Y \subset X$ is C-**semicompact** if (3.6) holds for t equal to $\leq_{\mathrm{cl}\, C}$. As a generalization of this notion, Luc ([243, Def. 2.1] and [244, Def. II.3.2]) says that Y is C-*complete* if (3.22) holds for $K = \mathrm{cl}\, C$.

Postolică [303] and Isac [178] say that Y is C-*bounded* if there exists a τ-bounded set $Y_0 \subset Y$ such that $Y \subset Y_0 - C$, Y is C-*closed* if $Y - C$ is closed, and Y is C-*semicompact* if Y is C-bounded and C-closed; in fact, these notions are also used in [175, Def. 3], but with Y_0 a singleton.

Dedieu [92] says that Y is **asymptotically compact** (a.c. for short) if there exist $\gamma > 0$ and U a neighborhood of $0 \in X$ such that $([0, \gamma]Y) \cap U$ is relatively compact; note that Y is a.c. iff $\mathrm{cl}\, Y$ is a.c. (see [389, Prop. 2.2(i)]). Of course, every subset of $\mathbb{R}^m$ is asymptotically compact. Several properties of a.c. sets can be found in [389, Prop. 2.2].

The **asymptotic cone** of the nonempty set $A \subset X$ is

$$A_\infty := \{x \in X \mid \exists\, (t_i)_{i \in I} \subset (0, \infty),\ t_i \to 0,\ \exists\, (a_i)_{i \in I} \subset A\ :\ t_i a_i \to x\};$$

if X is a normed space, in particular, for $X = \mathbb{R}^m$, one can use sequences instead of nets. Note that $A_\infty = (\mathrm{cl}\, A)_\infty$. If A is closed and convex, then A_∞ is given by the known formula from convex analysis $A_\infty = \bigcap_{t>0} t(A - a)$ for some fixed $a \in A$. The importance of this notion in our context is shown by the following result.

Proposition 3.2.26. *Let $C \subset X$ be a closed convex cone and $\emptyset \neq Y \subset X$ closed.*

(i) *If Y is asymptotically compact and $Y_\infty \cap C = \{0\}$, then $Y \cap (x + C)$ is compact for every $x \in X$ (hence Y is C-compact).*

(ii) *Suppose that there exists a compact set $Q \subset X$ such that $0 \notin Q$ and $C = [0, \infty) \cdot Q$ and $Y_\infty \cap C = \{0\}$. Then $Y \cap (x + C)$ is compact for every $x \in X$. Moreover, if C is pointed, then $Y - C$ is closed and $(Y - C) \cap (x + C)$ is compact for every $x \in X$.*
Conversely, if Y is convex and $Y \cap (x + C)$ is nonempty and compact for some $x \in X$, then $Y_\infty \cap C = \{0\}$.

PROOF. Suppose that $Y \cap (x + C) \neq \emptyset$. Both in (i) and (ii) we have that

$$\{0\} \subset [Y \cap (x + C)]_\infty \subset Y_\infty \cap (x + C)_\infty = Y_\infty \cap C = \{0\}.$$

If $C = [0, \infty) \cdot Q$ with Q compact and $0 \notin Q$, one obtains easily that C is locally compact. Using [389, Prop. 2.2(ii)] we get that C is an a.c. set. So, in both cases $Y \cap (x + C)$ is a.c. as a subset of such a set. Applying [389, Prop. 2.3] we obtain that $Y \cap (x + C)$ is relatively compact; since the set is closed, it is compact, too.

Consider now case (ii) and suppose that C is pointed. Applying [389, Cor. 3.12] we have that $Y - C$ is closed and $(Y - C)_\infty = Y_\infty - C$. Let $z \in (Y - C)_\infty \cap C$. It follows that there are $y \in Y_\infty$ and $c \in C$ such that $z = y - c$. Since $z \in C$, we obtain that $y = z + c \in C \cap Y_\infty = \{0\}$. Since C is pointed, we get $z = 0$. By what was obtained above we have that $(Y - C) \cap (x + C)$ is compact for every $x \in X$.

Now let Y be convex and assume that $Y_\infty \cap C \neq \{0\}$. Since Y is also closed, fixing $\bar{y} \in Y$ and $0 \neq u \in Y_\infty \cap C$, we have that $\bar{y} + tu \in Y$ for every $t \geq 0$. Therefore $\{\bar{y} + tu \mid t \geq 0\} \subset Y \cap (\bar{y} + C)$, whence $Y \cap (\bar{y} + C)$ is not compact. □

Corollary 3.2.27. *Let $C \subset X$ be a closed convex cone and $\emptyset \neq Y \subset X$ closed. Assume that $Y_\infty \cap C = \{0\}$. If Y is asymptotically compact or $C = [0, \infty) \cdot Q$ for some compact set $Q \subset X$ with $0 \notin Q$, then Y has the domination property.*

PROOF. By the preceding proposition $Y_C^+(y)$ is compact for every $y \in Y$; by Proposition 3.2.20 Max $\left(Y_C^+(y); C\right) \neq \emptyset$. Therefore Y has the domination property. □

3.2.5 Classification of Existence Results for Efficient Points

In the sequel we mention several existence results for efficient points that can be found in the literature in equivalent formulations (some of them were stated for minimization problems, or using other terminology). In all the results below $C \subset X$ is a convex cone and $Y \subset X$ is a nonempty set.

Results Corresponding to Proposition 3.2.14 (The Nontopological Case)

Theorem 3.2.28. (Corley [82, Th. 3.1]) *Let X be a separated t.v.s. and cl C pointed. Suppose that Y is C-semicompact in Corley's sense. Then* Max$(Y; C) \neq \emptyset$.

PROOF. By hypothesis, Y satisfies (3.6), and therefore (3.5), for $t = \leq_{\mathrm{cl}\,C}$. It follows that (iv) of Proposition 3.2.14 holds for $\mathrm{cl}\,C$ and Y, and so $\mathrm{Max}(Y; \mathrm{cl}\,C) \neq \emptyset$. Using Proposition 3.2.15 (iii) for $K = \mathrm{cl}\,C$, the conclusion follows.   □

We put this result here because applied to $K = \mathrm{cl}\,C$, the result is not topological.

Theorem 3.2.29. (Chew [77]) *Let X be real vector space. Suppose that the intersection of every nonempty chain (w.r.t. inclusion) of C-upper sections of Y is nonempty. Then $\mathrm{Max}(Y; C) \neq \emptyset$.*

PROOF. Taking into account Proposition 3.2.4, $\left(Y_C^+(y_i)\right)_{i \in I}$, with $(y_i)_{i \in I} \subset Y$, is a chain if and only if $(y_i)_{i \in I}$ is a chain in $(Y, \leq_C)$. So, using Proposition 3.2.14 (ii), the conclusion follows.   □

Note that the above formulation covers Proposition 4.6 (where it is supposed that C is pointed), Corollary 4.4, and Proposition 4.7 from [77] (in Proposition 4.7 it is assumed that X is a Hilbert space and $C \cap (-C)$ is closed).

Results Corresponding to Propositions 3.2.17 and 3.2.18

Theorem 3.2.30. (Jameson [199, Cor. 3.8.10]) *Let X be an H.t.v.s. and C well-based and closed. If Y is complete and bounded, then Y has the domination property.*

PROOF. By Theorem 3.2.23 condition (P4) is satisfied; so, apply Proposition 3.2.17 (iv).   □

Theorem 3.2.31. (Penot [288, Th. 3.3]) *Let X be an H.t.v.s. and let C be Daniell and closed. If Y is C-upper bounded and closed, then $\mathrm{Max}(Y; C) \neq \emptyset$.*

PROOF. Apply Proposition 3.2.17 (i).   □

Theorem 3.2.32. (Cesari and Suryanarayana [60, Lemma 4.1]) *Let X be a Banach space and C a closed (π)-cone. Suppose that Y is C-upper bounded. Then $\mathrm{Max}(w - \mathrm{cl}\,Y; C) \neq \emptyset$.*

PROOF. We noted in the preceding section that C is Daniell and complete w.r.t. w. It follows that C and $w - \mathrm{cl}\,Y$ satisfy condition (i) of Proposition 3.2.17 for $\tau = w$.   □

The above result also follows from Corollary 3.2.27 because $Y \subset y_0 - C$ implies $Y_\infty \cap C = \{0\}$.

Theorem 3.2.33. (Borwein [39, Th. 1]) *Let X be an l.c.s. and let C be Daniell and closed. If (i) there exists $y \in Y$ such that $Y_C^+(y)$ is C-upper bounded and closed, or (ii) Y is τ-bounded and closed, while C is boundedly order complete, then $\mathrm{Max}(Y; C) \neq \emptyset$.*

PROOF. In the first case apply Proposition 3.2.17 (i) for $Y_C^+(y)$ and then Proposition 3.2.15 (i). In the second case (P3) holds [see also (3.19)], so that the conclusion follows by applying Proposition 3.2.17 (iii). □

Theorem 3.2.34. (Isac [175, Th. 2]) *Let X be an l.c.s. and let C be super-normal. Suppose that there exists $\emptyset \neq Z \subset Y$ such that $Y_C^+(Z)$ is complete and τ-bounded. Then $\mathrm{Max}(Y;C) \neq \emptyset$.*

PROOF. By Proposition 3.2.24 $\mathrm{cl}\,C$ satisfies (P4). Using also Proposition 3.2.15 (iii) for $K = \mathrm{cl}\,C$ and (i) we have

$$\emptyset \neq \mathrm{Max}\left(Y_C^+(Z); \mathrm{cl}\,C\right) \subset \mathrm{Max}\left(Y_C^+(Z); C\right) \subset \mathrm{Max}\left(Z; C\right). \qquad \square$$

Note that Postolică [304, Th. 3.2] obtained a slightly more general variant: X, C, and Y being as in Isac's theorem, he supposes that there exists Z such that $Y \subset Z \subset Y - C$ and $\emptyset \neq Y_0 \subset Y$, so that $Z_C^+(Y_0)$ is complete and τ-bounded; then $\mathrm{Max}(Y;C) \neq \emptyset$. The conclusion follows from Isac's theorem applied to Z and Proposition 3.2.15 (iv).

Theorem 3.2.35. (Postolică [304, Cor. 3.2.1]) *Let X be an l.c.s. and C have a complete bounded base. Suppose that Y is bounded and closed. Then Y has the domination property.*

PROOF. As observed in the previous section, C is complete and supernormal; therefore C satisfies (P3). Applying Proposition 3.2.17 (iii), the conclusion follows. □

Theorem 3.2.36. (Attouch and Riahi [10, Th. 2.5]) *Let X be a Banach space and C a closed convex cone satisfying (2.21) for some $x^* \in X^*$. Suppose that Y is closed and $x^*(Y)$ is bounded above (in $\mathbb{R}$). Then Y has the domination property.*

PROOF. To obtain the conclusion we use Proposition 3.2.14, but the ideas are those from Proposition 3.2.17; in fact, when Y is bounded the result follows from either of the preceding two theorems.

Let $(y_i)_{i \in I} \subset Y$ be a C-increasing net. It follows that $(\langle y_i, x^* \rangle)_{i \in I} \subset [0, \infty)$ is increasing and bounded, and therefore is a Cauchy net. Taking $\varepsilon > 0$, there exists $i_\varepsilon \in I$ such that $|\langle y_i, x^* \rangle - \langle y_j, x^* \rangle| < \varepsilon/2$ for $i_\varepsilon \preceq i, j$. From (2.21) we obtain that for $i, j \in I$, $i_\varepsilon \preceq i, j$,

$$\|y_i - y_j\| \leq \|y_i - y_{i_\varepsilon}\| + \|y_j - y_{i_\varepsilon}\| \leq |\langle y_i - y_{i_\varepsilon}, x^* \rangle| + |\langle y_j - y_{i_\varepsilon}, x^* \rangle| < \varepsilon.$$

Therefore (y_i) is a Cauchy net, and so $y_i \to y \in X$. Since Y is closed, $y \in Y$. From Proposition 3.2.11 we have that y is a C-upper bound for (y_i). Applying Proposition 3.2.14 (iv), the conclusion follows. □

Theorem 3.2.37. (Ha [148, 149]) *Let X be a quasi-complete H.t.v.s. Assume that $\mathrm{cl}\,C$ satisfies (P5) and $(A + \mathrm{cl}\,C) \cap (A - \mathrm{cl}\,C)$ is bounded for every bounded set $A \subset X$. If $Y - \mathrm{cl}\,C$ is closed and $Y \subset Y_0 - C$ for some bounded $Y_0 \subset X$, then $\mathrm{Max}(Y;C) \neq \emptyset$.*

PROOF. The set $\operatorname{cl} C$ satisfies $(P3)$ by what was observed after the introduction of conditions (P1)–(P5). Let $y_0 \in Y$ and consider $Y_1 = (Y - \operatorname{cl} C)^+_{\operatorname{cl} C}(y_0)$; Y_1 is closed and bounded, being included in $((Y_0 \cup \{y_0\}) + \operatorname{cl} C) \cap ((Y_0 \cup \{y_0\}) - \operatorname{cl} C$. By Proposition 3.2.17 (iii), $\operatorname{Max}(Y_1; \operatorname{cl} C) \neq \emptyset$, whence, by Proposition 3.2.15 (i) and (ii), $\operatorname{Max}(Y; C) \neq \emptyset$.   □

Theorem 3.2.38. (Ng and Zheng [272, Th. 3.1]) *Let X be an H.t.v.s. and let C be well-based and closed. If there exists $y_0 \in Y$ such that $Y_C^+(y_0)$ is sequentially C-complete and bounded, then $\operatorname{Max}(Y; C) \neq \emptyset$.*

PROOF. By Theorem 3.2.23 condition (SP4) holds. The conclusion follows using Propositions 3.2.25 and 3.2.18 (iv).   □

Even if the next result is a special case of the preceding theorem, we state it mainly because we are in a position to furnish a direct and constructive proof of it.

Theorem 3.2.39. *Let X be an H.l.c.s. and let C be well-based and closed. If there exists $y_0 \in Y$ such that $Y_C^+(y_0)$ is sequentially C-complete and bounded, then $\operatorname{Max}(Y; C) \neq \emptyset$.*

PROOF. Because C is well-based, there exists $y^* \in Y^*$ such that $B := \{y \in C \mid \langle y, y^* \rangle = 1\}$ is a bounded base of C. Because $A := Y_C^+(y_0)$ is bounded,

$$\alpha_0 := \sup\{\langle y, y^* \rangle \mid y \in A\} = \sup\{\langle y, y^* \rangle \mid y \in A \cap (y_0 + C)\} \in \mathbb{R}.$$

Take $y_1 \in A \cap (y_0 + C)$ such that $\langle y_1, y^* \rangle \geq \alpha_0 - 2^{-1}$. Then $\alpha_1 := \sup\{\langle y, y^* \rangle \mid y \in A \cap (y_1 + C)\} \leq \alpha_0$; take $y_2 \in A \cap (y_1 + C)$ such that $\langle y_2, y^* \rangle \geq \alpha_1 - 2^{-2}$. Continuing in this way we obtain the C-increasing sequence $(y_n)_{n \geq 0} \subset A$ and the decreasing sequence $(\alpha_n) \subset \mathbb{R}$ such that $y_{n+1} \in A \cap (y_n + C)$, $\alpha_n = \sup\{\langle y, y^* \rangle \mid y \in A \cap (y_n + C)\}$ and $\langle y_{n+1}, y^* \rangle \geq \alpha_n - 2^{-n-1}$ for $n \geq 0$. It follows that

$$\alpha_{n+1} \geq \langle y_{n+1}, y^* \rangle \geq \alpha_n - 2^{-n-1}, \quad \langle y_{n+1}, y^* \rangle \geq \langle y_n, y^* \rangle \quad \forall n \geq 0. \quad (3.23)$$

Because $y_{n+1} - y_n \in C$, we have that $y_{n+1} - y_n = \lambda_n b_n$ with $b_n \in B$ and $\lambda_n = \langle y_{n+1} - y_n, y^* \rangle \geq 0$. From (3.23) we obtain that $\lambda_n = \langle y_{n+1}, y^* \rangle - \langle y_n, y^* \rangle \leq \alpha_{n+1} + 2^{-n} - \alpha_{n-1} \leq 2^{-n}$ for $n \geq 1$. Let $p : X \to \mathbb{R}$ be a continuous seminorm. Because (b_n) is bounded, there exists $M_p > 0$ such that $p(b_n) \leq M_p$ for every $n \geq 0$. Then

$$p(y_{n+m} - y_n) \leq \sum_{k=n}^{n+m-1} p(y_{k+1} - y_k) = \sum_{k=n}^{n+m-1} \lambda_k p(b_k) \leq 2^{-n+1} M_p$$

for all $n, m \geq 1$. Hence the sequence $(y_n) \subset A$ is Cauchy. Because A is sequentially C-complete, (y_n) converges to some $y \in A$. Since $y_{n+m} \in A \cap (y_n + C)$ for $m \geq 0$ and C is closed, we have that $y \in y_n + C$ for every $n \geq 0$. In particular, $y \geq y_0$. Take $z \in A \cap (y + C)$. Then $\alpha_n \geq \langle z, y^* \rangle \geq \langle y, y^* \rangle$. Because, by (3.23), $\lim \alpha_n = \langle y, y^* \rangle$, we obtain that $\langle z, y^* \rangle = \langle y, y^* \rangle$. Since $y^* \in C^\#$, we obtain that $z = y$, and so $y \in \operatorname{Max}(A, C) \subset \operatorname{Max}(Y, C)$.   □

Remark 3.2.40. When C is well-based and Y is C-bounded from above, or more generally, when y^* is bounded above on Y for some $y^* \in C^\#$ with $B := \{y \in C \mid \langle y, y^* \rangle = 1\}$ bounded, $Y_C^+(y)$ is bounded for every $y \in Y$.

Indeed, if $z \in Y_C^+(y)$, then $z - y = \lambda b$ for some $\lambda \geq 0$ and $b \in B$. Thus, $0 \leq \lambda = \langle z - y, y^* \rangle \leq \sup y^*(Y) - \langle y, y^* \rangle =: \gamma$. Hence $Y_C^+(y) \subset [0, \gamma] \cdot B$.

Results Corresponding to Proposition 3.2.20 and Its Corollary

Theorem 3.2.41. (Bitran and Magnanti [32, Prop. 3.1]) *Let $X = \mathbb{R}^m$, let C be closed with $C^\# \neq \emptyset$, and Y closed and convex. Then $\mathrm{Max}(Y; C) \neq \emptyset$ if and only if $Y_\infty \cap C = \{0\}$.*

PROOF. The necessity follows from the last part of Proposition 3.2.26, while the sufficiency follows from both parts of Corollary 3.2.27.   □

Note that the convexity of Y is used only for the sufficiency part.

Theorem 3.2.42. (Nieuwenhuis [273, Th. I-14]) *Let X be a Banach space and C have nonempty interior. If Y is compact, then $\mathrm{Max}(Y; C_0) \neq \emptyset$, where $C_0 = \mathrm{int}\, C \cup \{0\}$.*

PROOF. By Proposition 3.2.20, $\mathrm{Max}(Y; \mathrm{cl}\, C) \neq \emptyset$. The conclusion follows from Proposition 3.2.15 (iii) because $\mathrm{cl}\, C \cap (-C_0) = \{0\} \subset C_0$.   □

Theorem 3.2.43. (Henig [162, Th. 2.1]) *Let $X = \mathbb{R}^m$ and let $C^\#$ be nonempty. Suppose that there exists Z such that $Y \subset Z \subset Y - \mathrm{cl}\, C$, Z is closed, and $Z_\infty \cap \mathrm{cl}\, C = \{0\}$. Then $\mathrm{Max}(Y; C) \cap (y_0 + \mathrm{cl}\, C) \neq \emptyset$ for every $y_0 \in Y$.*

PROOF. As in (or from) Bitran–Magnanti's theorem, $\mathrm{Max}(Z_{\mathrm{cl}\, C}^+(y_0); \mathrm{cl}\, C) \neq \emptyset$. The conclusion follows from Proposition 3.2.15 (iii) taking $K = \mathrm{cl}\, C$.   □

Theorem 3.2.44. (Borwein [39, Th. 1]) *Let X be an l.c.s. and C closed. If $Y_C^+(y)$ is compact for some $y \in Y$, then $\mathrm{Max}(Y; C) \neq \emptyset$.*

PROOF. Apply Corollary 3.2.21 for $Z = \{y\}$.   □

Theorem 3.2.45. (Penot and Sterna-Karwat [291, Rem. 3.3]) *Let X be an H.t.v.s., Y a closed convex set, and $C \subset X$ have a compact base. If $\mathrm{Max}(Y; C) \neq \emptyset$, then Y has the domination property.*

PROOF. Taking $y \in \mathrm{Max}(Y; C)$, $Y_C^+(y)$ is nonempty and compact, being a singleton. Using Proposition 3.2.26, $Y_\infty \cap C = \{0\}$, whence, by Corollary 3.2.27, Y has the domination property.   □

Theorem 3.2.46. (Jahn [194, Th. 2.3(b), Th. 2.6(b)]) *Let $(X, \|\cdot\|)$ be a reflexive Banach space and $Y \subset X$ have a nonempty C-upper section $Y_C^+(y)$ that is weakly closed and C-upper bounded.*

(i) *Suppose that $\|x\| < \|x + y\|$ for all $x, y \in C \setminus \{0\}$. Then $\mathrm{Max}(Y; C) \neq \emptyset$.*

(ii) *Suppose that* aint $C \neq \emptyset$ *and* $\|x\| < \|x+y\|$ *for all* $x \in C$, $y \in$ aint C.

Then $\mathrm{Max}(Y;C) \neq \emptyset$.

PROOF. Note first that when A (subset of an H.t.v.s.) is convex and cor $A \neq \emptyset$, then cl $A = $ cl(aint A). Indeed, let $u \in$ aint A and $x \in$ cl A. There exists $(x_i)_{i \in I} \subset A$ with $x_i \to x$. For every $\lambda \in (0,1)$ and $i \in I$ we have that $(1-\lambda)u + \lambda x_i \in$ cor A, and so $(1-\lambda)u + \lambda x \in$ cl(aint A). Taking $\lambda \to 1$, we get $x \in$ cl(aint A). Therefore cl $A = $ cl(aint A).

Let $x, y \in$ cl $C \setminus \{0\}$. There exist $(x_n)_{n \in \mathbb{N}}$, $(y_n)_{n \in \mathbb{N}}$ sequences from $C \setminus \{0\}$ in case (i) and from aint C in case (ii) such that $x_n \to x$, $y_n \to y$. Since $\|x_n\| < \|x_n + y_n\|$ in both cases, it follows that $\|x\| \le \|x + y\|$. Therefore $\|x\| \le \|x + y\|$ for all $x, y \in$ cl C in both cases. In particular, C is an acute (normal) cone.

From the normality of C (or directly from the monotonicity conditions), $Y_C^+(y)$ is also bounded. It follows that $Y_C^+(y)$ is w-compact. Since (3.21) is satisfied, C being acute, the conclusion follows by applying Corollary 3.2.21. □

Note that Jahn stated a similar result when X is the dual of a normed space and $Y_C^+(y)$ is w^*-closed.

Theorem 3.2.47. (Tanaka [350, Lemma 2.4]) *Let* $X = \mathbb{R}^m$ *and let* $C^\#$ *be nonempty. Suppose that* Y *is* C*-compact. Then* $\mathrm{Max}(Y;C) \neq \emptyset$*. Moreover, if* C *is correct, then* Y *has the domination property.*

PROOF. Note that by taking $K = $ cl C, condition (3.17) is satisfied, while when C is correct, (3.18) holds, too. Since Y is C-compact, $Y_{\mathrm{cl}\,C}^+(y)$ is compact for every $y \in Y$. From Corollary 3.2.21 we have that Max $\left(Y_{\mathrm{cl}\,C}^+(y); \mathrm{cl}\,C \right) \neq \emptyset$ for every $y \in Y$, and so Y has the domination property w.r.t. cl C. In particular, by Proposition 3.2.15 (iii), $\mathrm{Max}(Y;C) \neq \emptyset$, while when C is correct we obtain that Y has (DP) using assertion (v) of the same proposition. □

Results Corresponding to Proposition 3.2.22

Theorem 3.2.48. (Luc [243, Th. 2.6, Cor. 2.12], [244, Th. II.3.3]) *Assume that* C *is correct and* Y *satisfies (3.22) for* $K = $ cl C*. Then* Y *has the domination property.*

PROOF. The conditions of Proposition 3.2.22 are satisfied in this case. □

Theorem 3.2.49. (Malivert [248, Th. 3.5]) *Let* C *be a Daniell cone with* int $C \neq \emptyset$*. Suppose that* Y *is closed and for every* $y \in Y \setminus \mathrm{Max}(Y;C_0)$ *there exists* $y_0 \in (y + \mathrm{int}\,C) \cap Y$ *such that* $(y_0 + \mathrm{int}\,C) \cap Y$ *is* C*-upper bounded. Then* Y *has the domination property w.r.t.* C_0*, where* $C_0 = \{0\} \cup \mathrm{int}\,C$.

PROOF. Let $Y_0 = (y_0 + \text{int}\,C) \cap Y$, and consider $(z_i)_{i \in I} \subset Y_0$ be strictly C_0-increasing. It follows that (z_i) is C-increasing and C-upper bounded. Since C is Daniell, $z_i \to z \in X$ and $z_i \leq_C z$ for every $i \in I$. Since Y is closed, $z \in Y$. Since $(z_i) \subset Y_0$ is strictly C_0-increasing, it follows that $z \in Y_0$. So, we obtain that $z \in Y_0 \cap \left(\bigcap_{i \in I} (z_i + C) \right)$. Therefore the hypotheses of Proposition 3.2.22 are satisfied by C, K, Y replaced by C_0, C, Y_0, respectively. $\square$

Note that Propositions 3.2.17–3.2.22 have different fields of application, although these fields are not disjoint. For example, even in finite-dimensional spaces there exist convex cones whose closure is pointed that are not correct; for such a cone and a compact set one can apply Corollary 3.2.21 but not Proposition 3.2.22 in Luc's version.

Also note that there are other results concerning existence of efficient points w.r.t. cones that do not enter into our classification. To our knowledge, among these results are Theorem 4.1 of Hartley [158], Proposition 4.10 and Proposition 4.11 of Chew [77], and Theorem 2.2 of Sterna-Karwat [330] (see also the paper of Turinici [359] for a discussion related to the use of ordinals). The presentation of this section follows that in the paper by Sonntag and Zălinescu [327]; however, we modified slightly Proposition 3.2.17 and introduced Propositions 3.2.18 and 3.2.25 in order to cover Theorem 3.2.30 by Jameson [199] and Theorem 3.2.38 by Ng and Zheng [272].

3.2.6 Some Density and Connectedness Results

Let $\emptyset \neq Y \subset X$, and $C \subset X$ a convex cone. We say that $y_0 \in Y$ is *properly maximal w.r.t.* C if there exists $x^* \in C^\#$ such that $\langle y_0, x^* \rangle \geq \langle y, x^* \rangle$ for every $y \in Y$; we denote the set of properly maximal points of Y w.r.t. C by $\text{PrMax}(Y; C)$.

Another notion of proper maximality is due to Henig [161]; namely, $y_0 \in Y$ is *Henig proper maximal w.r.t.* C if there exists a proper convex cone $K \subset X$ such that $C \setminus \{0\} \subset \text{int}\,K$ and $y_0 \in \text{Max}(Y; K)$. We denote by $\text{HMax}(Y; C)$ the set of Henig proper maximal points of Y w.r.t. C. We have that

$$\text{PrMax}(Y; C) \subset \text{HMax}(Y; C) \subset \text{Max}(Y; C). \tag{3.24}$$

For the first inclusion in (3.24) take $y_0 \in \text{PrMax}(Y; C)$ and $K := \{x \in X \mid \langle x, x^* \rangle \geq 0\}$, where $x^* \in C^\#$ is the element in the definition of proper maximality, while the second inclusion in (3.24) is obvious. Note that when one of the first two sets in (3.24) is nonempty, then C is a pointed cone.

Lemma 3.2.50. *If the set Y is convex, then* $\text{PrMax}(Y; C) = \text{HMax}(Y; C)$.

PROOF. If $y_0 \in \text{HMax}(Y; C)$, then there exists a proper convex cone $K \subset X$ such that $C \setminus \{0\} \subset \text{int}\,K$ and $(Y - y_0) \cap K = \{0\}$; hence $(Y - y_0) \cap \text{int}\,K = \emptyset$. Applying a separation theorem, we get $x^* \in X^*$ such that

$$\langle y - y_0, x^* \rangle \leq 0 < \langle x, x^* \rangle \quad \forall y \in Y, \ \forall x \in \text{int}\,K.$$

It follows that $x^* \in C^\#$ and $\langle y, x^* \rangle \leq \langle y_0, x^* \rangle$ for every $y \in Y$. Therefore $y_0 \in \mathrm{PrMax}(Y; C)$. $\qquad\square$

The elements of $\mathrm{PrMax}(Y; C)$ are obtained by linear scalarization (compare with Section 3.1.1), and so they can be obtained (quite) easily. So, for practical purposes, it is important to have the possibility to approximate each element in $\mathrm{Max}(Y; C)$ by elements of $\mathrm{PrMax}(Y; C)$. The first result in this direction was obtained by Arrow, Barankin, and Blackwell [9] for $X = \mathbb{R}^n$, $C = \mathbb{R}^n_+$, and Y a compact convex set. This result was later extended in several directions. One of these extensions was realized by Henig [161, Th. 5.1], who proved that $\mathrm{Max}(Y; C) \subset \mathrm{cl}\left(\mathrm{HMax}(Y; C)\right)$ when $X = \mathbb{R}^n$, C is a closed pointed convex cone, and there exists $A \subset C$ such that $0 \in A$, $Y - A$ is closed, and $(Y - A)_\infty \cap C = \{0\}$. In the sequel we extend this result to the case in which $(X, \|\cdot\|)$ is a normed vector space. When no mention of the topology is made for a topological notion, this is considered w.r.t. the norm topology. First we prove an auxiliary result; this is due essentially to Gong [136, Lem. 2.1] (see also Sterna-Karwat [334, Th. 3.1]).

Lemma 3.2.51. *Let $C \subset (X, \|\cdot\|)$ be a based closed convex cone with base B and take $\delta := d(0, B) > 0$. For $\varepsilon \in \,]0, \delta[$ consider $B_\varepsilon := \{x \in X \mid d(x, B) \leq \varepsilon\}$ and $C_\varepsilon := [0, \infty[\cdot B_\varepsilon$ the cone generated by B_ε. Then*

(i) *C_ε is a closed convex cone for every $\varepsilon \in \,]0, \delta[$;*
(ii) *if $0 < \gamma < \varepsilon < \delta$, then $C \setminus \{0\} \subset C_\gamma \setminus \{0\} \subset \mathrm{int}\, C_\varepsilon$;*
(iii) *$C = \cap_{\varepsilon \in \,]0,\delta[} C_\varepsilon = \cap_{n \in \mathbb{N}} C_{\varepsilon_n}$, where $(\varepsilon_n) \subset \,]0, \delta[$ converges to 0.*

PROOF. (i) Fix $\varepsilon \in \,]0, \delta[$. Because B is convex, $d(\cdot, B)$ is a convex (continuous) function, and so B_ε is a closed convex set with $0 \notin B_\varepsilon$. We have to show that C_ε is closed. So, let $y \in \mathrm{cl}\, C_\varepsilon$. Fix $(\varepsilon_n) \subset \,]\varepsilon, \infty[$ converging to ε. Because $B_\varepsilon \subset B + \varepsilon' U_X$ for $\varepsilon' > \varepsilon$, there exist $(\lambda_n) \subset [0, \infty[$, $(b_n) \subset B$ and $(u_n) \subset U_X$ such that $y_n := \lambda_n(b_n + \varepsilon_n u_n) \to y$. The sequence (λ_n) is bounded above. Otherwise, there exists a subsequence (λ_{n_k}) with limit ∞. Taking $b'_n := b_n + \varepsilon_n u_n$, it follows that $b'_{n_k} \to 0$. Since $\delta = d(0, B) \leq \|b_{n_k}\| = \|\varepsilon_{n_k} u_{n_k} - b'_{n_k}\| \leq \varepsilon_{n_k} + \|b'_{n_k}\|$, we get the contradiction $\delta \leq \varepsilon$. So, we may assume that $\lambda_n \to \lambda \in [0, \infty[$. If $\lambda = 0$, then $y = \lim \lambda_n b_n \in C \subset C_\varepsilon$, because C is closed. Let $\lambda > 0$. Then $b'_n \to \lambda^{-1} y$ and $d(\lambda^{-1} y, B) \leq \|\lambda^{-1} y - b_n\| = \|\lambda^{-1} y - b'_n + \varepsilon_n u_n\| \leq \varepsilon_n + \|\lambda^{-1} y - b'_n\|$. It follows that $d(\lambda^{-1} y, B) \leq \varepsilon$, and so $\lambda^{-1} y \in B_\varepsilon$, whence $y \in C_\varepsilon$.

(ii) Let $0 < \gamma < \varepsilon < \delta$. The first inclusion being obvious, let us prove the second. Let $y \in C_\gamma \setminus \{0\}$; i.e., $y = \lambda x$ with $\lambda > 0$ and $x \in B_\gamma$. Because $B_\gamma \subset B + \varepsilon\, \mathrm{int}\, U_X \subset \mathrm{int}\, C_\varepsilon$, the conclusion follows.

(iii) The second equality being obvious, taking into account (ii), it is sufficient to prove that $\cap_{n \in \mathbb{N}} C_{\varepsilon_n} \subset C$ for a fixed $(\varepsilon_n) \subset \,]0, \delta[$ convergent to 0. So let $y \in \cap_{n \in \mathbb{N}} C_{\varepsilon_n}$. There exist $(\lambda_n) \subset [0, \infty[$, $(b_n) \subset B$, and $(u_n) \subset U_X$ such that $y_n := \lambda_n(b_n + 2\varepsilon_n u_n) \to y$. As in (i) we obtain that (λ_n) is bounded above. It follows that $y = \lim \lambda_n b_n \in C$. $\qquad\square$

The cone C_ε introduced in Lemma 3.2.51 is called a *Henig dilating cone*.

Remark 3.2.52. Borwein and Zhuang [42] say that y_0 is a **super efficient point** of Y w.r.t. C, written $y_0 \in \mathrm{SE}(Y;C)$, if $(U_X + C) \cap \mathbb{R}_+(Y - y_0) \subset \mu U_X$ for some $\mu > 0$. Under the hypothesis that C has a bounded base B, in [42, Prop. 2.5] one proves that $y_0 \in \mathrm{SE}(Y;C)$ if and only if $y_0 \in \mathrm{Max}(Y;C_\varepsilon)$ for some $0 < \varepsilon < d(0,B)$.

Theorem 3.2.53. *Let X be a normed vector space, $C \subset X$ a based closed convex cone, and τ a locally convex topology on X compatible with the dual system (X, X^*). Assume that Y is a nonempty τ-closed and τ-asymptotically compact subset of X with $Y_\infty \cap C = \{0\}$, the asymptotic cone being taken w.r.t. τ. Then*

$$\mathrm{Max}(Y;C) \subset \tau\text{-}\mathrm{cl}\left(\mathrm{HMax}(Y;C)\right). \tag{3.25}$$

PROOF. Taking $\bar{y} \in \mathrm{Max}(Y;C)$, we must show that $\bar{y} \in \tau\text{-}\mathrm{cl}\left(\mathrm{HMax}(Y;C)\right)$. Without loss of generality, we take $\bar{y} = 0$; hence $Y \cap C = \{0\}$.

Let B be a base for C; because C is closed, we (may) assume that B is closed; of course, B is convex, too.

Because Y is τ-a.c., by [389, Prop. 2.2 (i)], so is Y_∞. Using [389, Cor. 3.12] we obtain that $B - Y_\infty$ is τ-closed, and so $B - Y_\infty$ is (norm) closed. Because $Y_\infty \cap C = \{0\}$, we have that $Y_\infty \cap B = \emptyset$, whence $0 \notin B - Y_\infty$. It follows that

$$\gamma := d(B, Y_\infty) := \inf\{\|b - y\| \mid b \in B,\ y \in Y_\infty\} = d(0, B - Y_\infty) > 0. \tag{3.26}$$

Let $B_n = \{x \in X \mid d(x,B) \leq \frac{\gamma}{2n+1}\}$ for $n \in \mathbb{N}^*$; B_n is a closed convex set with $0 \notin B_n$. From Lemma 3.2.51 we have that $C_n := [0,\infty) \cdot B_n$ is a based closed convex cone such that $C \setminus \{0\} \subset \mathrm{int}\, C_n$.

Consider $Y_n := Y \cap C_n$. Since $Y_\infty \cap B_n = \emptyset$, we have that $Y_\infty \cap C_n = \{0\}$, and so, by Proposition 3.2.26, Y_n is τ-compact for every $n \geq 1$. Now using Proposition 3.2.20 we get the existence of $y_n \in \mathrm{Max}(Y_n; C_n)$. From Proposition 3.2.15 (i) we have that $y_n \in \mathrm{Max}(Y; C_n)$, and so $y_n \in \mathrm{HMax}(Y; C_n)$. Of course, $(y_n) \subset Y_1$; since Y_1 is τ-compact, (y_n) is bounded. Because $y_n \in C_n$, there exist $(\lambda_n) \subset [0,\infty[$, $(b_n) \subset B$, and $(x_n) \subset X$ such that $\|x_n\| \leq \gamma/2n$ and $y_n = \lambda_n(b_n + x_n)$ for every $n \geq 1$.

Let us prove that $\lambda_n \to 0$. In the contrary case there exists a subnet $(\lambda_{\theta(j)})_{j \in J}$ with $0 < \lambda_{\theta(j)} \to \lambda \in]0,\infty]$. Because Y_1 is τ-compact, we may assume that $(y_{\theta(j)})_{j \in J}$ τ-converges to some $y \in Y_1$. If $\lambda = \infty$, from $b_{\theta(j)} = \lambda_{\theta(j)}^{-1} y_{\theta(j)} - x_{\theta(j)}$ we obtain the contradiction $0 \in B$. Hence $\lambda < \infty$. If $\lambda \in]0,\infty[$, then $\lambda^{-1}y \in B$, whence $y \in C \cap Y$. It follows that $y = 0$, and so we get the contradiction $0 \in B$. Therefore $\lambda_n \to 0$.

Let us prove that $y_n \to^\tau 0$. In the contrary case there exists a τ-neighborhood U of 0 such that the set $P := \{n \in \mathbb{N}^* \mid y_n \notin U\}$ is infinite. As above, there exists a subnet $(y_{\theta(j)})_{j \in J}$ of $(y_n)_{n \in P}$ τ-convergent to some $y \in Y_1 \subset Y$. Then $y = \tau - \lim \lambda_{\theta(j)} b_{\theta(j)} \in C \cap Y$, and so $y = 0$. It follows that there exists some $j_0 \in J$ such that $y_{\theta(j)} \in U$ for $j \geq j_0$; this is a contradiction because $\theta(j_0) \in P$. Hence $0 = \tau - \lim y_n \in \tau\text{-}\mathrm{cl}\left(\mathrm{HMax}(Y;C)\right)$. □

The convergence in the preceding theorem is strong if the classes of weak (or even τ) and norm neighborhoods of $\overline{y}$ in Y coincide (that is, $\overline{y}$ is a continuity point of Y); this happens, for example, if $Y - \overline{y}$ is contained in a Phelps cone. The next result furnishes other conditions for having strong convergence at every $y \in \operatorname{Max}(Y;C)$.

Theorem 3.2.54. *Let X be a normed vector space, τ a locally convex topology on X compatible with the dual system (X, X^*), $C \subset X$ a closed convex cone, and $\emptyset \neq Y \subset X$. Then*

$$\operatorname{Max}(Y;C) \subset \operatorname{cl}\big(\operatorname{HMax}(Y;C)\big), \tag{3.27}$$

provided one of the following conditions is satisfied:

(i) *C has a bounded base, Y is τ-closed, τ-asymptotically compact, and $Y_\infty \cap C = \{0\}$, the asymptotic cone being taken w.r.t. τ;*

(ii) *C has a w-compact base, Y is τ-complete, and $Y \subset A - C$ for some bounded set $A \subset X$.*

(iii) *Y is a closed convex set and either (a) C is based and Y is locally compact, (b) C has a bounded base and Y is τ-locally compact, or (c) C has a w-compact base and the τ-closed bounded subsets of Y are τ-complete.*

PROOF. We follow the lines of the proof of the preceding theorem, and we point out only the respective differences. Let $\overline{y} = 0 \in \operatorname{Max}(Y;C)$.

(i) The base B of C is, moreover, bounded. With the same notation as in the proof of Theorem 3.2.53, we have that $\lambda_n \to 0$. Because (b_n) is bounded, we obtain that (y_n) strongly converges to 0.

(ii) The base B of C is w-compact, and so bounded. This time $Y_\infty \subset (A - C)_\infty = -C$, because A is bounded, and so $B - Y_\infty \subset B + C = [1, \infty[\cdot B$ (the asymptotic cone being taken w.r.t. w). The last set being closed, $\gamma > 0$. With the notation of Theorem 3.2.53 let us show that Y_n is bounded. Indeed,

$$Y_n \subset C_n \cap (A - C) \subset (A_0 + C_n) \cap (A_0 - C_n),$$

where $A_0 = A \cup \{0\}$. Since C_n is well-based, C_n is normal, whence the boundedness of Y_n follows. Moreover, Y_n is τ-complete as a τ-closed subset of a τ-complete set. Applying Propositions 3.2.24 and 3.2.17(iv), we find that $(y_n) \subset Y_1 \cap \operatorname{HMax}(Y;C)$, as in Theorem 3.2.53. Exploiting this time the w-compactness of B we obtain that $\lambda_n \to 0$, and so $y_n \to 0$ for the strong topology.

(iii) (a) Let U be a convex neighborhood of 0 such that $Y_0 := Y \cap U$ is compact. Of course, $0 \in \operatorname{Max}(Y_0;C)$. By Theorem 3.2.53 for the norm topology, there exists a sequence $(y_n) \subset \operatorname{HMax}(Y_0;C)$ strongly converging to 0. There exists n_0 such that $y_n \in \operatorname{int} U$ for $n \geq n_0$. Because Y_0 is convex, by Lemma 3.2.50, $y_n \in \operatorname{PrMax}(Y_0;C)$, and so y_n maximizes some $x_n^* \in C^\#$ on Y_0. Let $y \in Y$. Then there exists $t \in]0,1[$ such that $(1-t)y_n + ty \in Y_0$, and so $\langle (1-t)y_n + ty, x_n^* \rangle \leq \langle y_n, x_n^* \rangle$, whence $\langle y, x_n^* \rangle \leq \langle y_n, x_n^* \rangle$. Hence $y_n \in \mathrm{PrMax}(Y; C)$.

In case (b) we take U to be a τ-closed and convex neighborhood of 0 such that $Y_0 := Y \cap U$ is τ-compact, while in case (c) one takes U_X instead of U (then Y_0 is a τ-complete convex and bounded set). The proof continues as for (a) but using assertion (i) or (ii) above instead of Theorem 3.2.53. □

Note that for a fixed $\bar{y} \in \mathrm{Max}(Y; C)$, when Y is a convex set, the fact that $\bar{y} \in \mathrm{cl}\left(\mathrm{PrMax}(Y; C)\right)$ can be obtained as above, for (a) C has a bounded base and $Y \cap (\bar{y} + r U_X)$ is τ-compact, or (b) C has a w-compact base and $Y \cap (\bar{y} + r U_X)$ is τ-complete, where r is some positive number.

Remark 3.2.55. In the statements of Theorems 3.2.53 and 3.2.54 besides Y we can consider another set Z satisfying $Y \subset Z \subset Y - C$ and ask that this set Z satisfy the conditions imposed on Y, the conclusion remaining the same for the set Y. Moreover, when Y (or Z) is closed and convex, there is no need for the hypothesis $Y_\infty \cap C = \{0\}$ in Theorems 3.2.53 and 3.2.54(i), because in the contrary case, $\mathrm{Max}(Y; C) = \emptyset$.

Density results like those proved in Theorems 3.2.53 and 3.2.54 can be used to establish the connectedness of the set $\mathrm{Max}(Y; C)$. Connectedness of an efficient point set provides a possibility of moving continuously from one optimal solution to any other along optimal alternatives only.

Theorem 3.2.56. *Let X be a normed vector space, $C \subset X$ a closed convex cone, and $\emptyset \neq Y, Z \subset X$ such that $Y \subset Z \subset Y - C$ and $Y - C$ is convex.*

(i) *If Z is w-compact and C is based, then $\mathrm{PrMax}(Y; C)$ and $\mathrm{Max}(Y; C)$ are connected w.r.t. the weak topology w.*
(ii) *If Z is compact and C is well-based, then $\mathrm{PrMax}(Y; C)$ and $\mathrm{Max}(Y; C)$ are connected.*

PROOF. (i) Because $\mathrm{PrMax}(Y; C) = \mathrm{PrMax}(Y - C; C) = \mathrm{PrMax}(Z; C)$ and $\mathrm{Max}(Y; C) = \mathrm{Max}(Y - C; C) = \mathrm{Max}(Z; C)$, we (may) assume that Y is w-compact.

Consider the multifunction

$$\Gamma : C^\# \rightrightarrows Y, \quad \Gamma(x^*) := \{y \in Y \mid \langle y, x^* \rangle \geq \langle y', x^* \rangle \ \forall y' \in Y\}.$$

It is clear that $\mathrm{PrMax}(Y; C) = \mathrm{Im}\,\Gamma = \Gamma(C^\#)$. Because Y is w-compact, $\Gamma(x^*)$ is a nonempty w-compact set for every $x^* \in C^\#$, and so $\mathrm{dom}\,\Gamma = C^\#$. Observe that

$$\sup\{\langle y', x^* \rangle \mid y' \in Y\} = \sup\{\langle y', x^* \rangle \mid y' \in Y - C\},$$

and so $\Gamma(x^*)$ is convex for every $x^* \in C^\#$. In particular, $\Gamma(x^*)$ is w-connected for every $x^* \in C^\#$. Because $C^\#$ is a convex set (in fact, $C^\# \cup \{0\}$ is a convex cone), and so a (norm) connected set, in order to obtain that $\Gamma(C^\#)$ is w-connected, by Proposition 2.5.8(ii), it is sufficient to show that Γ is (norm–w)

u.c. Assume that Γ is not $\|\cdot\|$-w u.c. at $x^* \in C^\#$. Then there exist a weak open set $U \subset X$ and a sequence $(x_n^*) \subset C^\#$ converging to x^* such that $\Gamma(x^*) \subset U$ and $\Gamma(x_n^*) \not\subset U$ for every $n \in \mathbb{N}$. Take $y_n \in \Gamma(x_n^*) \setminus U \subset Y$ for $n \in \mathbb{N}$. Because Y is weakly compact, by Eberlein–Šmulian theorem (see [168]), there exists a subsequence $(y_{n_p})_{p \in \mathbb{N}}$ converging weakly to $\overline{y} \in Y$. But

$$\langle y_{n_p}, x_{n_p}^* \rangle \geq \langle y, x_{n_p}^* \rangle \quad \forall y \in Y,$$

and so, taking the limit (for $y \in Y$ fixed), we obtain that $\overline{y} \in \Gamma(x^*) \subset U$. Since U is w-open, we get the contradiction that $y_{n_p} \in U$ for $p \geq p_0$, for some $p_0 \in \mathbb{N}$. Now using Theorem 3.2.53 we have that $\mathrm{PrMax}(Y;C) \subset \mathrm{Max}(Y;C) \subset w\text{-}\mathrm{cl}\left(\mathrm{PrMax}(Y;C)\right)$. We obtain that $\mathrm{Max}(Y;C)$ is connected w.r.t. w by a well-known result in topology.

(ii) The proof is similar. $\qquad\qquad\qquad\qquad\qquad\qquad\qquad\qquad\qquad\qquad$ □

Note that in general, $\mathrm{Max}(Y;C)$ is not connected, as the following example shows.

Example 3.2.57. Let $Y = \mathbb{R}^3$, $C = \mathbb{R}_+^3$, $C_0 = \{0\} \cup \mathrm{int}\, C$, and

$$Y = \mathrm{conv}\left(\bigcup_{n \in \mathbb{N}} \{(0,0,k),(0,1,k),(1,0,k),(1-2^{-k},1-2^{-k},k)\}\right).$$

Then $\mathrm{Max}(Y;C) = \emptyset$, but $\mathrm{Max}(Y;C_0) = \{(1,0,z) \mid z \geq 0\} \cup \{(0,1,z) \mid z \geq 0\}$ is obviously not connected.

The most recent density results in the literature, for normed vector spaces, are those of Ferro [116] and Daniilidis [89] (stated for Y convex). So, in [116, Th. 3.1] one obtains Theorem 3.2.53 for Y weakly compact, Theorem 3.2.54(ii) for C well-based and Y w-compact, and Theorem 3.2.54(iii) for X a Banach space, C with w-compact base, and Y closed, while in [89, Th. 2] one obtains Theorem 3.2.53 for Y τ-compact. Theorems 3.2.53 and 3.2.54 also subsume the density results obtained by Jahn [197, Th. 3.1], Petchke [295, Th. 4.1], Borwein and Zhuang [42, Th. 4.2] (see also Remark 3.2.52), Zhuang [393, Th. 3.3], and Gong [136, Th. 3.1]. There are several density results established for topological vector spaces; we mention only the paper by Makarov and Rachkovski [247].

Theorem 3.2.56(i) was established by Song in [325, Th. 3.1], while Theorem 3.2.56(ii) was essentially established by Gong [136, Th. 4.1].

3.3 Continuity Properties with Respect to a Scalarization Parameter

Let Y be a topological vector space and $C \subset Y$ a proper convex cone with nonempty interior and $\emptyset \neq A \subset Y$. Let $k_0 \in \mathrm{int}\, C$. For every $p \in Y$ consider the minimization problem

$P(p)$ minimize t subject to $t \in \mathbb{R}$, $y \in A$, $p \in tk_0 - y - C$.

Consider also the **marginal function** $m : Y \to \overline{\mathbb{R}}$ associated with problems $P(p)$ defined by

$$m(p) := \inf\{t \in \mathbb{R} \mid y \in A, \ p \in tk_0 - y - C\} = \inf\{t \in \mathbb{R} \mid p \in tk_0 - (A + C)\}.$$

We have the following relations between the weak efficient points of A w.r.t. C and the solutions of $P(p)$. Let us denote the set $B \setminus \operatorname{int} B$ by ∂B; note that ∂B differs from the usual boundary $\operatorname{bd} B$ of B.

Proposition 3.3.1. (i) *If* $(\overline{t}, \overline{y}) \in \mathbb{R} \times A$ *is a solution of* $P(p)$, *then* $\overline{y} \in w\operatorname{Eff}(A; C)$ *and* $\overline{t}k_0 - p \in \partial(A + C)$.
(ii) *If* $\overline{y} \in w\operatorname{Eff}(A; C)$, *then* $(0, \overline{y})$ *is a solution of* $P(-\overline{y})$.

PROOF. (i) Let $(\overline{t}, \overline{y}) \in \mathbb{R} \times A$ be a solution of $P(p)$ and suppose that $\overline{y} \notin w\operatorname{Eff}(A; C)$. Then there exists $\overline{k} \in \operatorname{int} C$ such that $y := \overline{y} - \overline{k} \in A$. There exists $\delta > 0$ such that $\overline{k} - \delta k_0 \in C$. So

$$p - (\overline{t} - \delta)k_0 \in -\overline{y} + \delta k_0 - C = -y - (\overline{k} - \delta k_0) - C \subset -y - C,$$

contradicting the optimality of $(\overline{t}, \overline{y})$. Therefore $\overline{y} \in w\operatorname{Eff}(A; C)$. If $\overline{t}k_0 - p \notin \partial(A + C)$, then $\overline{t}k_0 - p \in \operatorname{int}(A + C)$. There exists $\delta > 0$ such that $(\overline{t} - \delta)k_0 - p \in A + C$, which shows that $(\overline{t}, \overline{y})$ is not a solution of $P(p)$, a contradiction. Therefore $\overline{t}k_0 - p \in \partial(A + C)$.
(ii) Let $\overline{y} \in w\operatorname{Eff}(A; C)$; of course, $(0, \overline{y})$ is an admissible solution of $P(-\overline{y})$. Suppose that $(0, \overline{y})$ is not a solution of $P(-\overline{y})$. Then there exist $t < 0$ and $y \in A$ such that $-\overline{y} \in tk_0 - y - C$, whence $y \in \overline{y} - \operatorname{int} C$, a contradiction. □

We say that the **segment** $[y, z] \subset Y$ with $y \neq z$ is **parallel to** C if $y \leq_C z$ or $z \leq_C y$.

Lemma 3.3.2. *Let* $\emptyset \neq A, V \subset Y$ *be such that* $\partial(A + C) \cap V$ *contains no segments parallel to* C. *Then* $\partial(A + C) \cap \operatorname{aint} V \subset \operatorname{Eff}(A; C)$.

PROOF. Let $\overline{y} \in [(A + C) \setminus \operatorname{int}(A + C)] \cap \operatorname{aint} V$. Suppose that $\overline{y} - \overline{k} \in A$ for some $\overline{k} \in C \setminus \{0\}$. Since $\overline{y} \in \operatorname{aint} V$, there exists $\delta \in]0, 1[$ such that $\overline{y} + t\overline{k} \in V$ for every $t \in [-\delta, \delta]$. Of course, for $t \in [0, \delta]$ we have that $\overline{y} - t\overline{k} = \overline{y} - \overline{k} + (1 - t)\overline{k} \subset A + C$. Suppose that $\overline{y} - t\overline{k} \in \operatorname{int}(A + C)$ for some $t \in]0, \delta]$; i.e. $\overline{y} - t\overline{k} + W \subset A + C$ for some neighborhood W of 0 in Y. Then $\overline{y} + W \subset A + C + t\overline{k} \subset A + C$, which shows that $\overline{y} \in \operatorname{int}(A + C)$, a contradiction. Therefore $\overline{y} - t\overline{k} \in \partial(A + C) \cap V$ for every $t \in [0, \delta]$. Since the segment $[\overline{y}, \overline{y} - \delta\overline{k}]$ is parallel to C contained in $\partial(A + C) \cap V$, we get a contradiction. It follows that $A \cap (\overline{y} - C) \subset \{\overline{y}\}$. Since $\overline{y} \in A + C$, we obtain that $\overline{y} \in A$ and, furthermore, $\overline{y} \in \operatorname{Eff}(A; C)$. □

As application of the preceding lemma we have the following result.

Proposition 3.3.3. *Let $p \in Y$ be such that the problem $P(p)$ has at least a solution. If there exists an open set $V \subset Y$ such that $m(p)k_0 - p \in V$ and $\partial(A + C) \cap V$ contains no segments parallel to C, then $y \in A$ obtained by solving $P(p)$ coincides with $m(p)k_0 - p$ and belongs to $\mathrm{Eff}(A; C)$.*

PROOF. The conclusion follows immediately by using Proposition 3.3.1 (i) and Lemma 3.3.2. □

Concerning the continuity properties of the marginal function m we have the following results.

Proposition 3.3.4. *The marginal function m is upper continuous at every $p \in \mathrm{dom}\, m$.*

PROOF. Indeed, let $p_0 \in \mathrm{dom}\, m$ and $\lambda \in \mathbb{R}$ such that $m(p_0) < \lambda$. Then there exists $t < \lambda$ such that $p_0 \in tk_0 - (A + C)$. It follows that

$$p_0 \in \lambda k_0 - [(\lambda - t)k_0 + A + C] \subset \lambda k_0 - (A + \mathrm{int}\, C) \subset \lambda k_0 - \mathrm{int}(A + C).$$

Therefore $U := \lambda k_0 - (A + C)$ is a neighborhood of p_0. Of course, for $p \in U$ we have that $m(p) \leq \lambda$. Hence m is upper continuous at p_0. □

Let $M := \{p \in Y \mid m(p) \in \mathbb{R}\}$ and consider the function $g : M \to Y$ defined by $g(p) := m(p)k_0 - p$.

Corollary 3.3.5. *The function g is C-u.c. on M.*

PROOF. Let $p_0 \in M$ be fixed and $V \in \mathcal{V}_Y$. Consider $V_1 \in \mathcal{V}_Y$ such that $V_1 + V_1 \subset V$. There exists $\varepsilon > 0$ such that $\varepsilon k_0 \in V_1$. Since m is upper continuous at p_0, there exists $U_1 \in \mathcal{V}_Y$ such that $m(p) < m(p_0) + \varepsilon$ for every $p \in p_0 + U_1$. Consider $U := U_1 \cap V_1$ and $p \in M \cap (p_0 + U)$. Then $m(p) - m(p_0) = \varepsilon - \gamma$ with $\gamma > 0$, whence, for $p \in M \cap (p_0 + U)$,

$$(m(p) - m(p_0))\, k_0 - (p - p_0) \subset (\varepsilon - \gamma)k_0 + V_1 \subset V_1 + V_1 - \gamma k_0 \subset V - C.$$

It follows that $g(p) \in g(p_0) + V - C$, which shows that g is C-u.c. at p_0. □

Proposition 3.3.6. *Suppose that $A + C$ is closed. Then m is lower continuous on Y and continuous on $\mathrm{dom}\, m$.*

PROOF. Taking $T : Y \times \mathbb{R} \to Y$ defined by $T(p,t) := tk_0 - p$, T is a continuous linear operator. Therefore the set $T^{-1}(A + C)$ is closed. It is also a set of epigraph type. It follows that $\mathrm{epi}\, m = T^{-1}(A + C)$, and so m is lower continuous. The continuity of m on $\mathrm{dom}\, m$ follows now from the preceding proposition. □

Note that in the case in which $A + C$ is closed, M is the set of those $p \in Y$ for which problem $P(p)$ has solutions; in this case, by Proposition 3.3.1 (i), we have that $g(p) \in w\,\mathrm{Eff}(A; C)$ for every $p \in M$.

Corollary 3.3.7. *Suppose that $A + C$ is closed. Then g is C-continuous on M. Moreover, if C is normal, then g is continuous on M.*

PROOF. The proof of the C-lower continuity of g is similar to that of Corollary 3.3.5, so we omit it. Therefore g is C-continuous on M. When C is normal, C-continuity and usual continuity are equivalent. □

Combining Proposition 3.3.3 and Corollary 3.3.7 we obtain the following result.

Corollary 3.3.8. *Suppose that $A + C$ is closed and that there exists an open set $V \subset Y$ such that $g(p_0) \in V$ for some $p_0 \in M$, and $\partial(A + C) \cap (V - C)$ contains no segments parallel to C. Then there exists a neighborhood U of p_0 in M such that g is C-continuous on U with values in $\mathrm{Eff}(A; C)$.*

PROOF. By the preceding corollary we have that $g(p) \in V - C$ for p in a neighborhood U of p_0 in M. Applying now Proposition 3.3.3 to each $g(p)$ with $p \in U$, we obtain that $g(p) \in \mathrm{Eff}(A; C)$. The C-continuity of g is stated in the preceding corollary. □

Consider now $f : X \to Y$, where (X, τ) is a topological space, and consider $A := f(X)$. We associate with f the multifunction $\Sigma : Y \rightrightarrows X$ defined by

$$\Sigma(p) := \begin{cases} \{x \in X \mid f(x) \in m(p)k_0 - p - C\} & \text{if } p \in M, \\ \emptyset & \text{if } p \in Y \setminus M. \end{cases}$$

Of course, $\Sigma(p)$ is the set of weak efficient solutions of the problem to minimize $f(x)$ w.r.t. C, by solving $P(p)$. The next result holds.

Proposition 3.3.9. *Suppose that f is continuous, C is closed, and m is continuous at $p_0 \in M$. Then Σ is closed at p_0.*

PROOF. Let $((p_i, x_i))_{i \in I} \subset \mathrm{gr}\, \Sigma$ converging to (p_0, x). Then $m(p_i)k_0 - f(x_i) - p_i \in C$ for every $i \in I$. Taking the limit, we obtain that $m(p_0)k_0 - f(x) - p_0 \in C$, which shows that $x \in \Sigma(p_0)$. □

Of course, if Σ has some compactness properties at p_0, then Σ is even u.c. at p_0.

Note that Pascoletti and Serafini [286] considered Y to be a finite-dimensional space and a supplementary parameter q in span C; also, Sterna-Karwat considered this supplementary parameter q in the case in which the intrinsic core of C is nonempty, Y being an arbitrary topological vector space. Our motivation to consider a fixed parameter q $(= k_0)$ is that with the help of it one can find all the weak efficient elements of the set A. The results stated above correspond to similar ones in [331].

3.4 Well-Posedness of Vector Optimization Problems

Throughout this section X is a separated topological space, Y is a separated topological vector space, and $C \subset Y$ is a convex cone with $C \neq Y$.

Recall that for $A \subset Y$ a nonempty set,

$$\mathrm{Eff}(A; C) := \{a \in A \mid A \cap (a - C) \subset a + C\}.$$

Of course, $\mathrm{Eff}(\emptyset; C) = \emptyset$. We say that A has the **domination property** (DP) (w.r.t. C) if $A \subset \mathrm{Eff}(A; C) + C$. When $\mathrm{int}\, C \neq \emptyset$ we denote by $w\,\mathrm{Eff}(A; C)$ the set $\mathrm{Eff}(A; \{0\} \cup \mathrm{int}\, C)$.

Remark 3.4.1. If $A \subset Y$ and $\mathrm{int}\, C \neq \emptyset$, then

$$A \cap \mathrm{cl}\,(\mathrm{Eff}(A; C)) \subset w\,\mathrm{Eff}(A; C) = A \cap \mathrm{cl}\,(w\,\mathrm{Eff}(A; C)).$$

In particular, if A is closed, then $w\,\mathrm{Eff}(A; C)$ is closed, too.

Indeed, let $w\,\mathrm{Eff}(A; C) \supset (y_i)_{i \in I} \to y \in A$. Suppose that $y \notin w\,\mathrm{Eff}(A; C)$. Then there exists $y' \in A \cap (y - \mathrm{int}\, C)$. It follows that $y_i - y' \to y - y' \in \mathrm{int}\, C$, whence $y_i - y' \in \mathrm{int}\, C$ for $i \succeq i_0$, contradicting the fact that $y_{i_0} \in w\,\mathrm{Eff}(A; C)$. The other assertions are obvious.

Let $f : X \to Y$ be a function and $A \subset X$ a nonempty set. The **optimal value** of the vector optimization problem

(P) C-minimize $f(x)$ subject to $x \in A$

is $\Omega = \Omega(f, A; C) := \mathrm{Eff}(f(A); C)$, while its **solution set** is $\Sigma(f, A; C) := A \cap f^{-1}(\Omega)$; an element of $\Sigma(f, A; C)$ is called an **optimal solution**, or simply *solution*, of problem (P). We associate with (P) and $\eta \in \Omega$ the multifunction $\Pi^\eta : Y \rightrightarrows X$ defined by $\Pi^\eta(\varepsilon) := \{x \in A \mid f(x) \leq_C \eta + \varepsilon\}$ for $\varepsilon \in C$ and $\Pi^\eta(\varepsilon) := \emptyset$ otherwise. We also consider the multifunction $\Pi : Y \rightrightarrows X$ defined by $\Pi(\varepsilon) := \bigcup_{\eta \in \Omega} \Pi^\eta(\varepsilon)$. In the case $X = Y$ and $f = \mathrm{Id}_X$ we denote Π^η and Π by Π_0^η and Π_0, respectively. An element of $\Pi(\varepsilon)$ is called an ε-**optimal solution** of (P).

Definition 3.4.2. *Let $f : X \to Y$ be a function and $A \subset X$ a nonempty set. We say that*

(i) (P) *is η-**well-posed** if $\Omega \neq \emptyset$ and Π^η is u.c. at 0 for every $\eta \in \Omega$;*
(ii) (P) *is **well-posed** if $\Omega \neq \emptyset$ and Π is u.c. at 0;*
(iii) (P) *is **weakly well-posed** if X is a topological vector space, $\Omega \neq \emptyset$, and Π is H-u.c. at 0.*

Before stating the next result, let us introduce another notion. We say that the nonempty set $A \subset Y$ has the **containment property** (CP) if

$$\forall W \in \mathcal{V}_Y, \ \exists V \in \mathcal{V}_Y \ \ : \ [A \setminus (\mathrm{Eff}(A; C) + W)] + V \subset \mathrm{Eff}(A; C) + C.$$

If $\operatorname{int} C \neq \emptyset$, then (CP) is equivalent to

$$\forall W \in \mathcal{V}_Y, \ \exists V \in \mathcal{V}_Y, \ \forall y \in A \setminus (\operatorname{Eff}(A; C) + W), \ \exists \bar{y} \in \operatorname{Eff}(A; C),$$
$$\exists k \in C \ : \ y = \bar{y} + k, \quad k + V \subset C. \tag{3.28}$$

It is obvious that (3.28) is sufficient for (CP) (and implies that $\operatorname{int} C \neq \emptyset$ if $A \not\subset \operatorname{cl}(\operatorname{Eff}(A; C))$). Suppose that (CP) holds and $\operatorname{int} C \neq \emptyset$. For $W \in \mathcal{V}_Y$ let $V_0 \in \mathcal{V}_Y$ be such that $[A \setminus (\operatorname{Eff}(A; C) + W)] + V_0 \subset \operatorname{Eff}(A; C) + C$. Consider $\bar{k} \in V_0 \cap \operatorname{int} C$; there exists $V \in \mathcal{V}_Y$ such that $\bar{k} + V \subset C$. Let $y \in A \setminus (\operatorname{Eff}(A; C) + W)$. It follows that $y - \bar{k} = y_0 + k'$ for some $y_0 \in \operatorname{Eff}(A; C)$ and $k' \in C$. Hence $y = y_0 + k_0$ with $k_0 = \bar{k} + k'$. Since $k_0 + V = \bar{k} + k' + V \subset C + C = C$, (3.28) holds.

Generally the condition (DP) does not imply and is not implied by (CP). However, if (CP) holds and $\operatorname{int} C \neq \emptyset$, then $w \operatorname{Eff}(A; C) = A \cap \operatorname{cl}(\operatorname{Eff}(A; C))$ and $A \subset w \operatorname{Eff}(A; C) + C$.

Indeed, suppose that (CP) holds and $y \in A \setminus \operatorname{cl}(\operatorname{Eff}(A; C))$. Then there exists $W \in \mathcal{V}_Y$ such that $y \in A \setminus (\operatorname{Eff}(A; C) + W)$. By (3.28), $y = \bar{y} + k$ with $\bar{y} \in \operatorname{Eff}(A; C)$ and $k \in \operatorname{int} C$. Therefore $y \notin w \operatorname{Eff}(A; C)$. Hence $w \operatorname{Eff}(A; C) = A \cap \operatorname{cl}(\operatorname{Eff}(A; C))$. The preceding argument shows also that $A \subset w \operatorname{Eff}(A; C) + C$.

We have the following result.

Proposition 3.4.3. *Let $A \subset Y$ be nonempty and compact and let C be closed and with nonempty interior. If $\operatorname{Eff}(A; C) = w \operatorname{Eff}(A; C)$, then (CP) holds for A.*

PROOF. It is known that A has (DP) in this case. Since $C \cap (-C_0) \subset C_0$ and $C + (C_0 \setminus \{0\}) \subset C_0$, where $C_0 := \{0\} \cup \operatorname{int} C$, we have that (DP) holds w.r.t. C_0, too. Therefore, taking into account that $\operatorname{Eff}(A; C) = w \operatorname{Eff}(A; C)$,

$$A \subset \operatorname{Eff}(A; C_0) + C_0 = \operatorname{Eff}(A; C) \cup (\operatorname{Eff}(A; C) + \operatorname{int} C).$$

Let $W \in \mathcal{V}_Y$. From the above inclusion we have that every $y \in A \setminus \operatorname{Eff}(A; C)$ has a representation $y = e_y + k_y$ with $e_y \in \operatorname{Eff}(A; C)$ and $k_y \in \operatorname{int} C$; let $V_y \in \mathcal{V}_Y$ be such that $k_y + V_y \subset C$. There exists $V_y^1 \in \mathcal{V}_Y$ open such that $V_y^1 + V_y^1 \subset V_y$. It follows that the family $\{y + V_y^1 \mid y \in A \setminus (\operatorname{Eff}(A; C) + \operatorname{int} W)\}$ represents an open cover of the compact set $A \setminus (\operatorname{Eff}(A; C) + \operatorname{int} W)$. Therefore there exist $y_1, \ldots, y_n \in A \setminus (\operatorname{Eff}(A; C) + \operatorname{int} W)$ such that

$$A \setminus (\operatorname{Eff}(A; C) + \operatorname{int} W) \subset \bigcup_{i=1}^{n} (y_i + V_{y_i}^1).$$

Let $V := \bigcap_{i=1}^{n} V_{y_i}^1 \in \mathcal{V}_Y$ and consider $y \in A \setminus (\operatorname{Eff}(A; C) + \operatorname{int} W)$. Then there exists $1 \leq i \leq n$ such that $y \in y_i + V_{y_i}^1$. So, using the above notation, $y = e_{y_i} + k_{y_i} + v_i$, with $v_i \in V_{y_i}^1$. Taking $k_y' := k_{y_i} + v_i$, we have that $k_y' + V = k_{y_i} + v_i + V \subset k_{y_i} + V_{y_i}^1 + V \subset k_{y_i} + V_{y_i} \subset C$. Therefore condition (3.28) holds, and so (CP) holds, too. □

Proposition 3.4.4. *Let $A \subset Y$ be a nonempty set. Then*

(i) *Π_0 is $(-C)$-H-u.c. at 0.*
(ii) *If (DP) holds for A, then Π_0 is C-u.c. at 0.*
(iii) *If (CP) holds for A, then Π_0 is H-u.c. at 0.*
(iv) *If C is normal and closed, $\text{Eff}(A;C)$ is compact, and (DP) holds for A, then Π_0 is u.c. at 0.*

PROOF. (i) Let $V \in \mathcal{V}_Y$. Consider $\varepsilon \in V \cap C$ and $y \in \Pi_0(\varepsilon)$. By definition, there exists $\eta \in \Pi_0(0)$ with $y \leq_C \eta + \varepsilon$, whence $y \in \Pi_0(0) + V - C$. Therefore $\Pi_0(V) \subset \Pi_0(0) + V - C$, which shows that Π_0 is $(-C)$-H-u.c. at 0.

(ii) Since (DP) holds, $\Pi_0(\varepsilon) \subset A \subset A + C = \Pi_0(0) + C \subset D + C$ for every $\varepsilon \in X$ and every open set $D \subset Y$ with $\Pi_0(0) \subset D$. Therefore Π_0 is C-u.c. at 0.

(iii) Let $W \in \mathcal{V}_Y$. Since (CP) holds for A, there exists $V \in \mathcal{V}_Y$ such that $[A \setminus (\Pi_0(0) + W)] + V \subset \Pi_0(0) + C$. Consider $\varepsilon \in C \cap (V \cap W)$ and $y \in \Pi_0(\varepsilon)$. Suppose that $y \notin \Pi_0(0) + W$; then $y + V \subset \Pi_0(0) + C$. Since $y \in \Pi_0(\varepsilon)$, $y = \eta + \varepsilon - k$ for some $\eta \in \Pi_0(0)$ and $k \in C$. But $y - \varepsilon \in y + V$, whence $y - \varepsilon = \eta - k = \eta' + k'$ with $\eta' \in \Pi_0(0)$ and $k' \in C$. Therefore $\eta = \eta' + (k + k')$. Since $\eta, \eta' \in \text{Eff}(A;C)$ and C is pointed, it follows that $k = k' = 0$, and so $y \in \Pi_0(0) + W$, a contradiction. Hence $\Pi_0(V \cap W) \subset \Pi_0(0) + W$, which shows that Π_0 is H-u.c. at 0.

(iv) Suppose that Π_0 is not u.c. at 0; then there exist $D_0 \subset Y$ an open set with $\Pi_0(0) \subset D_0$ and a net $(\varepsilon_i)_{i \in I} \subset C$ with $(\varepsilon_i) \to 0$ and $\Pi_0(\varepsilon_i) \not\subset D_0$ for every $i \in I$. Therefore for every $i \in I$ there exists $y_i \in \Pi_0(\varepsilon_i) \setminus D_0$. It follows that for every i there exist $\eta_i \in \Pi_0(0)$ and $k_i \in C$ such that $y_i = \eta_i + \varepsilon_i - k_i$. Since (DP) holds for A, $y_i = \eta_i' + k_i'$ with $\eta_i' \in \Pi_0(0)$ and $k_i' \in C$ for every i. It follows that $\eta_i + \varepsilon_i = \eta_i' + (k_i' + k_i)$ for $i \in I$. Since $\Pi_0(0)$ is compact, passing to subnets if necessary, we may suppose that (η_i) and (η_i') converge to η and η' from $\Pi_0(0)$, respectively. We obtain that $(k_i' + k_i) \to \eta - \eta' \in C$. Since $\eta, \eta' \in \text{Eff}(A;C)$, we obtain that $\eta = \eta'$, and so $(k_i' + k_i) \to 0$. Since C is normal, it follows that $(k_i) \to 0$, and so $(y_i) \to \eta \in \Pi_0(0) \subset D_0$. We get that $y_i \in D_0$ for $i_0 \preceq i$ (for some $i_0 \in I$), a contradiction. Therefore Π_0 is u.c. at 0. $\square$

In the general case we have the following result concerning the relations between η-well-posedness and well-posedness.

Proposition 3.4.5. *Let $f : X \to Y$, and $A \subset X$ a nonempty set. Suppose that $\text{Eff}(f(A);C)$ is compact, $\text{int}\, C \neq \emptyset$, and (P) is η-well-posed. Then (P) is well-posed.*

PROOF. Consider $D \subset X$ an open set such that $\Pi(0) \subset D$. Since Π^η is u.c. at 0 and $\Pi^\eta(0) \subset \Pi(0)$ for every $\eta \in \text{Eff}(f(A);C)$, for each $\eta \in \text{Eff}(f(A);C)$ there exists $V_\eta \in \mathcal{V}_Y$ such that $\Pi^\eta(V_\eta) = A \cap f^{-1}(\eta + V_\eta \cap C - C) \subset D$. Of course, for every $\eta \in \text{Eff}(f(A);C)$ there exists an open neighborhood V_η' of $0 \in Y$ such that $V_\eta' + V_\eta' \subset V_\eta$. Since the family

$$\{\eta + (V'_\eta \cap \operatorname{int} C) - (V'_\eta \cap \operatorname{int} C) \mid \eta \in \operatorname{Eff}(f(A); C)\}$$

represents an open cover of the compact set $\operatorname{Eff}(f(A); C)$, there exist $\eta_1, \ldots, \eta_n$ in the set $\operatorname{Eff}(f(A); C)$ such that

$$\operatorname{Eff}(f(A); C) \subset \bigcup_{i=1}^{n} \left(\eta_i + (V'_{\eta_i} \cap \operatorname{int} C) - (V'_{\eta_i} \cap \operatorname{int} C) \right). \qquad (3.29)$$

Consider $V := \bigcap_{i=1}^{n} V'_{\eta_i} \in \mathcal{V}_Y$. Let $\varepsilon \in V \cap C$ and $x \in \Pi(\varepsilon)$. There exist $\eta \in \operatorname{Eff}(f(A); C)$ and $k \in C$ such that $f(x) = \eta + \varepsilon - k$. From (3.29) we have that $\eta = \eta_i + k_i - k'_i$ for some $k_i, k'_i \in V'_{\eta_i} \cap \operatorname{int} C$. It follows that $\varepsilon + k_i \in V_{\eta_i} \cap C$, whence $f(x) \in \eta_i + V_{\eta_i} - C$, and so $x \in \Pi^{\eta_i}(V_{\eta_i}) \subset D$. Therefore $\Pi(V) \subset D$. The proof is complete. □

The notion of well-posedness for vector optimization problems, as well as that of the containment property, was introduced by Bednarczuk (see [17], [18]) in the framework of topological vector spaces ordered by closed and pointed convex cones with nonempty interior. In this framework, assertions (ii) and (iii) of Theorem 3.4.4 are proved in [17], [18], while assertion (iv) is proved in [18] for $\dim Y < \infty$. Proposition 3.4.5 is stated by Bednarczuk [18], too. Proposition 3.4.3 is proved in [19].

3.5 Continuity Properties

3.5.1 Continuity Properties of Optimal-Value Multifunctions

As in the preceding section we consider X a topological space, Y a topological vector space, $C \subset Y$ a convex cone with $C \neq Y$, and the multifunction $\Gamma : X \rightrightarrows Y$.

With Γ we associate the **optimal-value multifunctions** $\Omega, \overline{\Omega} : X \rightrightarrows Y$ defined by

$$\Omega(x) := \operatorname{Eff}(\Gamma(x); C), \qquad \overline{\Omega}(x) := \operatorname{Eff}(\overline{\Gamma}(x); C).$$

When $\operatorname{int} C \neq \emptyset$ we also consider the **weak optimal-value multifunctions** $\Omega_w, \overline{\Omega}_w$ obtained from Ω and $\overline{\Omega}$, respectively, by replacing C with $\{0\} \cup \operatorname{int} C$.

We say that Γ has the domination property (around $x_0 \in X$) if $\Gamma(x)$ has the domination property at every $x \in X$ (at every $x \in U_0$, where U_0 is some neighborhood of x_0). Note that when Γ has the domination property, then $\Gamma_C = \Omega_C$, and so Ω is C-(nearly) convex if X is a topological vector space and Γ is C-(nearly) convex.

An important special case occurs when $\mathcal{U}$ is another topological (vector) space and Γ is replaced by $f\Lambda$, where $f : X \times \mathcal{U} \to Y$ and $\Lambda : \mathcal{U} \rightrightarrows X$, i.e., $\Gamma(u) := f(\Lambda(u) \times \{u\})$ and $\Omega(u) := \operatorname{Eff}(f(\Lambda(u) \times \{u\}); C)$ (and similarly for $\Omega_w(u)$).

In the first two results we give continuity properties of Ω under convexity assumptions for Γ.

Proposition 3.5.1. *Let X be a topological vector space and let Γ be C-nearly convex and have the domination property. If Γ is C-l.c. at $(x_0, y_0) \in \operatorname{gr}\Gamma$ and $\Gamma(x_0) \subset B + C$ for some bounded set $B \subset Y$, then Ω is C-H-continuous on $\operatorname{int}(\operatorname{dom}\Gamma)$.*

PROOF. Since Γ has the domination property, we have that $\Gamma_C = \Omega_C$. Because Γ is C-l.c. at $(x_0, y_0) \in \operatorname{gr}\Gamma \subset \operatorname{gr}\Gamma_C = \operatorname{gr}\Omega_C$, Ω_C is C-l.c. at (x_0, y_0). Of course, $\Omega_C(x_0) \subset B + C$. Applying Theorem 2.6.4 we obtain that Ω_C is C-H-continuous on $\operatorname{dom}\Omega_C$, whence Ω is C-H-continuous on $\operatorname{dom}\Gamma = \operatorname{dom}\Omega$. $\square$

Proposition 3.5.2. *Let X be a topological vector space and let Γ be C-nearly convex and have the domination property. Suppose that there exist $x_0 \in \operatorname{int}(\operatorname{dom}\Gamma)$, $B \subset Y$ a bounded set, and $U \in \mathcal{V}_X$ such that $\Gamma(x_0) \subset B + C$ and $\Gamma(x_0 + u) \cap (B - C) \neq \emptyset$ for every $u \in U$. Then Ω is C-H-continuous on $\operatorname{int}(\operatorname{dom}\Gamma)$.*

PROOF. Since Γ has the domination property, it follows that $\Omega(x_0) \subset B + C$ and $\Omega(x_0 + u) \cap (B - C) \neq \emptyset$ for every $u \in U$. Since Ω is C-nearly convex, we obtain that Ω is C-H-continuous on $\operatorname{int}(\operatorname{dom}\Gamma) = \operatorname{int}(\operatorname{dom}\Omega)$, by applying Theorem 2.6.6. $\square$

In the sequel we study the continuity properties of the optimal multi-function Ω without convexity assumptions. We begin with a result concerning the lower semicontinuity of the optimal multifunction.

Proposition 3.5.3. *Suppose that Γ is C-l.s.c. and (DP) holds for $\Gamma(x)$ for every $x \in X$. Then Ω is C-l.s.c., too.*

PROOF. Let $y \in Y$ and $(x_i)_{i \in I} \subset \operatorname{lev}_\Omega(y)$, $(x_i)_{i \in I} \to x \in X$. Then $(x_i)_{i \in I} \subset \operatorname{lev}_\Gamma(y)$. Since Γ is C-lsc, $x \in \operatorname{lev}_\Gamma(y)$; therefore $\Gamma(x) \cap (y - C) \neq \emptyset$. Since $\Gamma(x) \subset \Omega(x) + C$, it follows that $\Omega(x) \cap (y - C) \neq \emptyset$, whence $x \in \operatorname{lev}_\Omega(y)$. $\square$

We say that the set $A \subset Y$ is C-**complete** if every C-decreasing net of A is C-lower bounded by an element of A.

Corollary 3.5.4. *Let $G : X \rightrightarrows Y$ be C-l.s.c. and $\Lambda : \mathcal{U} \rightrightarrows X$ be compact at every $u \in \mathcal{U}$. If C is closed and $G(x) + C$ is C-complete for every $x \in X$, then $\Gamma = G\Lambda$ has (DP) at every $u \in \mathcal{U}$ and Ω is C-l.s.c.*

PROOF. From Proposition 2.5.28 (v) we have that Γ is C-l.s.c.; in order to apply the preceding result we must show that $\Gamma(u)$ has (DP) for every $u \in \mathcal{U}$. In order to apply Proposition 3.2.14 (iv), let $(y_i)_{i \in I} \subset \Gamma(u)$ be a C-decreasing net. For every $i \in I$ there exists $x_i \in \Lambda(u)$ such that $y_i \in G(x_i)$. Since $\Lambda(u)$ is compact, there exists the subnet $(x_{\varphi(j)})_{j \in J}$ converging to $x \in \Lambda(u)$. Let $i \in I$; there exists $j_i \in J$ such that $\varphi(j) \succeq i$ for every $j \succeq j_i$. It follows that

$y_{\varphi(j)} \leq_C y_i$ for all $j \succeq j_i$. Hence $(x_{\varphi(j)})_{j \succeq j_i} \subset \mathrm{lev}_G(y_i)$. Since G is C-l.s.c., $x \in \mathrm{lev}_G(y_i)$; i.e., $(y_i)_{i \in I} \subset G(x) + C$. By hypothesis, there exist $y_0 \in G(x)$ and $k_0 \in C$ such that $y_0 + k_0 \leq_C y_i$ for every $i \in I$. Therefore $y_0 \in \Gamma(u)$, and it is a minorant for $(y_i)_{i \in I}$. By the above-mentioned result we have that $\Gamma(u)$ has (DP). $\qquad\square$

We continue with a closedness result for the weak optimal multifunction.

Theorem 3.5.5. *Suppose that* $\mathrm{int}\, C \neq \emptyset$ *and* Γ *is* C-*l.c. at* x_0. *If* Γ *or* $\overline{\Gamma}$ *is closed at* x_0, *then* Ω_w *or* $\overline{\Omega}_w$ *is closed at* x_0, *respectively.*

PROOF. Suppose first that Γ is closed at x_0. Let $y \notin \Omega_w(x_0)$. If $y \notin \Gamma(x_0)$, since Γ is closed at x_0, there exist $U \in \mathcal{V}_X(x_0)$ and $V \in \mathcal{V}_Y(y)$ such that $\Gamma(U) \cap V = \emptyset$, whence $\Omega_w(U) \cap V = \emptyset$.

Suppose now that $y \in \Gamma(x_0)$. Then there exists $z \in \Gamma(x_0) \cap (y - \mathrm{int}\, C)$. Consider $W \in \mathcal{V}_Y$ such that $W + W \subset z - y + \mathrm{int}\, C$. Since Γ is C-l.c. at (x_0, z), there exists $U \in \mathcal{V}_X(x_0)$ such that $\Gamma(x) \cap (z + W - C) \neq \emptyset$; i.e., $z \in \Gamma'(x) + W + C$, for every $x \in U$. Let $z' \in y + W$. It follows that

$$z' \in y + W - z + \Gamma(x) + W + C \subset \Gamma(x) + \mathrm{int}\, C + C \subset \Gamma(x) + \mathrm{int}\, C,$$

which means that $z' \notin \Omega_w(x)$. Therefore $\Omega_w(U) \cap W = \emptyset$. Hence Ω_w is closed at x.

If $\overline{\Gamma}$ is closed at x_0, since in our conditions $\overline{\Gamma}$ is C-l.c. at x_0, by applying the first part we get that $\overline{\Omega}$ is closed at x_0. $\qquad\square$

Corollary 3.5.6. *Suppose that* $\mathrm{int}\, C \neq \emptyset$, $\Lambda : \mathcal{U} \rightrightarrows X$ *is l.c. at* u_0, *and* $f : X \times \mathcal{U} \to Y$ *is* C-*u.c. on* $\Lambda(u_0) \times \{u_0\}$. *If* $f\Lambda$ *is closed at* u_0, *then* Ω_w *is closed at* x_0.

PROOF. Using Proposition 2.5.28 (iii) we have that $f\Lambda$ is C-l.c. at x_0. The conclusion follows from the preceding theorem. $\qquad\square$

Using the preceding theorem we get also an upper continuity result for Ω_w.

Corollary 3.5.7. *Suppose that* $\mathrm{int}\, C \neq \emptyset$, Γ *is u.c., and* C-*l.c. at* x_0 *and* $\Gamma(x_0)$ *or* $\mathrm{cl}\,\Gamma(x_0)$ *is compact. Then* Ω_w *or* $\overline{\Omega}_w$ *is u.c. at* x_0, *respectively.*

PROOF. Suppose that $\Gamma(x_0)$ is compact. Using Proposition 2.5.10 (iii) we have that Γ is closed at x_0. Then using Theorem 3.5.5 we have that Ω_w is closed at x_0. Since $\Omega_w(x) = \Omega_w(x) \cap \Gamma(x)$ for every $x \in X$, using Proposition 2.5.11 (i), we obtain that Ω_w is u.c. at x_0.

Suppose now that $\mathrm{cl}\,\Gamma(x_0)$ is compact. Using Proposition 2.5.20 (iii) and (iv) we obtain that $\overline{\Gamma}$ is C-l.c. and u.c. at x_0. The conclusion follows from the first part. $\qquad\square$

The following examples show that the hypotheses on Γ in Theorem 3.5.5 and Corollary 3.5.7 are essential.

Example 3.5.8. ([356]) Let $X =\,]-\infty, 0[$, $Y = \mathbb{R}^2$, $C = \mathbb{R}_+^2$, and let the multi-function Γ be defined by

$$\Gamma(x) = \begin{cases} \{(y_1, y_2) \in \mathbb{R}^2 \mid y_2 \geq xy_1, \, y_1 \leq 1, \, y_2 \leq 1\} & \text{if } x \neq -1, \\ \{(y_1, y_2) \in \mathbb{R}^2 \mid y_2 \geq -y_1 + 1, \, y_1 \leq 1, \, y_2 \leq 1\} & \text{if } x = -1. \end{cases}$$

Then Γ is l.c. (in particular, C-l.c.), but neither closed nor u.c., at $x_0 = -1$; moreover, $\Gamma(x_0)$ is compact. We have that

$$\Omega(x) = \Omega_w(x) = \begin{cases} \{(y_1, y_2) \in \mathbb{R}^2 \mid y_2 = xy_1, \, x^{-1} \leq y_1 \leq 1\} & \text{if } x \neq -1, \\ \{(y_1, y_2) \in \mathbb{R}^2 \mid y_2 = -y_1 + 1, \, 0 \leq y_1 \leq 1\} & \text{if } x = -1; \end{cases}$$

Ω is not closed at x_0.

Example 3.5.9. ([356]) Let $X =\,]-\infty, 0[$, $Y = \mathbb{R}^2$, $C = \mathbb{R}_+^2$, and let the multi-function Γ be defined by

$$\Gamma(x) = \begin{cases} \{(y_1, y_2) \in \mathbb{R}^2 \mid y_2 \geq xy_1, \, y_1 \leq 1, \, y_2 \leq 1\} & \text{if } x \neq -1, \\ \{(y_1, y_2) \in \mathbb{R}^2 \mid y_2 \geq -y_1 - 1, \, y_1 \leq 1, \, y_2 \leq 1\} & \text{if } x = -1. \end{cases}$$

Then Γ is is closed and u.c., but not C-l.c., at $x_0 = 1$; moreover, $\Gamma(x_0)$ is compact. We have that

$$\Omega(x) = \Omega_w(x) = \begin{cases} \{(y_1, y_2) \in \mathbb{R}^2 \mid y_2 = xy_1, \, x^{-1} \leq y_1 \leq 1\} & \text{if } x \neq -1, \\ \{(y_1, y_2) \in \mathbb{R}^2 \mid y_2 = -y_1 - 1, \, -2 \leq y_1 \leq 1\} & \text{if } x = -1; \end{cases}$$

Ω is not closed at x_0.

Proposition 3.5.10. *Suppose that Γ is C-u.c. at x_0 and (DP) holds for $\Gamma(x_0)$. Then Ω is C-u.c. at x_0.*

PROOF. Let $D \subset Y$ be an open set such that $\Omega(x_0) \subset D$. Then $\Gamma(x_0) \subset \Omega(x_0) + C \subset D + C$. Since Γ is C-u.c. at x_0 and $D + C$ is open, there exists $U \in \mathcal{V}_X(x_0)$ such that $\Gamma(U) \subset D + C + C$, whence $\Omega(U) \subset D + C$. Therefore Ω is C-u.c. at x_0.   □

In Example 3.5.8 (DP) holds for $\Gamma(x_0)$, but Ω is not C-u.c. at x_0, showing that the C-upper continuity of Γ at x_0 is essential.

Corollary 3.5.11. *Let $\Lambda : \mathcal{U} \rightrightarrows X$ be u.c. at $u_0 \in \mathcal{U}$.*

(i) *If $g : X \to Y$ is C-l.c. on $\Lambda(u_0)$ and (DP) holds for $g\Lambda(u_0)$, then Ω is C-u.c. at u_0.*

(ii) *Assume that $f : X \times \mathcal{U} \to Y$ is C-l.c. on $\Lambda(u_0) \times \{u_0\}$, (DP) holds for $f\Lambda(u_0)$, and either the set $\Lambda(u_0)$ is compact or the multifunction $\widetilde{\Lambda} : \mathcal{U} \rightrightarrows X \times \mathcal{U}$, $\widetilde{\Lambda}(u) = \Lambda(u) \times \{u\}$, is u.c. at u_0; then Ω is C-u.c. at u_0.*

PROOF. (i) Using Proposition 2.5.28 (i), we have that $g\Lambda$ is C-u.c. at u_0. The conclusion follows from the preceding proposition.

(ii) If $\Lambda(u_0)$ is compact, then as seen in the proof of Proposition 2.5.28 (ii), $\tilde{\Lambda}$ is u.c. at u_0. So, in both cases the conclusion follows from (i) applied with g and Λ replaced by f and $\tilde{\Lambda}$, respectively. □

In order to obtain the P-H-upper continuity of Ω for a convex cone $P \subset Y$ we introduce the multifunction

$$\Gamma^P : X \rightrightarrows Y, \qquad \Gamma^P(x) := \begin{cases} \Gamma(x_0) & \text{if } x = x_0, \\ \Gamma(x) \setminus (\Omega(x_0) + P) & \text{if } x \neq x_0. \end{cases}$$

For $P = \{0\}$ the next result establishes another sufficient condition for Ω to be H-u.c. at u_0.

Theorem 3.5.12. *Suppose that* $\operatorname{int} C \neq \emptyset$ *and (CP) holds for* $\Gamma(x_0)$. *If* Γ *is uniformly C-l.c. at x_0 on $\Omega(x_0)$ and Γ^P is H-u.c. at x_0, then Ω is H-P-u.c. at x_0.*

PROOF. When $x_0 \notin \operatorname{dom} \Gamma$, the conclusion is obvious. Let $x_0 \in \operatorname{dom} \Gamma$ and suppose that Ω is not H-P-u.c. at x_0. Then there exists $W \in \mathcal{V}_Y$ such that

$$\forall U \in \mathcal{V}_X(x_0), \ \exists x_U \in U, \ \exists y_U \in \Omega(x_U) \setminus (\Omega(x_0) + W + P). \tag{3.30}$$

Consider $W_1 \in \mathcal{V}_Y$ such that $W_1 + W_1 \subset W$. Since (CP) holds for $\Gamma(x_0)$, there exists $V \in \mathcal{V}_Y$ open such that

$$\forall y \in \Gamma(x_0) \setminus (\Omega(x_0) + W_1), \ \exists e_y \in \Omega(x_0), \ k_y \in C : \ y = e_y + k_y, \ k_y + V \subset C.$$

Let $V_1 \in \mathcal{V}_Y$ be such that $V_1 + V_1 \subset V$. Since Γ^P is H-u.c. at x_0, there exists $U_1 \in \mathcal{V}_X(x_0)$ such that

$$\Gamma(x) \setminus (\Omega(x_0) + P) \subset \Gamma(x_0) + V_1 \cap W_1 \quad \forall x \in U_1. \tag{3.31}$$

Since Γ is uniformly C-l.c. at x_0 on $\Omega(x_0)$, there exists $U_2 \in \mathcal{V}_X(x_0)$ such that

$$\Omega(x_0) \subset \Gamma(x) + V_1 + C \quad \forall x \in U_2. \tag{3.32}$$

From (3.30) we obtain $x' \in U_1 \cap U_2$ and $y' \in \Omega(x') \setminus (\Omega(x_0) + W + P) \subset \Gamma(x') \setminus (\Omega(x_0) + P)$. By (3.31), $y' = y_0 + w'$ for some $y_0 \in \Gamma(x_0)$ and $w' \in V_1 \cap W_1$. If $y_0 \in \Omega(x_0) + W_1$, then $y' \in \Omega(x_0) + W + P$, a contradiction. Therefore $y_0 \notin \Omega(x_0) + W_1$, whence $y_0 = e_0 + k_0$ with $e_0 \in \Omega(x_0)$ and $k_0 + V \subset \operatorname{int} C$. Since $x' \in U_2$, by (3.32), there exist $y'' \in \Gamma(x')$, $v' \in V_1$, and $k' \in C$ such that $e_0 = y'' + v' + k'$. Hence $y' = y'' + v' + k' + k_0 + w' \in \Gamma(x') + k_0 + V + C \subset \Gamma(x') + \operatorname{int} C$, contradicting the fact that $y' \in \Omega(x')$. The proof is complete. □

Note that for $P = \{0\}$, Γ^P is H-u.c. at x_0 if and only if Γ is H-u.c. at x_0.

3.5 Continuity properties   127

Corollary 3.5.13. *Suppose that Γ is H-u.c. at $x_0 \in X$, and C is proper, closed, with nonempty interior. If $\Gamma(x_0)$ is compact, $\mathrm{Eff}(\Gamma(x_0); C) = w\,\mathrm{Eff}(\Gamma(x_0); C)$, and Γ is C-l.c. at (x_0, y) for all $y \in \Omega(x_0)$, then Ω is u.c. at x_0.*

PROOF. We may suppose that $x_0 \in \mathrm{dom}\,\Gamma$. Using Proposition 3.4.3, we have that (CP) holds for $\Gamma(x_0)$, while from Remark 2.5.19 we have that Γ is uniformly C-l.c. at x_0 on $\Omega(x_0)$. Then, from the preceding theorem for $P = \{0\}$, we obtain that Ω is H-u.c. at x_0. Since $\mathrm{Eff}(A; C) = w\,\mathrm{Eff}(A; C)$, $\Omega(x_0)$ is closed, and so compact. Hence Ω is actually u.c. at x_0. □

In Examples 3.5.8 and 3.5.9 all the conditions in Theorem 3.5.12 (for $P = \{0\}$) and Corollary 3.5.13 are satisfied, except the fact that Γ is H-u.c. at x_0 in Example 3.5.8 and Γ is uniformly C-l.c. at x_0 on $\Omega(x_0)$ in Example 3.5.9, showing that these two conditions are essential in those results. The next example shows that the condition (CP) for $\Gamma(x_0)$ is also essential in Theorem 3.5.12.

Example 3.5.14. Let $X = [0, 1]$, $Y = \mathbb{R}^2$, $C = \mathbb{R}_+^2$, and let Γ be defined by

$$\Gamma(x) = \begin{cases} \{(y_1, y_2) \mid y_1 \geq 0,\ y_2 = \min(y_1^2, e^{-y_1+1})\} & \text{if } x = 0, \\ \{(y_1, y_2) \mid x \leq y_1 \leq 2 - 2\ln x,\ y_2 = \min(y_1^2, e^{-y_1+1})\} & \text{if } x \neq 0. \end{cases}$$

Then Γ is H-u.c. and l.c. at 0, whence Γ^P is H-u.c. at x_0 for every convex cone $P \subset Y$, $\Gamma(0)$ has the domination property, but (CP) does not hold for $\Gamma(0)$. Moreover,

$$\Omega(x) = \begin{cases} \{(0,0)\} & \text{if } x = 0, \\ \{(x, x^2)\} \cup \{(y, e^{1-y}) \mid 1 - 2\ln x < y \leq 2 - 2\ln x\} & \text{if } x \neq 0. \end{cases}$$

So Γ is uniformly C-l.c. at x_0 on $\Omega(x_0)$, but Ω is not H-P-u.c. at x_0 for $P \subset -C$.

Note that asking Γ to be C-l.c. at x_0 makes the preceding corollary a particular case of Corollary 3.5.7 because $\Omega(x_0) = \Omega_w(x_0)$ and $\Omega(x) \subset \Omega_w(x)$ for $x \in X$.

Corollary 3.5.15. *Suppose that C is closed, proper, with nonempty interior, $\Lambda : \mathcal{U} \rightrightarrows X$ is u.c. and l.c. at u_0, $\Lambda(u_0)$ is compact, and $f : X \times \mathcal{U} \to Y$ is continuous on $\Lambda(u_0) \times \{u_0\}$. If $w\,\mathrm{Eff}\,(f\Lambda(u_0); C) = \mathrm{Eff}\,(f\Lambda(u_0); C)$, then Ω is u.c. at u_0.*

PROOF. Since $\Lambda(u_0)$ is compact and f is continuous on $\Lambda(u_0) \times \{u_0\}$, using Proposition 2.5.28 (ii) and (vi), $f\Lambda$ is u.c. and l.c. at u_0; moreover, $f\Lambda(u_0)$ is compact. Using the preceding corollary we obtain the conclusion. □

In the following results we establish sufficient conditions for the lower continuity of the optimal multifunction Ω.

Proposition 3.5.16. *Suppose that Γ is C-l.c. at (x_0, y_0) with $y_0 \in \Omega(x_0)$. If (DP) holds for Γ around x_0, then Ω is C-l.c. at (x_0, y_0).*

PROOF. Take $U_0 \in \mathcal{V}_X(x_0)$ such that $\Gamma(x) \subset \Omega(x) + C$ for $x \in U$. Let $W \in \mathcal{V}_Y$. Since Γ is C-l.c. at (x_0, y_0), there exists $U_1 \in \mathcal{V}_X(x_0)$ such that $\Gamma(x) \cap (W - C) \neq \emptyset$ for every $x \in U_1$. Let $x \in U_0 \cap U_1$. Then there exists $y \in \Gamma(x) \cap (W - C)$; it follows that $y = y' + k$ with $y' \in \Omega(x)$ and $k \in C$. Therefore $y' \in W - C - C = W - C$, whence $\Omega(x) \cap (W - C) \neq \emptyset$. Hence Ω is C-l.c. at (x_0, y_0). □

In Example 3.5.8 (DP) holds for $\Gamma(x)$ for every $x \in X$, but Ω is not C-l.c. at (x_0, y) for every $y \in \Omega(x_0)$; Γ is not C-l.c. at (x_0, y) for every $y \in \Omega(x_0)$. In the next example (DP) does not hold for Γ around x_0.

Example 3.5.17. ([356]) Let $X = [0, \infty[$, $Y = \mathbb{R}^2$, $C = \mathbb{R}_+^2$, and let Γ be defined by

$$\Gamma(x) = \begin{cases} \{(y_1, y_2) \in \mathbb{R}^2 \mid y_1^2 + y_2^2 < x^2\} & \text{if } x \neq 1, \\ \{(y_1, y_2) \in \mathbb{R}^2 \mid y_1^2 + y_2^2 \leq 1\} & \text{if } x = 1. \end{cases}$$

Then Γ is l.c. (and so C-l.c.) at every $x \in X$ and

$$\Omega(x) = \begin{cases} \emptyset & \text{if } x \neq 1, \\ \{(y_1, y_2) \in \mathbb{R}^2 \mid y_1^2 + y_2^2 = 1, \, y_1 \leq 0, \, y_2 \leq 0\} & \text{if } x = 1. \end{cases}$$

So (DP) does not hold for Γ around $x_0 = 1$, and Ω is not C-l.c. at (x_0, y) for every $y \in \Omega(x_0)$.

Corollary 3.5.18. *Suppose that $\Lambda : \mathcal{U} \rightrightarrows X$ is l.c. at $u_0 \in \mathcal{U}$, and $f : X \times \mathcal{U} \to Y$ is C-u.c. on $\Lambda(u_0) \times \{u_0\}$. If (DP) holds for $f\Lambda$ around u_0, then Ω is C-l.c. at u_0.*

PROOF. Using Proposition 2.5.28 (iii), $f\Lambda$ is C-l.c. at u_0. The conclusion follows from the preceding proposition. □

In order the obtain the next result we need a uniform (CP) condition. So, we say that (CP) **holds for Γ uniformly around** $x_0 \in X$ if there exists $U \in \mathcal{V}_X(x_0)$ such that

$$\forall W \in \mathcal{V}_Y, \ \exists V \in \mathcal{V}_Y, \ \forall x \in U \ : \ [\Gamma(x) \setminus (\Omega(x) + W)] + V \subset \Omega(x) + C.$$

When $\operatorname{int} C \neq \emptyset$, one can write a similar condition to (3.28).

Theorem 3.5.19. *Suppose that $\operatorname{int} C \neq \emptyset$, (CP) holds for Γ uniformly around $x_0 \in X$, and Γ is H-C-u.c. at x_0. If Γ is l.c. at (x_0, y_0), where $y_0 \in \Omega(x_0)$, then Ω is l.c. at (x_0, y_0). Moreover, if Γ is uniformly $\{0\}$-l.c. at x_0 on $\Omega(x_0)$, then Ω is H-l.c. at x_0.*

PROOF. Let $U_0 \in \mathcal{V}_X(x_0)$ be such that (CP) holds uniformly for $\Gamma(x)$ with $x \in U$. Let us fix $W \in \mathcal{V}_Y$ and consider $W_1 \in \mathcal{V}_Y$ such that $W_1 + W_1 \subset W$. There exists $V \in \mathcal{V}_Y$ open such that

$$\forall x \in U_0, \ \forall y \in \Gamma(x) \setminus (\Omega(x) + W_1), \ \exists e \in \Omega(x), \ k \in C : y = e + k, \ k + V \subset C. \tag{3.33}$$

There exists $V_1 \in \mathcal{V}_Y$ such that $V_1 + V_1 \subset V$. Since Γ is l.c. at (x_0, y_0), there exists $U_1 \in \mathcal{V}_X(x_0)$ such that

$$y_0 \in \Gamma(x) + V_1 \cap W_1 \quad \forall x \in U_1. \tag{3.34}$$

Now, since Γ is H-C-u.c. at x_0, there exists $U_2 \in \mathcal{V}_X(x_0)$ such that

$$\Gamma(x) \subset \Gamma(x_0) + V_1 + C \quad \forall x \in U_2. \tag{3.35}$$

Let $U := U_0 \cap U_1 \cap U_2$ and take $x \in U$. From (3.34) we have that $y_0 = y_x + w_x$ with $y_x \in \Gamma(x)$ and $w_x \in V_1 \cap W_1$. Suppose that $y_x \notin \Omega(x) + W_1$. From (3.33) we get $e_x \in \Omega(x)$ and $k_x \in C$ such that $y_x = e_x + k_x$ and $k_x + V \subset \operatorname{int} C$. From (3.35) we have that $e_x = y_x^0 + v_x' + k_x'$ with $y_x^0 \in \Gamma(x_0)$, $v_x' \in V_1$, and $k_x' \in C$. So, $y_0 = y_x^0 + v_x' + k_x' + k_x + w_x$, whence

$$y_0 \in y_x^0 + k_x + V_1 + V_1 + C \subset y_x^0 + \operatorname{int} C + C \subset y_x^0 + \operatorname{int} C,$$

contradicting the fact that $y_0 \in \Omega(x_0)$. Therefore $y_x \in \Omega(x) + W_1$, whence $y_0 \in \Omega(x) + W$. Hence Ω is l.c. at (x_0, y_0).

If Γ is uniformly $\{0\}$-l.c. at x_0 on $\Omega(x_0)$, relation (3.34) holds for every $y_0 \in \Omega(x_0)$, and so $\Omega(x_0) \subset \Omega(x) + W$ for all $x \in U$. □

In Examples 3.5.8, 3.5.9, and 3.5.17 all the hypotheses of the preceding theorem are satisfied at x_0 and $y_0 \in \Omega(x_0)$, except, respectively, the fact that Γ is H-C-u.c. at x_0, Γ is l.c. at (x_0, y_0), and (CP) holds for Γ, uniformly around x_0; Ω is not l.c. at (x_0, y_0).

The assumption that (CP) holds uniformly for Γ around x_0 is sufficiently strong. This condition may be relaxed if $\Gamma(x_0)$ or $\Omega(x_0)$ is compact.

Theorem 3.5.20. *Suppose that C is closed and normal, Γ satisfies (DP) around $x_0 \in X$, Γ is H-C-u.c. at x_0, and either $\Gamma(x_0)$ is closed and $\Omega(x_0)$ is relatively compact (for example, if $\Gamma(x_0)$ is compact) or $\Omega(x_0)$ is compact. Then, if Γ is l.c. at $(x_0, y_0) \in \operatorname{gr} \Omega$, Ω is l.c. at (x_0, y_0), too. Moreover, if Γ is l.c. at (x_0, y) for every $y \in \Omega(x_0)$, then Ω is l.c. at x_0 (even H-l.c. at x_0 if $\Omega(x_0)$ is compact).*

PROOF. We give the proof for X, Y first-countable spaces; in the general case one uses the equivalence of (i) and (ii) of Proposition 2.5.6 and assertion (ii) of Proposition 2.5.14 instead of the equivalence of (i) and (iii), and assertion (iv), respectively.

Let us note first that C is necessarily pointed because C is normal and Y is separated. Let $X \supset (x_n) \to x_0$. Since Γ is l.c. at (x_0, y_0), using Proposition

2.5.6, there exists $Y \supset (y_n) \to y_0$ such that $y_n \in \Gamma(x_n)$ for every $n \geq n_0$. We may suppose that (DP) holds for $\Gamma(x_n)$ for $n \geq n_0$. Hence $y_n = \overline{y}_n + k_n$ with $\overline{y}_n \in \Omega(x_n) \subset \Gamma(x_n)$ and $k_n \in C$ for $n \geq n_0$. Since Γ_C is H-u.c. at x_0, using Proposition 2.5.14, there exist $(y_n^0) \subset \Gamma(x_0)$ and $(k_n^0) \subset C$ such that $w_n := \overline{y}_n - (y_n^0 + k_n^0) \to 0$. Since $\Gamma(x_0)$ has (DP), we have that $y_n^0 = \overline{y}_n^0 + \overline{k}_n^0$ with $\overline{y}_n^0 \in \Omega(x_0)$ and $\overline{k}_n^0 \in C$. So, for $n \geq n_0$,

$$y_n = \overline{y}_n + k_n = y_n^0 + w_n + (k_n^0 + k_n) = \overline{y}_n^0 + w_n + (\overline{k}_n^0 + k_n^0 + k_n),$$

where $(y_n^0) \subset \Gamma(x_0)$, $(\overline{y}_n^0) \subset \Omega(x_0)$, $(\overline{k}_n^0)$, (k_n^0), $(k_n) \subset C$ and $(w_n) \to 0$.

Since $\Omega(x_0)$ is relatively compact, there exists a subsequence $(\overline{y}_{n_p}^0) \to y^0$; in both cases $y^0 \in \Gamma(x_0)$, and so $(\overline{k}_n^0 + k_{n_p}^0 + k_{n_p}) \to y_0 - y^0 \in C$. Since $y_0 \in \Omega(x_0)$ and C is pointed, we have that $y_0 = y^0$, and so $(\overline{k}_n^0 + k_{n_p}^0 + k_{n_p}) \to 0$. Since C is normal, we obtain that $(k_{n_p}) \to 0$, whence $(\overline{y}_{n_p}) \to y_0$. Arguing by contradiction we obtain, in fact, that $(\overline{y}_n) \to y_0$.

Using again Proposition 2.5.6 we obtain that Ω is l.c. at (x_0, y_0). The second part is an immediate consequence of the first one. $\qquad\square$

In Examples 3.5.8, 3.5.9, and 3.5.17 all the hypotheses of the preceding theorem are satisfied at x_0 and $y_0 \in \Omega(x_0)$, except, respectively, the fact that Γ is H-C-u.c. at x_0, Γ is l.c. at (x_0, y_0), and (DP) holds for Γ around x_0; Ω is not l.c. at (x_0, y_0).

In the next result we suppose that Y is a normed vector space.

Theorem 3.5.21. *Suppose that Y is a normed vector space, C is closed and normal, and Γ is l.c. and closed at x_0. If there exists a neighborhood U_0 of x_0 such that $\Gamma(x)$ satisfies (DP) for $x \in U_0^\bullet := U_0 \setminus \{x_0\}$, $\Gamma(U_0^\bullet)$ is C-bounded from below (i.e., $\Gamma(U_0^\bullet) \subset B_0 + C$ for some bounded set $B_0 \subset Y$), and the bounded subsets of $\Omega(U_0^\bullet)$ are relatively compact, then Ω is l.c. at x_0.*

PROOF. Consider $y_0 \in \Omega(x_0)$ and $X \supset (x_i)_{i \in I} \to x_0$. We may suppose that $(x_i)_{i \in I} \subset U_0^\bullet$. Since Γ is l.c. at x_0, by Proposition 2.5.6, there exist a subnet $(x_{\varphi(j)})_{j \in J}$ of (x_i) and a net $(y_j)_{j \in J} \to y_0$ with $y_j \in \Gamma(x_{\varphi(j)})$ for every $j \in J$. Because $\Gamma(x_i)$ satisfies (DP) for $i \in I$, there exist $\overline{y}_j \in \Omega(x_{\varphi(j)})$ and $k_j \in C$ such that $y_j = \overline{y}_j + k_j$ for all $j \in J$. Since $(y_j) \to y_0$, there exists a bounded set $B_1 \subset Y$ such that $(y_j)_{j \succeq j_1} \subset B_1$ for some $j_1 \in J$. Taking $B := B_0 \cup B_1$, we obtain that $\overline{y}_j \in (B + C) \cap (B - C)$ for $j \succeq j_1$. Since C is normal, the set $(B + C) \cap (B - C)$ is bounded, whence $(\overline{y}_j)_{j \succeq j_1} \subset \Omega(U_0^\bullet)$ is bounded. By hypothesis, $(\overline{y}_j)_{j \succeq j_1}$ contains a subnet $(\overline{y}_{\psi(p)})_{p \in P}$ converging to $\overline{y} \in Y$. Since Γ is closed at x_0, it follows that $\overline{y} \in \Gamma(x_0)$. Since $k_j = y_j - \overline{y}_j \in C$ and C is closed, we obtain that $y_0 - \overline{y} \in C$, whence $y_0 = \overline{y}$ (C being pointed and $y_0 \in \Omega(x_0)$). Therefore $(\overline{y}_{\psi(p)})_{p \in P} \to \overline{y}$. Using again Proposition 2.5.6 we have that Ω is l.c. at x_0. $\qquad\square$

Examples 3.5.8, 3.5.9, and 3.5.17 show that the conditions Γ is closed at x_0, Γ is l.c. at x_0, and (DP) holds for Γ around x_0 are essential in the preceding theorem. Also, the hypothesis that $\Gamma(U_0^\bullet)$ is bounded from below for some $U \in \mathcal{V}_X(x_0)$ is important, as the next example shows.

Example 3.5.22. Let $X = [0, \infty[$, $Y = \mathbb{R}^2$, $C = \mathbb{R}_+^2$, and let Γ be defined by

$$\Gamma(x) := \begin{cases} \{(0,0)\} \cup]-\infty, 1] \times \{1\} & \text{if } x = 0, \\ \{(0,0)\} \cup \{(y_1, \frac{x}{x+1}y_1 + \frac{1}{x+1}) \mid y_1 \in [-\frac{1}{x}, 1]\} & \text{if } x > 0. \end{cases}$$

Then Γ is closed and l.c. at $x_0 = 0$, $\Gamma(x)$ has the domination property for every $x \neq 0$, but $\Gamma(U^\bullet)$ is not bounded from below for every $U \in \mathcal{V}_X(x_0)$. Moreover,

$$\Omega(x) = \begin{cases} \{(0,0)\} & \text{if } x = 0, \\ \{(-\frac{1}{x}, 0)\} & \text{if } x > 0. \end{cases}$$

Of course, Ω is not l.c. at x_0.

Another version of the preceding theorem is given in the next result.

Theorem 3.5.23. *Suppose that Y is a normed vector space, C is closed and $C \subset \{y \in Y \mid \|y\| \leq \langle y, \overline{y}^* \rangle\}$ for some $\overline{y}^* \in Y^*$, and Γ is l.c. and closed at $x_0 \in X$. If there exists a neighborhood U_0 of x_0 such that $\Gamma(x)$ is complete for $x \in U_0^\bullet$, $\overline{y}^*\left(\Gamma(U_0^\bullet)\right)$ is bounded from below, and the bounded subsets of $\Omega(U_0^\bullet))$ are relatively compact, then Ω is l.c. at x_0.*

PROOF. Fix $x \in U_0^\bullet$. Since $\Gamma(x)$ is complete and $\overline{y}^*\left(\Gamma(x)\right)$ is bounded from below, using Theorem 3.2.36 of Attouch–Riahi (see also [10, Th. 2.5]; it is stated for Y a Banach space, but it is sufficient for the set to be complete), $\Gamma(x)$ has the domination property. Using the same notation as in the proof of Theorem 3.5.21, we have that $\|y_j - \overline{y}_j\| \leq \langle y_j - \overline{y}_j, \overline{y}^* \rangle \leq \langle y_j, \overline{y}^* \rangle - \sup \overline{y}^*\left(\Gamma(U_0^\bullet)\right)$. It follows that $(\overline{y}_j)_{j \succeq j_1}$ is bounded. The conclusion follows as in the proof of Theorem 3.5.21. □

The same examples, as for Theorem 3.5.21, show that all the conditions on Γ in the preceding theorem are essential.

Corollary 3.5.24. *Suppose that X is a topological vector space, C is closed and normal, $\Lambda : \mathcal{U} \rightrightarrows X$ is H-u.c. and l.c. at u_0, and either $f : X \times \mathcal{U} \to Y$ is equicontinuous on $\Lambda(u_0) \times \{u_0\}$ or f is continuous on $\Lambda(u_0) \times \{u_0\}$ and $\widetilde{\Lambda} : \mathcal{U} \rightrightarrows X \times \mathcal{U}$, $\widetilde{\Lambda}(u) := \Lambda(u) \times \{u\}$, is u.c. at u_0. If (DP) holds for $f\Lambda$ around u_0 and either $\Lambda(u_0)$ is compact or $\Omega(u_0)$ is compact, then Ω is l.c. at u_0.*

PROOF. From Proposition 2.5.28 (iii) we have that $f\Lambda$ is l.c. at u_0, while from (v) if f is equicontinuous on $\Lambda(u_0) \times \{u_0\}$ or (i) if $\widetilde{\Lambda}$ is u.c. at u_0 and f is continuous on $\Lambda(u_0) \times \{u_0\}$ (of the same proposition), we have that $f\Lambda$ is H-u.c., or u.c. at u_0, respectively. Moreover, $f\Lambda(u_0)$ is compact if $\Lambda(u_0)$ is so. The conclusion follows by applying the preceding theorem. □

One can obtain the lower continuity of Ω at (x_0, y_0) without using condition (CP) or compactness conditions if one knows that y_0 is a strong proper efficient point of $\Gamma(x_0)$. One says that y is a **strong proper efficient point** of $A \subset Y$, the class of such points being denoted by $\mathrm{SPEff}(A; C)$, if $y \in \mathrm{Eff}(A; P)$, where P is a proper convex cone such that the pair (C, P) has the property

$$\forall W \in \mathcal{V}_Y, \ \exists V \in \mathcal{V}_Y \ : \ (C \setminus W) + V \subset P. \tag{3.36}$$

Proposition 3.5.25. *Let $P \subset Y$ be a convex cone such that (3.36) holds. Then $(\mathrm{cl}\, C, P)$ satisfies (3.36), $\mathrm{cl}\, C \setminus \{0\} \subset \mathrm{int}\, P$, and*

$$\forall W \in \mathcal{V}_Y, \ \exists V \in \mathcal{V}_Y, \ \forall y \in Y \setminus (W \cup P) \ : \ (y + V) \cap C = \emptyset. \tag{3.37}$$

If C has a bounded base B, then (C, P) satisfies (3.36) if and only if there exists $V \in \mathcal{V}_Y$ such that $B + V \subset P$.

So, if C has a bounded base and Y is a locally convex space, then there exists a proper convex cone $P \subset Y$ such that (3.36) holds. The converse is true if Y is a normed vector space.

PROOF. Let (C, P) satisfy (3.36) and take $W \in \mathcal{V}_Y$. There exists $W_0 \in \mathcal{V}_Y$ such that $W_0 + W_0 \subset W$. By (3.36) there exists $V \in \mathcal{V}_Y$ such that $(C \setminus W_0) + V \subset P$. Consider $V_0 \in \mathcal{V}_Y$ such that $V_0 + V_0 \subset V$. Then $(\mathrm{cl}\, C \setminus W) + V_0 \subset P$. Indeed, let $x = \bar{x} + v$ with $\bar{x} \in \mathrm{cl}\, C \setminus W$ and $v \in V_0$. Since $\bar{x} \in \mathrm{cl}\, C \subset C + (V_0 \cap W_0)$, $\bar{x} = k + w$ for some $k \in C$ and $w \in V_0 \cap W_0$. Then $k \notin W_0$; otherwise, $\bar{x} \in W_0 + W_0 \subset W$. Moreover, $w + v \in V_0 + V_0 \subset V$, whence $x \in (C \setminus W_0) + V \subset P$.

Consider now $y \in \mathrm{cl}\, C \setminus \{0\}$; there exists $W \in \mathcal{V}_Y$ such that $y \notin W$. Since $(\mathrm{cl}\, C, P)$ satisfies (3.36), there exists $V \in \mathcal{V}_Y$ such that $(\mathrm{cl}\, C \setminus W) + V \subset P$, whence $y + V \subset P$. Therefore $y \in \mathrm{int}\, P$.

In order to prove (3.37), let $W \in \mathcal{V}_Y$ and take $W_1 \in \mathcal{V}_Y$ with $W_1 + W_1 \subset W$. There exists $V_1 \in \mathcal{V}_Y$ such that $(C \setminus W_1) + V_1 \subset P$. Consider $V := V_1 \cap W_1$. Suppose that for some $y \in Y \setminus (W \cup P)$ there exists $y' \in (y + V) \cap C$; then $y' \notin W_1$ (otherwise, $y \in W_1 - V \subset W$). Hence $y \in y' + V \subset P$, a contradiction.

Suppose now that B is a (convex) base of C. Suppose first that (3.36) holds. Since $0 \notin \mathrm{cl}\, B$, there exists $W \in \mathcal{V}_Y$ such that $B \cap W = \emptyset$. There exists $V \in \mathcal{V}_Y$ such that $B + V \subset (C \setminus W) + V \subset P$.

Suppose now that $V_0 \in \mathcal{V}_Y$ and $B + V_0 \subset P$. By contradiction, suppose that (3.36) does not hold. Then there exists $W_0 \in \mathcal{V}_Y$ such that $(C \setminus W_0) + V \not\subset P$ for every $V \in \mathcal{V}_Y$. Of course, we may suppose that $B \cap W_0 = \emptyset$. Since B is bounded, there exists $\lambda > 0$ such that $B \subset \lambda W_0$. It follows that

$$\forall V \in \mathcal{V}_Y, \ \exists t_V > 0, \ \exists b_V \in B, \ \exists y_V \in V \ : \ t_V b_V \notin W_0, \ t_V b_V + y_V \notin P.$$

Therefore $b_V + t_V^{-1} y_V \notin P$, whence $t_V^{-1} y_V \notin V_0$ for every $V \in \mathcal{V}_Y$. Since $t_V b_V \notin W_0$ and $B \subset \lambda W_0$ we get $t_V^{-1} < \lambda$ for every $V \in \mathcal{V}_Y$. Since $(y_V)_{V \in \mathcal{V}_Y} \to 0$, it follows that $(t_V^{-1} y_V) \to 0$, contradicting the fact that $t_V^{-1} y_V \notin V_0$ for every $V \in \mathcal{V}_Y$. Therefore (3.36) holds.

Now let Y be a locally convex space and C have a bounded base B. Then there exists a convex $V \in \mathcal{V}_Y$ such that $B \cap V = \emptyset$. Then $0 \notin B + V$. Taking $P = [0, \infty[\cdot (B + V)$, P is a proper convex cone and $B + V \subset P$.

Suppose now that Y is a normed vector space and (3.36) holds for some proper convex cone $P \subset Y$. Let $B_0 = \{y \in C \mid \|y\| = 1\}$. From (3.36) for $W = B(0, 1) = \{y \in Y \mid \|y\| < 1\}$, there exists $r > 0$ such that $B_0 + B(0, r) \subset P$. Then $B = \operatorname{conv} B_0$ is a bounded base for C. Indeed, it is obvious that $C = [0, \infty[\cdot B$. But, $B + B(0, r) = \operatorname{conv}(B_0 + B(0, r)) \subset P$, and so $\operatorname{cl} B + B(0, r/2) \subset P$. If $0 \in \operatorname{cl} B$, then $B(0, r/2) \subset P$, whence $P = Y$, a contradiction. It follows that B is a (bounded) base for C. □

In the definition of a strong proper efficient point we may suppose that P is pointed (otherwise, replace P by $\{0\} \cup \operatorname{int} P$).

Theorem 3.5.26. *Consider $y_0 \in \operatorname{SPEff}(\Gamma(x_0); C)$ and suppose that Γ is C_0-H-u.c. at x_0 and l.c. at (x_0, y_0), where $C_0 = C$ if C is normal and $C_0 = \{0\}$ otherwise. Then Ω is l.c. at (x_0, y_0) if for some $U_0 \in \mathcal{V}_X(x_0)$ one of the following conditions holds:*

(i) *$\Gamma(x)$ has the domination property for every $x \in U_0$;*
(ii) *Y is locally compact, C is closed, and $\Gamma(u)$ is closed for every $x \in U_0$.*

PROOF. Let P be the pointed convex cone with the property (3.37) for which $y_0 \in \operatorname{Eff}(\Gamma(x_0); P)$. It follows that

$$\Gamma(x_0) \subset (y_0 - P)^c \cup \{y_0\}, \tag{3.38}$$

where A^c denotes $Y \setminus A$. Consider $W \in \mathcal{V}_Y$ such that $(W + C_0) \cap (W - C_0) = W$, and W compact if Y is locally compact. Let $W_1 \in \mathcal{V}_Y$ be such that $W_1 + W_1 \subset W$. From (3.37) we get $V \in \mathcal{V}_Y$ such that $(y + V) \cap C = \emptyset$ for every $y \in Y \setminus (W \cup P)$. It follows that

$$[((y_0 - P)^c \setminus (y_0 + W_1)) + V] \cap (y_0 - C) = \emptyset.$$

Taking $V_1 \in \mathcal{V}_Y$ such that $V_1 + V_1 \subset V$, we obtain that

$$[((y_0 - P)^c \setminus (y_0 + W_1)) + V_1] \cap [(y_0 + V_1) - C] = \emptyset. \tag{3.39}$$

From (3.38), taking into account that $(y_0 - P)^c + C = (y_0 - P)^c$, we obtain

$$\Gamma(x_0) + V_1 \cap W_1 + C_0 \subset [((y_0 - P)^c \setminus (y_0 + W_1)) + V_1 \cap W_1] \cup (y_0 + W + C_0).$$

Since Γ is C_0-H-u.c. at x_0, there exists $U_1 \in \mathcal{V}_X(x_0)$ such that

$$\Gamma(x) \subset \Gamma(x_0) + V_1 \cap W_1 + C_0 \quad \forall x \in U_1.$$

Therefore

$$\Gamma(x) \subset [((y_0 - P)^c \setminus (y_0 + W_1)) + V_1 \cap W_1] \cup (y_0 + W + C_0) \quad \forall x \in U_1. \tag{3.40}$$

Since Γ is l.c. at (x_0, y_0), there exists $U_2 \in \mathcal{V}_X(x_0)$ such that $(y_0 + V_1 \cap W_1) \cap \Gamma(x) \neq \emptyset$ for every $x \in U_2$. For $x \in U := U_0 \cap U_1 \cap U_2$ let $y_x \in (y_0 + V_1 \cap W_1) \cap \Gamma(x)$. Then, by applying (3.39), we obtain that

$$(y_x - C) \cap [((y_0 - P)^c \setminus (y_0 + W_1)) + V_1 \cap W_1] = \emptyset.$$

Thus, from (3.40), we obtain that

$$(y_x - C) \cap \Gamma(x) \subset (y_0 + W + C_0) \cap (y_0 + V_1 \cap W_1 - C) \subset y_0 + W \quad \forall x \in U.$$

For $x \in U$, if (i) holds, then there exists $\bar{y}_x \in \Omega(x) \cap (y_x - C)$, while if (ii) holds, $(y_x - C) \cap \Gamma(x)$ is a compact set (as a closed subset of a compact set), and we obtain again $\bar{y}_x \in \Omega(x) \cap (y_x - C)$; in both cases we have that $\Omega(x) \cap (y_0 + W) \neq \emptyset$ for all $x \in U$. The proof is complete.   □

As a consequence of the preceding theorem we get the following result.

Corollary 3.5.27. *Let $Y = \mathbb{R}^m$ and let C be closed and pointed. Suppose that y_0 is an element of $\mathrm{SPEff}(\Gamma(x_0); C)$ and Γ is closed-valued, H-C-u.c. at x_0, and l.c. at (x_0, y_0). Then Ω is l.c. at (x_0, y_0).*

PROOF. Note that C is normal in this case. Applying Theorem 3.5.26 (ii) we get the conclusion.   □

Examples 3.5.8, 3.5.9 (with $y_0 \in \Omega(x_0)$), and 3.5.17 (with $y_0 = (-2^{-1/2}, -2^{-1/2})$) show that the hypotheses Γ is H-C-u.c. at x_0, Γ is l.c. at (x_0, y_0), and Γ is closed-valued, respectively, are essential for Theorem 3.5.26 and Corollary 3.5.27.

Corollary 3.5.28. *Suppose that $\Lambda : \mathcal{U} \rightrightarrows X$ is l.c. and H-u.c. at u_0, $f : X \times \mathcal{U} \to Y$ is equicontinuous on $\Lambda(u_0) \times \{u_0\}$, and $f\Lambda$ satisfies (DP) around u_0. Then Ω is l.c. at (u_0, y_0) for all $y_0 \in \mathrm{SPEff}(f\Lambda(u_0); C)$.*

PROOF. Applying again Proposition 2.5.28 (iii) and (vii), $f\Lambda$ is l.c. and H-u.c. at u_0. The conclusion follows by applying the preceding theorem.   □

Note that Sterna-Karwat [333] established Proposition 3.5.1 for Γ with the domination property, C-mid-convex, bounded-valued, and H-C-l.c. at $x_0 \in \mathrm{dom}\,\Gamma$, and Proposition 3.5.2 for Γ with the domination property, C-mid-convex, bounded-valued, and C-upper bounded, i.e., $\mathrm{Im}\,\Gamma \subset B - C$ for some bounded set $B \subset Y$.

Corollary 3.5.4 was established by Ferro [114, Th. 4.1] under the following additional assumptions: $\mathcal{U}$ is a locally convex space ordered by closed convex cone D with nonempty interior, G is a function, $\Lambda(u_1) \subset \Lambda(u_2)$ if $u_2 - u_1 \in \mathrm{int}\,D$.

Theorem 3.5.5, Propositions 3.5.10, 3.5.16, and Corollaries 3.5.11, 3.5.7, 3.5.18 were obtained by Penot and Sterna-Karwat in [291] and [292] even for more general relations then $\leq_C$.

Theorem 3.5.12 for $P = \{0\}$ is stated by Bednarczuk in [17], while Theorem 3.5.12 for $P = -C$ and Corollary 3.5.13 are stated in [19]. Theorem 3.5.19 with the stronger conditions that Γ is H-u.c. and l.c. at x_0, Theorem 3.5.20 under somewhat different conditions (for Y first countable, without asking that C be normal, but we doubt that this is true), Theorem 3.5.26 (except the case when C is normal) and its corollary are obtained in [19]. Also, Corollaries 3.5.15, 3.5.24, and 3.5.28 are from [19]. Theorem 3.5.23 was obtained by Attouch and Riahi [10] for $X = \mathbb{N} \cup \{\infty\}$, $x_0 = \infty$, and Y a Banach space.

3.5.2 Continuity Properties for the Optimal Multifunction in the Case of Moving Cones

As in the preceding section, X is a topological space, Y is a topological vector space, and $\Gamma : X \rightrightarrows Y$. In this section instead of a single convex cone $C \subset Y$ we consider a multifunction $\mathcal{C} : X \rightrightarrows Y$ whose values are pointed convex cones. We associate also the multifunction $\mathcal{C}^c : X \rightrightarrows Y$ defined by $\mathcal{C}^c(x) := Y \setminus (\mathcal{C}(x) \setminus \{0\}) = (Y \setminus \mathcal{C}(x)) \cup \{0\}$. In this section $\Omega, \Omega_w : X \rightrightarrows Y$, $\Omega(x) := \mathrm{Eff}(\Gamma(x); \mathcal{C}(x))$, and $\Omega_w(x) := w \, \mathrm{Eff}(\Gamma(x); \mathcal{C}(x))$.

Theorem 3.5.29. *Suppose that*

$$\Gamma(x_0) \subset \liminf_{x \to x_0} (\Gamma(x) + \mathcal{C}(x)) \quad \text{and} \quad \limsup_{x \to x_0} \mathcal{C}^c(x) \subset \mathcal{C}^c(x_0). \qquad (3.41)$$

(i) *If* $\limsup_{x \to x_0} \Gamma(x) \subset \Gamma(x_0) + \mathcal{C}(x_0)$, *then* $\limsup_{x \to x_0} \Omega(x) \subset \Omega(x_0)$.
(ii) *If* Γ *is closed at* x_0 *and* $\mathcal{C}(x_0) \setminus \{0\}$ *is open, then* Ω *is closed at* x_0.
(iii) *If* Γ *is compact at* x_0, *then* Ω *is u.c. at* x_0.

PROOF. (i) Let $y \in \limsup_{x \to x_0} \Omega(x)$. By Proposition 2.5.3 (i), there exists a net $((x_i, y_i))_{i \in I}$ in $\mathrm{gr} \, \Omega$ converging to (x_0, y) with $x_i \neq x_0$ for $i \in I$. Since $\mathrm{gr} \, \Omega \subset \mathrm{gr} \, \Gamma$, from the hypothesis, we obtain that $y \in \Gamma(x_0) + \mathcal{C}(x_0)$. Suppose that $y \notin \Omega(x_0) = \mathrm{Eff} \, (\Gamma(x_0) + \mathcal{C}(x_0); \mathcal{C}(x_0))$; then there exists $k \in \mathcal{C}(x_0) \setminus \{0\}$ such that $y' := y - k \in \Gamma(x_0)$. Since $\Gamma(x_0) \subset \liminf_{x \to x_0} (\Gamma(x) + \mathcal{C}(x))$, using again Proposition 2.5.3, there exist a subnet $(x_{\varphi(j)})_{j \in J}$ of (x_i) and the nets (y'_j), (k_j) with $y'_j \in \Gamma(x_{\varphi(j)})$, $k_j \in \mathcal{C}(x_{\varphi(j)})$ for $j \in J$ and $(y'_j + k_j) \to y'$. Since $(y_{\varphi(j)} - y'_j - k_j) \to y - y' \neq 0$, we may suppose that $y_{\varphi(j)} - y'_j - k_j \neq 0$ for all $j \in J$. Taking into account that $y_{\varphi(j)} \in \Omega(x_{\varphi(j)})$, we have that $y_{\varphi(j)} - y'_j - k_j \in \mathcal{C}^c(x_{\varphi(j)})$ for $j \in J$. Since $\limsup_{x \to x_0} \mathcal{C}^c(x) \subset \mathcal{C}^c(x_0)$, we obtain that $k \in \mathcal{C}^c(x_0)$, a contradiction. Therefore $y \in \Omega(x_0)$.

(ii) If Γ is closed at x_0, then $\Gamma(x_0)$ is closed and $\limsup_{x \to x_0} \Gamma(x) \subset \Gamma(x_0)$. From (i) we obtain that $\limsup_{x \to x_0} \Omega(x) \subset \Omega(x_0)$. Since $\mathcal{C}(x_0) \setminus \{0\}$ is open, $\Omega(x_0)$ is closed. Using Proposition 2.5.10 we obtain that Ω is closed at x_0.

(iii) Suppose that Ω is not u.c. at x_0. Then there exist an open set $D_0 \subset Y$ such that $\Omega(x_0) \subset D_0$ and a net $((x_i, y_i))_{i \in I} \subset \mathrm{gr} \, \Omega \, (\subset \mathrm{gr} \, \Gamma)$ with $(x_i) \to x_0$ and $y_i \notin D_0$ for every $i \in I$. Since Γ is compact at x_0, there exists a subnet $(y_{\varphi(j)})_{j \in J} \to y \in \Gamma(x_0)$. Of course, $y \notin D_0$, and so $y \notin \Omega(x_0)$. Hence there exists $k \in \mathcal{C}(x_0) \setminus \{0\}$ such that $y' = y - k \in \Gamma(x_0)$. Proceeding as in (i) we obtain a contradiction. Therefore Ω is u.c. at x_0. $\qquad \square$

In the next example all the hypotheses of the preceding theorem are satisfied at $x_0 = -1$, except $\Gamma(x_0) \subset \liminf_{x \to x_0} (\Gamma(x) + C(x))$, and the conclusions of (i)–(iii) fail.

Example 3.5.30. Let $X =]-\infty, 0[$, $Y = \mathbb{R}^2$, $C(x) = \{(0,0)\} \cup]0, \infty[\times]0, \infty[$, and let Γ be defined by

$$\Gamma(x) = \begin{cases} \{(y_1, y_2) \in \mathbb{R}^2 \mid y_2 \geq xy_1,\, y_1 \leq 1,\, y_2 \leq 1\} & \text{if } x \neq -1, \\ \{(y_1, y_2) \in \mathbb{R}^2 \mid y_2 \geq -y_1 - 1,\, y_1 \leq 1,\, y_2 \leq 1\} & \text{if } x = -1. \end{cases}$$

Then

$$\Omega(x) = \begin{cases} \{(y_1, y_2) \in \mathbb{R}^2 \mid y_2 = xy_1,\, x^{-1} \leq y_1 \leq 1\} & \text{if } x \neq -1, \\ \{(y_1, y_2) \in \mathbb{R}^2 \mid y_2 = -y_1 - 1,\, -2 \leq y_1 \leq 1\} & \text{if } x = -1. \end{cases}$$

In the next example all the hypotheses of the preceding theorem are satisfied at $x_0 = 1$, except $\limsup_{x \to x_0} C^c(x) \subset C^c(x_0)$, and the conclusions of (i)–(iii) fail.

Example 3.5.31. ([356], modified) Let $X =]0, \infty[$, $Y = \mathbb{R}^2$, and Γ and C be defined by

$$\Gamma(x) = \{(y_1, y_2) \in \mathbb{R}^2 \mid y_1^2 + y_2^2 \leq 1\},$$
$$C(x) = \begin{cases} \{(0,0)\} \cup \{(y_1, y_2) \in \mathbb{R}^2 \mid y_2 > 0,\, y_1 + y_2 > 0\} & \text{if } x = 1, \\ \{(0,0)\} \cup]0, \infty[\times]0, \infty[& \text{if } x \neq 1. \end{cases}$$

Then

$$\Omega(x) = \begin{cases} \{(y_1, y_2) \in \mathbb{R}^2 \mid y_1^2 + y_2^2 = 1,\, y_1 \leq 0,\, y_1 - y_2 \geq 0\} & \text{if } x = 1, \\ \{(y_1, y_2) \in \mathbb{R}^2 \mid y_1^2 + y_2^2 = 1,\, y_1 \leq 0,\, y_2 \leq 0\} & \text{if } x \neq 1. \end{cases}$$

In the next example condition (3.41) holds, but the hypotheses of (i) and (ii) fail at $x_0 = 0$, as well as their conclusions.

Example 3.5.32. ([356], modified) Let $X =]0, \infty[$, $Y = \mathbb{R}^2$, and let Γ and C be defined by

$$\Gamma(x) = \{(y_1, y_2) \in \mathbb{R}^2 \mid y_1^2 + y_2^2 \leq 1\} \setminus \{(-1, 0)\},$$
$$C(x) = \{(0,0)\} \cup \{(y_1, y_2) \in \mathbb{R}^2 \mid y_2 > 0,\, y_1 + xy_2 > 0\}.$$

Then

$$\Omega(x) = \begin{cases} \{(y_1, y_2) \in \mathbb{R}^2 \mid y_1^2 + y_2^2 = 1,\, y_1 \leq 0,\, y_2 \leq xy_1\} & \text{if } x \neq 0, \\ \{(y_1, y_2) \in \mathbb{R}^2 \mid y_1^2 + y_2^2 = 1,\, y_1 \leq 0,\, y_2 < 0\} & \text{if } x = 0. \end{cases}$$

Theorem 3.5.33. *Consider $(x_0, y_0) \in \operatorname{gr} \Omega$. Suppose that $\Gamma + C$ is l.c. at (x_0, y_0) and $\Gamma(x)$ has (DP) w.r.t. $C(x)$ for $x \in U^\bullet$, where $U \in V_X(x_0)$. Then*

(i) $y_0 \in \liminf_{x \to x_0} (\Omega(x) + C(x))$.

(ii) *Moreover, assume that* $\limsup_{x \to x_0} \Gamma(x) \subset \Gamma(x_0)$ *(or more generally,* $\limsup_{x \to x_0} \Omega(x) \subset \Gamma(x_0) + \mathcal{C}(x_0)$ *) and* $\limsup_{x \to x_0} \mathcal{C}(x) \subset \mathcal{C}(x_0)$ *and for all nets* $((x_i, y_i))_{i \in I} \subset \operatorname{gr} \Gamma$, $((x_i, z_i))_{i \in I} \subset \operatorname{gr} \Omega$ *with* $x_i \neq x_0$ *and* $z_i \leq_{\mathcal{C}(x_i)} y_i$ *for* $i \in I$, (z_i) *contains a convergent subnet. Then* Ω *is l.c. at* (x_0, y_0).

PROOF. Let $X \setminus \{x_0\} \supset (x_i) \to x_0$. We may suppose that $(x_i) \subset U^\bullet$. Since $\Gamma + \mathcal{C}$ is l.c. at (x_0, y_0), there exist a subnet $(x_{\varphi(j)})_{j \in J}$ and a net $(y_j + k_j)_{j \in J} \to y_0$ with $y_j \in \Gamma(x_{\varphi(j)})$ and $k_j \in \mathcal{C}(x_{\varphi(j)})$ for every $j \in J$. Since $\Gamma(x_{\varphi(j)})$ has (DP) w.r.t. $\mathcal{C}(x_{\varphi(j)})$, there exist $z_j \in \Omega(x_{\varphi(j)})$ and $k'_j \in \mathcal{C}(x_{\varphi(j)})$ such that $y_j = z_j + k'_j$.

(i) It follows that $y_j + k_j \in \Omega(x_{\varphi(j)}) + \mathcal{C}(x_{\varphi(j)})$ for every $j \in J$. Therefore y_0 is in $\liminf_{x \to x_0} (\Omega(x) + \mathcal{C}(x))$.

(ii) By hypothesis, there exists a subnet $(z_{\psi(l)})_{l \in L}$ convergent to some $z \in Y$. It follows that $z = z_0 + k_0$ with $z_0 \in \Gamma(x_0)$, $k_0 \in \mathcal{C}(x_0)$, and $(k_{\psi(l)} + k'_{\psi(l)}) \to y_0 - z =: k \in \mathcal{C}(x_0)$. Therefore $y_0 = z + k = z_0 + k_0 + k$. Since $y_0 \in \Omega(x_0)$ and $\mathcal{C}(x_0)$ is pointed, we obtain that $z = z_0 = y_0$. Using Proposition 2.5.6, we have that Ω is l.c. at y_0. $\qquad\square$

Note that the last condition of the hypothesis of (ii) in the preceding theorem holds if $\Omega(U^\bullet)$ is relatively compact for some $U \in \mathcal{V}_X(x_0)$.

Taking $\mathcal{C}(x) = C$ and $y_0 \in \Omega(x_0)$ in Examples 3.5.9, 3.5.17, 3.5.8, and 3.5.22, we observe that the conditions of Theorem 3.5.33, $\Gamma + \mathcal{C}$ is l.c. at (x_0, y_0), $\Gamma(x)$ has (DP) w.r.t. $\mathcal{C}(x)$ for $x \in U^\bullet$, $\limsup_{x \to x_0} \Gamma(x) \subset \Gamma(x_0)$, and the second hypothesis of (ii) are essential for obtaining (i) and the conclusion of (ii), respectively.

In the following example all the hypotheses of Theorem 3.5.33 (ii) for $x_0 = 1$ and $y_0 = (-1, 0)$ hold, except $\limsup_{x \to x_0} \mathcal{C}(x) \subset \mathcal{C}(x_0)$; Ω is not l.c. at (x_0, y_0).

Example 3.5.34. Let $X =]0, \infty[$, $Y = \mathbb{R}^2$, and let the multifunctions Γ and $\mathcal{C}$ be defined by

$$\Gamma(x) = \{(y_1, y_2) \in \mathbb{R}^2 \mid y_1^2 + y_2^2 \leq 1\},$$
$$\mathcal{C}(x) = \begin{cases} \{(y_1, y_2) \in \mathbb{R}^2 \mid y_2 \geq 0, \, y_1 + y_2 \geq 0\} & \text{if } x \neq 1, \\ \mathbb{R}_+^2 & \text{if } x = 1. \end{cases}$$

Then

$$\Omega(x) = \begin{cases} \{(y_1, y_2) \in \mathbb{R}^2 \mid y_1^2 + y_2^2 = 1, \, y_1 \leq 0, \, y_1 - y_2 \geq 0\} & \text{if } x \neq 1, \\ \{(y_1, y_2) \in \mathbb{R}^2 \mid y_1^2 + y_2^2 = 1, \, y_1 \leq 0, \, y_2 \leq 0\} & \text{if } x = 1. \end{cases}$$

Before stating the next result we extend the notion of lower semicontinuous multifunction to the case of moving cones. So, if $F : X \times \mathcal{U} \rightrightarrows Y$ $[G : \mathcal{U} \rightrightarrows Y]$, we say that F $[G]$ is $\mathcal{C}$-**lower semicontinuous** ($\mathcal{C}$-l.s.c. for short), where this time $\mathcal{C} : \mathcal{U} \rightrightarrows Y$ if the sets $\operatorname{lev}_{F,\mathcal{C}}(y) := \{(x, u) \in X \times \mathcal{U} \mid F(x, u) \cap (y - \mathcal{C}(u)) \neq \emptyset\}$ $[\operatorname{lev}_{G,\mathcal{C}}(y) := \{u \in \mathcal{U} \mid G(u) \cap (y - \mathcal{C}(u)) \neq \emptyset\}]$ are closed for every $y \in Y$. We have the following extension of Corollary 3.5.4.

Theorem 3.5.35. *Let* $F : X \times \mathcal{U} \rightrightarrows Y$ $\mathcal{C}$-*l.s.c. and* $\Lambda : \mathcal{U} \rightrightarrows X$ *compact at every* $u \in \mathcal{U}$. *If* $F(x, u) + \mathcal{C}(u)$ *is* $\mathcal{C}(u)$-*complete for every* $u \in \mathcal{U}$ *and* $x \in \Lambda(u)$, *then* $\Gamma(u) := F(\Lambda(u) \times \{u\})$ *has the domination property w.r.t.* $\mathcal{C}(u)$ *for every* $u \in \mathcal{U}$, *and the optimal multifunction* Ω *is* $\mathcal{C}$-*l.s.c.*

PROOF. Let us note first that $\Gamma(u)$ has (DP) w.r.t. $\mathcal{C}(u)$ for every $u \in \mathcal{U}$. The proof is the same as that of the corresponding part in the proof of Corollary 3.5.4; just replace $G(x)$ by $F(x, u)$ and C by $\mathcal{C}(u)$. Let $y \in Y$, and suppose that the net $(u_i) \subset \text{lev}_{\Omega, \mathcal{C}}(y)$ converge to $u \in \mathcal{U}$. Therefore, for every $i \in I$ there exist $x_i \in \Lambda(u_i)$ and $y_i \in F(x_i, u_i)$ such that $y_i \in y - \mathcal{C}(u_i)$. It follows that $((x_i, u_i))_{i \in I} \subset \text{lev}_{F, \mathcal{C}}(y)$. Since Λ is compact at u, there exists the subnet $(x_{\varphi(j)})_{j \in J}$ of $(x_i)_{i \in I}$ converging to $x \in \Lambda(u)$. Since F is $\mathcal{C}$-l.s.c. we get $(x, u) \in \text{lev}_{F, \mathcal{C}}(y)$, whence $u \in \text{lev}_{\Gamma, \mathcal{C}}(y)$. Since $\Gamma(u)$ has (DP) w.r.t. $\mathcal{C}(u)$, we obtain that $u \in \text{lev}_{\Gamma, \mathcal{C}}(y)$. Therefore Ω is $\mathcal{C}$-l.s.c.   □

Theorem 3.5.29 (i), (iii) was essentially obtained by Dolecki and Malivert [98], while (ii) was obtained by Penot and Sterna-Karwat [291]. From this theorem one obtains easily Theorem 2.1 of [245] (established for Y a normed vector space, $X = \mathbb{N} \cup \{\infty\}$, $x_0 = \infty$, and and $\mathcal{C}(x) \setminus \{0\}$ open) and Theorem 2.2 of [245] (established for Y a reflexive Banach space endowed with the weak topology and $X = \mathbb{N} \cup \{\infty\}$, $x_0 = \infty$). Theorem 3.5.33 (i) extends a result of Luc [244], while (ii) is obtained, essentially, in [291] for $\mathcal{C}$ a constant multifunction, in [98] for Ω compact at x_0, and in [245] for $X = \mathbb{N} \cup \{\infty\}$, $x_0 = \infty$. From this theorem one obtains easily Theorems 3.3 and 3.4 from [245]. Under the condition $\limsup_{x \to x_0} \Omega(x) \subset \Gamma(x_0) + \mathcal{C}(x_0)$ with $\mathcal{C}(x_0) = C$ a pointed closed convex cone, $X = \mathbb{N} \cup \{\infty\}$, and $x_0 = \infty$, Theorem 3.5.33 (ii) is obtained in [242, Th. 2.2]; note that [242, Th. 2.3] can be obtained from Theorem 3.5.33 (ii), too.

The notion of lower semicontinuity w.r.t. moving cones is given by Ferro [115]. Theorem 3.5.35 was established by Ferro [115, Th. 4.1], too; in [115, Th. 4.2] a slightly different result is also stated.

3.5.3 Continuity Properties for the Solution Multifunction

Consider $\mathcal{U}$, X two topological spaces and Y a topological vector space. As in the preceding section, consider a multifunction $\mathcal{C} : \mathcal{U} \rightrightarrows Y$ whose values are pointed convex cones. With $\mathcal{C}$ we associate the multifunction $\mathcal{C}^c : \mathcal{U} \rightrightarrows Y$ defined by $\mathcal{C}^c(u) := (Y \setminus \mathcal{C}(u)) \cup \{0\}$. Take also $\Lambda : \mathcal{U} \rightrightarrows X$ and $f : X \times \mathcal{U} \to Y$. For a fixed $u_0 \in \mathcal{U}$ we consider the problem

(P) $\mathcal{C}(u_0)$-minimize $f(x, u_0)$ subject to $x \in \Lambda(u_0)$,

and its perturbed problems

(P_u) $\mathcal{C}(u)$-minimize $f(x)$ subject to $x \in \Lambda(u)$.

Thus the initial problem is just (P_{u_0}). We associate the **optimal-value multifunction** $\Omega : \mathcal{U} \rightrightarrows Y$ and the **solution multifunction** $\Sigma : \mathcal{U} \rightrightarrows X$ defined, respectively, by

$$\Omega(u) := \mathrm{Eff}\,(f(\Lambda(u) \times \{u\}); \mathcal{C}(u)),\qquad \Sigma(u) := \{x \in \Lambda(u) \mid f(x,u) \in \Omega(u)\}.$$

It is obvious that $\mathrm{dom}\,\Sigma = \mathrm{dom}\,\Omega$.

In the preceding section we obtained continuity properties of Ω in Corollaries 3.5.15, 3.5.24, and 3.5.28. In this section we are interested in continuity properties for the solution multifunction Σ. We begin with a closedness condition for Σ_w.

Theorem 3.5.36. *Suppose that f is continuous on $\Lambda(u_0) \times \{u_0\}$,*

$$\Lambda(u_0) \subset \liminf_{u \to u_0} \Lambda(u),\ \text{and}\ \limsup_{u \to u_0} \mathcal{C}^c(u) \subset \mathcal{C}^c(u_0). \tag{3.42}$$

(i) *If $\limsup_{u \to u_0} \Lambda(u) \subset \Lambda(u_0)$, then $\limsup_{u \to u_0} \Sigma(u) \subset \Sigma(u_0)$.*
(ii) *If $\mathcal{C}(u_0) \setminus \{0\}$ is open and Λ is closed at u_0, then Σ is closed at u_0.*
(iii) *If $\mathcal{C}(u_0) \setminus \{0\}$ is open and Λ is compact at u_0 ($\Lambda(u_0)$ is compact and Λ is u.c. at u_0), then Σ is compact at u_0 (Σ is u.c. at u_0).*

PROOF. (i) Let $x \in \limsup_{u \to u_0} \Sigma(u)$. This means that there exists a net $((u_i, x_i))_{i \in I} \subset \mathrm{gr}\,\Sigma\ (\subset \mathrm{gr}\,\Lambda)$ converging to (u_0, x_0), with $u_i \neq u_0$ for every $i \in I$. From the hypothesis we have that $x_0 \in \Lambda(u_0)$. Suppose that $x_0 \notin \Sigma(u_0)$; then there exists $x \in \Lambda(u_0)$ such that $k := f(x_0, u_0) - f(x, u_0) \in \mathcal{C}(u_0) \setminus \{0\}$. Since $x_0 \in \Lambda(u_0)$, from (3.42) there exist a subnet $(u_{\varphi(j)})_{j \in J}$ of (u_i) and a net $(x'_{\varphi(j)})_{j \in J}$ converging to x. Since f is continuous on $\Lambda(u_0) \times \{u_0\}$, $f(x'_{\varphi(j)}, u_{\varphi(j)}) - f(x_{\varphi(j)}, u_{\varphi(j)}) \to k \neq 0$, so we may suppose that $f(x'_{\varphi(j)}, u_{\varphi(j)}) - f(x_{\varphi(j)}, u_{\varphi(j)}) \neq 0$ for every $j \in J$. Since $x_{\varphi(j)} \in \Sigma(u_{\varphi(j)})$, we have that $f(x'_{\varphi(j)}, u_{\varphi(j)}) - f(x_{\varphi(j)}, u_{\varphi(j)}) \in \mathcal{C}^c(u_{\varphi(j)})$ for $j \in J$. From (3.42) we obtain that $k \in \mathcal{C}^c(u_0)$, a contradiction. Therefore $x_0 \in \Sigma(u_0)$.

(ii) Using the fact that $\Lambda(u_0)$ is closed (Λ being closed at u_0) and $\mathcal{C}(u_0) \setminus \{0\}$ is open, one obtains easily that $\Sigma(u_0)$ is closed. The conclusion then follows from (i).

(iii) If Λ is compact at u_0, the proof is similar to that of (ii). If $\Lambda(u_0)$ is compact and Λ is u.c. at u_0, then Λ is closed at u_0 by Proposition 2.5.10 (iv). Noting that $\Sigma(u) = \Sigma(u) \cap \Lambda(u)$ for every $u \in \mathcal{U}$, the upper continuity of Σ follows from (ii) and Proposition 2.5.11 (i). $\qquad\square$

Note that the preceding result can be stated for a multifunction $F : X \times \mathcal{U} \rightrightarrows Y$ instead of the function f; the conclusions remain valid if we suppose that F is compact and l.c. at (x, u_0) for every $x \in \Lambda(u_0)$.

In the sequel in this section $f(x, u) = g(x)$, where $g : X \to Y$ and $\mathcal{C}(u) = C$ for every $u \in \mathcal{U}$.

Theorem 3.5.37. *Suppose that $\mathrm{int}\,C \neq \emptyset$, g is continuous, Λ is u.c. at u_0, (P) is well-posed, and Ω is $(-C)$-H-u.c. at u_0. Then Σ is u.c. at u_0.*

PROOF. If $\Sigma(u_0) = \emptyset$, then $\Omega(u_0) = \emptyset$. Since Ω is $(-C)$-H-u.c. at u_0, it follows that $u_0 \in \mathrm{int}\,(\mathcal{U} \setminus \mathrm{dom}\,\Omega) = \mathrm{int}\,(\mathcal{U} \setminus \mathrm{dom}\,\Sigma)$, and so Σ is u.c. at u_0.

Let $\Sigma(u_0) \neq \emptyset$. Consider an open set $D \subset Y$ such that $\Sigma(u_0) \subset D$. Since (P) is well-posed, there exists $W \in \mathcal{V}_Y$ such that

$$\bigcup_{\eta \in \Omega(u_0)} \{x \in \Lambda(u_0) \mid g(x) \leq_C \eta + \varepsilon\} \subset D \quad \forall \varepsilon \in W \cap C.$$

Let $\mathcal{L}(A) := \Omega(u_0) + A - C$ for $A \subset Y$. The above relation shows that $\Lambda(u_0) \subset D \cup [X \setminus g^{-1}(\mathcal{L}(W \cap C))]$. Fix $\varepsilon \in W \cap \mathrm{int}\, C$. There exists $W_1 \in \mathcal{V}_Y$ with $\varepsilon + W_1 \subset C$. It follows that $W_1 \subset \varepsilon - C \subset (W \cap C) - C$, and so $\Lambda(u_0) \subset D \cup [X \setminus g^{-1}(\mathcal{L}(W_1))]$. Let $W_2 \in \mathcal{V}_Y$ be such that $W_2 + W_2 \subset W_1$. Since $\mathrm{cl}\,\mathcal{L}(W_2) \subset W_2 + \mathcal{L}(W_2) \subset \mathcal{L}(W_1)$, we have that

$$\Lambda(u_0) \subset D \cup [X \setminus g^{-1}(\mathrm{cl}\,\mathcal{L}(W_2))] =: D_1.$$

Because g is continuous, $g^{-1}(\mathrm{cl}\,\mathcal{L}(W_2))$ is closed, and so D_1 is open. Taking into account that Λ is u.c. at u_0, there exists $U_1 \in \mathcal{V}_{\mathcal{U}}(u_0)$ such that $\Lambda(U_1) \subset D_1$. Since Ω is $(-C)$-H-u.c. at u_0, there exists $U_2 \in \mathcal{V}_{\mathcal{U}}(u_0)$ such that $\Omega(U_2) \subset \Omega(u_0) + W_2 - C = \mathcal{L}(W_2)$. Let $u \in U_1 \cap U_2$ and $x \in \Sigma(u)$ ($\subset \Lambda(u)$). Then $g(x) \in \Omega(u) \subset \mathcal{L}(W_2)$, which means that $x \in g^{-1}(\mathcal{L}(W_2)) \subset g^{-1}(\mathrm{cl}\,\mathcal{L}(W_2))$. It follows that $x \in D$. Therefore $\Sigma(U_1 \cap U_2) \subset D$, which shows that Σ is u.c. at u_0. □

Corollary 3.5.38. *Suppose that* $\mathrm{int}\, C \neq \emptyset$, g *is continuous,* Λ *is u.c. at* u_0, $\Omega(u_0)$ *is compact,* (P) *is* η-*well-posed, and* Ω *is* $(-C)$-*H-u.c. at* u_0. *Then* Σ *is u.c. at* u_0.

PROOF. By Proposition 3.4.5 the problem (P) is well-posed. The conclusion follows from the preceding theorem. □

The next result establishes sufficient conditions for Hausdorff upper continuity of Σ.

Theorem 3.5.39. *Suppose that* X *is a topological vector space,* $\mathrm{int}\, C \neq \emptyset$, g *is uniformly continuous on* X, Λ *is H-u.c. at* u_0, (P) *is weakly well-posed, and* Ω *is* $(-C)$-*H-u.c. at* u_0. *Then* Σ *is H-u.c. at* u_0.

PROOF. Let $V \in \mathcal{V}_X$ be fixed and consider $V_1 \in \mathcal{V}_X$ such that $V_1 + V_1 \subset V$. Since (P) is weakly well-posed, there exists $W \in \mathcal{V}_Y$ such that

$$\bigcup_{\eta \in \Omega(u_0)} \{x \in \Lambda(u_0) \mid g(x) \leq_C \eta + \varepsilon\} \subset \Sigma(u_0) + V_1 \quad \forall \varepsilon \in W \cap C;$$

i.e., with the notation from the proof of the preceding theorem,

$$\Lambda(u_0) \cap g^{-1}(\mathcal{L}(W \cap C)) \subset \Sigma(u_0) + V_1.$$

Taking ε, W_1, and W_2 as in the proof of the preceding theorem, we obtain that

$$\Lambda(u_0) \subset (\Sigma(u_0) + V_1) \cup [X \setminus g^{-1}(\mathcal{L}(W_1))]. \qquad (3.43)$$

By the uniform continuity of g there exists $V_2 \in \mathcal{V}_X$ such that

$$g\left((X \setminus g^{-1}(\mathcal{L}(W_1))) + V_2\right) \subset g\left(X \setminus g^{-1}(\mathcal{L}(W_1))\right) + W_2 \subset (Y \setminus \mathcal{L}(W_1)) + W_2.$$

Since $(Y \setminus \mathcal{L}(W_1)) + W_2 \subset Y \setminus \mathcal{L}(W_2)$, we obtain that

$$\left(X \setminus g^{-1}(\mathcal{L}(W_1))\right) + V_2 \subset g^{-1}(Y \setminus \mathcal{L}(W_2)) = X \setminus g^{-1}(\mathcal{L}(W_2)). \qquad (3.44)$$

Since Λ is H-u.c. at u_0, there exists $U_1 \in \mathcal{V}_{\mathcal{U}}(u_0)$ such that $\Lambda(U_1) \subset \Lambda(u_0) + V_1 \cap V_2$. From (3.43) we have that

$$\Lambda(u_0) + V_1 \cap V_2 \subset (\Sigma(u_0) + V_1 + V_1 \cap V_2) \cup \left[(X \setminus g^{-1}(\mathcal{L}(W_1))) + V_1 \cap V_2\right].$$

Taking into account (3.44) and that $V_1 + V_1 \subset V$, we get

$$\Lambda(u_0) + V_1 \cap V_2 \subset (\Sigma(u_0) + V) \cup \left(X \setminus g^{-1}(\mathcal{L}(W_2))\right).$$

Therefore

$$\Lambda(U_1) \subset (\Sigma(u_0) + V) \cup \left(X \setminus g^{-1}(\mathcal{L}(W_2))\right). \qquad (3.45)$$

Since Ω is $(-C)$-H-u.c. at u_0, there exists $U_2 \in \mathcal{V}_{\mathcal{U}}(u_0)$ such that $\Omega(U_2) \subset \Omega(u_0) + W_2 - C = \mathcal{L}(W_2)$. So, for $u \in U_1 \cap U_2$ and $x \in \Sigma(u)$ we have that $g(x) \in \mathcal{L}(W_2)$. From (3.45) we obtain that $x \in \Sigma(u_0) + V$. This shows that $\Sigma(U_1 \cap U_2) \subset \Sigma(u_0) + V$, and so Σ is u.c. at u_0. $\qquad \square$

Theorem 3.5.36 was stated, essentially, by Penot and Sterna-Karwat in [291] and Dolecki and Malivert [98]. Bednarczuk stated Theorems 3.5.37 and 3.5.39 and Corollary 3.5.38 in [18].

3.6 Sensitivity of Vector Optimization Problems

Throughout this section X, Y are normed vector spaces and $\{0\} \neq C \subset Y$ is a pointed closed convex cone; when we consider weak minimal points of a set we assume also that $\mathrm{int}\, C \neq \emptyset$. Let $D \subset Y$; recall that the efficient set of D w.r.t. C is

$$\mathrm{Eff}(D; C) = \{y \in D \mid D \cap (y - C) \subset y - C\},$$

the set of **Henig-proper efficient points** of D w.r.t. C is

$$\mathrm{HEff}(D; C) = \bigcup \{\mathrm{Eff}(D; K) \mid K \text{ convex cone}, C \setminus \{0\} \subset \mathrm{int}\, K \neq Y\},$$

the set of **Benson-proper efficient points** of D w.r.t. C is

$$\mathrm{BEff}(D; C) = \{y \in D \mid \overline{\mathrm{cone}}(D + C - y) \cap (-C) = \{0\}\},$$

and the set of **weakly efficient points** of D w.r.t. C is the set $w\,\text{Eff}(D;C) = \text{Eff}(D;C_0)$, where $C_0 = \{0\} \cup \text{int}\, C$.

It is obvious that

$$\text{SPEff}(D;C) \subset \text{HEff}(D;C) \subset \text{BEff}(D;C) \subset \text{Eff}(D;C) \subset w\,\text{Eff}(D;C),$$

the last inclusion if $\text{int}\,C \neq \emptyset$. Note that if C has a compact base, then $\text{SPEff}(D;C) = \text{HEff}(D;C) = \text{BEff}(D;C)$. The first equality follows from Proposition 3.5.25. Let us prove the second one. Suppose that B is a compact base for C (i.e., B is compact, convex, $0 \notin B$, and $C = \text{cone}\,B$), and $y \in \text{BEff}(D;C)$. It follows that $\overline{\text{cone}}(D - y) \cap (-B) = \emptyset$, and so $r := d(-B, \text{cone}(D - y)) > 0$. Otherwise, there are $(y_n) \subset \overline{\text{cone}}(D - y)$ such that $b_n + y_n \to 0$. There exists $b_{n_k} \to b \in B$ since B is compact, and so $y_{n_k} \to -b$, which shows that $\overline{\text{cone}}(D - y) \cap (-B) \neq \emptyset$. Taking $C := \text{cone}(B + \frac{r}{2}B_Y)$, we have that $C \setminus \{0\} \subset \text{int}\,C$ and $(D - y) \cap (-C) \subset \overline{\text{cone}}(D - y) \cap (-\tilde{C}) = \{0\}$. Therefore $y \in \text{HEff}(D;C)$. The equality of $\text{HEff}(D;C) = \text{BEff}(D;C)$ in the case of finite-dimensional spaces is proven by Henig [161].

Consider the multivalued mapping $\Gamma : X \rightrightarrows Y$ and the following associated multivalued mappings: $\Omega, H\Omega, \Omega_w : X \rightrightarrows Y$, given by

$$\Omega(x) := \text{Eff}(\Gamma(x);C), \quad H\Omega(x) := \text{HEff}(\Gamma(x);C),$$
$$\Omega_w(x) := w\,\text{Eff}(F(u);C).$$

In the sequel we shall also use the multivalued mapping Γ_C defined by $\Gamma_C(u) := \Gamma(u) + C$ for $u \in X$. Of course, $\text{gr}\,\Gamma_C = \text{gr}\,\Gamma + \{0\} \times C$.

Throughout this section we denote by $D\Gamma(\overline{x},\overline{y})(u)$ the set $\overline{D\Gamma}(\overline{x},\overline{y})(u)$.

Shi [319] introduced the following **derivative** of Γ at $(\overline{x},\overline{y}) \in \text{gr}\,\Gamma$ in the direction u:

$$S\Gamma(\overline{x},\overline{y})(u) := \{y \in Y \mid \exists\,(t_n) \subset (0,\infty),\ ((x_n, y_n)) \subset \text{gr}\,\Gamma\ :$$
$$x_n \to \overline{x},\ t_n(x_n - \overline{x}, y_n - \overline{y}) \to (u, y)\}.$$

It is obvious that for $(\overline{x},\overline{y}) \in \text{gr}\,\Gamma$,

$$\text{gr}\,D\Gamma(\overline{x},\overline{y}) \subset \text{gr}\,S\Gamma(\overline{x},\overline{y}) \subset \overline{\text{cone}}\,(\text{gr}\,\Gamma - (\overline{x},\overline{y})),$$

with equality if $\text{gr}\,\Gamma$ is a convex set, and

$$S\Gamma(\overline{x},\overline{y})(u) = D\Gamma(\overline{x},\overline{y})(u) \quad \forall u \in X \setminus \{0\}.$$

In the sequel we shall use several times the following condition:

$$S\Gamma(\overline{x},\overline{y})(0) \cap (-C) = \{0\}. \tag{3.46}$$

Note that if (3.46) holds, then $\overline{\text{cone}}(\Gamma(\overline{x}) - \overline{y}) \cap (-C) = \{0\}$, while if C has a compact base, then $\overline{y} \in \text{BEff}\,(\Gamma(\overline{x});C)$, and therefore $\overline{y} \in H\Omega(\overline{x})$.

In the following theorem we give sufficient conditions for (3.46).

Theorem 3.6.1. *Let $\overline{y} \in \Gamma(\overline{x})$. Then each of the following conditions is sufficient for (3.46):*

(i) $\overline{y} \in \mathrm{BEff}\,(\Gamma(\overline{x}); C)$, Γ *is upper Lipschitz at $\overline{x}$; i.e., there exist $L, \rho > 0$ such that $\Gamma(x) \subset \Gamma(\overline{x}) + L\,\|x - \overline{x}\|\,B_Y$ for all $x \in B(\overline{x}, \rho)$;*

(ii) $\overline{y} \in \mathrm{BEff}\,(\Gamma(\overline{x}); C)$, X, Y *are finite-dimensional, $\overline{x} \in \mathrm{raint}(\mathrm{dom}\,F)$, and Γ is C-convex;*

(iii) $\overline{y} \in \mathrm{HEff}\,(\Gamma(\overline{x}); C)$, X, Y *are Banach spaces, $\overline{x} \in \mathrm{int}(\mathrm{dom}\,F)$, and Γ is C-convex and C-closed (gr Γ_C is closed).*

PROOF. (i) Let $y \in S\Gamma(\overline{x}, \overline{y})(0) \cap (-C)$. Then there exist $(t_n) \subset (0, \infty)$, $((x_n, y_n)) \subset \mathrm{gr}\,\Gamma$ such that $x_n \to 0$ and $t_n\,((x_n, y_n) - (\overline{x}, \overline{y})) \to (0, y)$. We may suppose that $x_n \in B(\overline{x}, \rho)$ for every $n \in \mathbb{N}$. It follows that for every n there exist $\overline{y}_n \in \Gamma(\overline{x})$ and $v_n \in B_Y$ such that $y_n = \overline{y}_n + L\,\|x_n - \overline{x}\| \cdot v_n$. Therefore $\|t_n(\overline{y}_n - \overline{y}) - t_n(y_n - \overline{y})\| \leq L t_n\,\|x_n - \overline{x}\|$ for every $n \in \mathbb{N}$, whence $t_n(\overline{y}_n - \overline{y}) \to y$. So $y \in \overline{\mathrm{cone}}(\Gamma(\overline{x}) - \overline{y}) \cap (-C) = \{0\}$, a contradiction.

(ii) Taking $(\overline{x}, \overline{y}) = (0, 0)$ (replacing Γ by $\Gamma - (\overline{x}, \overline{y})$) and replacing X by $\mathrm{span}(\mathrm{dom}\,\Gamma)$ if necessary, we may suppose that $\overline{x} \in \mathrm{int}(\mathrm{dom}\,\Gamma)$. Since $\dim Y < \infty$ and $\overline{y} \in \mathrm{BEff}\,(\Gamma(\overline{x}); C)$, we have that $\overline{y} \in \mathrm{HEff}\,(\Gamma(\overline{x}); C)$; therefore there exists a convex cone C such that $C \setminus \{0\} \subset \mathrm{int}\,C \neq Y$ and $\Gamma(\overline{x}) \cap (-C) \subset C$. Since $\mathrm{gr}\,\Gamma_C = \mathrm{gr}\,\Gamma + \{0\} \times C + \{0\} \times C$, the set $\mathrm{gr}\,\Gamma_C$ is convex. We have that $(\overline{x}, \overline{y}) \notin \mathrm{raint}\,\mathrm{gr}\,\Gamma_C$; in the contrary case for $k \in C \setminus \{0\}$ there exists some $\lambda > 0$ such that $(\overline{x}, \overline{y}) - \lambda(0, k) = (\overline{x}, \overline{y} - \lambda k) \in \mathrm{gr}\,\Gamma + \{0\} \times C$; i.e., $-\lambda k - c' \in \Gamma(\overline{x})$ for some $c' \in C$. This contradicts the hypothesis, since $\lambda k + c' \in C \setminus \{0\} + C \subset \mathrm{int}\,C$. Using a separation theorem, there exists $(u^*, y^*) \in X^* \times Y^* \setminus \{(0, 0)\}$ such that

$$\langle \overline{x}, u^* \rangle + \langle \overline{y} + k, y^* \rangle \leq \langle x, u^* \rangle + \langle c + k, y^* \rangle \quad \forall\, (x, y) \in \mathrm{gr}\,\Gamma,\ c \in C. \quad (3.47)$$

If $y^* = 0$, then $\langle \overline{x}, u^* \rangle \leq \langle x, u^* \rangle$ for every $u \in \mathrm{dom}\,\Gamma$, and so $u^* = 0$. Therefore $y^* \neq 0$. From (3.47) we obtain that $\langle c, y^* \rangle \geq 0$, whence $\langle c, y^* \rangle > 0$ for every $c \in \mathrm{int}\,C \supset C \setminus \{0\}$. Suppose that there exists $y \in S\Gamma(\overline{x}, \overline{y})(0) \cap (-C) \setminus \{0\}$. Then there are $(t_n) \subset (0, \infty)$ and $((x_n, y_n)) \subset \mathrm{gr}\,\Gamma$ such that $x_n \to 0$ and $t_n\,((x_n, y_n) - (\overline{x}, \overline{y})) \to (0, y)$. From (3.47) we obtain that $\langle t_n(x_n - \overline{x}), u^* \rangle + \langle t_n(y_n - \overline{y}), y^* \rangle \geq 0$ for every n, whence $\langle -y, y^* \rangle \leq 0$, a contradiction.

(iii) Let C be a convex cone such that $C \setminus \{0\} \subset \mathrm{int}\,C \neq Y$ and $\overline{y} \in \mathrm{Eff}(F(\overline{u}); C)$. If we show that $\mathrm{int}\,\mathrm{gr}\,\Gamma_C \neq \emptyset$, arguments like those in the proof of (ii) show that (3.46) holds. For this aim consider the relation $\mathcal{R} := \{(y, x) \mid (x, y) \in \mathrm{gr}\,\Gamma_C\}$; $\mathcal{R}$ is closed, convex, and $0 \in \mathrm{int}(\mathrm{Im}\,\mathcal{R})$. Applying the Robinson–Ursescu theorem, we have that $\mathcal{R}(V)$ is a neighborhood of $\overline{x}$ for every neighborhood V of $\overline{y}$. Let $c_0 \in \mathrm{int}\,C$. There exists $\rho > 0$ such that $B(c_0, \rho) \subset C$. It follows that for some $\delta > 0$ we have $B(\overline{x}, \delta) \subset \mathcal{R}\,(B(\overline{y}, \rho/2))$. Let us show that $B(\overline{x}, \delta) \times B(c_0 + \overline{y}, \rho/2) \subset \mathrm{gr}\,\Gamma_C$. Indeed, let (u, y) belong to the first set. There exists $y' \in \mathcal{R}^{-1}(u)$ such that $\|y' - \overline{y}\| \leq \rho/2$. It follows that $y = y' + (y - y')$ and

$$\|c_0 - (y - y')\| \leq \|(c_0 + \overline{y} - y)\| + \|y' - \overline{y}\| \leq \rho/2 + \rho/2 = \rho,$$

which shows that $y - y' \in B(c_0, \rho) \subset C$. Therefore $(x, y) \in \mathcal{R}^{-1} + \{0\} \times C =$ gr Γ_C. The proof is complete. $\qquad\qquad\qquad\qquad\qquad\qquad\qquad\qquad\qquad\qquad\square$

Remark 3.6.2. Part (i) of the preceding theorem is proven in [319, Prop. 3.2].

The condition that Γ is upper Lipschitz at $\bar{x}$ in Theorem 3.6.1(i) is important, as the next example shows.

Example 3.6.3. ([354]) Let $X = Y = \mathbb{R}$, $C = \mathbb{R}_+$, and let $\Gamma : \mathbb{R} \rightrightarrows \mathbb{R}$ be defined by

$$\Gamma(x) := \begin{cases} \{0\} & \text{if } x \leq 0, \\ \{-\sqrt{x}, 0\} & \text{if } x > 0. \end{cases}$$

Then

$$\Gamma_C(x) = \begin{cases} [0, \infty[& \text{if } x \leq 0, \\ [-\sqrt{x}, \infty[& \text{if } x > 0, \end{cases} \qquad \Omega(x) = \begin{cases} \{0\} & \text{if } x \leq 0, \\ \{-\sqrt{x}\} & \text{if } x > 0. \end{cases}$$

Take $\bar{x} = 0$ and $\bar{y} = 0$. Then

$$S\Gamma(\bar{x}, \bar{y})(u) = D\Gamma(\bar{x}, \bar{y})(u) = \begin{cases} \{0\} & \text{if } u \neq 0, \\ -C & \text{if } u = 0, \end{cases}$$

$$D\Gamma_C(\bar{x}, \bar{y})(u) = \begin{cases} C & \text{if } u < 0, \\ \mathbb{R} & \text{if } u \geq 0, \end{cases} \qquad D\Omega(\bar{x}, \bar{y})(u) = \begin{cases} \{0\} & \text{if } u < 0, \\ -C & \text{if } u = 0, \\ \emptyset & \text{if } u > 0, \end{cases}$$

$$\text{Eff}\,(D\Gamma(\bar{x}, \bar{y})(u); C) = \begin{cases} \{0\} & \text{if } u \neq 0, \\ \emptyset & \text{if } u = 0, \end{cases}$$

$$\text{Eff}\,(D\Gamma_C(\bar{x}, \bar{y})(u); C) = \begin{cases} \{0\} & \text{if } u < 0, \\ \emptyset & \text{if } u \geq 0. \end{cases}$$

The condition $\bar{y} \in \text{BEff}\,(\Gamma(\bar{x}); C)$ in the preceding theorem cannot be replaced by $\bar{y} \in \text{Eff}\,(\Gamma(\bar{x}); C)$. In the next example Γ is upper Lipschitz at $\bar{x} \in \text{int}(\text{dom}\,\Gamma)$, C-convex, and C-closed.

Example 3.6.4. Let $X = \mathbb{R}$, $Y = \mathbb{R}^2$, $C = \mathbb{R}_+^2$, and $\Gamma : X \rightrightarrows Y$ be defined by $\Gamma(x) := \{(y_1, y_2) \mid y_2 \geq y_1^2\}$. Take $\bar{x} = 0$ and $\bar{y} = 0$; then $\bar{y} \in \text{Eff}\,(\Gamma(\bar{x}); C) \backslash \text{BEff}\,(\Gamma(\bar{x}); C)$. Moreover, $S\Gamma(\bar{x}, \bar{y})(u) = D\Gamma(\bar{x}, \bar{y})(u) = \mathbb{R} \times \mathbb{R}_+$, and so $S\Gamma(\bar{x}, \bar{y})(u) \cap (-C) =]-\infty, 0] \times \{0\} \neq \{0\}$.

If $\dim X = \infty$ in Theorem 3.6.1(ii), the conclusion may be not true.

Example 3.6.5. Consider X an infinite-dimensional normed vector space, $Y := \mathbb{R}$, $\phi : X \to \mathbb{R}$ a noncontinuous linear functional, let $\Gamma : X \rightrightarrows Y$ be defined by $\Gamma(x) := \{\phi(x)\}$ and $C := \mathbb{R}_+$. One obtains easily that $S\Gamma(\bar{x}, \bar{y})(u) = D\Gamma(\bar{x}, \bar{y})(u) = \mathbb{R}$ for all $(\bar{x}, \bar{y}) \in \text{gr}\,\Gamma$ and $u \in X$.

Theorem 3.6.6. *Let $(\bar{x}, \bar{y}) \in \text{gr}\,\Gamma$. Then*

(i) *for every $u \in X$ we have $D\Gamma(\overline{x}, \overline{y})(u) + C \subset D\Gamma_C(\overline{x}, \overline{y})(u)$.*

(ii) *If the set $\{y \in C \mid \|y\| = 1\}$ is compact, then for every $u \in X$,*

$$\text{Eff}\left(D\Gamma_C(\overline{x}, \overline{y})(u); C\right) \subset \text{Eff}\left(D\Gamma(\overline{x}, \overline{y})(u); C\right) \subset D\Gamma(\overline{x}, \overline{y})(u). \qquad (3.48)$$

Moreover, if $\Gamma(x)$ is convex for $x \in V$, where V is some neighborhood of $\overline{x}$, then

$$\text{Eff}\left(D\Gamma(\overline{x}, \overline{y})(u); C\right) = \text{Eff}\left(D\Gamma_C(\overline{x}, \overline{y})(u); C\right) \quad \forall u \in X. \qquad (3.49)$$

(iii) *If the set $\{y \in C \mid \|y\| = 1\}$ is compact and (3.46) holds, then*

$$D\Gamma(\overline{x}, \overline{y})(u) + C = D\Gamma_C(\overline{x}, \overline{y})(u) \quad \forall u \in X. \qquad (3.50)$$

PROOF. (i) Let $y \in D\Gamma(\overline{x}, \overline{y})(u)$ and $k \in C$. Then there exist $(t_n) \to 0_+$ and $((u_n, y_n)) \to (u, y)$ such that $(\overline{x}, \overline{y}) + t_n(u_n, y_n) \in \text{gr}\,\Gamma$ for every $n \in \mathbb{N}$. It follows that $(\overline{x}, \overline{y}) + t_n(u_n, y_n + k) \in \text{gr}\,\Gamma_C$ and $(u_n, y_n + k) \to (u, y + k)$. Therefore $y + k \in D\Gamma_C(0, 0)(u)$.

In the rest of the proof we suppose that $Q := \{y \in C \mid \|y\| = 1\}$ is compact. Of course, $C = \text{cone}\,Q$.

(ii) Let $y \in \text{Eff}\,(D\Gamma_C(\overline{x}, \overline{y})(u); C)$; there exists $(t_n) \to 0_+$, $(\lambda_n) \subset [0, \infty)$, $(q_n) \subset Q$, $((u_n, y_n)) \subset X \times Y$ such that $(u_n, y_n + \lambda_n q_n) \to (u, y)$ and $(\overline{x}, \overline{y}) + t_n(u_n, y_n) \in \text{gr}\,\Gamma$ for every $n \in \mathbb{N}$. Since Q is compact, we may suppose that $q_n \to q \in Q$. Assume that there are a subsequence (λ_{n_k}) and $\gamma > 0$ such that $\lambda_{n_k} \geq \gamma$ for every $k \in \mathbb{N}$. Then $(\overline{x}, \overline{y}) + t_{n_k}(u_{n_k}, y_{n_k} + (\lambda_{n_k} - \gamma)q_{n_k}) \in \text{gr}\,\Gamma_C$ and $(u_{n_k}, y_{n_k} + (\lambda_{n_k} - \gamma)q_{n_k}) \to (u, y - \gamma q)$. It follows that $y - \gamma q \in D\Gamma_C(\overline{x}, \overline{y})(u) \cap (y - C) \setminus \{y\}$, a contradiction. Therefore $\lambda_n \to 0$, and so $(u_n, y_n) \to (u, y)$. It follows that $y \in D\Gamma(\overline{x}, \overline{y})(u)$. Using also (i), we have that for every $u \in X$,

$$\text{Eff}\left(D\Gamma_C(\overline{x}, \overline{y})(u); C\right) \subset D\Gamma(\overline{x}, \overline{y})(u) \subset D\Gamma_C(\overline{x}, \overline{y})(u).$$

From the above inclusions we obtain immediately that (3.48) holds.

Assume, furthermore, that $\Gamma(x)$ is convex near $\overline{x}$ and let y be an element of $\text{Eff}\,(D\Gamma(\overline{x}, \overline{y})(u); C)$ $(\subset D\Gamma(\overline{x}, \overline{y})(u) \subset D\Gamma_C(\overline{x}, \overline{y})(u))$. Suppose that $y \notin \text{Eff}\,(D\Gamma_C(\overline{x}, \overline{y})(u); C)$. Then there exists $k \in C \setminus \{0\}$ such that $y - k \in D\Gamma_C(\overline{x}, \overline{y})(u)$. It follows that there exists $((t_n, u_n, y_n))_{n \in \mathbb{N}} \to (0_+, u, y - k)$ such that $\overline{y} + t_n y_n \in \Gamma(\overline{x} + t_n u_n) + C$ for every $n \in \mathbb{N}$; i.e., $\overline{y} + t_n(y_n - \lambda_n q_n) \in \Gamma(\overline{x} + t_n u_n)$ for every n, with $(q_n) \subset Q$, $(\lambda_n) \subset [0, \infty[$. There exists $n_0 \in \mathbb{N}$ such that $\Gamma(\overline{x} + t_n u_n)$ is convex for every $n \geq n_0$. Suppose that $\limsup \lambda_n > 0$. Then there exist $(n_p)_{p \in \mathbb{N}} \subset \mathbb{N}$ an increasing sequence and $\alpha > 0$ such that $\lambda_{n_p} \geq \alpha$ for every $p \in \mathbb{N}$. Since Q is compact, we may suppose that $(q_{n_p}) \to q \in Q$. Since $\Gamma(\overline{x} + t_{n_p} u_{n_p})$ is convex, we have that

$$\overline{y} + t_{n_p}\left(y_{n_p} - \frac{\alpha}{\lambda_{n_p}} q_{n_p}\right) = \frac{\alpha}{\lambda_{n_p}}\left(\overline{y} + t_{n_p}(y_{n_p} - \lambda_{n_p} q_{n_p})\right)$$
$$+ \left(1 - \frac{\alpha}{\lambda_{n_p}}\right)(\overline{y} + t_{n_p} y_{n_p}) \in \Gamma(\overline{x} + t_{n_p} u_{n_p})$$

for every p. Since $\left(y_{n_p} - \frac{\alpha}{\lambda_{n_p}} q_{n_p}\right) \to y - k - \alpha q$, we obtain that $y - (k + \alpha q) \in D\Gamma(\bar{x}, \bar{y})(u)$, with $k + \alpha q \in C \setminus \{0\}$, contradicting the choice of y. Hence $(\lambda_n) \to 0$, whence the contradiction $y - k \in D\Gamma(\bar{x}, \bar{y})(u)$. Therefore $y \in \mathrm{Eff}\left(D\Gamma_C(\bar{x}, \bar{y})(u); C\right)$. It follows that (3.49) holds.

(iii) The inclusion $\supset$ being proven in (i), let us prove the converse one. Of course, we assume that (3.46) holds. Let $y \in D\Gamma_C(\bar{x}, \bar{y})(u)$; there exist $(t_n) \to 0_+$, $(\lambda_n) \subset [0, \infty)$, $(q_n) \subset Q$, $((u_n, y_n)) \subset X \times Y$ such that $(u_n, y_n + \lambda_n q_n) \to (u, y)$ and $(\bar{x}, \bar{y}) + t_n(u_n, y_n) \in \mathrm{gr}\,\Gamma$ for every $n \in \mathbb{N}$. We may suppose that $q_n \to q \in Q$. Taking also a subsequence if necessary, we may assume that $\lambda_n \to \lambda \in [0, \infty]$. Suppose that $\lambda = \infty$. Since $y_n + \lambda_n q_n \to y$, we obtain that $\lambda_n^{-1} y_n \to -q$. Of course, $\bar{x} + t_n u_n \to \bar{x}$ and

$$\frac{1}{\lambda_n t_n}\left((\bar{x} + t_n u_n, \bar{y} + t_n y_n) - (\bar{x}, \bar{y})\right) = \left(\frac{u_n}{\lambda_n}, \frac{y_n}{\lambda_n}\right) \to (0, -q).$$

It follows that $-q \in S\Gamma(\bar{x}, \bar{y})(0) \cap (-C) \setminus \{0\}$, a contradiction. Therefore $\lambda < \infty$, whence $(u_n, y_n) \to (u, y - \lambda q)$. Hence $y \in D\Gamma(\bar{x}, \bar{y})(u) + C$. The proof is complete. □

Remark 3.6.7. (i) and the first part of (ii) of Theorem 3.6.6 are proved by Tanino [354, Prop. 2.1, Th. 2.1] (see also [212, Th. 3.4]), while part (iii) is proved by Shi [319, Prop. 3.1].

The next example shows that the hypothesis "$\{y \in C \mid \|y\| = 1\}$ is compact" in the previous theorem is essential for having (3.48), (3.49), and (3.50).

Example 3.6.8. Let $X = \mathbb{R}$, $Y = \ell^2$, $C = \ell_+^2 = \{(x_n)_{n \in \mathbb{N}} \in \ell^2 \mid x_n \geq 0\ \forall n \in \mathbb{N}\}$, and $\Gamma : X \rightrightarrows Y$ be defined by

$$\Gamma(x) := \begin{cases} \{0\} & \text{if } x \in]-\infty, 0] \cup [1, \infty[, \\ \frac{1}{n}[-a - e_n, a] & \text{if } x \in [\frac{1}{n+1}, \frac{1}{n}[, \ n \in \mathbb{N}^*, \end{cases}$$

where $a \in \ell_+^2 \setminus \{0\}$ and $(e_n)_{n \in \mathbb{N}}$ is the canonical base of ℓ^2. Then

$$\Gamma_C(x) := \begin{cases} C & \text{if } x \in]-\infty, 0] \cup [1, \infty[, \\ -\frac{1}{n}(a + e_n) + C & \text{if } x \in [\frac{1}{n+1}, \frac{1}{n}[, \ n \in \mathbb{N}^*, \end{cases}$$

$$\Omega(x) := \begin{cases} \{0\} & \text{if } x \in]-\infty, 0] \cup [1, \infty[, \\ -\frac{1}{n}(a + e_n) & \text{if } x \in [\frac{1}{n+1}, \frac{1}{n}[, \ n \in \mathbb{N}^*. \end{cases}$$

Take $\bar{x} = 0$ and $\bar{y} = 0$. Then

$$D\Gamma(\bar{x}, \bar{y})(u) = \begin{cases} \{0\} & \text{if } u \leq 0, \\ \{ua\} & \text{if } u > 0, \end{cases}$$

$$D\Gamma_C(\bar{x}, \bar{y})(u) = \begin{cases} C & \text{if } u \leq 0, \\ -ua + C & \text{if } u > 0, \end{cases}$$

$$D\Omega(\bar{x}, \bar{y})(u) = \begin{cases} \{0\} & \text{if } u \leq 0, \\ \emptyset & \text{if } u > 0. \end{cases}$$

So (3.48), (3.49), and (3.50) do not hold for $u > 0$.

Let us prove the formulae for $D\Gamma(\overline{x}, \overline{y})(u)$ and $D\Gamma_C(\overline{x}, \overline{y})(u)$ for $u \geq 0$, those for $u < 0$ being obvious. So, let $u \geq 0$ and $v \in D\Gamma(\overline{x}, \overline{y})(u)$. Then there exist $(t_n) \subset]0, \infty[$ and gr $\Gamma \supset ((x_n, y_n)) \to (0, 0)$ such that $(t_n x_n) \to u$ and $(t_n y_n) \to v$. If $P := \{n \in \mathbb{N} \mid x_n \leq 0\}$ is infinite, then $y_n = 0$ for every $n \in P$, and so $(u, v) = (0, 0)$. Suppose that P is finite (this is the case when $u > 0$). Since $(x_n) \to 0$, there exists $n_0 \in \mathbb{N}^*$ such that $x_n \in]0, 1[$ for $n \geq n_0$. For every $n \geq n_0$ there exists $p_n \in \mathbb{N}^*$ such that $x_n \in [\frac{1}{p_n+1}, \frac{1}{p_n}[$; of course, $(p_n) \to \infty$. Then for every $n \geq n_0$ there exists $\lambda_n \in [0, 1]$ such that $y_n = \frac{1}{p_n}((1 - 2\lambda_n)a - \lambda_n e_n)$. It follows that $(\frac{t_n}{p_n}) \to u$ and $(\frac{t_n}{p_n}((1 - 2\lambda_n)a - \lambda_n e_{p_n})) \to v$. If $u = 0$, it is obvious that $v = 0$. Suppose that $u > 0$. Taking eventually a subsequence, we may assume that $(\lambda_n) \to \lambda \in [0, 1]$. We obtain that $(\lambda_n e_{p_n}) \to (1 - 2\lambda)a - u^{-1}v$. If $\lambda \neq 0$, we get the contradiction that (e_n) contains a norm-convergent subsequence. Therefore $(\lambda_n) \to 0$, whence $v = ua$. It is easy to get that $ua \in D\Gamma(\overline{x}, \overline{y})(u)$ for $u \geq 0$ (in fact, similar to the proof in the next paragraph). So the formula for $D\Gamma(\overline{x}, \overline{y})(u)$ holds.

Consider first $u > 0$ and $v = -ua + k$ with $k \in C$. Taking $x_n = \frac{1}{n+1}$, $y_n = \frac{1}{n}(-a + u^{-1}k) = \frac{1}{n}(-a - e_n + u^{-1}k + e_n)$ $(\in \Gamma_C(x_n))$, and $t_n = (n+1)u > 0$ for $n \in \mathbb{N}^*$, we have that $((x_n, y_n)) \to (0, 0)$ and $(t_n(x_n, y_n)) \to (u, v)$; if $u = 0$ and $v = k \in k$ just take $x_n = 0$, $y_n = \frac{1}{n}k$, and $t_n = n$. Therefore $v \in D\Gamma_C(\overline{x}, \overline{y})(u)$.

Let $u \geq 0$ and $v \in D\Gamma_C(\overline{x}, \overline{y})(u)$. Then there exist $(t_n) \subset]0, \infty[$ and gr $\Gamma_C \supset ((x_n, y_n)) \to (0, 0)$ such that $(t_n(x_n, y_n)) \to (u, v)$. As in the first part we may suppose that $x_n \in]0, 1[$ for $n \geq n_0$. So there exists $\mathbb{N}^* \supset (p_n) \to \infty$ with $x_n \in [\frac{1}{p_n+1}, \frac{1}{p_n}[$ for $n \geq n_0$. Of course, $y_n = -\frac{1}{p_n}(a + e_{p_n}) + k_n$ with $k_n \in C$. It follows that $(\frac{t_n}{p_n}) \to u$ and $(-\frac{t_n}{p_n}(a + e_{p_n}) + t_n k_n) \to v$. Since $0 = w\text{-}\lim e_n$, it follows that $v + ua = w\text{-}\lim(t_n k_n)$. Because C is weakly closed, $v + ua \in C$, whence $v \in -ua + C$.

The formula for $D\Omega(\overline{x}, \overline{y})(u)$ is obtained similarly.

In Example 3.6.3, (3.46) is not satisfied and $\Gamma(x)$ is not convex for $x > 0$; (3.50) and (3.49) do not hold for $u > 0$.

Corollary 3.6.9. *Assume that the conditions of Theorem 3.6.6 (iii) hold. Then the formulae (3.49) and (3.51) below hold:*

$$\text{HEff}\,(D\Gamma(\overline{x}, \overline{y})(u); C) = \text{HEff}\,(D\Gamma_C(\overline{x}, \overline{y})(u); C) \quad \forall u \in X. \qquad (3.51)$$

Moreover, if $\text{int}\, C \neq \emptyset$ *(so* $\dim Y < \infty$*), then*

$$w\,\text{Eff}\,(D\Gamma(\overline{x}, \overline{y})(u); C) \subset w\,\text{Eff}\,(D\Gamma_C(\overline{x}, \overline{y})(u); C) \quad \forall u \in X, \qquad (3.52)$$
$$w\,\text{Eff}\,(D\Gamma(\overline{x}, \overline{y})(u); C) = w\,\text{Eff}\,(D\Gamma_{\widetilde{C}}(\overline{x}, \overline{y})(u); C) \quad \forall u \in X, \qquad (3.53)$$

where $\widetilde{C}$ *is a closed convex cone such that* $\widetilde{C} \subset \{0\} \cup \text{int}\, C$.

PROOF. Using the preceding theorem, equality (3.50) holds. The conclusion then follows by using Proposition 3.2.15 (iv) for the pair (C, K) (there) replaced by the pairs (C, C) and (C, K), respectively, where the last K is from the definition of Henson proper efficiency.

Suppose now that $\operatorname{int} C \neq \emptyset$. Then (3.54) follows immediately from (3.50). Let $\widetilde{C}$ be a closed convex cone with $\widetilde{C} \subset \{0\} \cup \operatorname{int} C$. It is obvious that $\widetilde{C}$ satisfies the conditions of Theorem 3.6.6 (iii), and so (3.50) holds with C replaced by $\widetilde{C}$. Applying again Proposition 3.2.15 (iv), now for (C, K) replaced by $(\widetilde{C}, C_0)$ (with $C_0 = \{0\} \cup \operatorname{int} C$), we obtain (3.53), too.   □

Remark 3.6.10. When X, Y are finite-dimensional, formulae (3.49), (3.51), (3.54), and (3.53) are established in [220, Th. 2.1] under the conditions of Corollary 3.6.9, while formulae (3.51) and (3.53) are established in [221, Th. 2.1] for $\overline{x} \in \operatorname{int}(\operatorname{dom} \Gamma)$ and Γ locally C-convex near $\overline{x}$; i.e., $\Gamma|_V$ is C-convex for some neighborhood V of $\overline{x}$ [see also Theorem 3.6.1 (ii)].

In Example 3.6.3, (3.46) is not satisfied and (3.51), (3.52), (3.53) (for $\widetilde{C} = C$) do not hold for $u > 0$.

Theorem 3.6.11. *Let $(\overline{x}, \overline{y}) \in \operatorname{gr} \Gamma$. Assume that $\{y \in C \mid \|y\| = 1\}$ is compact and Γ is C-dominated by Ω near $\overline{x}$, i.e., $\Gamma(u) \subset \Omega(u) + C$ for every $u \in V$ for some neighborhood V of $\overline{x}$.*

(i) *If $\overline{y} \in \Omega(\overline{x})$, then for every $u \in X$,*

$$\operatorname{Eff}\left(D\Gamma_C(\overline{x}, \overline{y})(u); C\right) \subset \operatorname{Eff}\left(D\Omega(\overline{x}, \overline{y})(u); C\right) \subset D\Omega(\overline{x}, \overline{y})(u). \quad (3.54)$$

Moreover, if $\Gamma(x)$ is convex for $x \in V$, with V a neighborhood of $\overline{x}$, then

$$\operatorname{Eff}\left(D\Gamma(\overline{x}, \overline{y})(u); C\right) \subset \operatorname{Eff}\left(D\Omega(\overline{x}, \overline{y})(u); C\right) \quad \forall u \in X. \quad (3.55)$$

(ii) *If (3.46) holds, then $\overline{y} \in \Omega(\overline{x})$ and for every $u \in X$,*

$$\operatorname{Eff}\left(D\Gamma(\overline{x}, \overline{y})(u); C\right) = \operatorname{Eff}\left(D\Omega(\overline{x}, \overline{y})(u); C\right) \subset D\Omega(\overline{x}, \overline{y})(u), \quad (3.56)$$

$$\operatorname{HEff}\left(D\Gamma(\overline{x}, \overline{y})(u); C\right) = \operatorname{HEff}\left(D\Omega(\overline{x}, \overline{y})(u); C\right) \subset D\Omega(\overline{x}, \overline{y})(u). \quad (3.57)$$

PROOF. By hypothesis we have that $\Gamma_C(u) = \Omega_C(u)$ for every $u \in V$, where V is a neighborhood of $\overline{x}$. It follows that $D\Gamma_C(\overline{x}, \overline{y})(u) = D\Omega_C(\overline{x}, \overline{y})(u)$ for every $u \in X$.

(i) Using Theorem 3.6.6 (ii), we obtain for every $u \in X$,

$$\operatorname{Eff}\left(D\Gamma_C(\overline{x}, \overline{y})(u); C\right) = \operatorname{Eff}\left(D\Omega_C(\overline{x}, \overline{y})(u); C\right)$$
$$\subset \operatorname{Eff}\left(D\Omega(\overline{x}, \overline{y})(u); C\right) \subset D\Omega(\overline{x}, \overline{y})(u).$$

If $\Gamma(x)$ is convex for $x \in V$, with V a neighborhood of $\overline{x}$, by Theorem 3.6.6 (ii), (3.49) holds, whence the inclusion (3.55) follows immediately.

(ii) Assume that (3.46) holds. We already remarked that $\overline{y}$ is an element of $\mathrm{BEff}(\Gamma(\overline{x}); C)$, and so $\overline{y} \in \Omega(\overline{x})$. Since (3.46) holds for Γ, it holds for Ω, too. It follows that equality (3.50) and the corresponding one for Ω hold. The conclusion is immediate from Proposition 3.2.15 (iv), taking $K = C$ for (3.56) and (C, K) for (3.57), where K is the cone in the definition of Henig-proper efficient point. □

In Example 3.6.8 the set $\{y \in C \mid \|y\| = 1\}$ is not compact, but all the conditions on Γ in Theorem 3.6.11 hold, although the relations (3.54), (3.55), (3.56), and (3.57) do not hold for $u > 0$.

In Example 3.6.3, (3.46) is not satisfied and $\Gamma(x)$ is not convex for $x > 0$ (but $\Gamma(x)$ is C-dominated for every x); (3.55), (3.56), and (3.57) do not hold for $u > 0$.

In the next example $\Gamma(x)$ is not C-dominated for $x < 0$, the other conditions of Theorem 3.6.11 being satisfied; (3.55), (3.56), and (3.57) do not hold for $u < 0$.

Example 3.6.12. ([221]) Let $X = \mathbb{R}$, $Y = \mathbb{R}^2$, $C = \mathbb{R}^2_+$, and $\Gamma : \mathbb{R} \rightrightarrows \mathbb{R}^2$ be defined by

$$\Gamma(x) := \begin{cases} \{(y_1, y_2) \mid y_1 \geq 0, \ y_2 \geq y_1^2\} \text{ if } x \geq 0, \\ \{(y_1, y_2) \mid y_1 > 0, \ y_2 \geq y_1^2\} \text{ if } x < 0. \end{cases}$$

Then

$$\Omega(x) := \begin{cases} \{(0,0)\} \text{ if } x \geq 0, \\ \emptyset \qquad\quad \text{ if } x < 0. \end{cases}$$

Take $\overline{x} = 0$, $\overline{y} = (0,0)$. Then

$$D\Gamma(\overline{x},\overline{y})(u) = D\Gamma_C(\overline{x},\overline{y})(u) = C, \quad D\Omega(\overline{x},\overline{y})(u) = \Omega(u) \quad \forall\, u \in X.$$

Corollary 3.6.13. *Suppose that $\overline{y} \in \mathrm{BEff}\,(\Gamma(\overline{x}); C)$ and Γ is C-dominated by Ω near $\overline{x}$. Then (3.56) and (3.57) hold if one of the following conditions is satisfied:*

(i) *Γ is upper Lipschitz at $\overline{x}$ and $\{y \in C \mid \|y\| = 1\}$ is compact;*
(ii) *X, Y are finite-dimensional, $\overline{x} \in \mathrm{int}(\mathrm{dom}\,\Gamma)$, and Γ is C-convex near $\overline{x}$.*

PROOF. Applying Theorem 3.6.1, condition (3.46) holds if (i) or (ii) is satisfied. The conclusion follows by applying now part (ii) of the preceding theorem. □

Theorem 3.6.14. *Let $(\overline{x},\overline{y}) \in \mathrm{gr}\,\Gamma$. Assume that (3.46) holds, $\dim Y < \infty$, $\mathrm{int}\,C \neq \emptyset$, and Γ is $\widetilde{C}$-dominated by Ω_w near $\overline{x}$ for some closed convex cone $\widetilde{C}$ with $\widetilde{C} \subset \{0\} \cup \mathrm{int}\,C$. Then for every $u \in X$,*

$$w\,\mathrm{Eff}\,(D\Gamma(\overline{x},\overline{y})(u); C) = w\,\mathrm{Eff}\,(D\Omega_w(\overline{x},\overline{y})(u); C) \subset D\Omega_w(\overline{x},\overline{y})(u). \quad (3.58)$$

PROOF. Since (3.46) holds, we have that $\overline{y} \in \Omega(\overline{x}) \subset \Omega_w(\overline{x})$. Since the conditions of Corollary 3.6.9 hold, equality (3.53) holds, too. Since $\mathrm{gr}\,\Omega_w \subset \mathrm{gr}\,\Gamma$, condition (3.46) is satisfied with Γ replaced by Ω_w, whence (3.53) holds for the same substitution. Because Γ is $\widetilde{C}$-dominated by Ω_w near $\overline{x}$ we have that $\Gamma_{\widetilde{C}}(u) = (\Omega_w)_{\widetilde{C}}(u)$ for every u in a neighborhood of $\overline{x}$, whence $D\Gamma_{\widetilde{C}}(\overline{x},\overline{y}) = D(\Omega_w)_{\widetilde{C}}(\overline{x},\overline{y})$. The conclusion is obtained immediately by using again Proposition 3.2.15 (iv) for the pair of cones $(\widetilde{C}, C_0)$. □

In Examples 3.6.3 and 3.6.12 relation (3.58) does not hold; in Example 3.6.3, (3.46) is not satisfied, while in Example 3.6.12, $\Gamma(x)$ is not $\widetilde{C}$-dominated by $\Omega_w(x)$ for $x < 0$ for some closed convex cone $\widetilde{C}$ with $\widetilde{C} \subset \{0\} \cup \operatorname{int} C$.

Corollary 3.6.15. *Suppose that X, Y are finite-dimensional, $\overline{x} \in \operatorname{int}(\operatorname{dom} F)$, $\overline{y} \in \operatorname{BEff}(\Gamma(\overline{x}); C)$, and Γ is C-convex near $\overline{x}$. If Γ is $\widetilde{C}$-dominated by Ω_w near $\overline{x}$ for some closed convex cone $\widetilde{C}$ with $\widetilde{C} \subset \{0\} \cup \operatorname{int} C$, then (3.58) holds.*

PROOF. Applying Theorem 3.6.1, condition (3.46) holds. The conclusion follows from the preceding theorem. □

Besides Examples 3.6.3 and 3.6.12 mentioned above, we consider the following example where Γ is not C-convex (near $\overline{x}$); relation (3.58) does not hold for every $u \in X$.

Example 3.6.16. ([320]) Let $X = Y = \mathbb{R}$, $C = \mathbb{R}_+$, and let $\Gamma : \mathbb{R} \rightrightarrows \mathbb{R}$ be defined by

$$\Gamma(x) := \begin{cases} [-|x|, \infty[\cup \{-\sqrt{|x|}\} & \text{if } |x| \le 1, \\ [-\sqrt{|x|}, \infty[& \text{if } |x| > 1. \end{cases}$$

Then

$$\Gamma_C(x) = [-\sqrt{|x|}, \infty[, \quad \Omega(x) = \{-\sqrt{|x|}\} \quad \forall x \in X.$$

Take $\overline{x} = 0$, $\overline{y} = 0$. Then

$$D\Gamma(\overline{x},\overline{y})(u) = [-|u|, \infty[, \quad D\Gamma_C(\overline{x},\overline{y})(u) = X, \quad \forall u \in X,$$
$$D\Omega(\overline{x},\overline{y})(u) = \begin{cases}]-\infty, 0] & \text{if } u = 0, \\ \emptyset & \text{if } u \ne 0. \end{cases}$$

Corollary 3.6.17. *Suppose that Y is finite-dimensional, $\overline{y} \in \operatorname{BEff}(\Gamma(\overline{x}); C)$, $\Gamma(u)$ is C-closed for all u in a neighborhood of $\overline{x}$, and one of the following two conditions is satisfied:*

(i) *(3.46) holds and $\Gamma(u)$ is C-bounded (i.e., $(\Gamma(u))_\infty \cap (-C) = \{0\}$) for all u in a neighborhood of $\overline{x}$;*
(ii) *X is finite-dimensional, $\overline{x} \in \operatorname{int}(\operatorname{dom}\Gamma)$, and Γ is C-convex and C-dominated by Ω near $\overline{x}$.*

Then for every $u \in X$,

$$\operatorname{Eff}(D\Gamma(\overline{x},\overline{y})(u); C) = \operatorname{Eff}(D(H\Omega)(\overline{x},\overline{y})(u); C) \subset D(H\Omega)(\overline{x},\overline{y})(u).$$

PROOF. We have that (ii) ⇒ (i). Indeed, condition (3.46) holds by Theorem 3.6.1. Since Γ is C-dominated by Ω near $\overline{x} \in \operatorname{int}(\operatorname{dom} F)$, we have that $\Omega(u) \neq \emptyset$ for all u in a neighborhood V of $\overline{x}$; we may suppose that $\Gamma(u)$ is also C-closed and C-convex for $u \in V$. By a known result it follows that $\Gamma(u)$ is C-bounded for $u \in V$.

Suppose that (i) holds. Let V be a neighborhood of $\overline{x}$ such that $\Gamma(u)$ is C-dominated by $\Omega(u)$ and C-bounded for every $u \in V$. By [161, Th. 5.1] we have that

$$H\Omega(u) \subset \Omega(u) \subset \operatorname{cl}(H\Omega(u)) \quad \forall u \in V.$$

It follows that $\operatorname{gr} H\Omega|_V \subset \operatorname{gr} \Omega|_V \subset \operatorname{cl}(\operatorname{gr} H\Omega|_V)$, and so $D\Omega(\overline{x}, \overline{y}) = D(H\Omega)(\overline{x}, \overline{y})$. The conclusion follows by applying Theorem 3.6.11 (ii). □

In Example 3.6.16, $S\Gamma(\overline{x}, \overline{y})(0) = X$. So (3.46) does not hold and Γ is not C-convex near $\overline{x}$; the conclusion of the preceding corollary does not hold.

Remark 3.6.18. Formula (3.54) of Theorem 3.6.11 is proved by Tanino [354, Th. 3.1] for Y finite-dimensional and by Klose [212, Th. 3.4] for arbitrary Y; for X and Y finite-dimensional spaces, formula (3.55) is proved by Shi [320, Th. 4.1], Theorem 3.6.11 (ii) is proved by Shi [319, Th. 4.1], and Theorem 3.6.14 is proved by Kuk, Tanino, and Tanaka in [220, Th. 3.1] in the same conditions. Corollary 3.6.13 (i) was obtained by Tanino [354, Th. 3.2] for Y finite-dimensional, Shi [319, Cor. 4.1], and Klose [212, Th. 3.5], while part (ii) by Shi [320, Th. 4.2]. Corollary 3.6.15 was obtained by Shi in [320, Th. 5.2]. Formula (3.57) was obtained in [221, Th. 3.1] under the conditions of Corollary 3.6.13 (ii). Corollary 3.6.17 (i) and (ii) reinforce [220, Th. 3.2] and [221, Cor. 3.1] of Kuk, Tanino, and Tanaka.

Theorem 3.6.19. *Suppose that* $\operatorname{int} C \neq \emptyset$ *and consider* $\overline{y} \in \Omega_w(\overline{x})$. *If one of the following conditions holds,*

(i) Γ *is semidifferentiable at* $(\overline{x}, \overline{y})$;
(ii) Γ *is* C-convex *and* $\overline{x} \in \operatorname{int}(\operatorname{dom} \Gamma)$;
(iii) X, Y *are finite-dimensional,* Γ *is* C-convex *and* $\overline{x} \in \operatorname{raint}(\operatorname{dom} \Gamma)$,

then

$$D\Omega_w(\overline{x}, \overline{y})(u) \subset w\operatorname{Eff}(D\Gamma(\overline{x}, \overline{y})(u); C) \quad \forall u \in X. \tag{3.59}$$

Suppose that either (a) *(3.46) and* (i) *hold or* (b) $\overline{y} \in \operatorname{BEff}(\Gamma(\overline{x}); C)$ *and* (iii) *hold. If* $\{y \in C \mid \|y\| = 1\}$ *is compact and* Γ *is* $\widetilde{C}$-dominated *by* Ω_w *or* Ω *near* $\overline{x}$, *where* $\widetilde{C} \subset \{0\} \cup \operatorname{int} C$ *is a closed convex cone, then*

$$D\Omega_w(\overline{x}, \overline{y})(u) = w\operatorname{Eff}(D\Gamma(\overline{x}, \overline{y})(u); C) \quad \forall u \in X \tag{3.60}$$

or

$$D\Omega(\overline{x}, \overline{y})(u) = w\operatorname{Eff}(D\Gamma(\overline{x}, \overline{y})(u); C) \quad \forall u \in X, \tag{3.61}$$

respectively.

PROOF. Assume that (i) holds and consider $y \in D\Omega_w(\overline{x}, \overline{y})(u)$. If y does not belong to $w\,\mathrm{Eff}\,(D\Gamma(\overline{x}, \overline{y})(u); C)$, there exists $k \in \mathrm{int}\,C$ such that $y - k \in D\Omega_w(\overline{x}, \overline{y})(u)$. Therefore there are $(t_n) \to 0_+$ and $((u_n, y_n)) \to (u, y - k)$ such that $(\overline{x}, \overline{y}) + t_n(u_n, y_n) \in \mathrm{gr}\,\Omega_w \subset \mathrm{gr}\,\Gamma$. Since Γ is semidifferentiable at $(\overline{x}, \overline{y})$ and $(t_n) \to 0_+$, $u_n \to u$, there exists $(y_n') \to y - k$ such that $(\overline{x}, \overline{y}) + t_n(u_n, y_n') \in \mathrm{gr}\,\Gamma$ for every $n \geq n_0$. Therefore $\overline{y} + t_n y_n' \in \Gamma(\overline{x} + t_n u_n)$ and $\overline{y} + t_n y_n \in \Omega_w(\overline{x} + t_n u_n) \subset \Gamma(\overline{x} + t_n u_n)$. Since

$$((\overline{y} + t_n y_n') - (\overline{y} + t_n y_n))/t_n = y_n' - y_n \to y - k - y = -k \in -\,\mathrm{int}\,C,$$

there exists some $n_1 \geq n_0$ such that $(\overline{y} + t_n y_n') - (\overline{y} + t_n y_n) \in -\,\mathrm{int}\,C$ for every $n \geq n_1$, contradicting the fact that $\overline{y} + t_n y_n \in \Omega_w(\overline{x} + t_n u_n)$.

Assume now that (ii) holds; since $\mathrm{int}\,C \neq \emptyset$, $(\overline{x}, \overline{y}, \overline{k}) \in \mathrm{int}(\mathrm{gr}\,\Gamma_C)$ for $\overline{k} \in \mathrm{int}\,C$. By Theorem 2.7.6 (i) Γ_C is semidifferentiable at $(\overline{x}, \overline{y})$. Taking $y \in D\Omega_w(\overline{x}, \overline{y})(u)$, with the same argument as in the proof of (i), we obtain the same sequences, the sole difference being that $\overline{y} + t_n y_n' \in \Gamma(\overline{x} + t_n u_n) + C$ for $n \geq n_0$. The same contradiction is obtained.

If (iii) holds, by Theorem 2.7.6 (ii), Γ_C is semidifferentiable at $(\overline{x}, \overline{y})$. The proof is the same as for (ii).

(b) $\Rightarrow$ (a). Indeed, if (iii) holds, using Theorem 3.6.1 (ii) we get that (3.46) is satisfied.

Assume that (a) is satisfied and $\Gamma(u)$ is $\widetilde{C}$-dominated by $\Omega_w(u)$ for all $u \in V$, for some neighborhood V of $\overline{x}$. It follows that $\Gamma(u) + \widetilde{C} = \Omega_w(u) + \widetilde{C}$ for $u \in V$. So, $D\Gamma_{\widetilde{C}}(\overline{x}, \overline{y}) = D(\Omega_w)_{\widetilde{C}}(\overline{x}, \overline{y})$. Applying Theorem 3.6.6 for $(\Gamma, \widetilde{C})$ and $(\Omega_w, \widetilde{C})$ we obtain that $D\Gamma(\overline{x}, \overline{y})(u) + \widetilde{C} = D\Omega_w(\overline{x}, \overline{y})(u) + \widetilde{C}$ for every $u \in X$. Applying now Proposition 3.2.15 (iv) for the pair $(\widetilde{C}, C_0)$ we get for every $u \in X$,

$$w\,\mathrm{Eff}\,(D\Gamma(\overline{x}, \overline{y})(u); C) = w\,\mathrm{Eff}\,(D\Omega_w(\overline{x}, \overline{y})(u); C) \subset D\Omega_w(\overline{x}, \overline{y})(u).$$

From (3.59) and the above relation (3.60) follows.

If Γ is $\widetilde{C}$-dominated by Ω near $\overline{x}$, all we said above is true if we replace Ω_w by Ω. Therefore (3.61) holds, too. Note that $\overline{y} \in \Omega(\overline{x})$ because (3.46) holds. $\square$

In Example 3.6.3 Γ is neither semidifferentiable at $\overline{x}$, nor C-convex; relation (3.59) does not hold for $u = 0$. In Example 3.6.12 (a) and (b) are satisfied but Γ is not $\widetilde{C}$-dominated by Ω_w or Ω near $\overline{x}$, for some closed convex cone $\widetilde{C} \subset \{0\} \cup \mathrm{int}\,C$; relations (3.60) and (3.61) do not hold for $u < 0$.

In the next example neither (a) nor (b) is satisfied; relations (3.60) and (3.61) do not hold for $u > 0$.

Example 3.6.20. ([354]) Let $X = \mathbb{R}$, $Y = \mathbb{R}^2$, $C = \mathbb{R}_+^2$, and let $\Gamma : \mathbb{R} \rightrightarrows \mathbb{R}^2$ be defined by

$$\Gamma(x) := \{(y_1, -y_1) \mid y_1 \leq x\} \cup \{(y_1, -1 - y_1) \mid y_1 > 0\}.$$

Then

$$\Omega(x) = \Omega_w(x) = \{(y_1, -y_1) \mid y_1 \leq \min(0, x)\} \cup \{(y_1, -1 - y_1) \mid y_1 > 0\}.$$

Let $\overline{x} = 0$ and $\overline{y} = (0, 0)$. Then

$$D\Gamma(\overline{x}, \overline{y})(u) = \text{Eff}\,(D\Gamma(\overline{x}, \overline{y})(u); C) = \{(v_1, -v_1) \mid v_1 \leq u\},$$
$$D\Omega(\overline{x}, \overline{y})(u) = \text{Eff}\,(D\Gamma(\overline{x}, \overline{y})(u); C) = \{(v_1, -v_1) \mid v_1 \leq \min(0, u)\}.$$

Remark 3.6.21. For Ω_w replaced by Ω and X, Y finite-dimensional, (3.59) was obtained by Shi [320, Th. 5.1] under condition (iii) and by Kuk, Tanino, and Tanaka [220, Th. 3.3] under (i). Kuk, Tanino, and Tanaka obtained (3.59) under condition (iii) and (3.60) under condition (b) in [221, Th. 3.2] and (3.61) under condition (b) in [221, Th. 3.3].

Consider now another normed vector space $\mathcal{U}$, the multifunction $\Lambda : \mathcal{U} \rightrightarrows X$, the function $f : X \times \mathcal{U} \to Y$, and the multifunction $\Gamma : \mathcal{U} \rightrightarrows Y$ defined by $\Gamma(u) := f(\Lambda(u) \times \{u\})$.

Theorem 3.6.22. *Suppose that the following conditions hold:*

(i) Λ *is upper Lipschitz at* $\overline{u} \in \text{dom}\,\Lambda$ *and* $\Lambda(u)$ *is compact for* $u \in U$, *with* U *a neighborhood of* $\overline{u}$;
(ii) $\overline{x} \in \Lambda(\overline{u})$ *and* $\overline{y} = f(\overline{x}, \overline{u}) \in \text{BEff}(\Gamma(\overline{u}); C)$;
(iii) X *is finite-dimensional,* C *has a compact base,* f *is Lipschitz on bounded sets and Fréchet differentiable at* $(\overline{x}, \overline{u})$;
(iv) $\Lambda(\overline{u}) = \{\overline{x}\}$ *or the multifunction*

$$\widetilde{\Lambda} : \mathcal{U} \times Y \rightrightarrows X, \quad \widetilde{\Lambda}(u, y) := \{x \in \Lambda(u) \mid f(x, u) = y\},$$

is upper Lipschitz at $(\overline{u}, \overline{y})$ *and* $\widetilde{\Lambda}(\overline{u}, \overline{y}) = \{\overline{x}\}$.

Then for every $u \in X$,

$$\text{Eff}\,(\nabla_x f(\overline{x}, \overline{u})\,(D\Lambda(\overline{u}, \overline{x})(u)) + \nabla_u f(\overline{x}, \overline{u})(u); C) = \text{Eff}\,(D\Omega(\overline{u}, \overline{y})(u); C).$$

In order to prove this result we need the following lemma.

Lemma 3.6.23. *Let* Λ, f, Γ *be as above and* $\overline{u} \in \text{dom}\,\Lambda$. *If* f *is Lipschitz on bounded sets,* $\Lambda(\overline{u})$ *is bounded, and* Λ *is upper Lipschitz at* $\overline{u}$, *then* Γ *is upper Lipschitz at* $\overline{u}$.

PROOF. Consider the box norm on $\mathcal{U} \times X$. Since Λ is upper Lipschitz at $\overline{u}$, there exist $L, \rho > 0$ such that

$$\Lambda(u) \subset \Lambda(\overline{u}) + L\,\|u - \overline{u}\|\,B_X \quad \forall u \in B(\overline{u}, \rho). \tag{3.62}$$

It follows that the set $A := \bigcup_{u \in B(\overline{u}, \rho)} (\Lambda(u) \times \{u\})$ is bounded. From hypothesis, there exists $L' > 0$ such that f is L'-Lipschitz on A. Let $u \in B(\overline{u}, \rho)$

and $y \in \Gamma(u)$. Hence there exists $x \in \Lambda(u)$ such that $y = f(x, u)$. From (3.62), there exists $x' \in \Lambda(\bar{u})$ such that $\|x - x'\| \le L\|u - \bar{u}\|$. Taking $y' = f(x', \bar{u}) \in \Gamma(\bar{u})$, we obtain that

$$\|y - y'\| = \|f(x, u) - f(x', \bar{u})\| \le L' \|(x - x', u - \bar{u})\| \le L' \max(L, 1)\|u - \bar{u}\|.$$

Taking $L'' > L' \max(L, 1)$, we have that $\Gamma(u) \subset \Gamma(\bar{u}) + L''\|u - \bar{u}\| B_Y$ for every $u \in B(\bar{u}, \rho)$. Therefore Γ is upper Lipschitz at $\bar{u}$. □

Of course, a sufficient condition for f to be Lipschitz on bounded sets is that f be of class C^1 and $\mathcal{U}$ and X have finite dimension.

Proof of Theorem 3.6.22. From (i) and (iii), using Lemma 3.6.23, we obtain that Γ is upper Lipschitz at $\bar{u}$. Since $\bar{y} \in \mathrm{BEff}(\Gamma(\bar{u}); C)$, from Theorem 3.6.1(i) we obtain that (3.46) holds for Γ at $(\bar{u}, \bar{y})$. Since $\Lambda(u)$ is compact for $u \in U$ and f is continuous, $\Gamma(u)$ is compact for $u \in U$. Therefore $\Gamma(u) \subset \Omega(u) + C$ for $u \in U$. Now using Theorem 3.6.11(ii), we have that

$$\mathrm{Eff}\,(D\Gamma(\bar{u}, \bar{y})(u); C) = \mathrm{Eff}\,(D\Omega(\bar{u}, \bar{y})(u); C) \quad \forall u \in \mathcal{U}.$$

But from the second part of Theorem 2.7.8, we have that

$$D\Gamma(\bar{u}, \bar{y})(u) = \nabla_x f(\bar{x}, \bar{u})\,(D\Lambda(\bar{u}, \bar{x})(u)) + \nabla_u f(\bar{x}, \bar{u})(u) \quad \forall u \in \mathcal{U}.$$

The conclusion follows. □

Note that the preceding theorem (with a slightly weaker conclusion) is obtained by Tanino in [354, Th. 4.1] for $\mathcal{U}$, X, Y finite-dimensional and f of class C^1 and by Klose [212, Th. 4.4] (observe that the compactness of $\Lambda(u)$ for $u \in U$ was used for having that $\Gamma(u) \subset \Omega(u) + C$, while the Lipschitz condition of f on bounded sets for having that Γ is upper Lipschitz at $\bar{u}$, conditions written directly in [212]). Variants of Lemma 3.6.23 are stated and proved in [354, Lemma 4.1] and [212, Lemmas 4.2, 4.3].

3.7 Duality

It is an old idea to try to complement a given optimization problem ($f(x) \to$ min with minimal value I) by a dual problem ($g(y) \to$ max with supremal value S, $S \le I$); remember the dual variational principles of Dirichlet and Thompson (see Zeidler [391]) or, e.g., the paper of K.O. Friedrichs [121]. There are at least three reasons to look for useful dual problems:

- The dual problem has (under additional conditions) the same optimal value as the given "primal" optimization problem, but solving the dual problem could be done with other methods of analysis or numerical mathematics.
- An approximate solution of the given minimization problem gives an estimation of the minimal value I from above, whereas an approximate solution of the dual problem is an estimation of I from below, so that one gets intervals that contain I.

- Recalling the Lagrange method, saddle points, equilibrium points of two-person games, shadow prices in economics, perturbation methods or dual variational principles, it becomes clear that optimal dual variables often have a special meaning for the given problem.

Of course, the advantages just listed require a skillfully chosen dual program. Nevertheless, the mentioned points are motivation enough, to look for dual problems in multicriteria optimization too. There are many papers dedicated to that aim, as well as many survey papers (see Jahn [195], Luc [244]). So we shall concentrate firstly upon nonconvex **multicriteria** problems and derive theoretical duality assertions. These theoretical results will be used in order to derive duality assertions on the basis of the special structure of multicriteria **approximation** problems.

There are different approaches to duality:

Conjugation: Schönfeld [318], Breckner [43, 44], Zowe [394, 395], Nehse [268], Rosinger [311], Tanino and Sawaragi [355], Brumelle [50], Kawasaki [207, 208], Gerstewitz and Göpfert [128], Malivert [248], Sawaragi, Nakayama and Tanino [314], Luc [244], Zălinescu [386].

Lagrangian: Corley [83, 84], Bitran [31], Nehse [268], Gerstewitz and Iwanow [129], Göpfert and Gerth [138], Jahn [190, 193, 195], Iwanow and Nehse [188], Nakayama [265, 266], Sawaragi, Nakayama and Tanino [314], Luc [244], Dolecki and Malivert [99].

Axiomatic Duality: Luc [244].

3.7.1 Duality Without Scalarization

Let (Y, C) be a (with the cone C) partially ordered linear topological space, P, D nonempty subsets of Y, and let us consider the following multicriteria problems using the solution concepts introduced in Remark 2.1.3

$$(P) \qquad \mathrm{Eff}_{\mathrm{Min}}(P, C),$$
$$(D) \qquad \mathrm{Eff}_{\mathrm{Max}}(D, C).$$

We speak of a pair of **weakly dual problems**, if

$$P \cap (D - (C \setminus \{0\})) = \emptyset. \tag{3.63}$$

Since C is pointed, this is equivalent to

$$(P + (C \setminus \{0\})) \cap (D - (C \setminus \{0\})) = \emptyset.$$

(P) and (D) are called **strongly dual**, if (3.63) holds together with $0 \in \mathrm{cl}(P - D)$ or equivalently $(P + O) \cap (D + O) \neq \emptyset$ for all open neighborhoods O of zero in Y. So strong duality means that P and D touch each other. Otherwise, we speak of a pair of dual problems with a duality gap (in the aforementioned scalar case at the beginning of this chapter that would mean $I > S$).

Lemma 3.7.1. *Assume (P), (D) to be weakly dual. If there are $z^0 \in B, \zeta^0 \in D$ such that $z^0 = \zeta^0$, then z^0 is minimal-efficient for (P), ζ^0 maximal-efficient for (D), and (P), (D) are strongly dual.*

PROOF. z^0 not minimal-efficient means that there is $z^1 \in P$ such that $z^1 \in z^0 - (C \setminus \{0\}) = \zeta^0 - (C \setminus \{0\})$, which contradicts (3.63). For ζ^0 to be maximal-efficient follows similarly. From $z^0 = \zeta^0$ it follows that (P) and (D) are strongly dual.   □

As usual, if one deals with multicriteria optimization problems, the question is whether to apply scalarization methods. So at first we prove duality theorems completely without taking into account scalarization of the goal function.

Instead of (P) we consider the following multicriteria problem with side conditions

$$(P_1)   \text{Eff}_{\text{Min}}(f[\mathcal{A}], C)   \tag{3.64}$$

with

$$\mathcal{A} := \{x \in M, g(x) \in C_V\},$$

where X is a linear space, M a nonempty set in X, (V, C_V) a reflexively preordered linear topological space, (Y, C) as above, $f : X \rightarrow Y^\bullet$, dom $f = M$, $g : X \rightarrow V^\bullet$, and dom $g = M$. So a vector optimization problem is given as usual.

Remark 3.7.2. To deal with duality in vector optimization, we suppose that for C in Y we choose a closed convex pointed cone $B \subset Y$ such that $B+(C\setminus\{0\}) \subset B \setminus \{0\}$ and later, in Section 3.7.2, such that

$$B + (C \setminus \{0\}) \subset \text{int } B,   \tag{3.65}$$

and an element $k^0 \in B$ with $\mathbb{R}k^0 - B = Y$. If we use in the definition of the scalarizing functional (2.23) the set B with condition (3.65), then the functional (2.23) is strictly C- monotone (see Theorem 2.3.1, (g)). Such a property is important for the proof of duality assertions via scalarization.

We introduce the following **generalized Lagrangian**, having another set M^{0*},

$$L : M \times M^{0*} \longrightarrow Y \cup \{+\infty_Y\},$$

and assume

$$L(x, y) \in f(x) - B$$

for all $x \in \mathcal{A}$ and for all $y \in M^{0*}$.

Now we are able to write down a problem (D_1) that can be considered as a dual problem to (P_1):

$$(D_1)   \text{Eff}_{\text{Max}}(f^*[M^*], C),   \tag{3.66}$$

with
$$M^* := \{y \in M^{0*} \ : \ \mathrm{Eff}_{\mathrm{Min}}(L[M, y], B) \neq \emptyset\},$$
where is B as explained above, and $f^*(y) \in \mathrm{Eff}_{\mathrm{Min}}(\{L(x, y) \mid x \in M\}, B) \neq \emptyset \subset Y$. The problem (D_1) is a so-called **Lagrange dual problem** to (P_1), as is easy to see if we reduce to the finite-dimensional scalar case $(Y, C) = (\mathbb{R}^1, \mathbb{R}^1_+)$, $x \in \mathbb{R}^n$, $(V, C_V) = (\mathbb{R}^m, \mathbb{R}^m_+)$, $B = \mathbb{R}^1_+$, and if we take y as linear mapping λ. Then instead of (P_1) we get
$$f(x) \to \min \ \text{s.t.} \ x \in \mathcal{A} = \{x \in M, g(x) \in \mathbb{R}^m_+\}.$$
So, with $M^* = \{\lambda \in -\mathbb{R}^m_+ \mid \exists x_0 \in M : (f + \lambda^T g)(x) \geq (f + \lambda^T g)(x_0) \ \forall \, x \in M$ and $f^*(\lambda) = \min\limits_{x \in M} (f + \lambda^T g)(x) > -\infty\}$, (D_1) has the well-known maxmin form
$$\max_{\lambda \in M^*} f^*(\lambda) = \max_{\lambda \in M^*} \min_{x \in M}(f + \lambda^T g)(x).$$

Lemma 3.7.3. Weak duality. (P_1) *and* (D_1) *are weakly dual; i.e.,*
$$f[\mathcal{A}] \cap f^*[M^*] - (C \setminus \{0\}) = \emptyset. \tag{3.67}$$

PROOF. Take $\overline{y} \in M^*$; therefore $f^*(\overline{y}) \in \mathrm{Eff}_{\mathrm{Min}}(\{L(x, \overline{y}) \mid x \in M\}, B)$. This means that $\{L(x, \overline{y}) \mid x \in M\} \cap (f^*(\overline{y}) - (B \setminus \{0\})) = \emptyset$, and for $\mathcal{A} \subset M$ instead of M,
$$\{L(x, \overline{y}) \ : \ x \in \mathcal{A}\} \cap f^*(\overline{y}) - (B \setminus \{0\}) = \emptyset,$$
and regarding a property of B,
$$\{L(x, \overline{y}) \mid x \in \mathcal{A}\} \cap f^*(\overline{y}) - (B + (C \setminus \{0\})) = \emptyset$$
and
$$(\{L(x, \overline{y}) \mid x \in \mathcal{A}\} + B) \cap (f^*(\overline{y}) - (C \setminus \{0\})) = \emptyset.$$
By definition of the Lagrangian $L(x, \overline{y})$ we have for $x \in \mathcal{A}$ and $\overline{y} \in M^*$,
$$L(x, \overline{y}) \in f(x) - B,$$
and because $\overline{y} \in M^*$ is chosen arbitrarily, we can conclude that
$$f[\mathcal{A}]) \cap f^*[M^*] - (C \setminus \{0\})) = \emptyset,$$
which yields (3.67). $\qquad\qquad\qquad\qquad\qquad\qquad\qquad\qquad\qquad\quad \square$

A strong duality theorem holds as well. Therefore, we assume a condition (V1) (see Weidner [372]) that is related to the so-called domination property (cf. [244] and Section 3.4):
$$(f + y \circ g)[M] \subset \mathrm{Eff}_{\mathrm{Min}}((f + y \circ g)[M]), B) + B$$
for all $y \in M^*$. Since one can choose the wrapping cone B very close to $C \setminus \{0\}$, (V1) means that $(f + y \circ g)[M]$ cannot have improperly efficient elements with respect to C. This is reflected by the formulation of the next theorem, which takes into account only primarily efficient points with respect to a cone B.

Theorem 3.7.4. Strong duality theorem. *For* (P_1) *and* (D_1) *as given above, let* $(V1)$ *be fulfilled and assume* $f(\overline{x}) \in \mathrm{Eff}_{\mathrm{Min}}(f[\mathcal{A}], B)$ *to be in* $\mathrm{Eff}_{\mathrm{Min}}(f[\mathcal{A}], C)$. *Then* $f(\overline{x}) \in \mathrm{Eff}_{\mathrm{Max}}(f^*[M^*], C)$.

PROOF. This time we take into account that the Lagrangian $L : M \times M^* \longrightarrow Y \cup \{+\infty_Y\}$ for an element $\overline{y} \in M^*$ can be chosen as

$$L(x, \overline{y}) = \begin{cases} f(x) & : \quad g(x) \in C_V, \\ (+\infty)_Y & : \quad \text{otherwise.} \end{cases}$$

Therefore $\overline{y} \in M^*$ and $\mathrm{Eff}_{\mathrm{Min}}(\{L(x, \overline{y}) : x \in \mathcal{A}\}, B) = \mathrm{Eff}_{\mathrm{Min}}(f[\mathcal{A}], B)$. The second setting yields $\mathrm{Eff}_{\mathrm{Min}}(\{L(x, \overline{y}) : x \in M \setminus \mathcal{A}\}, B) = \emptyset$. Together these results give

$$\mathrm{Eff}_{\mathrm{Min}}(\{L(x, \overline{y}) \mid x \in M\}, B) = \mathrm{Eff}_{\mathrm{Min}}(f[\mathcal{A}], B), \qquad (3.68)$$

since $\mathrm{Eff}_{\mathrm{Min}}(\{L(x, y) \mid x \in M\}, B) = \mathrm{Eff}_{\mathrm{Min}}(\mathrm{Eff}_{\mathrm{Min}}(\{L(x, y) \mid x \in \mathcal{A}\}, B) \cup \mathrm{Eff}_{\mathrm{Min}}(\{L(x, y) \mid x \in M \setminus \mathcal{A}\}, B), B)$. As assumed, $f(\overline{x}) \in \mathrm{Eff}_{\mathrm{Min}}(f[\mathcal{A}], B)$, so (3.68) gives $f(\overline{x}) \in \mathrm{Eff}_{\mathrm{Min}}(\{L(x, \overline{y}) \mid x \in M\}, B)$; that is, $f(\overline{x}) \in f^*[M^*]$.

For $\overline{y} \in M^*$ we have

$$(\{L(x, \overline{y}) \mid x \in M\} + (B \setminus \{0\})) \cap \{f^*(\overline{y})\} = \emptyset$$

in the proof of Lemma 3.7.3. For all cones $B' \subset Y$ with $B + (B' \setminus \{0\}) \subset (B \setminus \{0\})$ we get $(\{L(x, \overline{y}) \mid x \in M\} + B + (B' \setminus \{0\})) \cap \{f^*(\overline{y})\} = \emptyset$, and along the lines of the proof of Lemma 3.7.3 we have

$$f[\mathcal{A}] \cap (f^*[M^*] - (B' \setminus \{0\})) = \emptyset. \qquad (3.69)$$

Because $\overline{x} \in \mathcal{A}$, we have that $(f(\overline{x}) + (B' \setminus \{0\})) \cap f^*[M^*] = \emptyset$, and so $f(\overline{x}) \in \mathrm{Eff}_{\mathrm{Max}}(f^*[M^*], B')$ for all $B' \subset Y$ as above. In particular,

$$f(\overline{x}) \in \mathrm{Eff}_{\mathrm{Max}}(f^*[M^*], C).$$

The proof is complete.   □

Sometimes results like Theorem 3.7.4 are called a strong **direct** duality theorem, because a primal optimal solution is shown to be dually optimal. The converse direction is also interesting. This leads to converse duality theorems. To get such a theorem for our pair of optimization problems, we state a condition (V2):

(V2) Every solution of (D_1) is to be dominated by a properly efficient solution of (P_1), which means that with a cone B as in (3.65), for all $\tilde{d} \in \mathrm{Eff}_{\mathrm{Max}}(f^*[M^*], C)$ there is an $f(\tilde{x}) \in \mathrm{Eff}_{\mathrm{Min}}(f[\mathcal{A}], B)$ such that $f(\tilde{x}) \in \tilde{d} + C$.

Theorem 3.7.5. Converse duality theorem. *Assume that* $(V1)$ *and* $(V2)$ *hold and* $\tilde{d} \in \mathrm{Eff}_{\mathrm{Max}}(f^*[M^*], C)$. *Then there are* $\overline{x} \in \mathcal{A}$, *and a cone* $B' \subset Y$ *with* $B + (B' \setminus \{0\}) \subset (B \setminus \{0\})$ *such that* $\tilde{d} = f(\overline{x}) \in \mathrm{Eff}_{\mathrm{Min}}(f[\mathcal{A}], B')$.

PROOF. For $\tilde{d}$ as assumed, (V2) leads to an $f(\tilde{x}) \in \mathrm{Eff}_{\mathrm{Min}}(f[\mathcal{A}], B)$ with $\tilde{d} \in f(\tilde{x}) - C$. Then $\tilde{d} \in f[\mathcal{A}] + C$. Otherwise, $\tilde{d} \notin f[\mathcal{A}] + C$, so that $\tilde{d} \in f(\tilde{x}) - C$ can be satisfied only if

$$\tilde{d} \in f(\tilde{x}) - (C \setminus \{0\}). \tag{3.70}$$

With $f(\tilde{x})$ as above and (3.68) we obtain $f(\tilde{x}) \in f^*[M^*]$. Because of $\tilde{d} \in \mathrm{Eff}_{\mathrm{Max}}(f^*[M^*], C)$ it follows that $f(\tilde{x}) \notin \tilde{d} + (C \setminus \{0\})$, which contradicts (3.70).

Since $\tilde{d} \in f[\mathcal{A}] + C$, because of weak duality, $\tilde{d} \in f[\mathcal{A}]$. Therefore, there is $\overline{x} \in \mathcal{A}$ such that $\tilde{d} = f(\overline{x})$. The last assertion follows from (3.69), which holds for $(P_1), (D_1)$ for all B'. This means that $f^*[M^*] \cap (f(\overline{x}) + B') = \emptyset$, or equivalently, $f(\overline{x}) \in \mathrm{Eff}_{\mathrm{Max}}(f^*[M^*], B')$, especially $f(\overline{x}) \in \mathrm{Eff}_{\mathrm{Max}}(f^*[M^*], C)$.   □

The duality in the last section is an abstract and not scalarized Lagrangian formalism for very general optimization problems. In the next section we will prove duality theorems with the help of scalarization. Here we again use the Lagrangian form of duality. For an equivalent method, using the Fenchel form of duality, see Wanka [366, 367].

3.7.2 Duality by Scalarization

To prove duality theorems we come back to the general case (P_1) in the following form:

$$(P_2) \quad \mathrm{Eff}_{\mathrm{Min}}(f[\mathcal{A}], C) \; with \mathcal{A} = \{x \in M \mid g(x) \in C_V\} \neq \emptyset,$$

where $f : M \to Y, \emptyset \neq M \subset X, g : M \to V, C_V$ a nonempty set in V and X, Y, V t.v.s., Y Hausdorff, Y partially ordered by C. Furthermore, we suppose that the assumptions mentioned in Remark 3.7.2 are satisfied. Then (compare Theorem 2.3.1), the set S of surjective continuous strictly monotone (with respect to C) functionals $s : Y \to \mathbb{R}$ is nonempty.

Again we introduce a **generalized Lagrangian** $L : M \times M^* \longrightarrow Y \cup \{+\infty_Y\}$ and assume that for $s \in S$,

$$\sup_{y \in M^*} s(L(x, y)) = s(f(x)) \quad if \quad g(x) \in C_V.$$

Using the functionals $s \in S$ we define a dual problem to (P_2):

$$(D_2) \quad \mathrm{Eff}_{\mathrm{Max}}(\mathcal{A}_D, C), \tag{3.71}$$

where

$$\mathcal{A}_D = \{h \in Y \mid \exists s \in S, \; \exists y \in M^* \; : \; s(h) = \inf\{s(L(x, y)) \mid x \in M\} > -\infty\}.$$

Lemma 3.7.6. Weak duality. *The pair* $(P_2), (D_2)$ *satisfies*

$$f[\mathcal{A}] \cap (\mathcal{A}_D - (C \setminus \{0\})) = \emptyset.$$

PROOF. Otherwise, there would be $x_0 \in \mathcal{A}, h_0 \in \mathcal{A}_D$ such that $f(x_0) - h_0 \in -C \setminus \{0\}$. With $s_0 \in S$ corresponding to h_0 we get

$$s_0(f(x_0)) < s_0(h_0)$$

because s is strictly C-monotone. On the other hand, we consider with h_0 a corresponding $y_0 \in M^*$ and derive regarding the definition of the dual feasible set $\mathcal{A}_D$,

$$s_0(h_0) = \inf_{x \in M} s_0(L(x, y_0)) \leq s_0(L(x_0, y_0)) \leq \sup_{y \in M^*} s_0(L(x_0, y)) = s_0(f(x_0)),$$

a contradiction. □

For **properly efficient elements** $z_0 = f(x_0)$, this means that there is an $s_0 \in S$ such that $s_0(f(x_0)) = \inf\{s_0(f(x)) \mid x \in \mathcal{A}\} > -\infty$, we can prove a strong duality theorem:

Theorem 3.7.7. Strong (direct) duality. *If* $z_0 = f(x_0)$ *with* $x_0 \in \mathcal{A}$ *is properly efficient for* (P_2), *then it is efficient for* (D_2) *i.e.,* $z_0 \in \mathrm{Eff}_{\mathrm{Max}}(D_2)$.

PROOF. Let $f(x_0)$ be a properly efficient element of the primal problem (P_2). With the corresponding element $s_0 \in S$ we can choose the Lagrangian $L : M \times M^*$ for an element $y_0 \in M^*$ such that

$$s_0(L(x, y_0)) = \begin{cases} s_0(f(x)) & : \quad g(x) \in C_V, \\ +\infty & : \quad \text{otherwise.} \end{cases}$$

Using this Lagrangian we get

$$s_0(f(x_0)) = \inf_{x \in \mathcal{A}} s_0(L(x, y_0)) = \inf \left\{ \inf_{x \in \mathcal{A}} s(L(x, y_0)) \, , \, \inf_{x \in M \setminus \mathcal{A}} s(L(x, y_0)) \right\}$$

$$= \inf_{x \in M} s(L(x, y_0)).$$

This yields $f(x_0) \in \mathcal{A}_D$. From the weak duality assertion in Lemma 3.7.6 we can conclude that $f(x_0)$ is an efficient element for (D_2). □

The point of the last proof is that there are functionals y_0 and s_0 such that a possibly nonlinear Lagrangian works. So an essential task is to find sufficiently practicable s_0 and y_0 for special optimization problems such as (P_3) or (P_4) below.

Next we would like to complete the study of (P_2) and (D_2) with a converse duality theorem. For this we restrict Y to be an l.c.s. At first we show that efficient elements $h \in \mathcal{A}_D$ can be characterized by scalarization.

Lemma 3.7.8. $h^0 \in \mathrm{Eff}_{\mathrm{Max}}(D_2)$ *iff* $s_h(h) \leq s_h(h^0) \, \forall h \in \mathcal{A}_D$ *where* s_h *corresponds to* h *according to the definition of* $\mathcal{A}_D$ *in* (D_2).

PROOF. (a) If $h^0 \in \mathrm{Eff}_{\mathrm{Max}}(D_2)$ and $h \in h^0 + C \setminus \{0\})$, then $h \notin \mathcal{A}_D$. Therefore $s(h) > \inf\{s(f(x)) + y(g(x)) \mid x \in M\}$ for all $s \in S$ and y. Applying $s_{\bar{h}}$ (belonging to $\bar{h} \in \mathcal{A}_D$) to h as above, we get

$$s_{\bar{h}}(h^0) = \inf\{s_{\bar{h}}(h) \mid h \in h^0 + (C \setminus \{0\})\}$$
$$\geq \inf\{s_{\bar{h}}(f(x)) + y_{\bar{h}}(g(x)) \mid x \in M\} = s_{\bar{h}}(\bar{h}),$$

as claimed, where the inequality follows from the strict inequality above.

(b) If $h^0 \in \mathcal{A}_D$ but $h^0 \notin \mathrm{Eff}_{\mathrm{Max}}(D_2)$, then there is $h \in \mathcal{A}_D$ such that $h \in h^0 + (C \setminus \{0\})$. Therefore $s(h) > s(h^0)$ for all $s \in S$, contrary to our assumption. □

Additionally, a strong converse duality statement holds.

Theorem 3.7.9. Strong converse duality. *Under the conditions given for the pair* (P_2), (D_2) *and additionally, if* $P := f[\mathcal{A}] + C$ *is closed and if for all* $s \in S$ *such that* $\inf\{s(f(x)) \mid x \in \mathcal{A}\} > -\infty$ *there is an* $x_0 \in \mathcal{A}$ *with* $\inf\{s(f(x)) \mid x \in \mathcal{A}\} = s(f(x_0))$, *then a dually efficient element is properly primal efficient.*

PROOF. Let h be an element of $\mathrm{Eff}_{\mathrm{Max}}(D_2)$. If h were not in P, there would be an open convex neighborhood U of h with $U \cap P = \emptyset$. Even $(U - C) \cap P = \emptyset$; otherwise, there would exist $u \in U, k \in C$ such that $u - k \in P = f[\mathcal{A}] + C$, which means that $u \in f[\mathcal{A}] + C + k = P$. Now we consider the set $B := U - C$. Obviously, B is open and convex and $B - C = B$, and so $\bar{B} - C \subset \overline{B - C} \subset \bar{B}$. Theorems 2.3.1 and 2.3.6 deliver an $s_1 \in S$ such that $s_1(p) \geq 0 > s_1(u) \; \forall p \in P \; \forall u \in B$. We have $h \in B$, since $0 \in C$, so there is a number γ with

$$s_1(h) < \gamma < 0 \leq s_1(p) \; \forall p \in P.$$

From weak duality we get for $s_2 \in S$,

$$s_2(h) \leq s_2(p) \; \forall p \in P.$$

Taking $\lambda \in (0,1)$ we consider $s_\lambda := \lambda s_2 + (1 - \lambda)s_1$. Then s_λ is again in S, and

$$s_\lambda(h) = s_1(h) + \lambda(s_2(h) - s_1(h)),$$
$$\forall x \in A : \; s_\lambda(f(x)) = \lambda(s_2(f(x))) + (1 - \lambda)(s_1(f(x))),$$
$$\geq \lambda s_2(f(x)) \geq \lambda s_2(h),$$

since $f(x) \in P$. Now we choose λ sufficiently small; then not only $s_1(h) < \gamma$ but $s_\lambda(h) < \gamma$, and from $\gamma < 0$ it follows that $\gamma < \lambda s_2(h)$, so $s_\lambda(h) < \gamma < \lambda s_2(h) \leq s_\lambda(f(x)) \; \forall x \in \mathcal{A}$. This means that $\inf\{s_\lambda(f(x)) : x \in \mathcal{A}\} \geq \gamma$, and the assumptions give an $x_0 \in \mathcal{A}$ such that

$$s_\lambda(h) < \gamma \leq \inf\{s_\lambda(f(x)) : x \in \mathcal{A}\} = s_\lambda(f(x_0)).$$

But strong duality now demands $f(x_0) \in \mathcal{A}_D$, and Lemma 3.7.8 requires $s_\lambda(f(x_0)) \leq s_\lambda(h)$, a contradiction. So $h \in P = f[\mathcal{A}] + C$. Weak duality gives $h \in f[\mathcal{A}]$. □

3.7.3 Duality for Approximation Problems

As an example for Section 3.7.2 and because of its practicability we now give duality theorems for a general **approximation problem** and a corresponding dual problem.

To this end we consider (P_1) or (P_2) in the following form:

$$(P_3) \quad \text{Eff}_{\text{Min}}(f[\mathcal{A}], C) \text{ with} \mathcal{A} := \{x \in C_X \mid B(x) - b \in C_V\},$$

where now X is a linear normed space, partially ordered by a closed convex cone C_X, V is a reflexive Banach space, C_V a cone in V that is convex and closed, $B \in L(X, V)$, $Y = \mathbb{R}^p$, partially ordered by a closed convex cone K such that $K + \mathbb{R}^p_+ \subset K$ and int $K^+ \neq \emptyset$, and f may have the form

$$f(x) = C(x) + \begin{pmatrix} \alpha_1 \|A^1(x) - a^1\| \\ \cdots \\ \alpha_p \|A^p(x) - a^p\| \end{pmatrix} \tag{3.72}$$

with $C \in L(X, \mathbb{R}^p), \alpha_i \geq 0$ real $(i = 1, \ldots, p)$, $a^i \in U$ a given real normed space, $A^i \in L(X, U), i = 1, \ldots, p$, and A^{*i} denotes the adjoint operator to A^i. For brevity and clarity we sometimes leave out the parenthesis in connection with A^{*i}.

On the one hand, (P_3), although a special case of (P_1) or (P_2), itself contains many important **special cases**:

(i) (P_3) is a semi-infinite linear problem if $\alpha_i = 0 \ \forall i = 1, \ldots, p$.
(ii) f can be interpreted as a Lipschitz perturbed linear problem.
(iii) $C = 0$ gives a multicriteria location problem.
(iv) Consider $C(x) \rightarrow \min$, $x \in \mathcal{A}_0 = \{x \in \mathcal{A} \neq \emptyset \mid A^i(x) = a^i, i = 1, \ldots, p\} = \emptyset$. Then (3.72) is a parameterized surrogate problem.
(v) The general multicriteria approximation problem or location problem dealt with in Sections 4.1 and 4.3.

On the other hand, (P_3) can be generalized considerably if we assume a^i to be variable in a nonempty set $W_i \subset U_i$ $(i = 1, \ldots, p), U_i$ instead of a fixed space U, so that we get an optimization problem (P_4):

$$(P_4) \quad \text{Eff}_{\text{Min}}(f[\mathcal{A}], K)$$

where

$$\mathcal{A} = \{(x, a) \mid x \in C_X, \ a = (a^1, \ldots, a^p), \ a^i \in W_i, \ i = 1, \ldots, p, B(x) - b \in C_V\},$$

and

$$f(x, a) = C(x) + \begin{pmatrix} \cdots \\ \alpha_i \|A^i(x) - a^i\|_{U_i} \\ \cdots \end{pmatrix}.$$

Using (3.71) we get for this multicriteria optimization problem a useful dual problem (D_4), which satisfies together with (P_4) the above defined conditions for weak and strong duality. Moreover, in Section 4.3 a practicable algorithmic procedure will be constructed and tested. In order to use the results of Section 3.7.2 we introduce a suitable Lagrangian L_{λ^0} to (P_4) for a given $\lambda^0 \in \operatorname{int} K^+$:

$$L_{\lambda^0}(x, a, Y, u^*) = \lambda^0 \left(\begin{pmatrix} \cdots \\ \alpha_i Y^i(a^i - A^i(x)) \\ \cdots \end{pmatrix} + \begin{pmatrix} \cdots \\ C^i(x) \\ \cdots \end{pmatrix} \right) + u^*(b - B(x)),$$

where

$$(x, a) \in M = \{(x, a) \mid x \in C_X, \ a^i \in W_i\}$$

and

$$(Y, u^*) \in M^{0*} := \{(Y, u^*) \mid Y = (Y^1, \ldots, Y^p), \ Y^i \in L(U_i, \mathbb{R}), u^* \in L(V, \mathbb{R}),$$

$$u^* \in C_V^+, \alpha_i \|Y^i\|_* \leq \alpha_i, \ i = 1, \ldots, p\}.$$

From this setting two results follow immediately:

$$\sup_{(Y, u^*) \in M^{0*}} L_{\lambda^0}(x, a, Y, u^*)$$

$$= \begin{cases} \sum_i \lambda_i^0 \, \alpha_i \, \|a^i - A^i(x)\| + \lambda^0 C(x) & \text{if } B(x) - b \in C_V, \\ +\infty & \text{otherwise,} \end{cases} \tag{3.73}$$

and

$$\inf_{(x, a) \in M} L_{\lambda^0}(x, a, Y, u^*)$$

$$= \begin{cases} \sum_i \lambda_i^0 \, \inf_{a^i}(\alpha_i \, Y^i(a^i)) + u^*(b) & \text{if } \sum_i \lambda_i^0(-\alpha_i A^{i*} Y^i + C^{i*}) - B^*(u^*) \in C_X^+, \\ -\infty & \text{otherwise.} \end{cases} \tag{3.74}$$

To obtain a dual problem to (P_4) we recall (D_2), (3.71), and define $M_{\lambda^0}^*$ to be

$$M_{\lambda^0}^* := \{(Y, u^*) \in M^{0*} \mid \sum_i \lambda_i^0(-\alpha_i A^{i*} Y^i + C^{i*}) - B^*(u^*) \in C_X^+\}.$$

So we are able to give the following dual problem $(D_4)'$:

$$\operatorname{Eff}_{\operatorname{Max}}(\mathcal{A}_D, K)$$

where

$$\mathcal{A}_D = \{d \in Y \mid \exists \lambda \in \operatorname{int} K^+, \exists (Y, u^*) \in M_\lambda^* : \lambda(d) = \inf_{(x, a) \in M} L_\lambda(x, a, Y, u^*)\}.$$

It follows that

$$\lambda(d) = \sum_i \lambda_i \inf_{a^i \in W_i}(\alpha_i Y^i(a^i)) + u^*(b).$$

In order to descalarize the last line we consider the set M^* and use Lemma 3.7.10 below, which is related to a result given by Jahn [195]:

$$M^* = \{(Y, Z) \mid Y = (Y^1, \ldots, Y^p), Y^i \in L(U_i, \mathbb{R}), i = 1, \ldots, p, Z \in L(V, \mathbb{R}^p),$$

$$\exists \lambda^* \in \operatorname{int} K^+ \quad \text{such that} \quad \sum_{i=1}^{p} \lambda_i^* (-\alpha_i A^{i*} Y^i + C^i) - (Z(B))^* (\lambda^*) \in C_X^+,$$

$$Z^*(\lambda^*) \in C_V^+, \alpha_i \|Y^i\|_* \leq \alpha_i, i = 1, \ldots, p\}.$$

Moreover, consider the sets D_1 and D_2:

- $D_1 := \{d \in \mathbb{R}^p \mid \exists \lambda^* \in \operatorname{int} K^+ \text{ and } \exists (Y, u^*) \in M_{\lambda^*}^* \text{ such that } \lambda^*(d) = \sum_{i=1}^{p} \lambda_i^* (\alpha_i Y^i(a^i)) + u^*(b)\}$,

- $D_2 := \left\{ d \in \mathbb{R}^p \mid \exists (Y, Z) \in M^* \text{ such that } d = \begin{pmatrix} \cdots \\ \alpha_i Y^i(a^i) \\ \cdots \end{pmatrix} + Z(b) \right\}$,

where d in D_2 satisfies a descalarized equation.

Lemma 3.7.10. *For the sets D_1, D_2 as defined above, $D_2 \subset D_1$. Moreover, if $b \neq 0$, even $D_2 = D_1$.*

PROOF. (a) Let d be in D_2. According to the definition of M^* there is $\lambda^* \in \operatorname{int} K^+$, $Y^i \in L(U_i, \mathbb{R})$ $(i = 1, \ldots, p)$, $Z \in L(V, \mathbb{R}^p)$, $Z^*(\lambda^*) \in C_V^+$, $\alpha_i \|Y^i\|_* \leq \alpha_i$ $(i = 1, \ldots, p)$, such that $t := d - \begin{pmatrix} \cdots \\ \alpha_i Y^i(a^i) \\ \cdots \end{pmatrix} = Z(b)$ and

$\sum_{i=1}^{p} \lambda_i^* (-\alpha_i A^{i*} Y^i + C^{i*}) - (Z(B))^* (\lambda^*) \in C_X^+$. Then $\lambda^*(t) = u^*(b)$ with $u^* = Z^*(\lambda^*) \in L(V, \mathbb{R})$, and so $\lambda^*(d) = \sum \lambda_i^* (\alpha_i Y^i(a^i)) + u^*(b)$, which means that $d \in D_1$.

(b) Let d be in D_1. Therefore there exist $\lambda^* \in \operatorname{int} K^+$, $Y^i \in L(U_i, \mathbb{R})$ $(i = 1, \ldots, p)$, $u^* \in L(V, \mathbb{R})$, $u^* \in C_V^+$ such that

$$\sum \lambda_i^* (-\alpha_i A^{i*} Y^i + C^{i*}) - B^*(u^*) \in C_X^+ \tag{3.75}$$

and

$$\lambda^*(d) = \sum_{i=1}^{p} \lambda_i^* (\alpha_i Y^i(a^i)) + u^*(b). \tag{3.76}$$

Since $b \neq 0$, (3.76) leads to the existence of $Z \in L(V, \mathbb{R}^p)$ with

$$d = \begin{pmatrix} \cdots \\ \alpha_i Y^i(a^i) \\ \cdots \end{pmatrix} + Z(b) \tag{3.77}$$

and $Z^*(\lambda^*) = u^*$. Then from (3.75) and $u^* \in C_V^+$ it is clear that $d \in D_2$. To understand where the existence of Z comes from, we consider once more

$\lambda^*(t) = u^*(b)$, which follows from (3.76). It is well known that if $b \neq 0$ and $\lambda^* \neq 0$, then one can define an operator $T \in L(V, \mathbb{R}^p)$ such that $T^*(\lambda^*) = u^*$ and $t = T(b)$: If $u^*(b) \neq 0$, take $T(\cdot) = \frac{u^*(\cdot)}{u^*(b)} t$ over V; if $u^*(b) = 0$, then consider $\lambda^* \neq 0$. For λ^* there exists t such that $\lambda^*(t_0) = 1$. Since $b \neq 0$, there is $\tilde{u}^* \in L(V, \mathbb{R})$ that separates b and 0; i.e., $\tilde{u}^*(b) = 1$. Then take

$$T(\cdot) = u^*(\cdot) t_0 + \tilde{u}^*(\cdot) t \quad \text{over } V. \qquad \square$$

If $b \neq 0$, we obtain from $(D_4)'$, regarding Lemma 3.7.10, the following dual problem (D_4) to (P_4):

$$(D_4): \quad \text{Eff}_{\text{Max}}(f^*[M^*], K),$$

where

$$\left(\begin{array}{c} \cdots \\ \inf_{a^i \in W_i} \alpha_i Y^i(a^i) \\ \cdots \end{array} \right) + Z(b) =: f^*(Y, Z) \qquad \text{for} \quad (Y, Z) \in M^*.$$

For an example consider Section 4.3.2 or recall the dual problem for a common finite-dimensional linear multicriteria problem.

Example 3.7.11. Consider instead of (P_3) the following scalar optimization problem:

(P_5) $c(x) + \sum_{i=1}^p \alpha_i \|A^i(x) - a^i\| \to \min$, s.t. $x \in C_X, B(x) - b \in C_V$,

where $X, V, C_X, C_V, A^i, a^i (i = 1, \ldots, p)$ as above, a^i fixed, $c \in L(X, \mathbb{R}^1)$, and $\alpha_i > 0 \, \forall i = 1, \ldots, p$. Since the sum in (P_5) is really a norm, (P_5) is an example of (P_3). Then we get a dual problem (D_5) as a special case of (D_4):

(D_5) $\sum_{i=1}^p \alpha_i y^i(a^i) + z(b) \to \max$
  s.t. $y = (y^1, \ldots, y^p) \in L(U, \mathbb{R})^p, \|y^i\|_* \leq 1 \, \forall i, z \in L(V, \mathbb{R}), z \in C_V^+$,
  $\sum_{i=1}^p (-\alpha_i A^{i*}(y^i) + c) - B^*(z) \in C_X^+$.

The problem (D_5) follows immediately from (D_4), since $\|y\|_* \leq 1$ in (D_4) means that $\max_{i=1,\ldots,p} \|y^i\|_* \leq 1$ and so $\|y^i\|_* \leq 1$ for all i, because the maximum norm is suitable as a dual norm to a sum of norms.

Example 3.7.12. The ordinary scalar classic location problem is contained in (P_5): $X, \alpha_1, \ldots, \alpha_p, a_1, \ldots, a_p$ as above, $a_1, \ldots, a_p$ fixed,

(P_6) $\sum_{i=1}^p \alpha_i \|x - a^i\| \to \min_{x \in X}$ (Fermat's problem),
(D_6) $\sum_{i=1}^p \alpha_i y^i(a^i) \to \max$ s.t. $y^i \in X^*, \|y^i\|_* \leq 1 \, \forall i, \sum_1^p \alpha_i y^i = 0$.

Taking as $\mathcal{A} \subset X$ a closed linear subspace, we get

(P_7) $\|x - a\| \to \min$ s.t. $x \in \mathcal{A}$,
(D_7) $y(a) \to \max$ s.t. $y \in \mathcal{A}^\perp, \|y\|_* \leq 1$,

where $\mathcal{A}^\perp$ is the **annihilator** to $\mathcal{A}: \mathcal{A}^\perp = \{y \in X^* \mid y(x) = 0 \, \forall x \in \mathcal{A}\}$.

(P_5) can be generalized in another way, too (see Wanka [366, 367]). Wanka considers the optimization problem (P_8):

$$\sum_{i=1}^{p} \alpha_i \|x - a_i\|^{\beta_i} \rightarrow \min \ s.t. \ x \in U, \qquad (3.78)$$

where $\beta_i \geq 1 \ \forall i = 1, \ldots, p, \alpha_i > 0 \ \forall i = 1, \ldots, p, U$ a convex closed subset of a reflexive Banach space X, and $a^1, \ldots, a^p$ given elements of X. Then he studies the dual problem (D_8), which contains our problem (D_6), using Fenchel–Rockafellar theory instead of Lagrange theory. He also considers the vectorial variant of (3.78) together with a linear mapping as in (P_3). We come back to these general problems in Section 4.1.

$$(D_8): \quad \sum_{\substack{i=1 \\ \beta_i > 1}}^{p} \alpha_i (1 - \beta_i) \|y^i\|^{\frac{\beta_i}{1-\beta_i}} + \sum_{i=1}^{p} \alpha_i \beta_i y^i (a^i) - \sup_{x \in U} \sum_{i=1}^{m} \alpha_i \beta_i y^i (x) \rightarrow \max$$

$$(3.79)$$

s.t. $y^i \in X^* \ \forall i$ and $\|y^i\|_* \leq 1$ for i with $\beta_i = 1$.

Taking into account the fact that we constructed (D_4) as in Section 3.7.2, weak, strong, and converse duality follow immediately:

Theorem 3.7.13. Weak duality.

$$f(\mathcal{A}) \cap f^*(M^*) - (K \setminus \{0\}) = \emptyset. \qquad (3.80)$$

Now we derive a strong duality assertion. We use the chosen Lagrangian and the saddle point theorem of convex programming. For simplicity we assume $a^i \in W^i = \{a^i\}$; therefore, we write $x \in \mathcal{A}$ instead of $(x, a) \in \mathcal{A}$.

Theorem 3.7.14. Strong direct duality. *Let U be a reflexive Banach space and $b \neq 0$. Then, for every properly efficient $f(x^0)$ of (P_4) there exists an efficient element $f^*(Y^0, Z^0)$ of (D_4) such that*

$$f(x^0) = f^*(Y^0, Z^0) \qquad (3.81)$$

if the following condition is satisfied: There is a feasible element $\bar{x}$ of (P_4) with $B(\bar{x}) - b \in \operatorname{int} C_V$, and for every $\lambda^* \in \operatorname{int} K^+$ there is a feasible pair $(\bar{Y}, \bar{Z})$ of (D_4) such that the C_X^+-inclusion in (D_4) is satisfied with respect to $\operatorname{int} K^+$.

PROOF. Since $f(x^0)$ is properly efficient, there is $\lambda^0 \in \operatorname{int} K^+$ such that $\lambda^0(f(x^0)) = \inf\{\lambda^0(f(x)) : x \in \mathcal{A}\}$. Then (3.73) gives

$$\lambda^0(f(x^0)) = \inf_{x \in \mathcal{A}} \lambda^0(f(x)) = \inf_{x \in M} \sup_{(Y, u^*) \in M^{0*}} L_{\lambda^0}(x, Y, u^*).$$

We use the saddle point theorem of convex programming (see [391]) and have that there is (Y^0, u^{*0}) contained in $M_{\lambda^0}^*$ such that for $x \in M$ and $(Y, u^*) \in M_{\lambda^0}^*$,

$$L_{\lambda^0}(x, Y^0, u^{*0}) \geq L_{\lambda^0}(x^0, Y^0, u^{*0}) \geq L_{\lambda^0}(x_0, Y, u^*). \tag{3.82}$$

Then (3.74) gives $\inf_{x \in C_X} L_{\lambda^0}(x, Y^0, u^{*0}) = \sum \lambda_i^*(\alpha_i Y^{i0}(a^i)) + u^{*0}(b)$, so we get

$$\lambda^0(f(x^0)) = \sum_{i=1}^{p} \lambda_i^0(\alpha_i Y^{i0}(a^i)) + u^{*0}(b) \tag{3.83}$$

or using Lemma 3.7.10 its vectorial variant (3.81). Considering (D_4), we see that the proof is complete.   □

Our next aim is a converse duality theorem. To attempt this we use the scalarization Lemma 3.7.8. In our convex case it reads as follows:

Lemma 3.7.15. $d^0 \in \mathrm{Eff}_{\mathrm{Max}}(D_1, K)$ iff $\lambda_d(d^0) \geq \lambda_d(d)$ for all $d \in D_1$ and $\lambda_d \in \mathrm{int}\, K^+$ such that for every $(Y, u^*) \in M_{\lambda_d}^*$,

$$\lambda_d(d) = \sum \lambda_{d,i}(\alpha_i Y^i(a^i)) + u^*(b).$$

With the help of the last lemma a converse duality theorem follows. We again suppose $W^i = \{a^i\}, i = 1, \ldots, p$.

Theorem 3.7.16. Converse duality. We assume $b \neq 0, \mathrm{int}\, K \neq \emptyset, \exists \overline{x} \in \mathcal{A}$ with $B\overline{x} - b \in \mathrm{int}\, C_V^+$, for every $\lambda^* \in \mathrm{int}\, K^+$ with $\inf\{\lambda^*(f(x)) \mid x \in \mathcal{A}\} > -\infty$ there exists $(\overline{Y}, \overline{Z}) \in M^*$ such that

$$\sum \lambda_i^*(-\alpha_i A^{i*} \overline{Y}^i + C^{i*}) + (\overline{Z}b)^*(\lambda^*) \in \mathrm{int}\, C_X^+$$

and additionally an $x_\lambda \in \mathcal{A}$ with $\inf\{\lambda^*(f(x)) : x \in \mathcal{A}\} = \lambda^*(f(x_\lambda))$. Then to every $f^*(Y^0, Z^0)$ that is optimal with respect to (D_4) there is a properly efficient element $f(x^0)$ of (P_4) such that

$$f^*(Y^0, Z^0) = f(x^0). \tag{3.84}$$

We would like to emphasize that both strong duality Theorems 3.7.14 and 3.7.16 state duality in a nonscalarized form.

PROOF. Let $f^*(Y^0, Z^0)$ be maximal with respect to M^*. Then it is also maximal w.r.t. the set D_2, and since $b \neq 0$, also with respect to D_1; hence Lemma 3.7.15 yields

$$\lambda_d(f^*(Y^0, Z^0)) \geq \lambda_d(d) \tag{3.85}$$

$\forall d \in D_1$ with $\lambda_d \in \mathrm{int}\, K^+$ belonging to d. From (3.85) we get

$$d^0 := f^*(Y^0, Z^0) \in P := f(\mathcal{A}) + K. \tag{3.86}$$

Otherwise, $d^0 \notin f(\mathcal{A}) + K$. Now, separating P and d^0 by a continuous linear functional $\lambda_1^* \in \mathbb{R}^p \setminus \{0\}$, we proceed as in the proof of Theorem 3.7.9 and get

$$d^0 = f^*(Y^0, Z^0) = f(x^0)$$

for a suitable $x^0 \in \mathcal{A}$. We observe that $f(x^0)$ is even properly minimal w.r.t. (P_4) with a $\underline{\lambda}^* \in \operatorname{int} K^+$ that belongs to (Y^0, Z^0).   □

In Section 3.11 it is shown that the saddle point theory w.r.t. (P_4) can be extended considerably: We use sublinear functionals instead of norms and approximate efficiency instead of efficiency.

3.8 Vector Equilibrium Problems and Related Topics

It is well known that optimization and nonlinear analysis are two branches of modern mathematics much developed lately. The results obtained in these areas use certain kinds of differentiability (directional derivative, subgradient, generalized subgradients, etc.), certain generalizations of convexity, geometric methods (cone of feasible directions, normal and tangent cones, etc.), game theory, fixed point theory, topological degree, etc. The reader can find extensive expositions about these topics in a series of books of Zeidler [392] entitled *Nonlinear Analysis and Its Applications*, and those of Aubin and Ekeland [12], Aubin and Frankowska [13], and Rockafellar and Wets [309].

In the same way, equilibrium problems have come to play a central role in the development and unification of diverse areas of mathematics and physical sciences. Thus various problems of practical interest in optimization, variational inequalities, complementarity, economics, Nash equilibria, and engineering involve equilibrium in their description. The literature on equilibrium problems and treatment in these areas is vast; see [14], [34], [61], [384], [113], [203], [202], and so on. These sources provide extensive references and extended bibliographies for the reader wishing to explore the topic in depth.

Our object in this section is to present a developed investigation in generalized vector equilibrium problems, which embody at least vector optimization problems and vector variational inequalities. The vector variational inequalities have been widely developed in recent years, and various solutions have been characterized and computed. These were first introduced by Giannessi [132] and further developed by many authors; see, for instance, [64], [67], [322], [2], [229], [321], [75], and precisely, [133] in different areas.

Recent topics attracting considerable attention are equilibrium problems for vector-valued mappings, see [279], [280], [151]. Inspired by the scalar case, such problems have received different developments depending on the kind of order space where these have been considered. Some recent papers may be grouped in the following way:

1. Theory of vector optimization, vector equilibrium problems, and vector variational inequalities (Ansari [2, 3]; Ansari and Siddiqi [4]; Ansari, Siddiqi and Yao [5]; Bianchi, Hadjisvvas and Schaible [28]; Blum and Oettli [34, 33]; Chen and Craven [66, 67]; Fan [111]; Fu [122]; Kalmoun, Riahi

and Tanaka [201, 202, 203, 204, 205]; Konnov and Yao [215]; Lee and Kum [231]; Lee, G.M., Kim and Lee, B.S. [228, 234, 235, 227]; Lee, G.M., Kim and Lee, B.S., Cho [229]; Lee and Kum [232]; Li, Yang and Chen [238]; Lin, Yang, Yao [239]; Nagurney, N. and Zhao, L., [264], Oettli [278]; Samuelson [313], Siddiqi, Ansari, Ahmad [321]; Yang [379]; Yang and Chen [380]; Yang and Goh [381], Yu and Yao [383]).

2. Existence of solutions for generalized vector variational inequalities and complementarity problems (Chen [64]; Chen and Hou [70, 76]; Danilidis and Hadjisavvas [90]; Isac and Yuan [184]; Kazmi [209]; Lai and Yao [225]).
3. Vector variational inequalities and vector equilibrium problems with multi-functions (Ding and Tarafdar [95]; Kalmoun and Riahi [201, 202, 203]; Siddiqi, Ansari, Khan [323]; Song [326])
4. Vector variational inequalities, vector optimization, and scalarization (Chen, G.-Y., Chen, G.M. [65]; Chen and Craven [67]; Giannessi, Mastroeni and Pellegrini [134]; Goh and Yang [135]; Lee, G.M., Kim, Lee, B.S., Yen [230]).
5. Monotone vector variational inequalities (Chowdhury and Tan [78]; Hadjisavvas and Schaible [151], Ding and Tarafdar [97]).

We propose two approaches to establish the existence of solutions of equilibrium problems in the vector case. The first one directly uses a generalization of the well-known lemma of Knaster, Kuratowski, and Mazurkiewicz (KKM-Lemma) as proposed by Brézis, Nirenberg, and Stampacchia [47]. The second one, as proposed by Oettli [278], leads in a straightforward way to existence results for vector equilibrium problems from the results about scalar case. A key tool for the study of such problems is an appropriate gauge function.

We will see in subsequent sections that an overwhelming majority of vector problems of potential interest in nonlinear analysis can be cast in the form of vector equilibrium problems. In this section our attention is focused on the existence of a generalized vector equilibrium.

3.8.1 Vector Equilibrium Problems

Let us agree to define a standard **scalar (= real) equilibrium problem** as follows. Given a nonempty subset M of a real topological vector space X, and a real function f defined on $M \times M$,

(EP) find at least one point $\overline{x} \in M$ satisfying $f(\overline{x}, y) \leq 0$ for all $y \in M$.

Let us see whether this scalar equilibrium problem can be formulated for the case of vector-valued mappings.

Given the following:

- Y a real topological vector space;
- M a nonempty subset of X, called the set of feasible decisions;

- a multifunction $P : M \rightrightarrows Y$ that is used for an outcome y at a decision event x : for $y_1, y_2 \in Y$, y_1 dominates y_2 at the decision event $x \in M$ iff $y_1 - y_2 \in P(x)$;
- a multifunction $f : M \times M \rightrightarrows Y$, called the criterion mapping.

We shall assume for the multifunctions P and f that $\operatorname{dom} P = M$ and $\operatorname{dom} f = M \times M$.

Inspired by the scalar equilibrium problem, the generalized vector equilibrium problems (in short GVEP) we can consider can be generalized in several possible ways, for instance, to establish the existence of some feasible decision $\overline{x} \in M$ such that

1. $f(\overline{x}, y) \cap P(\overline{x}) \neq \emptyset$ for all $y \in M$;
2. $f(\overline{x}, y) \subset P(\overline{x})$ for all $y \in M$;
3. $f(\overline{x}, y) \cap - \operatorname{int} P(\overline{x}) = \emptyset$ for all $y \in M$;
4. $f(\overline{x}, y) \not\subset - \operatorname{int} P(\overline{x})$ for all $y \in M$;
5. $f(\overline{x}, y) \cap - (P(\overline{x}) \setminus \{0\}) = \emptyset$ for all $y \in M$;
6. $f(\overline{x}, y) \not\subset - (P(\overline{x}) \setminus \{0\})$ for all $y \in M$.

Proposition 3.8.1. *If we denote the solution set of the problem (GVEP$_i$) by $S(f, P)_i$, for $i = 1, \ldots, 6$, respectively, we have*

(1) $S(f, P)_2 \subset S(f, P)_1$, $S(f, P)_3 \subset S(f, P)_4$ and $S(f, P)_5 \subset S(f, P)_6$;

(2) $S(f, P)_5 \subset S(f, P)_3$ and $S(f, P)_6 \subset S(f, P)_4$ if $0 \notin \operatorname{int} P(\overline{x})$;

(3) $S(f, P)_2 \subset S(f, P)_3$ and $S(f, P)_1 \subset S(f, P)_4$, if $\forall x \in M$, $P(x)$ is w-pointed, i.e., $P(x) \cap - \operatorname{int} P(x) = \emptyset$;

(4) $S(f, P)_2 \subset S(f, P)_5$ and $S(f, P)_1 \subset S(f, P)_6$, if $\forall x \in M$, $P(x)$ is pointed, i.e., $P(x) \cap -P(x) = \{0\}$;

(5) $S(f, P)_4 \subset S(f, P)_1$ and $S(f, P)_3 \subset S(f, P)_2$ if $\forall x \in M$, the complement of $P(x)$ is included in $- \operatorname{int} P(x)$.

The proof of this proposition, which can be argued directly from the definition of the problems (GVEP$_i$), for $i = 1, \ldots, 6$, can be omitted.

Remark 3.8.2. (1) Combining (3) and (5) we can state, under the condition that the sets $P(x)$ are **connected**, i.e., $P(x) \cap - \operatorname{int} P(x) = \emptyset$ and $P(x) \cup - \operatorname{int} P(x) = Y$, that $S(f, P)_2 = S(f, P)_3$ and $S(f, P)_1 = S(f, P)_4$.

(2) Taking $M = Y = \mathbb{R}$, $P(x) = [0, \infty[$ and $f(x, y) = \{-1, 1\}$ for all $x, y \in Y$, one can see that $S(f, P)_1 = S(f, P)_4 = S(f, P)_6 = Y$ and $S(f, P)_2 = S(f, P)_3 = S(f, P)_5 = \emptyset$. This example contradicts the converse inclusions in (1).

(3) If we suppose that $P(x) \setminus \{0\}$ is open in Y, then the converse inclusions in (2) are valid; i.e., $S(f, P)_5 = S(f, P)_3$ and $S(f, P)_6 = S(f, P)_4$.

Remark 3.8.3. When f is a single-valued function, these problems (GVEP$_i$) can be reduced to

$(\text{GVEP}_1, \text{GVEP}_2) \quad f(\bar{x}, y) \in P(\bar{x}) \qquad\qquad \text{for all } y \in M;$

$(\text{GVEP}_3, \text{GVEP}_4) \quad f(\bar{x}, y) \notin -\operatorname{int} P(\bar{x}) \qquad \text{for all } y \in M;$

$(\text{GVEP}_5, \text{GVEP}_6) \quad f(\bar{x}, y) \notin -(P(\bar{x}) \setminus \{0\}) \quad \text{for all } y \in M.$

Remark 3.8.4. If f is as in the last remark, $P =]-\infty, 0]$, and $Y = \mathbb{R}$, all the problems (GVEP_i), for $i = 1, \ldots, 6$, are reduced to the scalar equilibrium problem (EP).

In order to emphasize the relationship to vector optimization we decide to consider here the fourth and sixth generalized **vector equilibrium problems**

(WGVEP) find $\bar{x} \in M$ such that $f(\bar{x}, y) \not\subset -\operatorname{int} P(\bar{x}) \quad \forall y \in M$

and

(SGVEP) find $\bar{x} \in M$ such that $f(\bar{x}, y) \not\subset -(P(\bar{x}) \setminus \{0\}) \quad \forall y \in M$. We mean by (WGVEP) (resp. (SGVEP)) weak (resp. strong) generalized vector equilibrium problem.

3.8.2 General Vector Monotonicity

Recently, much attention has been given to the development of different monotonicity notions. These **monotonicity notions** permit one to lighten the topological requirements needed to establish a solution of vector equilibrium problems, as compared with the nonmonotone case.

Let us recall some of them.

Definition 3.8.5. *Let X, Y be two vector spaces, M a nonempty subset of X. Let f and g be two multifunctions from $M \times M$ into Y. Let K and L be two multifunctions from M into Y.*

(a) *The pair (f, g) is said to be (K, L)-monotone if for all $x, y \in M$,*

$$f(x, y) \subset K(y) \Longrightarrow g(x, y) \subset L(y).$$

(b) *f is said to be K-monotone if for all $x, y \in M$, one has $f(x, y) + f(y, x) \subset K(x) \cap K(y)$.*

Remark 3.8.6. If $g(x, y) = f(y, x)$ and $K(y) + K(y) \subset K(y)$ for every $x, y \in M$, then f is K-monotone implies that the pair (f, g) is $(-K, K)$-monotone.

The definition of (K, L)-monotonicity of a pair (f, g) unifies several notions of monotonicity in the literature. Let us cite some of them:

Given a closed convex cone P with nonempty interior $\operatorname{int} P$;

(1) if $g(x, y) = f(y, x)$, then $(\operatorname{int} P, -\operatorname{int} P)$-monotonicity of the pair (f, g) is just the P-pseudomonotonicity of f, as introduced in [277], [47];

(2) in case $g(x,y) = f(y,x)$ is a single-valued function, the above definition of (int $P, -P$)-monotonicity of the pair (f,g) reduces to that of P-quasi-monotonicity of f, as introduced in [151], [28].

If we take in the last definition Y to be $\mathbb{R}$ and $P = \mathbb{R}_-$, we obtain the definitions, which have been introduced by Bianchi and Schaible in [29], of pseudomonotonicity, i.e., $f(x,y) \geq 0$ implies $f(y,x) \leq 0$ for each $x,y \in K$, and quasimonotonicity, i.e., $f(x,y) > 0$ implies $f(y,x) \leq 0$ for each $x,y \in K$.

3.8.3 Existence of Vector Equilibria by Use of the Generalized KKM Lemma

The classical **Knaster–Kuratowski–Mazurkiewicz lemma** (the KKM lemma) is of fundamental importance in nonlinear analysis, game theory, economics, optimization theory, and variational inequalities. Many generalizations of the KKM lemma, beginning with Fan's result [111] (Fan–KKM lemma), have been given (see [64], [358], and the quoted bibliography). Most of the results obtained are based on the assumption that the considered multifunctions have closed values.

If M is a nonempty subset and $x \in M$, we shall denote by $\mathcal{F}(M)$ (respectively, $\mathcal{F}(M,x)$) the family of all nonempty finite subsets of M (respectively, of M containing x).

If X and Y are two subsets of a vector space and S a multifunction defined from X into Y, then S is said to be a **KKM-mapping** if for every $A \in \mathcal{F}(X)$, conv$(A) \subset \cup_{x \in A} S(x)$.

Note that if S is a KKM-mapping, then $x \in S(x)$ for all $x \in X$.

We will next deal with the Fan–KKM lemma.

Lemma 3.8.7. (Fan–KKM lemma, [111, Lemma 1])
Let X be a nonempty convex subset of a Hausdorff topological vector space E, and S a KKM-mapping defined from X into E. Suppose that

(i) *(compactness condition)* $\exists x_0 \in X$ *such that* $S(x_0)$ *is compact in* Y;
(ii) *(closedness condition)* $\forall x \in X$, $S(x)$ *is closed in* Y.

Then $\bigcap_{x \in X} S(x) \neq \emptyset$.

Here, by relaxing the closedness condition, we begin by stating the following **generalized Fan's theorem**, which will play a crucial role in proving the main existence result.

Lemma 3.8.8. *Let X be an arbitrary nonempty convex subset of a Hausdorff topological vector space E, and T a KKM-mapping defined from X into E. Suppose that for some $x_0 \in X$ and each $A \in \mathcal{F}(X, x_0)$ one has:*

(i) $T(x_0) \cap X$ *is relatively compact in* X;

(ii) *for every* $x \in \mathrm{conv}(A)$, *the intersection* $T(x) \cap \mathrm{conv}(A)$ *is closed in* $\mathrm{conv}(A)$;

(iii) $\mathrm{cl}_X(X \cap (\cap_{x \in F} T(x))) \cap F = (\cap_{x \in F} T(x)) \cap F$, *where* $F = \mathrm{conv}(A)$.

Then the intersection of all subsets $T(x)$, *for* $x \in X$, *is nonempty.*

PROOF. For $A \in \mathcal{F}(X, x_0)$ consider $F_A = \mathrm{conv}(A) \subset X$ and S_A a multi-function defined on F_A by $S_A(x) := T(x) \cap F_A$. Note that $S_A(x)$ is closed, and so S_A is compact-valued. We claim that S_A is a KKM-mapping. Indeed, for $\emptyset \neq B \subset F_A$, B finite, we have that

$$\mathrm{conv}(B) \subset (\cup_{x \in B} T(x)) \cap F_A = \cup_{x \in B}(T(x) \cap F_A) = \cup_{x \in B} S_A(x).$$

Using the Fan–KKM lemma (Lemma 3.8.7), there exists $x_A \in \cap_{x \in F_A} S_A(x) \subset (\cap_{x \in F_A} T(x)) \cap X \subset T(x_0) \cap X$. Consider

$$U_A := \mathrm{cl}_X\{x_B \mid B \in \mathcal{F}(X, x_0), A \subset B\}$$
$$\subset \mathrm{cl}_X((\cap_{x \in F_A} T(x)) \cap X)$$
$$\subset \mathrm{cl}_X(T(x_0) \cap X).$$

The family $\{U_A \mid A \in \mathcal{F}(X, x_0\}$ has the finite intersection property. Since $\mathrm{cl}_X(T(x_0) \cap X)$ is compact, there exists some $\bar{x} \in \cap_{B \in \mathcal{F}(X, x_0)} U_B$. Let $x \in X$ and take $A := \{x, \bar{x}, x_0\} \in \mathcal{F}(X, x_0)$. Since $\bar{x} \in A \subset F_A$, then

$$\bar{x} \in U_A \cap F_A \subset \mathrm{cl}_X((\cap_{y \in F_A} T(y)) \cap X) \cap F_A = (\cap_{y \in F_A} T(y)) \cap F_A.$$

Since $x \in F_A$, we obtain $\bar{x} \in T(x)$. Therefore $\bar{x} \in \cap_{x \in X} T(x)$. $\qquad\square$

Remark 3.8.9. (1) Observe that in the case where T is closed-valued, this result becomes the Fan–KKM lemma.

(2) This result is in effect a generalization of the Brézis–Nirenberg–Stampacchia theorem; see [47, p. 295].

We are now in position to provide a first **existence result**.

Theorem 3.8.10. *Let* X *and* Y *be two real Hausdorff topological vector spaces. Let* M *be a nonempty convex subset of* X, *and consider two multi-functions* K *and* L *from* M *to* Y. *Suppose that* f *and* g *are two multifunctions from* $M \times M$ *to* 2^Y *such that*

(H1) *the pair* (f, g) *is* (K, L)-*monotone;*

(H2) $\bigcup_{i=1}^n g(x_i, z) \not\subset L(z) \; \forall \, (x_1, \ldots, x_n) \in M, \; \forall \, z \in \mathrm{conv}(x_1, \ldots, x_n)$;

(H3) *for every fixed* $x \in M$ *and every* $A \in \mathcal{F}(M)$, *the set* $\{y \in \mathrm{conv}(A) \mid f(x, y) \not\subset K(y)\}$ *is closed in* $\mathrm{conv}(A)$;

(H4) *for every* $A \in \mathcal{F}(M)$, *and every converging net* (y_α) *on* M *to* $y \in \mathrm{conv}(A)$

$$f(z, y_\alpha) \not\subset K(y_\alpha) \; \forall \, z \in \mathrm{conv}(A) \; \Rightarrow \; f(z, y) \not\subset K(y) \; \forall \, z \in \mathrm{conv}(A);$$
$$(3.87)$$

(H5) *there is a nonempty subset B of X and $x_0 \in M \cap B$ such that $M \cap B$ is relatively compact in M, and $f(x_0, y) \subset K(y)$ for all $y \in M \setminus B$.*

Then there exists $\bar{x} \in M \cap B$ such that $f(y, \bar{x}) \not\subset K(\bar{x})$ for all $y \in M$.

PROOF. For each $y \in M$, let $T(y) := \{x \in M \mid f(y, x) \not\subset K(x)\}$.
From (H5), $T(x_0) \subset M \cap B$ is relatively compact in M, and therefore (i) of Lemma 3.8.8 is satisfied. (H1) and (H2) lead clearly to the fact that T is a KKM-mapping, while property (ii) follows from (H3).

It remains to check (iii) of Lemma 3.8.8. For this, let F be the convex hull of a finite subset A of M. We have to show that

$$\mathrm{cl}_M \left(\cap_{z \in F} T(z) \right) \cap F = \left(\cap_{z \in F} T(z) \right) \cap F.$$

So let $x \in \mathrm{cl}_M \left(\cap_{z \in F} T(z) \right) \cap F$. Then there exists a net (x_α) on M that converges to $x \in F$ such that

$$f(z, x_\alpha) \not\subset K(x_\alpha) \ \forall z \in F.$$

From (H4) we get

$$f(z, x) \not\subset K(x) \ \forall z \in F,$$

which implies that $x \in \left(\cap_{z \in F} T(z) \right) \cap F$. Thus (iii) is satisfied. According to the above lemma we deduce that $\cap_{y \in M} T(y) \neq \emptyset$, which is the conclusion of Theorem 3.8.10. ☐

3.8.4 Existence by Scalarization of Vector Equilibrium Problems

In this subsection, we derive the existence theorem of (GVEP) by way of solving an appropriate scalar one. Very recently, Oettli [278] proposed an approach to solving vector equilibrium problems by using results in the scalar case. A key tool for the study of such problems is an appropriate Minkowski's gauge function.

Scalar problems

Before going into the study of vector equilibrium problems, let us consider the scalar case. Usually, we establish theorems in the theory of noncooperative games and in the theory of general equilibrium with the aid of the famous KKM lemma, which is equivalent to the Brouwer fixed point theorem. This method is more interesting; but since we have used this method in the first paragraph, we prefer for convenience to use partitions of unity and fixed point theorems to establish the existence of scalar equilibria.

Theorem 3.8.11. *Let X be a Hausdorff real topological vector space, M a nonempty closed convex subset. Consider two real functions φ and ψ defined on $M \times M$ such that:*

(1) $\phi(x, x) \leq 0$ *for each $x \in M$;*
(2) *for each $x, y \in M$, if $\phi(x, y) \leq 0$, then $\varphi(x, y) \leq 0$;*
(3) *for each $y \in M$, $\{x \in M \cap K \mid \varphi(x, y) \leq 0\}$ is closed for every compact subset K of M;*
(4) *for each $y \in M$, $\{x \in M \mid \phi(x, y) > 0\}$ is convex.*
(5) *(Compactness assumption) there exists a compact convex subset B of M such that for all $y \in M \setminus B$, there exists $x \in B$ such that $\varphi(x, y) > 0$.*

Then, there exists an equilibrium point $\overline{x} \in B$, i.e. $\varphi(y, \overline{x}) \leq 0$ for each $y \in M$.

PROOF. We proceed in two steps.

1. Case of $M = B$; i.e., M is assumed compact. Suppose that a solution does not exist. Then the subset M can be covered by the family of open subsets $O(y) := \{x \in M \mid \varphi(y, x) > 0\}$, where $y \in M$. Since M is assumed to be compact, let $\{O(y_i)\}$ be a finite subcover of M (we set $O_i = O(y_i)$).

Let $\{p_i \mid i = 1, \ldots, n\}$ be a continuous partition of unity subordinate to this finite covering. We introduce the function $p : M \longrightarrow M$ defined by

$$p(u) := \sum_{i=1}^{n} p_i(u) y_i.$$

This mapping is continuous; we obtain, by using Tychonoff's fixed point theorem, the existence of some $u_0 \in M$ such that $p(u_0) = u_0$.

Let I be the set of index $i \in \{1, \ldots, n\}$ for which $p_i(u_0) > 0$; then $u_0 \in O_i$ for all $i \in I$. From (2), we obtain $\phi(y_i, u_0) > 0$ for all $i \in I$. Therefore, by using assumption (4), we obtain

$$\phi(u_0, u_0) = \phi(p(u_0), u_0) = \phi\left(\sum_{i=1}^{n} p_i(u_0) y_i, u_0\right) > 0,$$

a contradiction with condition (1). Hence $\exists \overline{x} \in M$ such that $\varphi(y, \overline{x}) \leq 0$ for each $y \in M$.

2. General case. For every $y \in M$, consider $A(y) := \{x \in B \mid \varphi(y, x) \leq 0\}$. Let $\{y_1, \ldots, y_m\}$ be a fixed finite subset of M, and let M_0 be the convex hull of B and $\{y_1, \ldots, y_m\}$. Then M_0 is compact, since B is convex and compact and X is a Hausdorff topological vector space. By the first step, there exists some $x_0 \in M_0$ such that $\varphi(y, x_0) \leq 0$ for each $y \in M_0$.

Using assumption (5), we have $x_0 \in B$. Thus $x_0 \in \cap_{1 \leq i \leq m} A(y_i)$. Hence every finite subfamily of the sets $A(y)$ has nonempty intersection.

Observe that (3) justifies that $A(y)$ is compact for each $y \in M$, since B is compact. Thus by virtue of the finite intersection property in a compact set, we deduce that there exists $\overline{x} \in B$ that belongs to all the sets $A(y)$ for $y \in M$. This means that (EP) has a solution in B. $\qquad\square$

Vector problems

Now we prepare the treatment of the vector problem. First, let us recall the following propositions and notions on an ordered vector space.

Let Y be a Hausdorff l.c.s., and let Y^* denote its topological dual space. Let $P, P \neq Y$, be a closed convex proper cone with nonempty interior, and $P^+ := \{y^* \in Y^* \mid \langle y, y^* \rangle \geq 0 \ \forall y \in P\}$ its dual cone.

Since we suppose Y a Hausdorff locally convex space, $\operatorname{int} P \neq \emptyset$ and $P \neq Y$, the dual cone P^+ has a convex w^*-compact base B^*; i.e., $0 \notin B^*$ and $P^+ = \cup_{t \geq 0} t B^*$. As has been mentioned in Lemma 2.2.17, we can choose, for instance, $B^* = \{y^* \in P^+ \mid \langle b, y^* \rangle = 1\}$ for some arbitrary $b \in \operatorname{int} P$. We shall fix such a w^*-compact base B^* in what follows.

Let us consider the sup-support and inf-support functions

$$\alpha(y) := \min\{\langle y, y^* \rangle \mid y^* \in B^*\} \text{ and } \beta(y) := \max\{\langle y, y^* \rangle \mid y^* \in B^*\}.$$

Let us remark that $\beta(y) = -\alpha(-y)$, and that the infimum (resp. supremum) of linear and continuous functions α is superlinear and upper semicontinuous (resp. β is sublinear and lower semicontinuous) on the space Y.

Note that in Y, see Section 2.2, the order relations between two elements $x, y \in Y$ can be defined and characterized by

$$x \leq_P y \Leftrightarrow x \in y - P \Leftrightarrow \langle x - y, y^* \rangle \leq 0 \ \forall y^* \in P^+$$
$$\Leftrightarrow \langle x - y, y^* \rangle \leq 0 \ \forall y^* \in B^* \Leftrightarrow \beta(x - y) \leq 0$$
$$\Leftrightarrow \alpha(y - x) \geq 0;$$

and

$$x <_P y \Leftrightarrow x \in y - \operatorname{int} P \Leftrightarrow \langle x - y, y^* \rangle < 0 \ \forall y^* \in P^+ \setminus \{0\}$$
$$\Leftrightarrow \langle x - y, y^* \rangle < 0 \ \forall y^* \in B^* \Leftrightarrow \beta(x - y) < 0$$
$$\Leftrightarrow \alpha(y - x) > 0.$$

To produce the negation of each created relation for vector orders, it suffices to use negation of equivalent ones, i.e.,

$$x \not\leq_P y \Leftrightarrow x - y \notin -P \Leftrightarrow \beta(x - y) > 0 \Leftrightarrow \alpha(y - x) < 0$$

and

$$x \not<_P y \Leftrightarrow x - y \notin -\operatorname{int} P \Leftrightarrow \beta(x - y) \geq 0 \Leftrightarrow \alpha(y - x) \leq 0.$$

Now for a given vector-valued mapping f from $M \times M$ into Y, we define the real-valued functions

$$F(x, y) := \beta(f(x, y)). \tag{3.88}$$

Note that each inequality for F can be transformed into vector relations for f as follows:

$$f(x,y) \leq_P 0 \Longleftrightarrow F(x,y) \leq 0 \quad \text{and} \quad f(x,y) <_P 0 \Longleftrightarrow F(x,y) < 0;$$
$$f(x,y) \nleq_P 0 \Longleftrightarrow F(x,y) > 0 \quad \text{and} \quad f(x,y) \nless_P 0 \Longleftrightarrow F(x,y) \geq 0.$$

A direct application of the above equivalences between vector orders and real inequalities to Theorem 3.8.11 gives the following theorem.

Theorem 3.8.12. *Under the same setting of Theorem 3.8.11, we suppose in addition that Y is a Hausdorff locally convex topological vector space and P a closed convex cone with $\mathrm{int}\,P \neq \emptyset$ and $P \neq Y$. Consider $f, g : M \times M \longrightarrow Y$ satisfying*

(H1') $g(x,x) \notin -\mathrm{int}\,P$ *for all $x \in M$;*
(H2') $f(x,y) \in -\mathrm{int}\,P$ *implies $g(x,y) \in -\mathrm{int}\,P$ for every $x, y \in M$;*
(H3') *for every $y \in M$ and every compact subset K of X, the subset $\{x \in M \cap K \mid f(x,y) \notin -\mathrm{int}\,P\}$ is closed in M;*
(H4') *for every $y \in M$, the subset $\{x \in M \mid g(x,y) \in -\mathrm{int}\,P\}$ is convex;*
(H5') *there exists a compact subset B of M such that for every $x \in M \setminus B$, there exists some $y \in B$ for which $f(y,x) \in -\mathrm{int}\,P$.*

Then there exists $\overline{x} \in B$ such that $f(y, \overline{x}) \notin -\mathrm{int}\,P$.

3.8.5 Some Knowledge About the Assumptions

The conditions of Theorems 3.8.10 and 3.8.22 are not particularly restrictive. We will now show, in the following lemmas, how to use these assumptions.

Lemma 3.8.13. *Suppose that*

(a) $g(x,x) \not\subset L(x)$ *for all $x \in M$,*
(b) *for every fixed $y \in M$, the set $\{x \in M \mid g(x,y) \subset L(y)\}$ is convex.*

Then the hypothesis (H2) is satisfied.

PROOF. Suppose by contradiction that for each $i = 1, \ldots, n$ there exist $x_i \in M$ and $\lambda_i \in [0,1]$ such that $\sum_{i=1}^{n} \lambda_i = 1$ and $g(x_i, \sum_{j=1}^{n} \lambda_j x_j) \subset L(\sum_{j=1}^{n} \lambda_j x_j)$ for every $i = 1, \ldots, n$. Assumption (b) shows that if $z = \sum_{j=1}^{n} \lambda_j x_j$, then $g(z,z) \subset L(z)$. This contradicts (a), and the proposition is proved. $\square$

Lemma 3.8.14. *Condition (b) of Lemma 3.8.13, namely convexity of the subset $\{x \in M \mid g(x,y) \subset L(y)\}$, is satisfied if one of the following conditions is satisfied:*

(i) *$L(y)$ is convex and $g(\cdot, y)$ is concave; i.e., for all $u, v \in M$ and $t \in [0,1]$,*

$$g(tu + (1-t)v, y) \subset tg(u,y) + (1-t)g(v,y).$$

(ii) $L(y)$ *is a convex open cone and* $g(\cdot, y)$ *is* **quasi $\overline{L(y)}$-concave** *on* M; *i.e., for all* $u, v \in M$ *and* $t \in [0, 1]$,

$$g(tu + (1 - t)v, y) \subset \mathrm{conv}(g(u, y) \cup g(v, y)) + \overline{L(y)}.$$

PROOF. Under assumption (i), the condition is trivially satisfied. For the second one, it suffices to use the inclusion $L(y) + \overline{L(y)} \subset L(y)$. □

Remark 3.8.15. One can confirm that g satisfies condition (ii) of Lemma 3.8.14 if one supposes that g satisfies one of the following conditions:

(a) $g(\cdot, y)$ is $\overline{L(y)}$-concave: for all $u, v, y \in M$ and $t \in [0, 1]$, $g(tu + (1-t)v, y) \subset tg(u, y) + (1 - t)g(v, y) + \overline{L(y)}$;

(b) for all $u, v, y \in M$ and $t \in [0, 1]$, either $g(tu + (1 - t)v, y) \subset g(u, y) + \overline{L(y)}$ or $g(tu + (1 - t)v, y) \subset g(v, y) + \overline{L(y)}$.

The proof of this assertion is trivial.

Lemma 3.8.16. *Suppose that the set* $\{y \in M \mid f(x, y) \not\subset K(y)\}$ *is closed for every* $x \in M$; *then assumptions* (H3) *and* (H4) *are satisfied.*

The proof of this lemma is immediate.

Lemma 3.8.17. *If* $K(y) = K$ *is open and independent of* $y \in M$, *then the assumption* (H3) *is satisfied if* $f(x, \cdot)$ *is* $\overline{K}$-*upper continuous on the convex hull of every finite subset of* M.

PROOF. Let F be the convex hull of a finite subset of M, and set $U(x) = f(x, \cdot)^{+1}(K) \cap F := \{y \in F \mid f(x, y) \subset K\}$. Let us show that $U(x)$ is open in F. Fix $y_0 \in U(x)$. Then K is an open set in Y that contains $f(x, y_0)$. Since $f(x, \cdot)$ is $\overline{K}$-upper continuous on F, see Definition 2.33, it follows that for some $U_0 \in \mathcal{V}_F(y_0)$, $\forall y \in U_0$ $f(x, y) \subset K + \overline{K} \subset K$. Thus $U_0 \subset U(x)$, and the proof is complete. □

When we suppose $f(x, \cdot)$ to be upper continuous on the convex hull of every finite subset of M, then since K is an open subset of Y, one can immediately justify that $f(x, \cdot)^{+1}(K) \cap F$ is open in F.

Lemma 3.8.18. *The assumption* (H3) *is satisfied when we suppose that for every fixed* $x \in M$, $f(x, \cdot)$, *with compact values, is upper continuous on the convex hull of every finite subset of* M, *and the multifunction* K *has an open graph in* $M \times Y$.

PROOF. Let us fix arbitrarily $x \in M$ and F the convex hull of a finite subset A of M. We claim that $M(x) := \{u \in F \mid f(x, u) \not\subset K(u)\}$ is closed in F. Indeed, consider a net $\{u_i \mid i \in I\}$ in $M(x)$ that converges to some $\overline{u} \in F$; then for every $i \in I$, there exists some $z_i \in f(x, u_i)$ such that $z_i \notin K(u_i)$.

Suppose first that the net $\{z_i\}$ converges, and let $\overline{z}$ be the limit. Since K has an open graph in $M \times Y$, we deduce that $\overline{z} \notin K(\overline{u})$. We claim that $\overline{z} \in f(x, \overline{u})$; otherwise, from the Hausdorff property of Y there exists $V \in$

$\mathcal{N}_X(f(x,\overline{u}))$ such that $\overline{z} \notin V$. The multifunction $f(x,\cdot)$ is assumed to be upper continuous and $u_i \to \overline{u}$. Then there exists $i_0 \in I$ such that $\forall i \in I$, $i \succeq i_0 \Rightarrow f(x,u_i) \subset V$, and $z_i \notin V$, which is a contradiction. Thus $\overline{u} \in M(x)$.

Suppose now that the net $\{z_i\}$ doesn't converge. Then for every $z \in f(x,\overline{u})$ one can find an open set $V(z)$ and $I(z) \subset I$ such that $z \in V(z)$ and $\forall i \in I(z)$, $z_i \notin V(z)$. Since $f(x,\overline{u})$ is compact and $f(x,\overline{u}) \subset \cup_{z \in f(x,\overline{u})} V(z)$, there exist $z_1, \ldots, z_n \in f(x,\overline{u})$ such that $f(x,\overline{u}) \subset V(z_1) \cup \cdots \cup V(z_n)$. If we take $V_0 = \cup_{1 \leq k \leq n} V(z_i)$ and $I_0 = \cap_{1 \leq k \leq n} I(z_i)$, then one has $\forall i \in I_0$, $z_i \notin V_0$, which gives $f(x,u_i) \not\subset V_0$. This contradicts our assumption on upper continuity of $f(x,\cdot)$ on F, and completes the proof. $\qquad\square$

Lemma 3.8.19. *The assumption* (H4) *of Theorem 3.8.10 is equivalent to the following:*

(H8) *For every* $x,y \in M$ *and* $(y_\alpha) \subset M$ *converging to* y, *then*
$$f(tx + (1-t)y, y_\alpha) \not\subset K(y_\alpha) \quad \forall t \in [0,1] \quad \text{implies } f(x,y) \not\subset K(y).$$

PROOF. Let us first suppose (H4). Consider $x,y \in M$ and (y_α) a net on M that converges to y. By taking $D = \mathrm{conv}\{x,y\}$ and using (H4), we have $f(z,y) \not\subset K(y)$ for every $z \in D$; in particular, $z = x$ does the job.

For the converse, let $A \in \mathcal{F}(M)$ and (y_α) a net on M converging to $y \in \mathrm{conv}(A)$. Suppose that $f(z,y_\alpha) \not\subset K(y_\alpha)$ for all α and all $z \in \mathrm{conv}(A)$.

Fix $z \in \mathrm{conv}(A)$. Then $f(tz + (1-t)y, y_\alpha) \not\subset K(y_\alpha)$ for all α and all $t \in [0,1]$, since $tz + (1-t)y \in \mathrm{conv}(A)$. Using (H8), we deduce that $f(z,y) \not\subset K(y)$. This is true for each $z \in \mathrm{conv}(A)$, and so (H4) is satisfied. $\qquad\square$

Consider the following problems:

(I)   find $\overline{x} \in M$ such that $f(y,\overline{x}) \not\subset K(\overline{x})$ for all $y \in M$; and

(II) find $\overline{x} \in M$ such that $f(\overline{x},y) \not\subset L(\overline{x})$ for all $y \in M$.

Lemma 3.8.20. *(Generalized Minty's Linearization)*

(i) *If* $\overline{x}$ *solves Problem* (II), *then* $\overline{x}$ *is a solution of Problem* (I) *whenever the pair* (f,g) *is* (K,L)-*monotone with* $g(x,y) = f(y,x)$ *for every* $x,y \in M$.

(ii) *If* $\overline{x}$ *solves Problem* (I), *then* $\overline{x}$ *is a solution of Problem* (II) *whenever the following assumptions are satisfied:*
  (1) $f(x,x) \not\subset L(x)$ *for all* $x \in M$;
  (2) *if* $x,y \in M$, $x \neq y$, *then* $u \in]x,y[$ *and* $f(u,y) \subset L(x)$ *implies* $f(u,x) \subset K(x)$;
  (3) *for all* $x,y \in M$, *the set* $\{v \in [x,y] \mid f(v,y) \subset L(x)\}$ *is open in* $[x,y]$.

PROOF. (i) The first assertion is derived from (K,L)-monotonicity of the pair (f,g); let us prove (ii). Assume, for contradiction, that $f(\overline{x},z) \subset L(\overline{x})$ for some $z \in M$. By assumption (1), we have $z \neq \overline{x}$. According to assumption (3), there exists some $u \in]\overline{x},z[$ such that $f(u,z) \subset L(\overline{x})$. This implies, see assumption (2), $f(u,\overline{x}) \subset K(\overline{x})$. Since $f(u,\overline{x}) \not\subset K(\overline{x})$, we obtain a contradiction. $\qquad\square$

Remark 3.8.21. This lemma remains true if we suppose instead of (2) the following:

1. if $x \neq y \in M$, then $u \in]x, y[$, $f(u, x) \not\subset K(x)$, and $f(u, y) \subset L(x)$ imply that $f(u, v) \not\subset K(x)$ for all $v \in]x, y]$;
2. $f(x, x) \subset K(y)$ for all $x, y \in M$.

3.8.6 Some Particular Cases

Let us single out some particular cases of Theorem 3.8.10. First, Theorem 3.8.10 and Lemma 3.8.13 give the following result:

Theorem 3.8.22. *In addition to the assumptions* (H1) *and* (H2) *in Theorem 3.8.10, assume that M is closed and the following assumptions* (H6) *and* (H7). *Then the conclusion of Theorem 3.8.10 remains true.*

(H6) *For every fixed $x \in M$ and every convex compact subset K of X, the set*
$\{y \in K \cap M \mid f(x, y) \not\subset K(y)\}$ *is closed in $K \cap M$;*
(H7) *there is a convex compact subset B of X such that for all $y \in M \setminus B$, there exists $x \in M \cap B$ for which $f(x, y) \subset K(y)$.*

PROOF. Let us consider the multifunction S defined, for each $y \in M$, by $S(y) := \{x \in M \cap B \mid f(y, x) \not\subset K(x)\}$. Let $A \in \mathcal{F}(M)$ and consider M_0 the convex hull of $(B \cap M) \cup A$. Since B is compact and X is a Hausdorff topological vector space, M_0 is compact. Note that the multifunction f from $M_0 \times M_0$ into Y satisfies all assumptions of Theorem 3.8.10. Then there exists $x_1 \in M_0$ such that $f(y, x_1) \not\subset K(x_1)$ for every $y \in M_0$.

Using compactness assumption (H7), we have $x_1 \in B$, and therefore $x_1 \in \cap_{y \in A} S(y)$. We conclude that every finite-intersection subfamily of the set $\{S(y) \mid y \in M\}$ has nonempty intersection.

Observe that $(H6)$ justifies that $S(y)$ is compact for each $y \in M$. Thus by virtue of the finite-intersection property in a compact set, we deduce that there exists $\overline{x} \in B$ that belongs to the set $S(y)$ for all $y \in M$. This means that $\overline{x}$ is a suitable solution in B. □

Remark 3.8.23. It is clear that the condition (H7) of Theorem 3.8.22 is weaker than the corresponding condition (H5) of Theorem 3.8.10. Nevertheless, we restrict ourselves to the more stringent compactness condition in (H6).

Theorem 3.8.22 and Lemma 3.8.13 give the following result:

Theorem 3.8.24. *Under the same setting of Theorem 3.8.22, we suppose that* (H1), (H7) *are satisfied, and*

(i) $g(x, x) \not\subset L(x)$ *for all $x \in M$;*
(ii) *for every fixed $y \in M$, the set $\{x \in M \mid g(x, y) \subset L(y)\}$ is convex;*
(iii) *for every fixed $x \in M$, the set $\{y \in M \mid f(x, y) \not\subset K(y)\}$ is closed.*

Then we get the same conclusion of Theorem 3.8.10.

To consider another particular case, we assume that $K(x) = L(x)$ and $f(x, y) = g(x, y)$ for all $x, y \in M$. Then, from Theorem 3.8.10 and Lemma 3.8.13 we obtain the following theorem:

Theorem 3.8.25. *Assume that*

(i) $f(x, x) \not\subset K(x)$ *for all* $x \in M$;
(ii) *for every fixed* $y \in M$, *the set* $\{x \in M \mid f(x, y) \subset K(y)\}$ *is convex;*
(iii) *for every fixed* $x \in M$, *the set* $\{y \in M \mid f(x, y) \not\subset K(y)\}$ *is closed;*
(iv) *there is a compact subset* B *of* X *and* $x_0 \in M \cap B$ *such that* $f(x_0, y) \subset K(y)$ *for all* $y \in M \setminus B$.

Then we get the same conclusion of Theorem 3.8.10.

Remark 3.8.26. This theorem contains as a special case, when f is a single-valued function, Lemma 2.1 from [228].

Remark 3.8.27. When $K(x) = \mathbb{R}_+$ for all $x \in M$ and $f(x, y) = h(x, y) - \sup_{x \in M} h(x, x)$, we can state a version of the celebrated 1972 **Fan's minimax inequality**; see [111, Theorem 1] and [11, Theorem 8.5].

Theorem 3.8.28. *Under the same setting of Theorem 3.8.10. Assume that the multifunction* $f : M \times M \to Y$ *satisfies*

(1) $f(x, y) \not\subset -\operatorname{int} P(x)$ *implies* $f(y, x) \not\subset \operatorname{int} P(x)$ *for all* $x, y \in M$;
(2) $f(x, x) \not\subset -\operatorname{int} P(x)$ *for all* $x \in M$;
(3) *for every fixed* $x \in M$, $f(x, \cdot)$ *satisfies condition (ii) of Lemma 3.8.14;*
(4) *for every* $x \in M$ *and* $A \in \mathcal{F}(M)$, *the mapping* $f(x, \cdot)$ *is upper continuous on* $\operatorname{conv}(A)$;
(5) *for every* $x \in M$ *and every* $A \in \mathcal{F}(M)$ *whenever* $y \in \operatorname{conv}(A)$ *and* $(y_\alpha) \subset M$ *converges to* y, *then* $f(tx + (1-t)y, y_\alpha) \not\subset \operatorname{int} P(y_\alpha) \, \forall \alpha, \forall t \in [0, 1]$ *implies* $f(x, y) \not\subset \operatorname{int} P(y)$;
(6) *there is a compact set* $B \subset X$ *and* $x_0 \in M \cap B$ *such that* $f(x_0, y) \subset \operatorname{int} P(y) \, \forall y \in M \setminus B$;
(7) *if* $x \neq y \in M$ *and* $u \in]x, y[$, *then* $f(u, x) \subset \operatorname{int} P(x)$ *or* $f(u, y) \not\subset -\operatorname{int} P(x)$;
(8) *for all* $x, y \in M$, *the set* $\{v \in [x, y] \mid f(v, y) \subset -\operatorname{int} P(x)\}$ *is open in* $[x, y]$.

Then there exists $\overline{x} \in M$ *such that* $f(\overline{x}, y) \not\subset -\operatorname{int} P(\overline{x})$ *for all* $y \in M$.

PROOF. All conditions of Theorem 3.8.10 are satisfied by assumptions (1)–(6) if we set $K(x) = -L(x) = \operatorname{int} P(x)$ and $g(x, y) = f(y, x)$. Indeed, (H1) and (H5) follow from (1) and (6); combining Lemmas 3.8.13, 3.8.14 with assumptions (2)–(3) we get (H2); using Lemma 3.8.17 and assumption (3) we obtain (H3); and (H4) follows from Lemma 3.8.19 and assumption (5). Therefore there exists $\overline{x} \in M$ such that $f(y, \overline{x}) \not\subset K(\overline{x}) \, \forall \, y \in M$. Then we apply Lemma 3.8.20 to say that assumptions (1), (7), (8) lead to $f(\overline{x}, y) \not\subset L(\overline{x}) \ \forall \ y \in M$.   □

A generalized **Minty's linearization lemma** for single-valued functions can be formulated as follows.

Lemma 3.8.29. *Suppose that the single-valued function f is P-pseudomonotone; i.e., for all $x, y \in M$, $f(x, y) \not\subset - \operatorname{int} P$ implies $f(y, x) \notin \operatorname{int} P$; then*

$$f(\overline{x}, y) \not\subset - \operatorname{int} P \quad \forall y \in M \Longrightarrow f(y, \overline{x}) \notin \operatorname{int} P \quad \forall y \in M.$$

The converse in true whenever for every $x \in M$, $f(x, \cdot)$ satisfies condition (ii) of Lemma 3.8.14, $f(\cdot, x)$ is P-lower continuous on every line segment in M, and $f(x, x) \in P$.

PROOF. As mentioned previously, the proof of the first statement is straightforward. Let us prove the second one. Fix $y \in M$ and set $y_t := ty + (1 - t)\overline{x}$, which is in M for every $t \in [0, 1]$. Thus

$$f(y_t, \overline{x}) \notin \operatorname{int} P, \quad \forall t \in [0, 1]. \tag{3.89}$$

Taking into account condition (ii) of Lemma 3.8.14 and Remark 3.8.15, we get for some $s \in [0, 1[$,

$$-f(y_t, y_t) + (sf(y_t, \overline{x}) + (1 - s)f(y_t, y)) \in P. \tag{3.90}$$

Combining relations (3.89), (3.90), and $f(y_t, y_t) \in P$, and using $Y \setminus \operatorname{int} P - P \subset Y \setminus \operatorname{int} P$, we have

$$(1 - s)f(y_t, y) \notin - \operatorname{int} P \Leftrightarrow f(y_t, y) \notin - \operatorname{int} P. \tag{3.91}$$

We deduce that $f(\overline{x}, y) \notin - \operatorname{int} P$; otherwise, since $f(\cdot, y)$ is P-lower continuous on $[\overline{x}, y[$ and $V = -f(\overline{x}, y) - \operatorname{int} P \in \mathcal{V}_Y$, one can find t near zero such that

$$f(y_t, y) \in f(\overline{x}, y) + V - P = f(\overline{x}, y) - (f(\overline{x}, y) + \operatorname{int} P) - P \subset - \operatorname{int} P,$$

which contradicts (3.91); and the result follows.   □

Just after this lemma we shall give a theorem of **existence of vector equilibria** for vector single-valued functions.

Let us set $g(x, y) = f(y, x)$, $K(x) = \operatorname{int} P$ and $L(x) = - \operatorname{int} P$ for all $x, y \in M$, where P is a closed convex and w-pointed cone. Then by using Theorem 3.8.24, Remark 3.8.15 and Lemmas 3.8.13, 3.8.14, 3.8.17, and 3.8.29, we can state the following theorem:

Theorem 3.8.30. *If $f : M \times M \to Y$ satisfies*

(1) *f is P-pseudomonotone; i.e., $f(x, y) \not\subset - \operatorname{int} P$ implies $f(y, x) \notin \operatorname{int} P$;*
(2) *$f(x, x) = 0$ for all $x \in M$;*

(3) *for every $x \in M$, $f(x, \cdot)$ is P-upper continuous and quasi P-concave on every convex compact subset of M;*

(4) *for every $y \in M$, $f(\cdot, y)$ is P-lower continuous on every line segment in M;*

(5) *there exists a convex compact subset B of X such that for all $y \in M \setminus B$, there is $x \in M \cap B$ for which $f(x, y) \in \operatorname{int} P$.*

Then there exists $\overline{x} \in M$ such that $f(\overline{x}, y) \notin -\operatorname{int} P$ for all $y \in M$.

3.8.7 Mixed Vector Equilibrium Problems

In this section, we establish results for vector equilibrium problems where the monotone criterion mappings are supposed to be perturbed by nonmonotone mappings.

Definition 3.8.31. *Let X, Y be two Hausdorff real topological vector spaces, f a mapping from $M \times M$ to Y, P a convex closed cone of Y, and P^+ its dual cone. The mapping f is said to be P-**pseudomonotone in the topological sense** whenever (x_i) is a net on M converging to $x \in M$ such that $\forall V \in \mathcal{V}_Y$, $\exists \alpha_x$ satisfying $\forall i \succeq \alpha_x$, $(f(x_i, x) - V) \not\subset f(x, x) - \operatorname{int} P$, then $\forall y \in M$ and $\forall W \in \mathcal{V}_Y$, $\exists \alpha_{xy}$ such that $\forall i \succeq \alpha_{xy}$, $(f(x_i, y) - W) \cap (f(x, y) - P) \neq \emptyset$.*

In the case where $Y = \mathbb{R}$ and $P = \mathbb{R}_+$, the definition of P-pseudomonotonicity in the topological sense coincides with the classical Brézis–Browder pseudomonotonicity:

$$x_i \overset{M}{\to} x \text{ and } \liminf_i f(x_i, x) \leq f(x, x) \Rightarrow \limsup_i f(x_i, y) \geq f(x, y) \ \forall y \in M.$$

Observe that if $f(\cdot, y)$ is upper P-semicontinuous for all $y \in M$, then f is P-pseudomonotone in the topological sense.

We give now the following result concerning a vector equilibrium problem with a nonmonotone perturbation of a P-monotone mapping.

Theorem 3.8.32. *Let X, Y be two real Hausdorff topological vector spaces, M a nonempty convex subset of X, and let P be a closed convex w-pointed cone in Y with a nonempty interior. Consider two mappings f and g from $M \times M$ to Y such that*

(1) *g is $-P$-monotone;*

(2) *for every $x \in M$, $g(x, x) \in P$ and $f(x, x) = 0$;*

(3) *for every $x \in M$, $f(x, \cdot)$ and $g(x, \cdot)$ are P-convex;*

(4) *for every $x \in M$, $g(x, \cdot)$ is P-lower continuous;*

(5) *for every $y \in M$, $f(\cdot, y)$ is P-upper continuous on the convex hull of every finite subset of M;*

(6) *f is P-pseudomonotone in the topological sense;*

(7) *for every $y \in M$, $g(\cdot, y)$ is P-upper continuous on each line segment in M;*

(8) *there is a compact subset B of X and $x_0 \in M \cap B$ such that $\forall y \in M \setminus B$, $g(x_0, y) - f(y, x_0) \in \operatorname{int} P$.*

Then there exists $\bar{x} \in M$ such that $f(\bar{x}, y) + g(\bar{x}, y) \notin - \operatorname{int} P \quad \forall y \in M$, which is equivalent to $f(\bar{x}, y) - g(y, \bar{x}) \notin - \operatorname{int} P \quad \forall y \in M$.

PROOF. Set $\varphi(x, y) = g(x, y) - f(y, x)$, $\phi(x, y) = f(y, x) + g(y, x)$ and $K(x) = \operatorname{int} P$ and $L(x) = - \operatorname{int} P$ for all $x \in M$. We begin by proving that all assumptions of Theorem 3.8.10 are satisfied.

(H1): Let $x, y \in M$ with $g(x, y) - f(y, x) \in \operatorname{int} P$; then

$$f(y, x) + g(y, x) = f(y, x) - g(x, y) + g(x, y) + g(y, x)$$
$$\in g(x, y) + g(y, x) - \operatorname{int} P.$$

From assumption (1), we conclude that $f(y, x) + g(y, x) \in -P - \operatorname{int} P \subset - \operatorname{int} P$; this is our claim.

(H2): Let x_1, x_2, $y \in M$. Then by P-convexity of $f(y, \cdot)$ and $g(y, \cdot)$, for every $t \in]0, 1[$ one has

$$g(y, tx_1 + (1 - t)x_2) + f(y, tx_1 + (1 - t)x_2)$$
$$\in t\,[g(y, x_1) + f(y, x_1)] + (1 - t)\,[g(y, x_2) + f(y, x_2)] - P.$$

Since $- \operatorname{int} P$ is convex and $-P - \operatorname{int} P \subset - \operatorname{int} P$, we conclude that $\{x \in M : \phi(x, y) \in - \operatorname{int} P\}$ is convex for all $y \in M$. (H3): Since $g(x, \cdot)$ and $-f(\cdot, x)$ are P-lower continuous on the convex hull of every finite subset of M for all fixed $x \in M$, then so is $-f(\cdot, x) + g(x, \cdot)$. Proposition 2.5.23(ii) means that (H3) is satisfied.

(H4): Let $x, y \in M$ and $\{y_i\} \subset M$, which converges to y. Suppose that for every i,

$$g(tx + (1 - t)y, y_i) - f(y_i, tx + (1 - t)y) \notin \operatorname{int} P \; \forall t \in [0, 1]. \qquad (3.92)$$

Suppose, contrary to our claim, that $g(x, y) - f(y, x) \in \operatorname{int} P$. Hence $\operatorname{int} P$ is a neighborhood of $g(x, y) - f(y, x)$ in Y, and then we can find two neighborhoods $V_1, V_2 \in \mathcal{V}_Y(0)$ such that $(g(x, y) + V_1) - (f(y, x) + V_2) \subset \operatorname{int} P$. Since $g(x, \cdot)$ is P-lower continuous, there exists an index α_{xy} such that

$$g(x, y_i) \in g(x, y) + V_1 + P, \; \forall i \succeq \alpha_{xy}. \qquad (3.93)$$

For $t = 0$ in (3.92), one has

$$g(y, y_i) - f(y_i, y) \notin \operatorname{int} P. \qquad (3.94)$$

Let $V \in \mathcal{V}_Y$. Since $g(y, \cdot)$ is P-lower continuous, there exists α_y such that $g(y, y_i) \in g(y, y) + V + P$ for any $i \succeq \alpha_y$. Combining this with assumption (2), the relation (3.94) yields

$$(f(y_i, y) - V) \not\subset - \operatorname{int} P \; \forall i \succeq \alpha_y.$$

By topological P-pseudomonotonicity of f, for V_2 there exists α'_{xy} such that

$$f(y_i, x) \in f(y, x) + V_2 - P \;\; \forall i \succeq \alpha'_{xy}. \tag{3.95}$$

Setting $\alpha_0 = \max(\alpha_{xy}, \alpha_y, \alpha'_{xy})$ and combining relations (3.93) and (3.95), we deduce that $\forall i \succeq \alpha_0$,

$$g(x, y_i) - f(y_i, x) \in (g(x, y) + V_1) - (f(y, x) + V_2) + P \subset \operatorname{int} P + P \subset \operatorname{int} P,$$

which contradicts (3.92) for $t = 1$. Thus (H4) is satisfied.

All assumptions (H1)–(H5) of Theorem 3.8.10 are satisfied, so we deduce that there exists $\overline{x} \in M$ such that

$$g(y, \overline{x}) - f(\overline{x}, y) \notin \operatorname{int} P \quad \text{for all } y \in M. \tag{3.96}$$

Let $y \in M$ be a fixed point and set $y_t = ty + (1-t)\overline{x}$ for $t \in (0,1)$. Since $g(y_t, \cdot)$ is P-convex and $g(y_t, y_t) \in P$, then

$$tg(y_t, y) + (1-t)g(y_t, \overline{x}) \in P. \tag{3.97}$$

From relation (3.96), one has $(1-t)g(y_t, \overline{x}) - (1-t)f(\overline{x}, y_t) \notin \operatorname{int} P$. Combining with (3.97), it follows that $tg(y_t, y) + (1-t)f(\overline{x}, y_t) \notin -\operatorname{int} P$. Using the convexity of $f(\overline{x}, \cdot)$, we can write $tg(y_t, y) + t(1-t)f(\overline{x}, y) + (1-t)^2 f(\overline{x}, \overline{x}) \notin -\operatorname{int} P$, or else $g(y_t, y) + (1-t)f(\overline{x}, y) \notin -\operatorname{int} P$ because $f(\overline{x}, \overline{x}) = 0$. The P-upper continuity of $g(\cdot, y)$ on line segments allows us to write

$$g(\overline{x}, y) + f(\overline{x}, y) \notin -\operatorname{int} P.$$

The converse follows immediately from $-P$-monotonicity of g, and the proof is complete. $\qquad\square$

Remark 3.8.33. The P-monotonicity of g can be replaced by the following (f, P)-pseudomonotonicity: For all $x, y \in M$

$$g(x, y) + f(x, y) \notin \operatorname{int} P \Rightarrow g(y, x) - f(x, y) \notin \operatorname{int} P.$$

In that case, Theorem 3.8.32 appears as a generalization of Theorem 5.1 in [28].

Remark 3.8.34. The conclusion of Theorem 3.8.32 is valid (see Theorem 3.8.22) when instead of assumptions (5), (6), and (8) we only suppose that

(8′) there is $B \subset X$ convex and compact for which $\forall y \in M \setminus B$, $\exists x \in M \cap B$ such that $g(x_0, y) - f(y, x_0) \in \operatorname{int} P$;

in return, we must suppose that

(5′) for every $y \in M$, $f(\cdot, y)$ is P-lower continuous on every convex compact subset of M.

In the next section we present some typical situations in which the vector equilibrium structure occurs quite naturally.

3.9 Applications to Vector Variational Inequalities

As a first example of particular case of the (GVE) problem, let us consider the following **generalized vector variational inequality**:

(GVVI) find $\bar{x} \in K$ s.t. $\forall y \in K$, $\exists \bar{t} \in T(\bar{x})$: $\bar{t}(\eta(\bar{x}, y)) \notin -\operatorname{int} P(\bar{x})$.

Here T is a multifunction from X into the space $L(X, Y)$ of all linear continuous mappings from X into Y, $\eta : K \times K \to K$ is a mapping, and $\bar{t}(x)$ denotes the evaluation of the linear mapping $\bar{t}$ at x. Thus (GVVI) is a particular case of the (WGVEP) problem if we take $f(x, y) = T(x)(\eta(x, y))$. This problem is an extension of the single-valued (GVVI) problem introduced by Siddiqi, Ansari, and Khaliq [322].

Note that for $\eta(x, y) = y - g(x)$, (GVVI) is equivalent to the vector variational problem introduced and studied by Konnov and Yao [215]:

(VVI) find $\bar{x} \in K$ such that $T(\bar{x})(y - g(\bar{x})) \notin -\operatorname{int} P(\bar{x}) \ \forall y \in K$.

When $P(x) = P$ for all $x \in K$, the (GVVI) problem becomes the vector variational inequality problem introduced by Chen and Yang [75].

If $Y = \mathbb{R}$, $X = \mathbb{R}^n$, $L(X, Y) = X^*$, $P(x) = \mathbb{R}_+ \ \forall x \in K$, and g is the identity mapping, the problem (GVVI) becomes the usual **scalar variational inequality** considered and studied by Hartman and Stampacchia [159].

The vector variational inequalities (VVI) were introduced by Giannessi [132] in a finite-dimensional Euclidean space in 1980. Chen and several authors (see [65], [75], [66], [67], [383], [229], [64]) have intensively studied the vector variational inequalities in abstract infinite-dimensional spaces. Recently, the equivalence between a (VVI) and vector optimization problems and the equivalence between (VVI) and vector complementarity problems have been studied; see e.g., [75].

3.9.1 Vector Variational-Like Inequalities

We assume:

- X and Y are two Hausdorff topological vector spaces;
- $L(X, Y)$ is the set of all continuous linear operators from X into Y;
- $M \subseteq X$ is a nonempty convex set;
- $\{P(x) \mid x \in M\}$ is a family of closed convex cones in Y with $\operatorname{int} P(x) \neq \emptyset$ for all $x \in M$;
- $\eta : M \times M \to M$ is continuous and affine with $\eta(x, x) = 0$, for every $x \in M$.

For $\Pi \subseteq L(X, Y)$ we write $\Pi(x) := \{\pi(x) \mid \pi \in \Pi\}$, for all $x \in M$. We have to assume that $L(X, Y)$ is topologized in such a way that the bilinear form $(\pi, x) \mapsto \pi(x)$ is continuous from $L(X, Y) \times M$ into Y; see the next lemma for some examples.

Lemma 3.9.1. *Let X, Y be two Banach spaces, M a closed convex subset of X, and let $\{(\alpha_i, x_i)\}$ be a net in $L(X,Y) \times M$, and $\{(\alpha, x)\}$ in $L(X,Y) \times M$.*

(i) *If $\|\alpha_i - \alpha\| \to 0$ in $L(X,Y)$ and $x_i \overset{w}{\to} x$ in X, then $\{\alpha_i(x_i)\}$ converges weakly to $\alpha(x)$ in Y.*

(ii) *If $\langle(\alpha_i - \alpha)(u), y^*\rangle \to 0$ for every $u \in X$, $y^* \in Y^*$ and $\|x_i - x\| \to 0$ in X, then $\{\alpha_i(x_i)\}$ converges weakly to $\alpha(x)$ in Y.*

(iii) *If $\|\alpha_i - \alpha\| \to 0$ in $L(X,Y)$ and $\|x_i - x\| \to 0$ in X, then $\|\alpha_i(x_i)\} - \alpha(x)\| \to 0$ in Y.*

PROOF. (i) Fix $y^* \in Y^*$. We have

$$|\langle(\alpha_i - \alpha)(x_i), y^*\rangle| \le \|y^*\| \cdot \|\alpha_i - \alpha\| \cdot \|x_i\|.$$

Our assumption $x_i \overset{w}{\to} x$ in X just says that $x^*(x_i) \to x^*(x)$ for all $x^* \in X^*$. Then, according to [100, Theorem II.3.20], $\{x_i\}$ is bounded in X, and thus

$$\limsup_i |\langle\alpha_i(x_i) - \alpha(x), y^*\rangle|$$

$$\le \limsup_i |\langle\alpha_i(x_i) - \alpha(x_i), y^*\rangle| + \limsup_i |\langle\alpha(x_i) - \alpha(x), y^*\rangle|$$

$$\le \|y^*\| \cdot \sup_i \|x_i\| \cdot \limsup_i \|\alpha_i - \alpha\|$$

$$+ \limsup_i |\langle x_i - x, \alpha^* y^*\rangle| \quad (\alpha^* \text{ is the adjoint operator of } \alpha)$$

$$\le 0.$$

This shows that $\{\alpha_i(x_i)\}$ converges weakly to $\alpha(x)$ in Y.

(ii) Using [100, Corollary II.3.21] and $\sup_i\langle\alpha_i(u), y^*\rangle < +\infty$ for every $u \in X$, $y^* \in Y^*$, we have, that $\{\alpha_i\}$ is bounded in $L(X,Y)$, and

$$\limsup_i |\langle\alpha_i(x_i) - \alpha(x), y^*\rangle|$$

$$\le \limsup_i |\langle\alpha_i(x_i) - \alpha_i(x), y^*\rangle| + \limsup_i |\langle(\alpha_i - \alpha)(x), y^*\rangle|$$

$$\le \|y^*\| \cdot \sup_i \|\alpha_i\| \cdot \limsup_i \|x_i - x\| + \limsup_i |\langle(\alpha_i - \alpha)(x), y^*\rangle|$$

$$\le 0.$$

(iii) is obvious.   □

Definition 3.9.2. *Let T be a multifunction from M to $L(X,Y)$.*

(1) *T is said to be (P, η)-monotone if $Ty(\eta(x,y)) - Tx(\eta(x,y)) \subset P(x) \cap P(y)$.*

(2) *$T : M \to L(X,Y)$ is said to be (P, η)-pseudomonotone if $Tx(\eta(x,y)) \subset - \operatorname{int} P(y)$ implies $Ty(\eta(x,y)) \subset - \operatorname{int} P(y)$.*

(3) *T is said to be V-hemicontinuous if for any $x, y \in M$ $T(tx + (1-t)y) \to Ty$ as $t \to 0_+$ (i.e., $\forall z_t \in T(tx + (1-t)y) \exists z \in Ty$ such that $\forall a \in M$, $z_t(a) \to z(a)$ as $t \to 0_+$).*

Then, according to Theorem 3.8.32, we can formulate the following **existence result** concerning a **vector variational-like inequality**.

Theorem 3.9.3. *Let T be a multifunction from M to $L(X,Y)$. If one has*

(i) *T is compact-valued, (P,η)-pseudomonotone and V-hemicontinuous;*
(ii) *for some convex compact subset B of X, we have for every $y \in M \setminus B$ there exists $x \in M \cap B$ such that $Tx_0(\eta(x_0,y)) \subset -\operatorname{int} P(y)$.*

Then there exists $\bar{x} \in B$ satisfying $\quad T\bar{x}(\eta(y,\bar{x})) \not\subset -\operatorname{int} P(\bar{x}) \quad \forall y \in M$.

PROOF. Set $f(x,y) = -Tx(\eta(x,y))$, $g(x,y) = -Ty(\eta(x,y))$, and $K(x) = \operatorname{int} P(x)$ for $x, y \in M$. Then assumptions (H1), (i), (ii), and (H7) of Theorem 3.8.24 are easily satisfied. Let us show that (H6) of Theorem 3.8.24 is also satisfied. For this, let $x \in M$ be fixed and let (y_i) be a net on M converging to $y \in M$ such that $Tx(\eta(x,y_i)) \not\subset -\operatorname{int} P(y_i)$. Therefore there exists $z_i \in Tx$ satisfying $z_i(\eta(x,y_i)) \notin -\operatorname{int} P(y_i)$.

Since Tx is compact, then, passing to a subnet if necessary, we may assume that z_i converges to $z \in Tx$. By the continuity of the mapping η and the bilinear form on $L(X,Y) \times M$, we get $z_i(\eta(x,y_i)) \to z(\eta(x,y))$. Hence

$$z(\eta(x,y)) \notin -\operatorname{int} P(y) \text{ or else } Tx(\eta(x,y)) \not\subset -\operatorname{int} P(y).$$

Then all assumptions of Theorem 3.8.24 are satisfied; thus, there exists $\bar{x} \in M$ such that

$$Ty(\eta(y,\bar{x})) \not\subset -\operatorname{int} P(\bar{x}) \text{ for all } y \in M.$$

Let $y \in M$ be fixed and set $y_t = ty + (1-t)\bar{x}$. Then $Ty_t(\eta(y_t,\bar{x})) \not\subset -\operatorname{int} P(\bar{x})$. Hence $Ty_t(\eta(y,\bar{x})) \not\subset -\operatorname{int} P(\bar{x})$. By the V-hemicontinuity of T and the closedness of $Y \setminus (-\operatorname{int} P(\bar{x}))$, it follows that $T\bar{x}(\eta(y,\bar{x})) \not\subset -\operatorname{int} P(\bar{x})$. □

3.9.2 Perturbed Vector Variational Inequalities

Now, if we take $P(x) = P$ for all $x \in M$, then we obtain from Theorem 3.8.32 the following proposition:

Proposition 3.9.4. *Let $\varphi : M \to Y$ be a P-lower continuous and P-convex function, and let $S, T : M \longrightarrow L(X,Y)$ satisfy the following assumptions:*

(i) *S is continuous on M;*
(ii) *T is P-monotone and continuous on each line segment on M;*
(iii) *there is a compact subset B of X and $x_0 \in M \cap B$ such that*

$$(Sy + Tx_0)(y - x_0) + \varphi(y) - \varphi(x_0) \in \operatorname{int} P \quad \text{for all } y \in M \setminus B.$$

Then there exists $\bar{x} \in M$ such that

$$(S+T)\bar{x}(y - \bar{x}) + \varphi(y) - \varphi(\bar{x}) \notin -\operatorname{int} P \quad \text{for all } y \in M.$$

PROOF. It suffices to set $f(x,y) = Sx(y-x)$ and $g(x,y) = Tx(y-x) + \varphi(y) - \varphi(x)$ and to check that all assumptions of Theorem 3.8.32 are satisfied. We observe that assumptions (1)–(3) and (8) arise automatically from P-convexity of φ, P-monotonicity of T, and condition (iii). Using P-lower continuity of φ, (i)–(ii), and the condition on the bilinear form on $L(X,Y) \times M$, one deduces (4)–(7). □

3.9.3 Hemivariational Inequality Systems

In 1981, by using the generalized Clarke's gradient, Panagiotopoulos introduced the notion of nonconvex super-potential. In the absence of convexity, a new type of inequality was obtained, the so-called hemivariational inequalities. These look like a variational formulation of mechanic problems.

Let us take a brief look at this concept of nonsmooth analysis.

Let X be a Banach space, $\varphi : X \to \mathbb{R}$, and $x \in X$. We say that φ is locally Lipschitz at x if there is some $\varepsilon > 0$ and $k > 0$ such that

$$\|u - x\| < \varepsilon \text{ and } \|v - x\| < \varepsilon \Longrightarrow |\varphi(u) - \varphi(v)| \le k\|u - v\|.$$

The Clarke's generalized directional derivative of φ at x in the direction $h \in X$ is defined by

$$\varphi^o(x; h) := \limsup_{\substack{y \to x \\ t \searrow 0}} \frac{\varphi(y + th) - \varphi(y)}{t},$$

and the associated subdifferential is defined at x by

$$\partial\varphi(x) := \{x^* \in X^* \mid \varphi^o(x; h) \ge \langle h, x^* \rangle \ \forall h \in X\}.$$

Proposition 3.9.5. *[81, Prop. 1.1 and Prop. 1.5] Let $\varphi : X \to \mathbb{R}$ be locally Lipschitz of rank k at x. Then*

(i) *the function $h \longmapsto \varphi^o(x; h)$ is finite, positively homogeneous, subadditive on X, and for every $h \in X$, $|\varphi^o(x; h)| \le k\|h\|$;*

(ii) *the function $(y, h) \longmapsto \varphi^o(y; h)$ is upper semicontinuous at (x, h) for every $h \in X$;*

(iii) *$\partial\varphi(x)$ is nonempty, convex, weak*-compact, and $\|x^*\| \le k$ for every $x^* \in \partial\varphi(x)$;*

(iv) *for every $h \in X$ we have $\varphi^o(x; h) = \max\{\langle h, x^* \rangle \mid x^* \in \partial\varphi(x)\}$.*

Definition 3.9.6. *We say that a multifunction R from X to X^* is (Browder–Hess) **BH-pseudomonotone**, (**quasi-pseudomonotone**), or (**of class $(S)_+$**) if (x_n) converges weakly to x and $\liminf_{n \to \infty} \sup_{x_n^* \in Rx_n} \langle x - x_n, x_n^* \rangle \ge 0$ imply $\limsup_{n \to \infty} \sup_{x_n^* \in Rx_n} \langle y - x_n, x_n^* \rangle \le \sup_{x^* \in Rx} \langle y - x, x^* \rangle \ \forall y \in X$, $(\lim_{n \to \infty} \sup_{x_n^* \in Rx_n} \langle x - x_n, x_n^* \rangle = 0)$, or $(x_n \to x$ and (Rx_n) is bounded for n sufficiently large).*

Proposition 3.9.7. *Consider two multifunctions R_1 and R_2 from X to X^*. Then $R = R_1 + R_2$ is BH-pseudomonotone if either R_1 or R_2 is BH-pseudo-monotone or if one of the pair (R_1, R_2) is the Clarke's subdifferential $\partial\varphi$ of a locally Lipschitz function φ that is assumed quasi-pseudomonotone and the other one is of class $(S)_+$.*

PROOF. Let us suppose that $x_n \rightharpoonup x$ and $\liminf_{n\to\infty} \sup_{x_n^* \in Rx_n} \langle x - x_n, x_n^* \rangle \geq 0$. Then

$$\liminf_{n\to\infty} \left(\sup_{u_n^* \in R_1 x_n} \langle x - x_n, u_n^* \rangle + \sup_{v_n^* \in R_2 x_n} \langle x - x_n, v_n^*, \rangle \right) \geq 0. \qquad (3.98)$$

We claim that

$$\liminf_{n\to\infty} \sup_{u_n^* \in R_1 x_n} \langle x - x_n, u_n^* \rangle \geq 0 \text{ and } \liminf_{n\to\infty} \sup_{v_n^* \in R_2 x_n} \langle x - x_n, x_n^* \rangle \geq 0. \qquad (3.99)$$

To the contrary, suppose, for instance, that the first is false. Then for a subsequence $r := \lim_{p\to\infty} \sup_{u_{n_p}^* \in R_1 x_{n_p}} \langle x - x_{n_p}, u_{n_p}^* \rangle < 0$. Using (3.98), it follows

$$\liminf_{p\to\infty} \sup_{v_{n_p}^* \in R_2 x_{n_p}} \langle x - x_{n_p}, v_{n_p}^* \rangle \geq -r > 0. \qquad (3.100)$$

• If R_2 is BH-pseudomonotone, then $\lim_{p\to\infty} \sup_{v_{n_p}^* \in R_2 x_{n_p}} \langle x - x_{n_p}, v_{n_p}^* \rangle = 0$, a contradiction. Thus (3.99) is satisfied. We conclude from BH-pseudo-monotonicity of R_1 and R_2 that $\forall y \in X$,

$$\limsup_{n\to\infty} \sup_{x_n^* \in Rx_n} \langle y - x_n, x_n^* \rangle$$

$$\leq \limsup_{n\to\infty} \sup_{u_n^* \in R_1 x_n} \langle y - x_n, u_n^* \rangle + \limsup_{n\to\infty} \sup_{v_n^* \in R_2 x_n} \langle y - x_n, v_n^* \rangle$$

$$\leq \sup_{u^* \in R_1 x} \langle y - x, u^* \rangle + \sup_{v^* \in R_2 x} \langle y - x, v^* \rangle$$

$$= \sup_{x^* \in Rx} \langle y - x, x^* \rangle.$$

This gives BH-pseudomonotonicity of R.
• If R_2 is of class $(S)_+$, then (3.100) gives $x_{n_p} \to x$ and $(R_2 x_{n_p})$ is bounded. Thus $\lim_{p\to\infty} \sup_{v_{n_p}^* \in R_2 x_{n_p}} \langle x - x_{n_p}, v_{n_p}^* \rangle = 0$, which contradicts (3.100) and establishes (3.99).
Using that R_2 is of class $(S)_+$, then $x_n \to x$ and $(R_2 x_n)$ is bounded. We conclude that $\forall y \in X$, $\limsup_{n\to\infty} \sup_{v_n^* \in R_2 x_n} \langle y - x_n, v_n^* \rangle \leq \sup_{v^* \in R_2 x} \langle y - x, v^* \rangle$.

Using Proposition 3.9.5 (ii) (iv), we deduce

$$\limsup_{n\to\infty} \sup_{x_n^* \in Rx_n} \langle y - x_n, x_n^* \rangle$$

$$\leq \limsup_{n\to\infty} \varphi^0(x_n, y - x_n) + \limsup_{n\to\infty} \sup_{v_n^* \in R_2 x_n} \langle y - x_n, v_n^* \rangle$$

$$\leq \varphi^0(x, y - x) + \sup_{v^* \in R_2 x} \langle y - x, v^* \rangle$$

$$= \sup_{u^* \in R_1 x} \langle y - x, u^* \rangle + \sup_{v^* \in R_2 x} \langle y - x, v^* \rangle$$

$$= \sup_{x^* \in Rx} \langle y - x, x^* \rangle.$$

• The same proof remains valid when we suppose that $R_2 = \partial \varphi$ is quasi-pseudomonotone and R_1 is of class $(S)_+$. □

As the maximal monotone operators for the variational inequalities, the generalized pseudomonotone operators are a mathematically important tool for the formulation of **existence results concerning the hemivariational mappings**.

In this subsection, we are interested in some **hemivariational inequality systems**. Assume that we are given X a real reflexive Banach space, M a non-empty convex subset of X. Let $S_k, T_k : M \to X^*$, for $k = 1, \ldots, m$, and let $J : X \to \mathbb{R}^m$ be locally Lipschitz near M.

Let us introduce the following mappings: $S, T : M \longrightarrow X^* \times X^* \times \cdots \times X^*$ defined by $Sx := \prod_{k=1}^m S_k x$ and $Tx := \prod_{k=1}^m T_k x$, and set $J^o(x; h) := (J_1^o(x; h), \ldots, J_m^o(x; h))$.

We consider the following hemivariational inequality system (HVIS), which consists in finding $\bar{x} \in M$ such that $\forall y \in M, \exists k_y \in \{1, \ldots, m\}$ with

$$\langle y - \bar{x}, S_{k_y} \bar{x} \rangle + \langle y - \bar{x}, T_{k_y} \bar{x} \rangle + J_{k_y}^o(\bar{x}; y - \bar{x}) \geq 0. \tag{3.101}$$

By making use of Theorem 3.8.32 and Remark 3.8.34 we are able to give conditions that guarantee the existence of a solution to the inequality (HVIS).

Theorem 3.9.8. *Let us impose the following conditions:*

(i) *for every $k = 1, \ldots, m$, S_k is BH-pseudomonotone and locally bounded;*

(ii) *for every $k = 1, \ldots, m$, either ∂J_k is BH-pseudomonotone, or ∂J_k is quasi-pseudomonotone and S is of class $(S)_+$;*

(iii) *for every $k = 1, \ldots, m$, T_k is monotone and continuous from each line segment of M to the weak topology on X^*;*

(iv) *there is a weak-compact subset $B \subset X$ such that for all $y \in M \setminus B$ there exists $x \in M \cap B$ such that $\forall k \in \{1, \ldots, m\}$ $J_k^o(y, x - y) < \langle y - x, S_k y + Tx \rangle$.*

Then the problem (HVIS) has at least one solution.

PROOF. Let us first remark that setting, for each $k \in \{1, \ldots, m\}$ and $x, y \in M$,

$$f_k(x, y) := \langle y - x, S_k x \rangle + \max_{\xi \in \partial J_k(x)} \langle y - x, \xi \rangle \text{ and } g_k(x, y) := \langle y - x, T_k x \rangle,$$

the system (3.101) is equivalent to

$$f(\bar{x}, y) + g(\bar{x}, y) \notin -\operatorname{int} \mathbb{R}_+^m \ \forall y \in M.$$

Let $Y = \mathbb{R}^m$ and $P = \mathbb{R}_+^m$. We shall verify the conditions of Theorem 3.8.32 when X and X^* are respectively endowed with the weak and weak* topologies: (1), (2), (3), (4), and (8) are obviously satisfied; (6) holds by using Propositions 3.9.7 and 3.9.5 (iv), since J and S satisfy conditions (i) and (ii).

(5): Fix $y \in M$ and F a convex hull of a finite subset of M. We have to show that for every $k = 1, \ldots, m$, $f(\cdot, y)$ is $\mathbb{R}_+^m$-lower continuous on F, which is equivalent to $f_k(\cdot, y)$ is $\mathbb{R}_+$-upper continuous on F; i.e., $f_k(\cdot, y)$ is $\mathbb{R}_+$-upper semicontinuous[1]. Let $\{x_n\}$ be a sequence in F weak-converging to $x_0 \in F$. Since F is a closed subset of a finite-dimensional space, we assert that $\{x_n\}$ converges in norm to x_0. By the upper semicontinuity of J_k^o on $M \times X$, see Proposition 3.9.5 (ii), and the local boundedness of S_k we obtain that $f(\cdot, y)$ is $\mathbb{R}_+$-upper semicontinuous at x_0.

(7): holds by using a similar argument to that used in proving (5).   □

Remark 3.9.9. The hemivariational inequalities studied in Chapter 4 of [267] correspond to the case where $k = 1$ and for all $x \in M$, $Tx = l \in X^*$.

Remark 3.9.10. The study of variational–hemivariational inequalities is a particular case of the study of the inequalities system (3.101). It corresponds to the case in which $k = 1$ and

$$\langle y - x, Tx \rangle = \varphi(y) - \varphi(x) - \langle y - x, l \rangle \quad \text{for all } x, y \in M,$$

with $l \in X^*$ and φ is a real lower semicontinuous convex function on M. Therefore, a solution of this inequality exists without recourse to a condition of quasi- or strong quasi-boundedness on $\partial\varphi$, as was made in [267].

3.9.4 Vector Complementarity Problems

To introduce a vector complementarity problem, let X, Y be two topological vector spaces, $M \subset X$, and let T be a function from X into $L(X, Y)$. Suppose Y is ordered by the family of cones $\{\operatorname{int} P(x) : x \in M\} \cup \{0\}$. We define the following **weak** and **strong vector complementarity problems**:

(WVCP) find $\bar{x} \in M$ with $T\bar{x}(\bar{x}) \notin \operatorname{int} P(\bar{x})$, $T\bar{x}(y) \notin -\operatorname{int} P(\bar{x}) \ \forall y \in M$.

(SVCP) find $\bar{x} \in M$ with $T\bar{x}(\bar{x}) \notin \operatorname{int} P(\bar{x})$, $T\bar{x}(y) \in P(\bar{x}) \ \forall y \in M$. If $Y = \mathbb{R}$, $P(x) = \mathbb{R}_+$ for each $x \in M$, the vector complementarity problems (WVCP) and (SVCP) coincide with the scalar complementarity problem:

(CP) find $\bar{x} \in M$ such that $\langle \bar{x}, T\bar{x} \rangle = 0$ and $T\bar{x} \in M^+$,

where $M^+ := \{x^* \in X^* \mid \langle x, x^* \rangle \geq 0 \ \forall x \in M\}$.

---

[1] Let us recall that vector $\mathbb{R}_+$-upper (lower) semicontinuity of a real function coincides with the usual scalar upper (lower) semicontinuity.

If we restrict ourselves to $X = \mathbb{R}^n$ and $M = \mathbb{R}^n_+$, then the space $L(X, Y)$ becomes equal to $\mathbb{R}^n$, and (CP) becomes, find $\overline{x} \geq 0$ such that $\langle \overline{x}, T\overline{x} \rangle = 0$ and $T\overline{x} \geq 0$. In the following, conditions under which solutions of vector complementarity problems **exist** are presented. To derive these conditions, Lemma 3.9.11 will be used.

Lemma 3.9.11. *Let us consider the following vector variational inequality (VVI):*

$$\text{find } \overline{x} \in M \text{ such that } T\overline{x}(y - \overline{x}) \notin - \operatorname{int} P(\overline{x}) \quad \forall y \in M. \qquad (3.102)$$

(i) *Suppose that the cone $P(\overline{x})$ is w-pointed (i.e., $\operatorname{int} P(\overline{x}) \cap -P(\overline{x}) = \emptyset$) and T is a single-valued function; then (SVCP) $\Longrightarrow$ (VVI).*[2]
(ii) *Suppose that M is a convex cone; then (VVI) $\Longrightarrow$ (WVCP).*
(iii) *Suppose that $\operatorname{int} P(\overline{x}) \cup -P(\overline{x}) = X$; then (WVCP) $\Longrightarrow$ (SVCP).*

PROOF. (i): Since $P(\overline{x})$ is w-pointed, (SVCP) $\Longrightarrow$ (VVI) follows from the implication $T\overline{x}(\overline{x}) \notin \operatorname{int} P(\overline{x})$ and $T\overline{x}(y) \in P(\overline{x}) \Rightarrow T\overline{x}(y) - T\overline{x}(\overline{x}) \notin - \operatorname{int} P(\overline{x})$.
(ii): Let us take $y = 0$ in (3.102); we have $T\overline{x}(\overline{x}) \notin \operatorname{int} P(\overline{x})$. Together, fix $z \in M$ and take $y = z + \overline{x}$ ($\in M$, since M is a convex cone) in (3.102); then $T\overline{x}(z) = T\overline{x}(y - \overline{x}) \notin - \operatorname{int} P(\overline{x})$.
(iii): It is immediate from condition $\operatorname{int} P(\overline{x}) \cup -P(\overline{x}) = X$. $\qquad \square$

Theorem 3.9.12. *Suppose that*

(i) *for every $x, y \in M$, $Tx(y-x) \notin - \operatorname{int} P(x)$ implies $Ty(y-x) \notin - \operatorname{int} P(x)$;*
(ii) *the graph of $\operatorname{int} P$ is open on $M \times Y$;*
(iii) *T is continuous from each line segment of M to a topology on $L(X, Y)$ for which the bilinear form $(\pi, x) \mapsto \pi(x)$ is continuous from $L(X, Y) \times M$ into Y;*
(iv) *there is a convex weakly compact subset B of X such that for every $y \in M \setminus B$ there exists $x \in M \cap B$ such that $Tx(y - x) \in \operatorname{int} P(\overline{x})$.*

Then the problem (WVCP) has at least one solution.

PROOF. The assumptions (1)–(5), (7), and (8) of Theorem 3.8.28 hold when applied to $f(x, y) = Tx(y - x)$ for $x, y \in M$. To verify (6), fix $x, y \in M$ such that $x \neq y$, and suppose to the contrary that there exists some $u \in]x, y[$ such that $Tu(x - u) \notin \operatorname{int} P(x)$ and $Tu(y - u) \in - \operatorname{int} P(x)$. According to $Y \setminus \operatorname{int} P(x) - \operatorname{int} P(x) \subset Y \setminus P(x)$, we deduce that for every $t \in]0, 1[$,

$$Tu(tx + (1 - t)y - u) = tTu(x - u) + (1 - t)Tu(y - u) \notin P(x),$$

which is impossible for t such that $u = tx + (1 - t)y$. Therefore, (VVI) has a solution; and invoking Lemma 3.9.11, we finish the proof of theorem. $\qquad \square$

---

[2] This means: $\overline{x}$ is a solution of (SVCP) $\Longrightarrow \overline{x}$ is a solution of (VVI).

3.9.5 Application to Vector Optimization Problems

Consider $\phi : M \longrightarrow Y$ and the following vector optimization problems

(WVOP) find $\overline{x} \in M$ such that $\phi(y) \notin \phi(\overline{x}) - \operatorname{int} P \qquad \forall y \in M;$

and

(VOP)   find $\overline{x} \in M$ such that $\phi(y) \notin \phi(\overline{x}) - (P \setminus \{0\}) \; \forall y \in M.$

General vector mappings

We first obtain a direct existence result for the weak minimum of a vector optimization problem by setting $f(x, y) = \phi(x) - \phi(y)$. It is trivial to check that $x \in M$ is a solution of (GVEP) (and (WGVEP)) iff x is a solution of (VOP) (and (WVOP)).

Theorem 3.9.13. *Suppose that*

(i) *ϕ is quasi P-convex and P-lower semicontinuous;*
(ii) *there exists $B \subseteq M$ convex and compact such that $\forall y \in M \setminus B$, $\exists x \in B$ such that $\phi(y) \in \phi(x) + \operatorname{int} P$.*

Then (WVOP) admits at least one solution $\overline{x} \in B$.

PROOF. We apply Theorem 3.8.25 with $f(x, y) = \phi(x) - \phi(y)$ and $K(x) = -\operatorname{int} P$. □

Remark 3.9.14. In the particular case where $Y = \mathbb{R}^m$ and $P = \mathbb{R}^m_-$ we have under assumptions (i) and (ii) of Theorem 3.9.13 the existence of a weak Pareto optimum, i.e., $\varphi(M) \cap (\varphi(\overline{x}) + \operatorname{int} P \cup \{0\}) = \{\varphi(\overline{x})\}$.

Remark 3.9.15. Theorem 3.9.13 is a generalization, in the case of single-valued functions, of a result [244, Corollary 5-6 p. 59] on **existence of solutions of vector optimization problems**.

Smooth vector mappings

In this paragraph, by considering smooth vector mappings, we prove the existence of weak minimum for vector optimization problems by means of a vector variational-like inequality and preinvex mappings. For our analysis, the following concepts are necessary.

Definition 3.9.16. *We say that a Fréchet differentiable function[3] $\phi : M \to Y$ is P-**invex** with respect to $\eta : M \times M \to X$ if $\phi(x) - \phi(y) - \nabla\phi(y)(\eta(x, y)) \in P \; \forall x, y \in M$. Here $\nabla\phi$ denotes the Fréchet derivative of ϕ.*

---

[3] ϕ must be defined in a neighborhood of M.

Definition 3.9.17. *We say that* $\phi : M \to Y$ *is P-**preinvex** with respect to* $\eta : M \times M \to X$ *if for all* $x, y \in M, t \in [0, 1], \; y + t\eta(x, y) \in M$ *and*

$$t(\phi(x) - \phi(y)) + \phi(y) - \phi(y + t\eta(x, y)) \in P.$$

Remark 3.9.18. If we suppose that ϕ is Fréchet differentiable, then P-invexity of ϕ is satisfied whenever ϕ is P-preinvex with respect to the same η.

Indeed, suppose that $t(\phi(x) - \phi(y)) + \phi(y) - \phi(y + t\eta(x, y)) \in P$. Dividing by t and letting t goes to zero, we obtain $\phi(x) - \phi(y) - \nabla\phi(y)(\eta(x, y)) \in P$; that is, P-invexity of ϕ. □

Remark 3.9.19. We remark that if we take $\mathbb{R}$ and $\mathbb{R}_+$ respectively in place of Y and P, we obtain the definitions of scalar invex and preinvex functions.

Note that invex functions were first introduced by Hanson [157]. To characterize the invexity, Craven and Glover showed that the class of invex functions is equivalent to the class of functions whose starting points are global minima.

Now, we prove the following existence theorem of weak vector minimum points.

Theorem 3.9.20. *Let M be a nonempty closed, convex subset of a real topological vector space. Let $\phi : M \to Y$ be a Fréchet differentiable and P-preinvex mapping such that its derivative $\nabla\phi$ is continuous from each line segment of M to $L(X, Y)$, i.e., $\forall x, y \in M, \; \forall z \in X \; \nabla\phi(x + t(y - x))(z) \to \nabla\phi(x)(z)$ in Y if $t \to 0^+$. Let $\eta : M \times M \to X$ be such that $\eta(x, x) = 0$ for every $x \in K$. Suppose that, for every $x, y \in M$,*

(1) $\nabla\phi(x)(\eta(y, x)) \notin - \operatorname{int} P \Longrightarrow \nabla\phi(y)(\eta(y, x)) \notin - \operatorname{int} P$;
(2) η *is linear in the first argument and continuous in the second one;*
(3) *there exists a compact subset B of M, and $x_0 \in B$ such that* $\nabla\phi(x_0)(\eta(y, x_0)) \in - \operatorname{int} P$ *for all $y \in M \setminus B$.*

Then the vector optimization problem (WVOP) has a global minimum $\overline{x} \in B$.

PROOF. By Theorem 3.8.24, with $K(x) = L(x) = - \operatorname{int} P$, $f(x, y) = \nabla\phi(x)(\eta(x, y))$, and $g(x, y) = \nabla\phi(y)(\eta(x, y))$, there exists $\overline{x} \in B$ such that

$$\nabla\phi(y)(\eta(y, \overline{x})) \notin - \operatorname{int} P \quad \forall y \in M.$$

For $y \in M$ fixed, set $y_t = ty + (1-t)\overline{x}$, where $t \in [0, 1]$; then $\nabla\phi(y_t)(\eta(y_t, \overline{x})) \notin - \operatorname{int} P \;\; \forall t \in [0, 1]$.
Using (2) and $\eta(\overline{x}, \overline{x}) = 0$, we obtain

$$\nabla\phi(y_t)(\eta(y, \overline{x})) = 1/t\nabla\phi(y_t)(\eta(y_t, \overline{x})) \notin - \operatorname{int} P.$$

Since $\nabla\phi$ is continuous from each line segment of M to $L(X, Y)$, it follows that the vector variational-like inequality

$$\nabla \phi(\overline{x})(\eta(y, \overline{x})) \notin - \operatorname{int} P \text{ for all } y \in M \qquad (3.103)$$

has a solution.

Because of the P-preinvexity, which implies P-invexity, of ϕ we deduce

$$\phi(y) - \phi(\overline{x}) - \nabla \phi(\overline{x})(\eta(y, \overline{x})) \in P. \qquad (3.104)$$

Combining (3.103) and (3.104) we obtain for every $y \in M$,

$$\phi(y) - \phi(\overline{x}) = (\phi(y) - \phi(\overline{x}) - \nabla \phi(\overline{x})(\eta(y, \overline{x}))) + \nabla \phi(\overline{x})(\eta(y, \overline{x})) \notin - \operatorname{int} P,$$

which completes the proof. $\qquad \qquad \qquad \qquad \qquad \qquad \qquad \qquad \square$

3.9.6 Minimax Theorem for Vector-Valued Mappings

Let us first introduce the notions of **cone saddle point** of vector-valued mappings.

Assume that we are given $X = X_1 \times X_2$ a product of two real topological vector spaces, and U, V two nonempty subsets of X_1 and X_2, respectively. Let $F : M = U \times V \longrightarrow Y$ be a vector-valued mapping.

Definition 3.9.21. (1) *A point $(u_0, v_0) \in M$ is said to be a weak P-saddle point of F with respect to $M = U \times V$, a WVSP for short, if*

$$\begin{cases} F(u_0, V) \cap (F(u_0, v_0) + \operatorname{int} P \cup \{0\}) = \{F(u_0, v_0)\} \\ F(U, v_0) \cap (F(u_0, v_0) - \operatorname{int} P \cup \{0\}) = \{F(u_0, v_0)\}. \end{cases}$$

(2) *A point $(u_0, v_0) \in M$ is said to be a P-saddle point of F with respect to $M = U \times V$, a VSP for short, if*

$$\begin{cases} F(u_0, V) \cap (F(u_0, v_0) + P) = \{F(u_0, v_0)\} \\ F(U, v_0) \cap (F(u_0, v_0) - P) = \{F(u_0, v_0)\}. \end{cases}$$

It should be remarked that a VSP is also a WVSP. In the case where $\operatorname{int} P \cup \{0\} = P$, the two concepts are coincident.

Lemma 3.9.22. *Assume that we are given F a vector-valued mapping from $M = U \times V$ into Y, P a nonempty cone with $\operatorname{int} P \neq \emptyset$, and $f : M \times M \to Y$ defined, for each $(u_1, v_1), (u_2, v_2) \in M$, by*

$$f((u_1, v_1), (u_2, v_2)) := F(u_1, v_2) - F(u_2, v_1).$$

If (u_0, v_0) satisfies $f((u, v), (u_0, v_0)) \notin - \operatorname{int} P$ (resp. $-P \setminus \{0\}$) $\forall (u, v) \in M$, then (u_0, v_0) is a WVSP (resp. VSP) of F.

PROOF. First, we remark that

$$(u_0, v_0) \text{ is a WVSP} \quad \Leftrightarrow \quad \begin{cases} F(u_0, v) - F(u_0, v_0) \notin \operatorname{int} P \quad \forall v \in V, \\ F(u, v_0) - F(u_0, v_0) \notin - \operatorname{int} P \quad \forall u \in U, \end{cases}$$

and

$$(u_0, v_0) \text{ is a VSP} \iff \begin{cases} F(u_0, v) - F(u_0, v_0) \notin P \setminus \{0\} \quad \forall v \in V, \\ F(u, v_0) - F(u_0, v_0) \notin -P \setminus \{0\} \quad \forall u \in U. \end{cases}$$

The end of the proof is immediate from these equivalences and the definition of f. $\qquad\square$

This lemma, combined with Theorem 3.8.25, Lemma 3.8.17, and Remark 3.8.15, leads to the following **vector minimax theorem**.

Theorem 3.9.23. *Suppose that U and V are compact and convex subsets of two Hausdorff topological vector spaces, P a closed convex cone with nonempty interior in a Hausdorff locally convex topological vector space. Let $F : U \times V \to Y$ satisfy the following assumptions:*

(1) for every $u \in U$, $F(u, \cdot)$ is P-concave and P-upper semicontinuous;
(2) for every $u \in V$, $F(\cdot, v)$ is P-convex and P-lower semicontinuous.

Then F admits a weak vector saddle point in $U \times V$.

For further references see [112], [352], and [274].

3.10 Minimal-Point Theorems in Product Spaces and Corresponding Variational Principles

The importance of the Ekeland variational principle in nonlinear analysis is well known. Below we recall a versatile variant.

Proposition 3.10.1. Ekeland's variational principle. *Let (X, d) be a complete metric space and $f : X \to \mathbb{R}^\bullet := \mathbb{R} \cup \{\infty\}$ a proper, lower semicontinuous function bounded below. Consider $\varepsilon > 0$ and $x_0 \in X$ such that $f(x_0) \leq \inf f + \varepsilon$. Then for every $\lambda > 0$ there exists $\overline{x} \in \operatorname{dom} f$ such that*

$$f(\overline{x}) + \lambda^{-1}\varepsilon d(\overline{x}, x_0) \leq f(x_0), \qquad d(\overline{x}, x_0) \leq \lambda, \qquad (3.105)$$

and

$$f(\overline{x}) < f(x) + \lambda^{-1}\varepsilon d(\overline{x}, x) \quad \forall x \in X \setminus \{\overline{x}\}. \qquad (3.106)$$

This means that for $\lambda, \varepsilon > 0$ and x_0 an ε-approximate solution of the minimization problem

$$f(x) \to \min \text{ s.t. } x \in X, \qquad (3.107)$$

there exists a new point $\overline{x}$ that is not worse than x_0 and belongs to a λ-neighborhood of x_0, and especially, $\overline{x}$ satisfies the variational inequality (3.106). Relation (3.106) says, in fact, that $\overline{x}$ minimizes globally $f + \lambda^{-1}\varepsilon d(\overline{x}, \cdot)$, which is nothing else than a Lipschitz perturbation of f (for "smooth" principles, see [41]). Note that $\lambda = \sqrt{\varepsilon}$ gives a useful compromise in Proposition 3.10.1. For applications see Section 4.1 and 4.5 and, e.g., [116, 117, 310, 339, 343].

There are several statements that are equivalent to Ekeland's variational principle (EVP); see, e.g., [87, 126, 290, 297, 296, 48, 73, 154, 140, 142, 141, 144, 143, 71, 72, 124]. We mention explicitly a result of Attouch and Riahi [10], who showed that (in Banach spaces) EVP is equivalent to the existence of minimal points with respect to cones satisfying some additional conditions.

Proposition 3.10.2. Phelps minimal-point theorem. *Let X be a Banach space, $C \subset X$ a closed convex cone such that $C \subset K_{x^*} := \{x \in X \mid \|x\| \le \langle x, x^* \rangle\}$ for some $x^* \in X^*$, and $A \subset X$ a nonempty closed set such that $x^*(A)$ is bounded from below. Then for every $x_0 \in A$ there exists $\overline{x} \in A \cap (x_0 - C)$ such that $\overline{x}$ is a minimal element of A with respect to the partial order induced by C.*

Obviously, K_{x^*} is a pointed closed convex cone, and so C is pointed, too. Moreover, K_{x^*} is well-based, a bounded base being $B := \{x \in K_{x^*} \mid \langle x, x^* \rangle = 1\}$, and so C is well-based, too. Sometimes the cone K_{x^*} is called a **Phelps cone.**

Since in this book we are mainly interested in multicriteria problems, we look for multicriteria (or vector) variants of EVP and minimal-point theorems. Loridan [240], then Khanh [210], Nemeth [269], and Tammer [337], were the first to prove vectorial EVP. We illustrate this kind of result with the following one, which is very close to a result stated by Tammer [337] (see also Corollary 3.10.19).

Proposition 3.10.3. Vector EVP. *Let M be a nonempty closed subset of the Banach space $(X, \|\cdot\|)$, Y a t.v.s., $C \subset Y$ a proper closed convex cone, $k^0 \in \operatorname{int} C$, and $f : M \to Y$. Assume that $\{x \in M \mid f(x) \le_C rk^0\}$ is closed for every $r \in \mathbb{R}$ and for some $\varepsilon > 0$ and $x_0 \in \operatorname{dom} f$, we have that $f(M) \cap (f(x_0) - \varepsilon k^0 - (C \setminus \{0\})) = \emptyset$. Then for every $\lambda > 0$ there exists $\overline{x} \in \operatorname{dom} f$ such that*

$$f(M) \cap (f(\overline{x}) - \lambda^{-1}\varepsilon k^0 - \operatorname{int} C) = \emptyset, \qquad \|\overline{x} - x_0\| \le \lambda, \qquad (3.108)$$

and

$$f(x) + \lambda^{-1}\varepsilon k^0 \|x - \overline{x}\| \le_C f(\overline{x}) \Rightarrow x = \overline{x}. \qquad (3.109)$$

In particular, $\overline{x} \in \operatorname{Eff}(f_{\lambda^{-1}\varepsilon k^0, \overline{x}}, C)$, where $f_{k, \overline{x}} := f + \|\cdot - \overline{x}\| \cdot k$.

Note that for $Y = \mathbb{R}$ the conditions of the preceding proposition are very close to those of the Ekeland variational principle (Proposition 3.10.1).

Before starting with more general minimal-point theorems let us have a look at (3.106). If we write it in the form

$$f(x) - f(\overline{x}) + \varepsilon \lambda^{-1} d(\overline{x}, x) \le 0 \Rightarrow x = \overline{x},$$

then $(\overline{x}, f(\overline{x}))$ is a minimal point of epi f with respect to the binary relation defined by

$$(x_1, t_1) \succ (x_2, t_2) \Leftrightarrow t_1 - t_2 \geq \varepsilon \lambda^{-1} d(x_1, x_2). \tag{3.110}$$

If X is a normed space, this binary relation is determined by the cone

$$C := \{(x, t) \in X \times \mathbb{R} \mid t \geq \varepsilon \lambda^{-1} \|x\|\}.$$

The binary relations defined as in (3.110) will play a decisive role in this section.

It is worth mentioning that a weaker result than a full (= authentic) minimal-point theorem gives an EVP, as shown in Section 3.10.1 below.

3.10.1 Not Authentic Minimal-Point Theorems

Throughout the section (X, d) is a complete metric space, Y is a separated locally convex space, Y^* is its topological dual, $C \subset Y$ is a convex cone; as usual, $C^+ = \{y^* \in Y^* \mid y^*(y) \geq 0 \ \forall y \in C\}$ is the dual cone of C and $C^\# = \{y^* \in Y^* \mid y^*(y) > 0 \ \forall y \in C \setminus \{0\}\}$. Consider also $k^0 \in C \setminus (-\overline{C})$. We notice that there exists $z^* \in C^+$ such that $z^*(k^0) = 1$. Indeed, using a separation theorem for k^0 and $-\overline{C}$ we get $z_1^* \in C^+$ and a real $\alpha > 0$ such that $z_1^*(k^0) > \alpha > z_1^*(y)$ for all $y \in C$, then take $z^* := \alpha^{-1} z_1^*$. The cone C determines a preorder on Y denoted, as usual, by $\leq_C$; so, for $y_1, y_2 \in Y$, $y_1 \leq_C y_2$ iff $y_2 - y_1 \in C$. It is known (see Section 2.1) that $\leq_C$ is reflexive and transitive; $\leq_C$ is antisymmetric iff C is pointed, i.e., $C \cap -C = \{0\}$. Using the element k^0 we introduce a preorder on $X \times Y$, denoted by $\preceq_{k^0}$, in the following manner:

$$(x_1, y_1) \preceq_{k^0} (x_2, y_2) \iff y_1 + k^0 d(x_1, x_2) \leq_C y_2. \tag{3.111}$$

Note that $\preceq_{k^0}$ is reflexive and transitive; if $(x_1, y_1) \preceq_{k^0} (x_2, y_2)$ and $(x_2, y_2) \preceq_{k^0} (x_1, y_1)$, then $x_1 = x_2$. If C is pointed then $\preceq_{k^0}$ is antisymmetric, too.

Note that if $(X, \|\cdot\|)$ is a normed vector space (n.v.s.), then $\preceq_{k^0}$ is determined by the convex cone $C_{k^0} := \{(x, y) \in X \times Y \mid \|x\| \cdot k^0 \leq_C y\}$; C_{k^0} is pointed if C is.

Consider a nonempty set $A \subset X \times Y$. For $(x, y) \in A$ we denote by $A(x, y)$ the lower section of A with respect to $\preceq_{k^0}$: $A(x, y) = \{(x', y') \in A \mid (x', y') \preceq_{k^0} (x, y)\}$. In the sequel we shall use the following condition on A:

(H1) *for every $\preceq_{k^0}$-decreasing sequence $((x_n, y_n)) \subset A$ with $x_n \to x \in X$ there exists $y \in Y$ such that $(x, y) \in A$ and $(x, y) \preceq_{k^0} (x_n, y_n)$ for every $n \in \mathbb{N}$.*

A related condition is

(H2) *for every sequence $((x_n, y_n)) \subset A$ with $x_n \to x \in X$ and $(y_n) \leq_C$-decreasing there exists $y \in Y$ such that $(x, y) \in A$ and $y \leq_C y_n$ for every $n \in \mathbb{N}$.*

These two conditions are motivated by conditions used by Isac [179] and Nemeth [269], respectively, as we shall see below.

Note that if A satisfies (H2) and C has closed lower sections with respect to $\mathbb{R}_+ k^0$, i.e., $C \cap (y - \mathbb{R}_+ k^0)$ is closed for every $y \in C$, then (H1) is also satisfied.

Indeed, let $((x_n, y_n)) \subset A$ be a $\preceq_{k^0}$-decreasing sequence with $x_n \to x$. It is obvious that (y_n) is $\leq_C$-decreasing. By (H2), there exists $y \in Y$ such that $(x, y) \in A$ and $y \leq_C y_n$ for every $n \in \mathbb{N}$. It follows that

$$y + k^0 d(x_{n+p}, x_n) \leq_C y_{n+p} + k^0 d(x_{n+p}, x_n) \leq_C y_n \quad \forall n, p \in \mathbb{N}.$$

Fixing n and letting $p \to \infty$, by the closedness of C in the direction k^0, one obtains that $y + k^0 d(x, x_n) \leq_C y_n$, i.e., $(x, y) \preceq_{k^0} (x_n, y_n)$ for every $n \in \mathbb{N}$.

Note also that (H2) holds if A is closed and every $\leq_C$-decreasing sequence in C is convergent (i.e., C is a sequential Daniell cone). This is the case (even for nets) if Y is a Banach space and C has a closed (convex) and bounded base (see [38, Prop. 3.6] and Section 2.2 for bases of cones).

We establish now our first **minimal-point theorem**. In the sequel P_X and P_Y will denote the projections of $X \times Y$ onto X and Y, respectively; so $P_X(x, y) = x$ for every $(x, y) \in X \times Y$.

Theorem 3.10.4. *Let $A \subset X \times Y$ satisfy (H1) and suppose that there exists $\tilde{y} \in Y$ such that $P_Y(A) \subset \tilde{y} + C$. Then for every $(x_0, y_0) \in A$ there exists $(\bar{x}, \bar{y}) \in A$ such that $(\bar{x}, \bar{y}) \preceq_{k^0} (x_0, y_0)$ and if $(x', y') \in A$ is such that $(x', y') \preceq_{k^0} (\bar{x}, \bar{y})$ then $x' = \bar{x}$.*

PROOF. First of all, note that for $(x, y) \in A$ the set $P_X(A(x, y))$ is bounded. Indeed, as noticed at the beginning of this section, there exists $z^* \in C^+$ with $z^*(k^0) = 1$. Let $x' \in P_X(A(x, y))$; there exists $y' \in Y$ such that $(x', y') \in A$ and $(x', y') \preceq_{k^0} (x, y)$. Then $d(x, x') \leq z^*(y) - z^*(y') \leq z^*(y) - z^*(\tilde{y})$, which shows that $P_X(A(x, y))$ is bounded. Let us construct a sequence $((x_n, y_n))_{n \in \mathbb{N}} \subset A$ in the following way: Having $(x_n, y_n) \in A$, where $n \in \mathbb{N}$, by the above remark, there exists $(x_{n+1}, y_{n+1}) \in A(x_n, y_n)$ such that

$$d(x_{n+1}, x_n) \geq \tfrac{1}{2} \sup\{d(x, x_n) \mid x \in B_n\},$$

where $B_n := P_X(A(x_n, y_n))$. We obtain in this way the sequence $((x_n, y_n)) \subset A$, which is $\preceq_{k^0}$-decreasing. Since $A(x_{n+1}, y_{n+1}) \subset A(x_n, y_n)$, we have that $B_{n+1} \subset B_n$ for every $n \in \mathbb{N}$. Of course, $x_n \in B_n$. Let us show that $\operatorname{diam} B_n \to 0$. In the contrary case there exists $\delta > 0$ such that $\operatorname{diam} B_n \geq \delta$ for every $n \in \mathbb{N}$. From $\tfrac{1}{4}\delta k^0 \leq_C k^0 d(x_m, x_{m+1}) \leq_C y_m - y_{m+1}$ one gets $\tfrac{1}{4}\delta \leq z^*(y_m) - z^*(y_{m+1})$, and so, adding these relations for m from 0 to $n - 1$, we obtain

$$\tfrac{1}{4}\delta n \leq z^*(y_0) - z^*(y_n) \leq z^*(y_0) - z^*(\tilde{y}),$$

which yields a contradiction for $n \to \infty$. Thus we have that the sequence $(\overline{B_n})$ is a decreasing sequence of nonempty closed subsets of the complete metric space (X, d), whose diameters tend to 0. By Cantor's theorem, $\bigcap_{n \in \mathbb{N}} \overline{B_n} = \{\bar{x}\}$ for some $\bar{x} \in X$. Of course, $x_n \to \bar{x}$. Since $((x_n, y_n)) \subset A$ is a $\preceq_{k^0}$-decreasing sequence, from (H1) we get an $\bar{y} \in Y$ such that $(\bar{x}, \bar{y}) \preceq_{k^0} (x_n, y_n)$ for every $n \in \mathbb{N}$; $(\bar{x}, \bar{y})$ is the desired element. Indeed, $(\bar{x}, \bar{y}) \preceq_{k^0} (x_0, y_0)$; let $(x', y') \in A(\bar{x}, \bar{y})$. It follows that $(x', y') \in A(x_n, y_n)$, and so $x' \in B_n \subset \overline{B_n}$ for every n. Thus $x' = \bar{x}$. $\qquad \square$

We want to apply the preceding results to obtain two **vectorial EVP**. To envisage functions defined on subsets of X we add to Y an element ∞ not belonging to the space Y, obtaining thus the space $Y^\bullet$: $Y^\bullet = Y \cup \{\infty\}$. We consider that $y \leq_C \infty$ for all $y \in Y$. Consider now the function $f : X \to Y^\bullet$. As usual, the domain of f is $\operatorname{dom} f = \{x \in X \mid f(x) \neq \infty\}$; the epigraph of f is $\operatorname{epi} f = \{(x, y) \in X \times Y \mid f(x) \leq_C y\}$; the graph of f is $\operatorname{gr} f = \{(x, f(x)) \mid x \in \operatorname{epi} f\}$. Of course, f is proper if $\operatorname{dom} f \neq \emptyset$.

Corollary 3.10.5. *Let $f : X \to Y^\bullet$ be a proper function, bounded from below (i.e., there exists $\tilde{y} \in Y$ such that $\tilde{y} \leq_C f(x)$ for every $x \in X$), satisfying the condition*

(H3) $\{x' \in X \mid f(x') + k^0 d(x', x) \leq_C f(x)\}$ *is closed for every $x \in X$.*

Then for every $x_0 \in \operatorname{dom} f$ there exists $\bar{x} \in X$ such that

$$f(\bar{x}) + k^0 d(\bar{x}, x_0) \leq_C f(x_0) \tag{3.112}$$

and

$$\forall x \in X \ : f(x) + k^0 d(x, \bar{x}) \leq_C f(\bar{x}) \ \Rightarrow \ x = \bar{x}. \tag{3.113}$$

PROOF. Consider $A = \operatorname{gr} f \subset X \times Y$. Let us show that (H1) holds. Indeed, let $((x_n, y_n)) \subset A$ be a $\preceq_{k^0}$-decreasing sequence with $x_n \to x \in X$. Of course, since $y_n = f(x_n)$, for all $n, p \in \mathbb{N}$ we have that $x_{n+p} \in \{x \in X \mid f(x) + k^0 d(x, x_n) \leq_C f(x_n)\}$. Since the last set is closed, it also contains the limit x of the sequence $(x_{n+p})_{p \in \mathbb{N}}$. Therefore $(x, y) \preceq_{k^0} (x_n, y_n)$ for every n, where $y = f(x)$. Of course, $(x, y) \in A$. Applying Theorem 3.10.4 we obtain $\bar{x} \in X$ such that $(\bar{x}, f(\bar{x})) \in \operatorname{gr} f$ satisfies the conclusion of that theorem. This means that (3.112) and (3.113) hold. $\qquad \square$

Note that Isac [179, Th. 4] obtained the above result when $\operatorname{dom} f = X$, C is a normal cone, and $k^0 \in C \setminus \{0\}$ (by Theorem 2.1.22, $\overline{C}$ is pointed in this case, and so $k^0 \in C \setminus (-\overline{C})$).

Corollary 3.10.6. *Let $f : X \to Y^\bullet$ be a proper function, bounded from below and satisfying the condition*

(H4) *for every sequence $(x_n) \subset \operatorname{dom} f$, with $x_n \to x$ and $(f(x_n)) \leq_{C^-}$ decreasing, $f(x) \leq_C f(x_n)$ for all $n \in \mathbb{N}$.*

If C is closed in the direction k^0, then the conclusion of Corollary 3.10.5 holds.

PROOF. Consider $A = \operatorname{gr} f \subset X \times Y$. Let us show that (H2) holds. Indeed, let $((x_n, y_n)) \subset A$ be such that (y_n) is $\leq_C$-decreasing and $x_n \to x \in X$. Of course, since $y_n = f(x_n)$, by (H4) we have that $y = f(x) \leq_C y_n$ for every n. By a previous discussion, under our conditions (H1) holds. Using the same proof as for Corollary 3.10.5, we obtain the conclusion. □

Note that Nemeth [269, Prop. 1] obtained the above result when $\operatorname{dom} f = X$, Y is a Banach space, C is a regular (supposed to be closed) cone, and $k^0 \in C \setminus \{0\}$; observe that a regular cone is pointed (so that, once again, $k^0 \in C \setminus (-\overline{C})$).

Traditionally, in the statements of the EVP there appears an $\varepsilon > 0$ and an estimate of $d(\bar{x}, x_0)$ (see Propositions 3.10.1 and 3.10.3). For the first situation just replace k^0 by εk^0 or d by εd. For the second one, suppose that in the conditions of Corollary 3.10.5 or Corollary 3.10.6, $f(X) \cap (f(x_0) - \lambda k^0 - C \setminus \{0\}) = \emptyset$, where $\lambda \in (0, \infty)$. Then $d(\bar{x}, x_0) \leq \lambda$. Indeed, in the contrary case, by (3.112), for some $k \in C$, we have

$$f(\bar{x}) = f(x_0) - k^0 d(\bar{x}, x_0) - k = f(x_0) - \lambda k^0 - (d(\bar{x}, x_0) - \lambda)k^0 - k$$
$$\in f(x_0) - \lambda k^0 - C \setminus \{0\},$$

since $(d(\bar{x}, x_0) - \lambda)k^0 + k \in C \setminus \{0\}$.

Note that in taking $Y = \mathbb{R}$ and $C = \mathbb{R}_+$ from Corollary 3.10.6 one obtains the EVP for functions that are **not necessarily lower semicontinuous**. For example, the function $f : \mathbb{R} \to \mathbb{R}$, $f(x) = \exp(-|x|)$ for $x \neq 0$ and $f(0) = 2$, satisfies the hypothesis of Corollary 3.10.6.

Theorem 3.10.4 does not ensure, effectively, a minimal point. In the next section we shall derive an authentic minimal-point theorem.

Note that from both Corollary 3.10.5 and Corollary 3.10.6 one obtains Loridan's variant of EVP [240], taking $X = Q$ a closed subset of a Banach space and $Y = \mathbb{R}^p$ ordered by $C = \mathbb{R}_+^p$.

3.10.2 Authentic Minimal-Point Theorems

We take X, Y, C, k^0 as in Section 3.10.1. In addition to the element k^0 considered in the preceding section, let us take also an element $z^* \in C^+$ such that $z^*(k^0) = 1$. We have noticed that such an element exists in our conditions. We introduce the order relation $\preceq_{k^0, z^*}$ on $X \times Y$ by

$$(x_1, y_1) \preceq_{k^0, z^*} (x_2, y_2) \iff \begin{cases} (x_1, y_1) = (x_2, y_2) \text{ or} \\ (x_1, y_1) \preceq_{k^0} (x_2, y_2) \text{ and } z^*(y_1) < z^*(y_2). \end{cases}$$

It is easy to verify that $\preceq_{k^0, z^*}$ is reflexive, transitive, and antisymmetric. The next theorem is the **main result** of Section 3.10.

Theorem 3.10.7. Let $A \subset X \times Y$ satisfy (H1) and suppose that z^* is bounded from below on $P_Y(A)$. Then for every $(x_0, y_0) \in A$ there exists $(\bar{x}, \bar{y}) \in A$ a minimal element of A with respect to $\preceq_{k^0, z^*}$ such that $(\bar{x}, \bar{y}) \preceq_{k^0, z^*} (x_0, y_0)$.

PROOF. We construct a sequence $((x_n, y_n))_{n \geq 0} \subset A$ as follows: Having $(x_n, y_n) \in A$, we take $(x_{n+1}, y_{n+1}) \in A$, $(x_{n+1}, y_{n+1}) \preceq_{k^0, z^*} (x_n, y_n)$, such that

$$z^*(y_{n+1}) \leq \inf\{z^*(y) \mid (x, y) \in A, \ (x, y) \preceq_{k^0, z^*} (x_n, y_n)\} + 1/(n+1).$$

Of course, $((x_n, y_n))$ is $\preceq_{k^0, z^*}$-decreasing. It follows that $((x_n, y_n))$ is $\preceq_{k^0}$-decreasing and (y_n) is $\leq_C$-decreasing, whence $(z^*(y_n))$ is decreasing, too.

Suppose first that there exists $n_0 \in \mathbb{N}$ such that $z^*(y_{n_0}) = \lim z^*(y_n)$. It follows that $z^*(y_n) = z^*(y_{n_0})$, and, for $n \geq n_0$, since $(x_n, y_n) \preceq_{k^0, z^*} (x_{n_0}, y_{n_0})$, $(x_n, y_n) = (x_{n_0}, y_{n_0}) =: (\bar{x}, \bar{y})$. Then

$$z^*(y_n) = z^*(\bar{y}) \leq \inf\{z^*(y) \mid (x, y) \in A, \ (x, y) \preceq_{k^0, z^*} (\bar{x}, \bar{y})\} + 1/n \quad \forall n \geq n_0,$$

whence $z^*(\bar{y}) \leq z^*(y)$ for every $(x, y) \in A$ with $(x, y) \preceq_{k^0, z^*} (\bar{x}, \bar{y})$. Once again, by the definition of $\preceq_{k^0, z^*}$, we obtain that $\{(x, y) \in A \mid (x, y) \preceq_{k^0, z^*} (\bar{x}, \bar{y})\} = \{(\bar{x}, \bar{y})\}$; i.e., $(\bar{x}, \bar{y})$ is a minimal point of A with respect to $\preceq_{k^0, z^*}$, and of course, $(\bar{x}, \bar{y}) \preceq_{k^0, z^*} (x_0, y_0)$.

Suppose now that $\lim_{m \to \infty} z^*(y_m) < z^*(y_n)$ for every $n \in \mathbb{N}$. Because $(x_{n+p}, y_{n+p}) \preceq_{k^0, z^*} (x_n, y_n)$ for $n, p \in \mathbb{N}$, we obtain

$$d(x_{n+p}, x_n) \leq z^*(y_n) - z^*(y_{n+p}) \leq 1/n \quad \forall n, p \in \mathbb{N}, \ n \geq 1.$$

It follows that (x_n) is a Cauchy sequence in the complete metric space (X, d), and so (x_n) is convergent to some $\bar{x} \in X$. Since $((x_n, y_n))$ is $\preceq_{k^0}$-decreasing, by (H1) there exists some $\bar{y} \in Y$ such that $(\bar{x}, \bar{y}) \in A$ and $(\bar{x}, \bar{y}) \preceq_{k^0} (x_n, y_n)$ for every $n \in \mathbb{N}$. It follows that $z^*(\bar{y}) \leq \lim z^*(y_n)$, and so $z^*(\bar{y}) < z^*(y_n)$ for every $n \in \mathbb{N}$. Therefore $(\bar{x}, \bar{y}) \preceq_{k^0, z^*} (x_n, y_n)$ for every $n \in \mathbb{N}$. Let $(x', y') \in A$ be such that $(x', y') \preceq_{k^0, z^*} (\bar{x}, \bar{y})$. Since $(\bar{x}, \bar{y}) \preceq_{k^0, z^*} (x_n, y_n)$ for every $n \in \mathbb{N}$, we have

$$d(x', \bar{x}) \leq z^*(\bar{y}) - z^*(y') \leq z^*(y_n) - z^*(y') \leq 1/n \quad \forall n \geq 1,$$

whence $z^*(y') = z^*(\bar{y})$. Once again, by the definition of $\preceq_{k^0, z^*}$, we obtain that $(x', y') = (\bar{x}, \bar{y})$. □

An immediate consequence of the preceding theorem is the following weaker result.

Corollary 3.10.8. *Let $A \subset X \times Y$ satisfy (H1) and suppose that z^* is bounded from below on $P_Y(A)$. Then for every $(x_0, y_0) \in A$ there exists $(\bar{x}, \bar{y}) \in A$ such that $(\bar{x}, \bar{y}) \preceq_{k^0} (x_0, y_0)$, and if $(x', y') \in A$ and $(x', y') \preceq_{k^0} (\bar{x}, \bar{y})$, then $x' = \bar{x}$ and $z^*(y') = z^*(\bar{y})$.*

Note that a direct proof is possible. Just take in the proof of Theorem 3.10.7 $(x_{n+1}, y_{n+1}) \in A(x_n, y_n)$ such that

$$z^*(y_{n+1}) \leq \inf\{z^*(y) \mid (x, y) \in A(x_n, y_n)\} + 1/(n+1).$$

But it is not possible to obtain Theorem 3.10.7 from Corollary 3.10.8, as we shall see below by an example. Of course, Theorem 3.10.4 is an immediate consequence of Corollary 3.10.8.

Example 3.10.9. Consider $X = \{a, b\}$ with $d(a, b) = 1$, $Y = \mathbb{R}^3$, $C = \mathbb{R}^3_+$, $k^0 = (1, 1, 1)$, $z^* \in C^+ = C$, $z^*(u, v, w) = u$, $y_0 = (2, 2, 3)$, $y_1 = (2, 2, 1)$, $y_2 = (2, 1, 0)$, $y_3 = (2, 0, 0)$, $y_4 = (1, 2, 0)$, and $y_5 = (1, 1, 1)$. Let us take

$$A = \{(a, y_0), (a, y_1), (a, y_2), (a, y_3), (a, y_4), (b, y_5)\}.$$

We have that

$$(b, y_5) \preceq_{k^0, z^*} (a, y_0), \quad (a, y_4) \preceq_{k^0, z^*} (a, y_1),$$

and

$$(a, y_3) \preceq_{k^0} (a, y_2) \preceq_{k^0} (a, y_1) \preceq_{k^0} (a, y_0),$$

but $\preceq_{k^0}$ may not be replaced by $\preceq_{k^0, z^*}$ in the above listing. Taking $(x_0, y_0) = (a, y_0)$, the conclusion of Theorem 3.10.7 is satisfied by (b, y_5) and (a, y_4), the conclusion of Corollary 3.10.8 is satisfied by (b, y_5), (a, y_4), (a, y_2), and (a, y_3), while the conclusion of Theorem 3.10.4 is satisfied by all the elements of A except (a, y_0).

We prefer to formulate and give a direct proof of Theorem 3.10.4 because it has the advantage of not containing any reference to an element $z^* \in C^+$ and the proof is interesting in itself.

Note that if $z^* \in C^\#$, then the order relations $\preceq_{k^0}$ and $\preceq_{k^0, z^*}$ coincide. Indeed, let $(x, y) \preceq_{k^0} (x', y')$. Of course, $d(x, x') \le z^*(y') - z^*(y)$. If $z^*(y') - z^*(y) > 0$, then $(x, y) \preceq_{k^0, z^*} (x', y')$, while in the contrary case $(x, y) = (x', y')$, since $y' - y \in C$.

Remark 3.10.10. Taking $\preceq_C \preceq_{k^0}$ or $\preceq_C \preceq_{k^0, z^*}$ a reflexive and transitive relation, the conclusions of Theorem 3.10.4 and Theorem 3.10.7 remain valid for $\preceq_{k^0}$ and $\preceq_{k^0, z^*}$ replaced by $\preceq$, respectively, if in (H1) $\preceq_{k^0}$ is replaced by $\preceq$.

Corollary 3.10.11. *Suppose that $C^\#$ is nonempty and let $B \subset Y$ be a nonempty subset such that every $\le_C$-decreasing sequence in B is bounded below by an element of B. If there exists an element of $C^\#$ that is bounded below on B, then B has the* **domination property***; i.e., for every $y_0 \in B$ there exists a minimal element $\bar{y}$ of B such that $\bar{y} \le_C y_0$.*

PROOF. Let $x_0 \in X$ be fixed and consider $A = \{x_0\} \times B$. The hypothesis shows that there exists $z^* \in C^\#$ such that z^* is bounded below on B. Take $k^0 \in C$ such that $z^*(k^0) = 1$; of course, $k^0 \in C \setminus (-\bar{C})$. It is obvious that A satisfies (H1). Applying Theorem 3.10.7 we get the desired $\bar{y} \in B$. $\qquad\square$

3.10.3 Minimal-Point Theorems and Gauge Techniques

We shall establish now a minimal-point theorem by a gauge technique. As in the previous section, (X, d) is a complete metric space, Y is a separated l.c.s., $C \subset$ is a convex cone, and $k^0 \in C \setminus (-\overline{C})$.

Let $A \subset X \times Y$ be a nonempty set and consider for $x \in X$ and $y \in Y$ the sets

$$A_x := \{y \in Y \mid (x, y) \in A\}, \quad A_y := \{x \in X \mid (x, y) \in A\}.$$

Of course, $P_X(A) = \{x \in X \mid A_x \neq \emptyset\}$.

Theorem 3.10.12. *Let* $(x_0, y_0) \in A$ *be fixed. Assume that* C *is closed and the following conditions hold:*

(i) *there exists* $\tilde{t} \in \mathbb{R}$ *such that* $P_Y(A) \cap (y_0 - \tilde{t}k^0 - (C \setminus \{0\})) = \emptyset$;

(ii) *the set* $A^r := \{x \in X \mid \exists y \leq_C y_0 + rk^0, \ (x, y) \in A\}$ *is closed for every* $r \in \mathbb{R}$;

(iii) *every* C-*increasing, proper, l.s.c., and sublinear function* $\psi : Y \to \overline{\mathbb{R}}$ *attains its infimum on* $A_x - y_0$ *for every* $x \in P_X(A)$.

Then there exists $(\bar{x}, \bar{y}) \in A$ *such that*

$$\bar{y} + k^0 d(\bar{x}, x_0) \leq_C y_0 \tag{3.114}$$

and

$$(x', y') \in A, \ y' + k^0 d(x', \bar{x}) \leq_C \bar{y} \Rightarrow \begin{cases} x' = \bar{x} \text{ and} \\ y' \in (\bar{y} - C) \setminus (\bar{y} - (0, \infty) \cdot k^0 - C). \end{cases} \tag{3.115}$$

If $k^0 \in \operatorname{int} C$, *we may replace* (iii) *by*

(iv) *every* C-*increasing, continuous, and sublinear function* $\psi : Y \to \mathbb{R}$ *attains its infimum on* $A_x - y_0$ *for every* $x \in P_X(A)$,

and (3.115) *by*

$$(x', y') \in A, \ y' + k^0 d(x', \bar{x}) \leq_C \bar{y} \Rightarrow x' = \bar{x}, \ y' \notin \bar{y} - \operatorname{int} C. \tag{3.116}$$

PROOF. Consider the function

$$\xi : X \to \overline{\mathbb{R}}, \quad \xi(x) = \begin{cases} \inf\{\varphi(y - y_0) \mid y \in A_x\} & \text{for } x \in P_X(A), \\ +\infty & \text{for } x \notin P_X(A), \end{cases}$$

where $\varphi = \varphi_{C,k^0}$ is defined in (2.23). Because φ is C-increasing (see Theorem 2.3.1(d)), from (i) we obtain that ξ is bounded from below. Indeed, suppose that there exists $x \in X$ such that $\xi(x) < -\tilde{t}$; then there exists $y \in Y$ such that $(x, y) \in A$ and $\varphi(y - y_0) < -\tilde{t}$, whence $y - y_0 \in -\tilde{t}k^0 - tk^0 - C$ for some $t > 0$, which contradicts (i). The function ξ is also lower semicontinuous. For this let $r, r' \in \mathbb{R}$ be such that $r < r'$; then

$$A^r \subset \{x \in X \mid \xi(x) \leq r\} \subset A^{r'}.$$

If $x \in A^r$, there exists $y \leq_C y_0 + rk^0$ with $(x, y) \in A$, whence $\xi(x) \leq \varphi(y - y_0) \leq \varphi(rk^0) = r$. Now let $x \in X$ be such that $\xi(x) \leq r$. Since $r < r'$, from the definition of ξ, there exists $y \in A_x$ such that $\varphi(y - y_0) \leq r'$, whence, by the properties of φ, $y \in y_0 + r'k^0 - C$, i.e., $y \leq_C y_0 + r'k^0$. It follows that $x \in A^{r'}$. Therefore the inclusions mentioned above hold. So we get

$$\{x \in X \mid \xi(x) \leq r\} \subset \bigcap_{r' > r} A^{r'} \subset \bigcap_{r' > r} \{x \in X \mid \xi(x) \leq r'\}$$
$$= \{x \in X \mid \xi(x) \leq r\}.$$

Since by (ii), $A^{r'}$ is closed for every $r' \in \mathbb{R}$, it follows that $\{x \in X \mid \xi(x) \leq r\}$ is closed for every $r \in \mathbb{R}$; hence ξ is l.s.c. on X. Applying the EVP (for example, Corollary 3.10.5 or Corollary 3.10.6 for $Y = \mathbb{R}$) we find $\bar{x} \in X$ such that

$$\xi(\bar{x}) + d(\bar{x}, x_0) \leq \xi(x_0) \leq \varphi(y_0 - y_0) = 0 \qquad (3.117)$$

and

$$\xi(\bar{x}) < \xi(x) + d(x, \bar{x}) \quad \forall x \in X, \ x \neq \bar{x}. \qquad (3.118)$$

By (iii) there exists $\bar{y} \in A_{\bar{x}}$ such that $\xi(\bar{x}) = \varphi(\bar{y} - y_0)$. From (3.117), taking into account (2.25), one obtains that $\bar{y} - y_0 + k^0 d(\bar{x}, x_0) \leq_C 0$; i.e., (3.114) holds. Now let $(x', y') \in A$ be such that $y' + k^0 d(x', \bar{x}) \leq_C \bar{y}$. Then

$$\xi(x') + d(x', \bar{x}) \leq \varphi(y' - y_0) + d(x', \bar{x}) = \varphi(y' - y_0 + k^0 d(x', \bar{x}))$$
$$\leq \varphi(\bar{y} - y_0) = \xi(\bar{x}),$$

whence, from (3.118), we obtain that $x' = \bar{x}$. If $y' \in \bar{y} - (0, \infty) \cdot k^0 - C$, then $y' = \bar{y} - tk^0 - k$ for some $t > 0$ and $k \in C$. It follows that

$$\xi(\bar{x}) = \xi(x') \leq \varphi(y' - y_0) = \varphi(\bar{y} - y_0 - tk^0 - k) \leq \varphi(\bar{y} - y_0 - tk^0)$$
$$= \varphi(\bar{y} - y_0) - t < \xi(\bar{x}),$$

a contradiction. Therefore (3.115) holds.

The second part is obvious because when $k^0 \in \operatorname{int} C$, then φ is finite, continuous, and $\operatorname{int} C \subset (0, \infty)k^0 + C$. □

Notice also that this result is not an authentic minimal-point theorem. Note also that if condition (ii) is replaced by

(ii') the set $\{x \in X \mid \exists y \leq_C rk^0, \ (x, y) \in A\}$ is closed for every $r \in \mathbb{R}$,

then (3.114) must be replaced (when $\operatorname{int} C \neq \emptyset$) by the weaker condition

$$\bar{y} + k^0 d(\bar{x}, x_0) \notin y_0 + \operatorname{int} C.$$

Remark 3.10.13. In the conditions of the preceding theorem we have also that $d(\bar{x}, x_0) \leq \tilde{t}$.

Indeed, since $\xi(\bar{x}) \geq \tilde{t}$, from (3.117) we get $-\tilde{t} + d(\bar{x}, x_0) \leq 0$, i.e., our assertion.

Applying Theorem 3.10.12 to the epigraph of an operator we obtain the following vectorial EVP.

Corollary 3.10.14. *Let $f : X \to Y^\bullet$ be a proper function and let $x_0 \in \operatorname{dom} f$. Assume that C is closed and for every $r \in \mathbb{R}$ the set $\{x \in X \mid f(x) \leq_C f(x_0) + rk^0\}$ is closed. If for some $\varepsilon > 0$ we have $f(X) \cap (f(x_0) - \varepsilon k^0 - (C \setminus \{0\})) = \emptyset$, then for every $\lambda > 0$ there exists $\bar{x} \in \operatorname{dom} f$ such that*

$$f(\bar{x}) + \lambda^{-1}\varepsilon k^0 d(\bar{x}, x_0) \leq_C f(x_0), \quad d(\bar{x}, x_0) \leq \lambda, \qquad (3.119)$$

and

$$f(x) + \lambda^{-1}\varepsilon k^0 d(x, \bar{x}) \leq_C f(\bar{x}) \Rightarrow x = \bar{x}. \qquad (3.120)$$

PROOF. Consider $A = \operatorname{epi} f$ and take $y_0 = f(x_0)$. It is obvious that the hypotheses of Theorem 3.10.12 are satisfied for $\tilde{t} = \varepsilon$. It follows that there exists $(\bar{x}, \bar{y}) \in A$ such that (3.114) and (3.115) hold with $\lambda^{-1}\varepsilon d$ instead of d. The first relation of the conclusion is immediate from (3.114) and Remark 3.10.13. Assuming that $d(\bar{x}, x_0) = \lambda + t$ with $t > 0$, we obtain that for some $c \in C$,

$$f(\bar{x}) = f(x_0) - \lambda^{-1}\varepsilon k^0 d(\bar{x}, x_0) - c = f(x_0) - \varepsilon k^0 - (\lambda^{-1}\varepsilon t k^0 + c)$$
$$\in f(x_0) - \varepsilon k^0 - (C \setminus \{0\}).$$

This contradiction proves that $d(\bar{x}, x_0) \leq \lambda$. Let $x \in X$ be such that $f(x) + \lambda^{-1}\varepsilon k^0 d(x, \bar{x}) \leq_C f(\bar{x})$ $(\leq_C \bar{y})$; of course, $x \in \operatorname{dom} f$. Taking $y = f(x)$, from (3.115) one obtains that $x = \bar{x}$. $\qquad \square$

Notice also in this case that if the condition "$\{x \in X \mid f(x) \leq_C f(x_0) + rk^0\}$ is closed for every $r \in \mathbb{R}$" is replaced by "$\{x \in X \mid f(x) \leq_C rk^0\}$ is closed for every $r \in \mathbb{R}$," then the first relation in (3.119) must be replaced by "$f(\bar{x}) + \lambda^{-1}\varepsilon k^0 d(\bar{x}, x_0) \notin f(x_0) + \operatorname{int} C$" (if $\operatorname{int} C \neq \emptyset$).

Embedding $C \setminus \{0\}$ in the interior of another proper cone, we obtain a variant of Theorem 3.10.7 in slightly different conditions by using the **gauge technique**.

Theorem 3.10.15. *Assume that there exists a proper closed convex cone $B \subset Y$ such that $C \setminus \{0\} \subset \operatorname{int} B$. Assume also that the set $A \subset X \times Y$ satisfies the condition (H1) in Section 3.10.1 and $P_Y(A) \cap (\tilde{y} - \operatorname{int} B) = \emptyset$ for some $\tilde{y} \in Y$. Then for every $(x_0, y_0) \in A$ there exists $(\bar{x}, \bar{y}) \in A$, minimal with respect to $\preceq_{k^0}$, such that $(\bar{x}, \bar{y}) \preceq_{k^0} (x_0, y_0)$.*

PROOF. Let $\varphi := \varphi_{B,k^0}$ be defined by (2.23). By Theorem 2.3.1, φ is a continuous sublinear function for which (2.24), (2.25), (2.29), and (2.30) hold with D replaced by B; moreover, if $y_2 - y_1 \in C \setminus \{0\}$, then $\varphi(y_1) < \varphi(y_2)$. Observe that for $(x, y) \in A$ we have $\varphi(y - \tilde{y}) \geq 0$. Otherwise, for some $(x, y) \in A$ we have $\varphi(y-\tilde{y}) < 0$. It follows that there exists $\lambda > 0$ such that $y-\tilde{y} \in -\lambda k^0 - B$. Hence

$$y \in \tilde{y} - (\lambda k^0 + B) \subset \tilde{y} - (B + \text{int } B) \subset \tilde{y} - \text{int } B,$$

a contradiction. Since $0 \leq \varphi(y - \tilde{y}) \leq \varphi(y) + \varphi(-\tilde{y})$, it follows that φ is bounded from below on $P_Y(A)$. Let us construct a sequence $((x_n, y_n))_{n \geq 0} \subset A$ as follows: Having $(x_n, y_n) \in A$, we take $(x_{n+1}, y_{n+1}) \in A$, $(x_{n+1}, y_{n+1}) \preceq_{k^0} (x_n, y_n)$, such that

$$\varphi(y_{n+1}) \leq \inf\{\varphi(y) \mid (x, y) \in A, \ (x, y) \preceq_{k^0} (x_n, y_n)\} + 1/(n + 1).$$

Of course, the sequence $((x_n, y_n))$ is $\preceq_{k^0}$-decreasing. It follows that

$$y_{n+p} + k^0 d(x_{n+p}, x_n) \leq_C y_n \quad \forall n, p \in \mathbb{N}^*,$$

so that

$$d(x_{n+p}, x_n) \leq \varphi(y_n) - \varphi(y_{n+p}) \leq 1/n \quad \forall n, p \in \mathbb{N}^*.$$

It follows that (x_n) is a Cauchy sequence in the complete metric space (X, d), and so (x_n) is convergent to some $\bar{x} \in X$. By (H1) there exists $\bar{y} \in Y$ such that $(\bar{x}, \bar{y}) \in A$ and $(\bar{x}, \bar{y}) \preceq_{k^0} (x_n, y_n)$ for every $n \in \mathbb{N}$. Let us show that $(\bar{x}, \bar{y})$ is the desired element. Indeed, $(\bar{x}, \bar{y}) \preceq_{k^0} (x_0, y_0)$. Suppose that $(x', y') \in A$ is such that $(x', y') \preceq_{k^0} (\bar{x}, \bar{y})$ $(\preceq_{k^0} (x_n, y_n)$ for every $n \in \mathbb{N})$. Thus $\varphi(y') + d(x', \bar{x}) \leq \varphi(\bar{y})$, whence

$$d(x', \bar{x}) \leq \varphi(\bar{y}) - \varphi(y') \leq \varphi(y_n) - \varphi(y') \leq 1/n \quad \forall n \geq 1.$$

It follows that $d(x', \bar{x}) = \varphi(\bar{y}) - \varphi(y') = 0$. Hence $x' = \bar{x}$. As $y' \leq_C \bar{y}$, if $y' \neq \bar{y}$, then $\bar{y} - y' \in C \setminus \{0\}$, whence $\varphi(y') < \varphi(\bar{y})$, a contradiction. Therefore $(x', y') = (\bar{x}, \bar{y})$.   □

Comparing with Theorem 3.10.7, note that the present condition on C is stronger (because in this case $C^\# \neq \emptyset$), while the condition on A is weaker (A may be not contained in a half-space). Note that when C and k^0 are as in Theorem 3.10.15, Corollaries 3.10.5 and 3.10.6 may be improved.

Corollary 3.10.16. *Let $f : X \to Y^\bullet$. Assume that there exists a proper convex cone $B \subset Y$ such that $C \setminus \{0\} \subset \text{int } B$ and $f(X) \cap (\tilde{y} - B) = \emptyset$ for some $\tilde{y} \in Y$. Also assume that either (H3) (in Section 3.10.1) holds or C is closed in the direction k^0 and (H4) holds. Then the conclusion of Corollary 3.10.5 holds, too.*

As above, C is closed in the direction k^0 if $C \cap (y - \mathbb{R}_+ k^0)$ is closed for every $y \in C$. The proof of Corollary 3.10.16 is similar to those of Corollaries 3.10.5 and 3.10.6.

3.10.4 Minimal-Point Theorems and Cone-Valued Metrics

Nemeth [270] obtained a vectorial EVP using cone-valued metrics. The aim of this section is to obtain a minimal-point theorem in product spaces using such metrics. Applying the result to operators, we establish a vectorial EVP slightly more general than Nemeth's result.

As in the previous sections, Y is a separated locally convex space, while X is a nonempty set that will be endowed with a cone-valued metric. In this sense consider P a convex cone in Y and let $\leq_P$ be the preorder on Y determined by P.

We say that the mapping $r : X \times X \to P$ is a P-(valued) **metric** if r satisfies the usual conditions, i.e., for all $x_1, x_2, x_3 \in X$ one has $r(x_1, x_2) = 0 \Leftrightarrow x_1 = x_2$, $r(x_1, x_2) = r(x_2, x_1)$ and $r(x_1, x_3) \leq_P r(x_1, x_2) + r(x_2, x_3)$. The notions of convergent net and fundamental net are defined as usual; so the net $(x_i)_{i \in I} \subset X$ converges to $x \in X$ if $r(x_i, x) \to 0$ in Y, while $(x_i)_{i \in I}$ is fundamental (or Cauchy) if for every neighborhood V of the origin in Y there exists $i_V \in I$ such that $r(x_i, x_j) \in V$ for all $i, j \in I$ with $i, j \succeq i_V$. One sees easily that when P is normal, the limit of a convergent net is unique and every convergent net is fundamental. Of course, (X, r) is complete if every Cauchy net is convergent. For other details and comments on cone-valued metrics see [270]. An example of a cone-valued metric is furnished below. Consider d a scalar metric on X and y_0 a fixed element in $Y \setminus \{0\}$. Taking $P = \mathbb{R}_+ \cdot y_0$, the mapping $\bar{r} : X \times X \to P$, defined by $\bar{r}(x_1, x_2) = y_0 d(x_1, x_2)$, is a P-metric. Note that $(x_i)_{i \in I} \subset X$ is $\bar{r}$-convergent ($\bar{r}$-fundamental) iff (x_i) is d-convergent (d-fundamental); so $(X, \bar{r})$ is complete iff (X, d) is complete.

Now let $C, C_0 \subset Y$ be convex cones with $C_0 \subset C$. We say that C_0 is **sequentially** C-**bound regular** (in short C-seq-b-regular) if every C_0-increasing and C-bounded sequence in C_0 is Cauchy. Of course, if C_0 is C-seq-b-regular then C_0 is pointed (even more: $C_0 \cap (-C) = \{0\}$).

In the terminology of Nemeth [270], C_0 is sequentially C-bound regular if every C_0-increasing and C-bounded net in C_0 is convergent (so our requirement is weaker when C_0 is normal).

Note that if $k^0 \in C \setminus (-\overline{C})$ and $C_0 = \mathbb{R}_+ \cdot k^0$, then C_0 is C-seq-b-regular (even in the sense of Nemeth). Indeed, let $(t_n)_{n \in \mathbb{N}} \subset \mathbb{R}_+$ be such that $(t_n k^0)$ is C_0-increasing and C-bounded. Then (t_n) is increasing in $\mathbb{R}_+$. Assuming that (t_n) is not bounded, $t_n \to \infty$ (so we can consider that $t_n > 0$ for every $n \in \mathbb{N}$). Since $t_n k^0 \leq_C y$ for some $y \in Y$ and every $n \in \mathbb{N}$, it follows that $k^0 = \frac{1}{t_n} y - k_n$, with $(k_n) \subset C$. So we get that $k_n \to -k^0 \in \overline{C}$, a contradiction.

Let now $C_0 \subset C$ be a convex cone and r a C_0-metric on X. We consider the relation $\leq_r$ on $X \times Y$ defined by

$$(x_1, y_1) \leq_r (x_2, y_2) \quad \text{if} \quad y_1 + r(x_1, x_2) \leq_C y_2.$$

One obtains easily that $\leq_r$ is a preorder (even a partial order if C is pointed).

In order to establish the main result of this section we need a similar condition to (H1):

(H5) *For every $\leq_r$-decreasing net $((x_i, y_i))_{i \in I} \subset A$ with $x_i \to x \in X$ there exists $y \in Y$ such that $(x, y) \in A$ and $(x, y) \leq_r (x_i, y_i)$ for every $i \in I$.*

We shall see during the proof of the next theorem that one can consider that I is a totally ordered set in (H5).

A related condition, which is independent of the metric r, and seems to be natural enough, is the following:

(H6) *For every net $((x_i, y_i))_{i \in I} \subset A$, with $x_i \to x \in X$ and $(y_i)_{i \in I}$ C-decreasing, there exists $y \in Y$ such that $(x, y) \in A$ and $y \leq_C y_i$ for every $i \in I$.*

Note that if the lower sections of C with respect to C_0 are closed, i.e., $C \cap (y - C_0)$ is closed for every $y \in C$, and C_0 is normal, then (H6)$\Rightarrow$(H5). Indeed, let $((x_i, y_i))_{i \in I} \subset A$ be a $\leq_r$-decreasing net with $x_i \to x$. Of course, $(y_i)_{i \in I}$ is C-decreasing, and by (H6), there exists $y \in Y$ such that $(x, y) \in A$ and $y \leq_C y_i$ for every $i \in I$. Then

$$y + r(x_i, x_j) \leq_C y_j + r(x_i, x_j) \leq_C y_i \quad \forall i, j \in I, \ j \succeq i.$$

It follows that $y_i - y - r(x_i, x_j) \in C \cap (y_i - y - C_0)$, whence, by taking the limit with respect to $j \in I$, we obtain that $y_i - y - r(x_i, x) \in C$; i.e., $(x, y) \leq_C (x_i, y_i)$ for every $i \in I$. The normality of C_0 was used to obtain that $r(x_i, x_j) \to r(x_i, x)$ when taking the limit with respect to $j \in I$.

Theorem 3.10.17. *Let $C \subset Y$ be a pointed convex cone, $C_0 \subset C$ a normal convex cone that is C-seq-b-regular, r a C_0-metric on X such that (X, r) is complete, $A \subset X \times Y$ a nonempty set such that $P_Y(A)$ is C-lower bounded, and suppose that (H5) holds. Then for every $(x_0, y_0) \in A$ there exists a $\leq_r$-minimal point $(\overline{x}, \overline{y}) \in A$ such that $(\overline{x}, \overline{y}) \leq_r (x_0, y_0)$.*

PROOF. Let $B = \{(x, y) \in A \mid (x, y) \leq_r (x_0, y_0)\}$ and let $\mathcal{C}$ be a maximal chain in B. Of course, $(x_0, y_0) \in \mathcal{C}$. Consider the directed set $(I, \succeq)$ defined by $I = \mathcal{C}$ and $i_1, i_2 \in I$, $i_1 = (x_1, y_1)$, $i_2 = (x_2, y_2)$, $i_1 \succeq i_2$ if $(x_1, y_1) \leq_r (x_2, y_2)$. So $\mathcal{C}$ becomes a net indexed on I: For $i = (x, y) \in I$, $(x_i, y_i) = (x, y)$. The net $(x_i)_{i \in I}$ is Cauchy with respect to r. In the contrary case there exists a neighborhood V of the origin of Y such that

$$\forall i \in I, \ \exists j, k \in I, \ j, k \succeq i \ : \ r(x_i, x_j) \notin V. \qquad (3.121)$$

Taking into account that $(I, \succeq)$ is totally ordered, for $i_0 = (x_0, y_0)$ there exist $i_1, i_2 \in I$ such that $i_2 \succeq i_1 \succeq i_0$ and $r(x_{i_2}, x_{i_1}) \notin V$. Taking now $i = i_2$ in (3.121), there exist $i_3, i_4 \in I$ such that $i_4 \succeq i_3 \succeq i_2$ and $r(x_{i_4}, x_{i_3}) \notin V$. Continuing in this manner we obtain an increasing sequence $(i_n)_{n \in \mathbb{N}}$ in I such that $r(x_{i_{2n+2}}, x_{i_{2n+1}}) \notin V$ for every $n \in \mathbb{N}$. Since for every n, $i_n = (x_n, y_n)$, we obtain a $\leq_r$-decreasing sequence $((x_n, y_n)) \subset \mathcal{C}$ with $r(x_{2n+2}, x_{2n+1}) \notin V$ for every n. Since $y_{n+1} + r(x_{n+1}, x_n) \leq_C y_n$ for every n, we get that for all $m \in \mathbb{N}$,

$$s_m = \sum_{n=0}^{m} r(x_{n+1}, x_n) \leq_C \sum_{n=0}^{m}(y_n - y_{n+1}) = y_0 - y_{m+1} \leq_C y_0 - \tilde{y},$$

where $\tilde{y}$ is a C-lower bound for $P_Y(A)$. Therefore (s_m) is a C_0-increasing and C-bounded sequence in C_0. Since C_0 is C-seq-b-regular, it follows that (s_m) is a Cauchy sequence. In particular, for the neighborhood V obtained above, for some $m_0 \in \mathbb{N}$ and all $m, n \in \mathbb{N}$, $m, n \geq m_0$, we have that $s_m - s_n \in V$. In particular, $r(x_{2m_0+2}, x_{2m_0+1}) = s_{2m_0+1} - s_{2m_0} \in V$, a contradiction. Therefore $(x_i)_{i\in I}$ is a Cauchy net. Because (X, r) is complete, (x_i) r-converges to some $\overline{x} \in X$. Since (H5) holds, there exists $\overline{y} \in Y$ such that $(\overline{x}, \overline{y}) \in A$ and $(\overline{x}, \overline{y}) \leq_r (x_i, y_i)$ for all $i \in I$; that is, $(\overline{x}, \overline{y}) \leq_r (x, y)$ for all $(x, y) \in \mathcal{C}$. It follows that $(\overline{x}, \overline{y}) \in B$, and since $\mathcal{C}$ is a maximal chain in B, $(\overline{x}, \overline{y}) \in \mathcal{C}$. Therefore $(\overline{x}, \overline{y})$ is the least element of $\mathcal{C}$. The fact that $(\overline{x}, \overline{y})$ is a minimal element of A follows easily. □

Note that if C is not pointed, then $(\overline{x}, \overline{y})$ is such that $(x', y') \in A$, $(x', y') \leq_r (\overline{x}, \overline{y})$ imply $(\overline{x}, \overline{y}) \leq_r (x', y')$.

Consider now $k^0 \in C \setminus \{0\}$ and d a scalar metric on X and define the order relation $\leq_{k^0}$ by (3.111).

As a consequence of Theorem 3.10.17 we obtain the following weaker version of Theorem 3.10.7.

Corollary 3.10.18. *Let $C \subset Y$ be a pointed convex cone, $k^0 \in C \setminus \{0\}$, d a (scalar) metric on X such that (X, d) is complete, and $A \subset X \times Y$ a nonempty set such that $P_Y(A)$ is C-lower bounded and (H1), with "sequence" replaced by "net," holds. Then for every $(x_0, y_0) \in A$ there exists a $\leq_{k^0}$-minimal point $(\overline{x}, \overline{y}) \in A$ such that $(\overline{x}, \overline{y}) \leq_{k^0} (x_0, y_0)$.*

PROOF. Consider the convex cone $C_0 = \mathbb{R}_+ \cdot k^0$ and the C_0-metric $\overline{r}$ on X defined by $\overline{r}(x_1, x_2) = k^0 d(x_1, x_2)$. The discussion from the beginning of this section shows that the hypotheses of Theorem 3.10.17 are satisfied. Since the relations $\leq_{\overline{r}}$ and $\leq_{k^0}$ coincide, the conclusion follows. □

Another consequence of Theorem 3.10.17 is the following vectorial EVP.

Corollary 3.10.19. *Let $C \subset Y$ be a pointed convex cone, $C_0 \subset C$ a normal convex cone that is C-seq-b-regular, r a C_0-metric on X such that (X, r) is complete, and $f : X \to Y^\bullet$. Suppose that all the lower sections of C with respect to C_0 are closed, that f is C-lower bounded and for every net $(x_i)_{i\in I} \subset X$ converging to $x \in X$ such that $(f(x_i))$ is C-decreasing, $f(x) \leq_C f(x_i)$ for every $i \in I$. Then for every $x_0 \in \text{dom} f$ there exists $\overline{x} \in \text{dom} f$ such that*

$$f(\overline{x}) + r(\overline{x}, x_0) \leq_C f(x_0)$$

and

$$\forall x \in X : f(x) + r(x, \overline{x}) \leq_C f(\overline{x}) \Rightarrow x = \overline{x}.$$

PROOF. Let us consider the set $A = \operatorname{gr} f$. It is obvious that A satisfies (H6). Since the lower sections of C with respect to C_0 are closed, it follows that A satisfies (H5), too. Applying Theorem 3.10.17 for $(x_0, y_0) = (x_0, f(x_0)) \in A$, we get $(\bar{x}, \bar{y})$ a minimal point of A with respect to $\leq_r$ such that $(\bar{x}, \bar{y}) \leq_r (x_0, y_0)$. Of course, $\bar{x} \in \operatorname{dom} f$ and $\bar{y} = f(\bar{x})$. It is obvious that $\bar{x}$ satisfies the conclusion of the corollary. □

Note that Nemeth [270, Th. 6.1] obtained the same conclusion as in Corollary 3.10.19 under the supplementary hypotheses that C is closed and C_0 is complete.

For further references to vector-valued variational principles, see, e.g., [179, 73, 238, 169].

3.10.5 Fixed Point Theorems of Kirk–Caristi Type

The following existence result is close to a vectorial variant of a theorem by Takahashi [336] obtained by Tammer [341].

Theorem 3.10.20. *Let (X, d) be a complete metric space, Y a t.v.s., $B \subset Y$ a closed convex cone, $C \subset Y$ a proper convex cone such that $C \setminus \{0\} \subset \operatorname{int} B$, and $k^0 \in C \setminus \{0\}$. Consider $f : X \to Y^\bullet$ a proper function for which $\{x \in X \mid f(x) \leq_B rk^0\}$ is closed for every $r \in \mathbb{R}$. Assume that $f(X) \cap (tk^0 - B) = \emptyset$ for some $t \in \mathbb{R}$ and for every $x \in \operatorname{dom} f \setminus \operatorname{Eff}(f, C)$ there exists $y \neq x$ such that $f(y) + k^0 d(y, x) \leq_C f(x)$. Then $\operatorname{Eff}(f, C) \neq \emptyset$.*

PROOF. Note first that for every $x_0 \in \operatorname{dom} f$ there exists some $\varepsilon > 0$ such that $f(X) \cap (f(x_0) - \varepsilon k^0 - (B \setminus \{0\})) = \emptyset$; otherwise, for some $x_0 \in \operatorname{dom} f$ and every $n \in \mathbb{N}^*$ there exists $x_n \in \operatorname{dom} f$ such that $f(x_n) \in f(x_0) - nk^0 - B$. It follows that $\varphi_{B,k^0}(f(x_n) - f(x_0)) \leq -n$, whence $\varphi_{B,k^0}(f(x_n)) \leq \varphi_{B,k^0}(f(x_0)) - n \leq t$ for n sufficiently large. This yields the contradiction $f(x_n) \in tk^0 - B$. Applying Corollary 3.10.14 for $x_0 \in \operatorname{dom} f$, $\varepsilon > 0$ such that $f(X) \cap (f(x_0) - \varepsilon k^0 - (B \setminus \{0\})) = \emptyset$, $\lambda := 1/\varepsilon$ and C replaced by B (and taking into account the discussion after the proof of that result), there exists $\bar{x} \in \operatorname{dom} f$ such that $x = \bar{x}$ whenever $f(x) + k^0 d(x, \bar{x}) \leq_B f(\bar{x})$. Assume that $\bar{x} \notin \operatorname{Eff}(f, C)$. By hypothesis, there exists $x \neq \bar{x}$ such that $f(x) + k^0 d(x, \bar{x}) \in f(\bar{x}) - C \subset f(\bar{x}) - B$. So, we get the contradiction $x = \bar{x}$. □

The following result is close to a **vectorial variant** of the **Kirk–Caristi fixed point theorem** obtained by Tammer [341].

Theorem 3.10.21. *Let (X, d) be a complete metric space, $T : X \to X$, Y a t.v.s., $C \subset Y$ a proper closed convex cone, and $k^0 \in \operatorname{int} C$. Assume that there exists a proper function $f : X \to Y^\bullet$ for which $\{x \in X \mid f(x) \leq_C rk^0\}$ is closed for every $r \in \mathbb{R}$ and $f(X) \cap (tk^0 - C) = \emptyset$ for some $t \in \mathbb{R}$. If $f(Tx) + k^0 d(Tx, x) \leq_C f(x)$ for every $x \in \operatorname{dom} f$, then T has at least one fixed point.*

PROOF. As in the proof of the preceding theorem (applied for (C, C_0) instead of (B, C), where $C_0 = \{0\} \cup \text{int } C$), there exists $\bar{x} \in \text{dom } f$ such that $x = \bar{x}$ whenever $f(x) + k^0 d(x, \bar{x}) \leq_C f(\bar{x})$. But $f(T\bar{x}) + k^0 d(T\bar{x}, \bar{x}) \leq_C f(\bar{x})$, and so $T\bar{x} = \bar{x}$. $\qquad\square$

3.11 Optimality Conditions

3.11.1 Lagrange Multipliers and Saddle Point Assertions

Consider a convex vector minimization problem

$$(P) \qquad \text{Eff}(f[\mathcal{A}], C_Y),$$

where $f : M \longrightarrow Y$, $g : M \longrightarrow Z$, Y and Z are normed spaces, C_Z and C_Y are closed convex pointed cones in Z, Y, respectively,0 and

$$\mathcal{A} := \{x \in M \mid g(x) \in -C_Z\}.$$

As in ordinary scalar optimization, **Lagrange multipliers** can be used for different purposes such as duality, saddle point theory, sensitivity, and numerical approaches (compare Amahroq and Taa [1], Clarke [80], El Abdouni and Thibault [107], Li and Wang [237], Miettinen, [260], Minami [261], Tanaka [351, 352, 353], Thibault [357], Wang [365]).

In the following we derive **existence results** for Lagrange multipliers. These results extend well-known theorems (compare Kosmol [216]) on Lagrange multipliers in nonlinear programming considerably.

Lemma 3.11.1. *Let X be a linear space, M a convex subset of X, Y, and Z normed spaces, C_Z and C_Y closed convex pointed cones in Z, Y, respectively, and int $C_Y \neq \emptyset$.*

Assume that $f : M \longrightarrow Y$, $g : M \longrightarrow Z$ are C_Y-convex, C_Z-convex, respectively, mappings for which the following regularity assumptions are satisfied:

(A.1) int$\{(y, z) \in Y \times Z \mid \exists x \in M \ : \ y \in f(x) + C_Y \text{ and } z \in g(x) + C_Z\} \neq \emptyset$,
(A.2) $\exists y_0 \in \text{cl } f[\mathcal{A}]$ and $f[\mathcal{A}] \cap (y_0 - (C_Y \setminus \{0\})) = \emptyset$.

Then there exist $y_0^* \in C_{Y^*}^+$, $z_0^* \in C_{Z^*}^+$ with $(y_0^*, z_0^*) \neq (0, 0)$, and

$$y_0^*(y_0) = \inf\{y_0^*(f(x)) + z_0^*(g(x)) \mid x \in M\}.$$

PROOF. Consider the following sets:

$$A := \{(y, z) \in Y \times Z \mid \exists x \in M \ : \ y \in f(x) + C_Y, \ z \in g(x) + C_Z\}$$

and

$$B := \{(y, z) \in Y \times Z \mid y \in y_0 - C_Y, \ z \in -C_Z\}.$$

In order to apply a separation theorem for convex sets (compare Theorem 2.2.7) we show that the assumptions of the separation theorem are satisfied.

The set A is convex, since we get for $(y^1, z^1) \in A$, $(y^2, z^2) \in A$, $0 \le \lambda \le 1$, and corresponding elements $x^1, x^2 \in M$,

$$\lambda y^1 + (1 - \lambda)y^2 \in \lambda f(x^1) + C_Y + (1 - \lambda)f(x^2) + C_Y$$

$$\subset f(\lambda x^1 + (1 - \lambda)x^2) + C_Y + C_Y \subset f(\lambda x^1 + (1 - \lambda)x^2) + C_Y,$$

since C_Y is a convex cone and f a C_Y-convex mapping. Together with

$$\lambda z^1 + (1 - \lambda)z^2 \in \lambda g(x^1) + (1 - \lambda)g(x^2) + C_Z \subset g(\lambda x^1 + (1 - \lambda)x^2) + C_Z,$$

because C_Z is a convex cone and g a C_Z-convex mapping, we can conclude that

$$(\lambda y^1 + (1 - \lambda)y^2, \lambda z^1 + (1 - \lambda)z^2) \in A.$$

Moreover, B is convex with respect to the convexity of C_Y and C_Z.

Under the assumption $(A.1)$, int $A \ne \emptyset$.

In order to show that int $A \cap B = \emptyset$ we suppose $\exists\, (y, z) \in$ int $A \cap B$. This implies

$$\exists\, x \in M \quad \text{with} \quad g(x) \in z - C_Z \subset -C_Z \quad \text{and} \quad f(x) \in y - C_Y \subset y_0 - C_Y,$$

so that we get $y_0 = y = f(x)$ because of the definition of y_0 in $(A.2)$ and since C_Y is a pointed convex cone.

Regarding $(y, z) \in$ int A, it follows that there are an $\varepsilon > 0$ and $U_\varepsilon(y) \subset Y$, $V_\varepsilon(z) \subset Z$ with $U_\varepsilon(y) \times V_\varepsilon(z) \subset A$; especially for $k^0 \in C_Y \setminus \{0\}$, $\|k^0\| = 1$ we consider $(y - \frac{\varepsilon}{2}k^0, z) \in A$; i.e., for some $x' \in M$,

$$y_0 - \frac{\varepsilon}{2}k^0 = y - \frac{\varepsilon}{2}k^0 \in f(x') + C_Y$$

and

$$g(x') \in z - C_Z \subset -C_Z.$$

This means that $x' \in \mathcal{A}$ and $f(x') \in y_0 - (C_Y \setminus \{0\})$, in contradiction to the definition of y_0 in $(A.2)$.

Consider the set $A - B$. Under the given assumptions $A - B$ is convex and int$(A - B) \ne \emptyset$. Taking into account int $A \cap B = \emptyset$ we get $0 \notin$ int$(A - B)$. Now, it is possible to apply a separation theorem for convex sets (Theorem 2.2.7). This separation theorem implies the existence of $(y_0^*, z_0^*) \in (Y^* \times Z^*) \setminus \{0\}$ such that

$$z_0^*(z^1) + y_0^*(y^1) \ge z_0^*(z^2) + y_0^*(y^2) \quad \forall\, (y^1, z^1) \in A,\ \forall\, (y^2, z^2) \in B. \quad (3.122)$$

In the following we show that $y_0^* \in C_{Y^*}^+$ and $z_0^* \in C_{Z^*}^+$.

If we suppose $y_0^* \notin C_{Y^*}^+$, i.e., $y_0^*(\bar{y}) < 0$ for an element $\bar{y} \in C_Y$, we get for $y := -\bar{y} \in -C_Y$, regarding that C_Y is a cone,

$$\sup\{y_0^*(ny) \mid n \in N\} = \sup\{ny_0^*(y) \mid n \in N\} = \infty,$$

in contradiction to the separation property (3.122). Analogously, we can show that $z_0^* \in C_{Z^*}^+$.

For all $x \in M$, $(f(x), g(x)) \in A$, and with $(y_0, 0) \in B$ we get

$$\inf\{z_0^*(g(x)) + y_0^*(f(x)) \mid x \in M\} \geq y_0^*(y_0).$$

Now consider a sequence $\{x_n\}_{n \in N}$ in $\mathcal{A} = \{x \in M \mid g(x) \in -C_Z\}$ with

$$\lim_{n \to \infty} f(x_n) = y_0.$$

Then we get

$$\begin{aligned} \inf\{z_0^*(g(x)) + y_0^*(f(x)) \mid x \in M\} &\leq \inf\{z_0^*(g(x)) + y_0^*(f(x)) \mid x \in \mathcal{A}\} \\ &\leq \inf\{y_0^*(f(x)) \mid x \in \mathcal{A}\} \\ &\leq \lim_{n \to \infty} y_0^*(f(x_n)) = y_0^*(y_0), \end{aligned}$$

so that the equation holds. $\qquad\square$

Lemma 3.11.2. *Additionally to the assumptions of Lemma 3.11.1 we suppose*

(A.3) *There exists an element $x^1 \in M$ such that for all $z^* \in C_{Z^*}^+ \setminus \{0\}$, $z^*(g(x^1)) < 0$.*

(i) *Then there exist elements $y_0^* \in C_{Y^*}^+ \setminus \{0\}$ and $z_0^* \in C_{Z^*}^+$ with*

$$y_0^*(y_0) = \inf\{y_0^*(f(x)) + z_0^*(g(x)) \mid x \in M\}.$$

(ii) *If $x_0 \in \mathcal{A}$ and $f(x_0) \in \mathrm{Eff}(f[\mathcal{A}], C_Y)$, then there exist $y_0^* \in C_{Y^*}^+ \setminus \{0\}$ and $z_0^* \in C_{Z^*}^+$ such that x_0 is also a minimal solution of $y_0^*(f(\cdot)) + z_0^*(g(\cdot))$ on M, and*

$$z_0^*(g(x_0)) = 0.$$

PROOF. (i) From Lemma 3.11.1 we can conclude that there exist $y_0^* \in C_{Y^*}^+$, $z_0^* \in C_{Z^*}^+$ with $(y_0^*, z_0^*) \neq 0$, and

$$y_0^*(y_0) = \inf\{y_0^*(f(x)) + z_0^*(g(x)) \mid x \in M\}. \tag{3.123}$$

Under the assumption (A.3) we suppose $y_0^* = 0$. Then we get in (3.122) with $z^1 = g(x^1)$, $z^2 = 0$,

$$z_0^*(g(x^1)) \geq z_0^*(0) = 0. \tag{3.124}$$

Regarding $(y_0^*, z_0^*) \neq 0$, we have $z_0^* \neq 0$, and now together with the assumption (A.3) we obtain a contradiction,

$$0 > z_0^*(g(x^1)) \geq 0,$$

because of (3.124).

(ii) If $x_0 \in \mathcal{A}$ and $y_0 := f(x_0) \in \text{Eff}(f[\mathcal{A}], C_Y)$, then (3.123) implies

$$y_0^*(y_0) \leq y_0^*(f(x_0)) + z_0^*(g(x_0)) \leq y_0^*(f(x_0)) = y_0^*(y_0),$$

so that $y_0^*(f(x_0)) + z_0^*(g(x_0)) = \inf\{y_0^*(f(x)) + z_0^*(g(x)) \mid x \in M\}$ and $z_0^*(g(x_0)) = 0.$ □

Remark 3.11.3. Conversely, if $x_0 \in M$ is a minimal solution of the Lagrangian $y_0^*(f(\cdot)) + z_0^*(g(\cdot))$ with $g(x_0) \in -C_Z$ and $z_0^*(g(x_0)) = 0$, then

$$f(x_0) \in w \, \text{Eff}(f[\mathcal{A}], C_Y)$$

follows without regularity assumption:

$$y_0^*(f(x_0)) = y_0^*(f(x_0)) + z_0^*(g(x_0)) \leq y_0^*(f(x)) + z_0^*(g(x)) \leq y_0^*(f(x))$$

for all $x \in M$ with $g(x) \in -C_Z$ and $f(x_0) \in w \, \text{Eff}(f[\mathcal{A}], C_Y)$, regarding $y_0^* \in C_{Y^*}^+ \setminus \{0\}.$

Theorem 3.11.4. *Suppose that $(A.1)$, $(A.2)$, and $(A.3)$ are satisfied. Assume $x_0 \in M$. Then:*

(i) *If $f(x_0) \in \text{Eff}(f[\mathcal{A}], C_Y)$, then there exist $y_0^* \in C_{Y^*}^+ \setminus \{0\}$ and $z_0^* \in C_{Z^*}^+$ such that the following saddle point assertion is satisfied:*

$$y_0^*(f(x_0)) + z^*(g(x_0)) \leq y_0^*(f(x_0)) + z_0^*(g(x_0)) \leq y_0^*(f(x)) + z_0^*(g(x))$$
$$(3.125)$$
$$\forall x \in M, \quad \forall z^* \in C_{Z^*}^+.$$

(ii) *Conversely, if there are $y_0^* \in C_{Y^*}^+ \setminus \{0\}$ and $(x_0, z_0^*) \in M \times C_{Z^*}^+$ such that the saddle point assertion (3.125) is satisfied for all $x \in M$ and $z^* \in C_{Z^*}^+$, then $f(x_0) \in w \, \text{Eff}(f[\mathcal{A}], C_Y)$.*

PROOF. (i) Assume $f(x_0) \in \text{Eff}(f[\mathcal{A}], C_Y)$. Using Lemma 3.11.2, (ii), we get that there exist $y_0^* \in C_{Y^*}^+ \setminus \{0\}$ and $z_0^* \in C_{Z^*}^+$ with

$$y_0^*(f(x_0)) + z_0^*(g(x_0)) \leq y_0^*(f(x)) + z_0^*(g(x)) \ \forall x \in M.$$

Furthermore, regarding $-g(x_0) \in C_Z$, it follows, again with Lemma 3.11.2, (ii), that

$$z^*(g(x_0)) \leq 0 = z_0^*(g(x_0)) \quad \forall z^* \in C_{Z^*}^+.$$

This yields

$$y_0^*(f(x_0)) + z^*(g(x_0)) \leq y_0^*(f(x_0)) + z_0^*(g(x_0)) \ \forall z^* \in C_{Z^*}^+.$$

Then both inequalities are satisfied.

(ii) Suppose $y_0^* \in C_{Y^*}^+ \setminus \{0\}$ and assume that the saddle point assertion is satisfied for $(x_0, z_0^*) \in M \times C_{Z^*}^+$. Then the first inequality implies

$$z^*(g(x_0)) \le z_0^*(g(x_0)) \quad \forall z^* \in C_{Z^*}^+,$$

so that we get, regarding that $C_{Z^*}^+$ is a convex cone,

$$(z^* + z_0^*)(g(x_0)) \le z_0^*(g(x_0)) \quad \forall z^* \in C_{Z^*}^+,$$
$$z^*(g(x_0)) \le 0 \quad \forall z^* \in C_{Z^*}^+,$$

and $g(x_0) \in -C_Z$. This implies

$$0 \ge z_0^*(g(x_0)) \ge 0(g(x_0)) = 0,$$

since $0 \in C_{Z^*}^+$, and so $z_0^*(g(x_0)) = 0$. Consider now $x \in M$ with $g(x) \le 0$. Then we conclude from the second inequality in the saddle point assertion that

$$y_0^*(f(x_0)) = y_0^*(f(x_0)) + z_0^*(g(x_0)) \le y_0^*(f(x)) + z_0^*(g(x)) \le y_0^*(f(x))$$

and

$$y_0^*(f(x_0)) \le y_0^*(f(x)) \qquad \forall x \in \mathcal{A}.$$

This means that $f(x_0) \in w \operatorname{Eff}(f[\mathcal{A}], C_Y)$.   □

Remark 3.11.5. A point $(x_0, z_0^*) \in M \times C_{Z^*}^+$ satisfying the property (3.125) for an element $y_0^* \in C_{Y^*} \setminus \{0\}$ is called a y_0^*-**saddle point** of the **Lagrangian**

$$\Phi(x, z^*) := y_0^*(f(x)) + z^*(g(x)), \quad x \in M, \ z^* \in C_{Z^*}^+.$$

In the next subsection we will derive necessary and sufficient conditions for y^*-saddle points of a generalized Lagrangian.

The relation (3.125) can be described by

$$\Phi(x_0, z_0^*) \in \operatorname{Min}(\{\Phi(x, z_0^*) \mid x \in M\}, y_0^*),$$

$$\Phi(x_0, z_0^*) \in \operatorname{Max}(\{\Phi(x_0, z^*) \mid z^* \in C_{Z^*}^+\}, y_0^*)$$

(cf. Section 3.11.2).

Remark 3.11.6. Taking $M = \mathbb{R}_+^2 \subset Y = \mathbb{R}^2$, $C_Y = C_Z = \mathbb{R}_+^2$, $Y = Z = \mathbb{R}^2$, $f = I$ (identity), $g(x) = -x \ \forall x \in M$ we have $\mathcal{A} = \{x \in \mathbb{R}_+^2\}$, and all assumptions of Theorem 3.11.4 are satisfied. Then $x_0 = (0, 1)^T$, $y_0^* = (1, 0)^T$, $z_0^* = (0, 0)^T$ is a y_0^*-saddle point of the Lagrangian $\Phi \mid \Phi(x, z^*) = y_0^*(f(x)) + z^*(g(x))$, $x \in M$, $z^* \in C_{Z^*}^+$, since $0 + z^*(-x_0) \le 0 \le (x)_1$. The element x_0 is only weakly efficient, as proved in the theorem. So we cannot expect a symmetrical assertion of the kind "saddle-point iff efficiency."

3.11.2 ε-Saddle Point Assertions

In the following we derive ε-saddle point assertions for a special class of convex vector optimization problems.

We consider in this section a general class of vector-valued approximation problems that contains many practically important special cases and apply the concept of approximately efficient elements introduced in Section 3.1.1 to this problem. Approximate solutions of optimization problems are of interest from the computational as well as the theoretical point of view. Especially, the solution set of the approximation problem may be empty in the general noncompact case, whereas approximate solutions exist under very weak assumptions.

Valyi [361, 362] has developed Hurwicz-type saddle point theorems for different types of approximately efficient solutions of convex vector optimization problems (cf. a survey in the paper of Dauer and Stadler [91]). The aim of this section is to derive approximate saddle point assertions for vector-valued location and approximation problems using a generalized Lagrangian.

We introduce a generalized saddle function for the vector-valued approximation problem and use different concepts of approximate saddle points. Furthermore, we derive necessary and sufficient conditions for approximate saddle points, estimate the approximation error, and study the relations between the original problem and saddle point assertions under regularity assumptions.

All topological linear spaces that will occur are over the field $\mathbb{R}$ of real numbers. If X and U are linear spaces, then $\mathcal{L}(X, U)$ denotes the set of all continuous linear mappings from X into U.

Let X be a topological linear space. We recall that a function $f : X \to \mathbb{R}$ is said to be *sublinear* (cf. Section 2.2) if

$$f(x + y) \leq f(x) + f(y)$$

and

$$f(\alpha x) = \alpha f(x)$$

whenever $\alpha \in \mathbb{R}_+$ and $x, y \in X$. If $f : X \to \mathbb{R}$ is a sublinear function, then the set

$$\partial f(0) := \{y^* \in \mathcal{L}(X, \mathbb{R}) \mid y^*(x) \leq f(x) \, \forall \, x \in X\}$$

is called the *subdifferential of f at the origin of X*. It is well known (Hahn–Banach theorem) that for each continuous sublinear function $f : X \to \mathbb{R}$ the following formula holds:

$$f(x) = \max\{y^*(x) \mid y^* \in \partial f(0)\} \qquad \forall \, x \in X. \tag{3.126}$$

Generalizing the concept of a sublinear function, we call a mapping $f = (f_1, \ldots, f_p) : X \to \mathbb{R}^p$ *sublinear* if its components $f_1, \ldots, f_p$ are sublinear functions. The *subdifferential at the origin of X* of a sublinear mapping $f := (f_1, \ldots, f_p) : X \to \mathbb{R}^p$ is defined by

$$\partial f(0) := \partial f_1(0) \times \cdots \times \partial f_p(0).$$

Taking into account formula (3.126), it follows for a continuous sublinear mapping $f : X \to \mathbb{R}^p$ that

$$f(x) \in \Lambda(x) + \mathbb{R}_+^p \qquad \forall\, \Lambda \in \partial f(0) \quad \forall x \in X. \tag{3.127}$$

Let F be a subset of $\mathbb{R}^p$, and let y_0 be a point in $\mathbb{R}^p$. Given a subset B of $\mathbb{R}^p$ and an element $e \in \mathbb{R}^p$, in extension of the notation in Section 3.1.2 the point y_0 is called a (B, e)-*minimal* (resp. (B, e)-*maximal*) *element* of F (Definition 2.1.2) if $y_0 \in F$ and

$$F \cap (y_0 - e - (B \setminus \{0\})) = \emptyset \quad (\text{resp. } F \cap (y_0 + e + (B \setminus \{0\})) = \emptyset).$$

The set consisting of all (B, e)-minimal (resp. (B, e)-maximal) elements of F is denoted by

$$\mathrm{Eff}_{\mathrm{Min}}(F, B, e) \quad (\text{resp. } \mathrm{Eff}_{\mathrm{Max}}(F, B, e)).$$

If e is the origin of $\mathbb{R}^p$, then the (B, e)-minimal (resp. (B, e)-maximal) elements of F are simply called B-*minimal* (resp. B-*maximal*) *elements* of F, and their set is denoted by

$$\mathrm{Eff}_{\mathrm{Min}}(F, B) \quad (\text{resp. } \mathrm{Eff}_{\mathrm{Max}}(F, B)).$$

Given a function $y^* \in \mathcal{L}(\mathbb{R}^p, \mathbb{R})$ and an element $e \in \mathbb{R}^p$, the point y_0 is called a (y^*, e)-*minimal* (resp. (y^*, e)-*maximal*) *point (or element)* of F if $y_0 \in F$ and

$$y^*(y_0) - y^*(e) \leq y^*(y) \qquad \forall\, y \in F$$
$$(\text{resp. } y^*(y) \leq y^*(y_0) + y^*(e) \qquad \forall\, y \in F).$$

The set consisting of all (y^*, e)-minimal (resp. (y^*, e)-maximal) elements of F is denoted by

$$\mathrm{Min}(F, y^*, e) \quad (\text{resp. } \mathrm{Max}(F, y^*, e)).$$

If e is the origin of $\mathbb{R}^p$, then the (y^*, e)-minimal (resp. (y^*, e)-maximal) elements of F are simply called y^*-*minimal* (resp. y^*-*maximal*) *elements* of F and their set is denoted by

$$\mathrm{Min}(F, y^*) \quad (\text{resp. } \mathrm{Max}(F, y^*)).$$

Let M and N be nonempty sets, let X be a topological vector space, and let Φ be a mapping from $M \times N$ to X. Given a subset B of X and an element $e \in X$, a point $(x_0, y_0) \in M \times N$ is said to be a (B, e)-*saddle point* of Φ with *respect to* $M \times N$ if the following conditions are satisfied:

$$\Phi(x_0, y_0) \in \mathrm{Eff}_{\mathrm{Min}}\left(\{\Phi(x, y_0) \mid x \in M\}, B, e\right); \tag{3.128}$$
$$\Phi(x_0, y_0) \in \mathrm{Eff}_{\mathrm{Max}}\left(\{\Phi(x_0, y) \mid y \in N\}, B, e\right). \tag{3.129}$$

Given a function $y^* \in \mathcal{L}(X, \mathbb{R})$ and an element $e \in X$, a point $(x_0, y_0) \in M \times N$ is said to be a (y^*, e)-*saddle point of Φ with respect to $M \times N$* if the following conditions are satisfied:

$$\Phi(x_0, y_0) \in \text{Min} \left(\{\Phi(x, y_0) \mid x \in M\}, y^*, e \right); \tag{3.130}$$
$$\Phi(x_0, y_0) \in \text{Max} \left(\{\Phi(x_0, y) \mid y \in N\}, y^*, e \right). \tag{3.131}$$

In order to formulate this problem we suppose in the whole section that

- X, U and V are reflexive Banach spaces;
- $A : X \to U$, $B : X \to V$, $l : X \to \mathbb{R}^p$ are continuous linear mappings;
- $f : U \to \mathbb{R}^p$ is a continuous sublinear mapping;
- $b \in V \setminus \{0\}$;
- $A \subseteq U$, $\mathcal{X} \subseteq X$, $C_V \subseteq V$, and $C \subset \mathbb{R}^p$ are closed, pointed, and convex cones;
- $C + \mathbb{R}^p_+ \subseteq C$.

Defining $F : \mathcal{A} \times \mathcal{X} \to \mathbb{R}^p$ by

$$F(a, x) := l(x) + f(a - A(x)),$$

and

$$S := \{(a, x) \in U \times X \mid a \in \mathcal{A}, x \in \mathcal{X}, B(x) - b \in C_V\},$$

we consider the following vector optimization problem:

(P(C))   Compute the set $\text{Eff}_{\text{Min}}(F[S], C)$.

Remark 3.11.7. Special cases of the vector optimization problem (P(C)) are:

1. The vector approximation problem and the vector location problem by setting

$$f(a - A(x)) = \begin{pmatrix} \alpha_1 \|a_1 - A_1(x)\|_1 \\ \cdots \\ \alpha_p \|a_p - A_p(x)\|_p \end{pmatrix}$$

   (cf. Jahn [195], Gerth (Tammer), Pöhler [130], Wanka [367]).
2. Linear vector optimization problems if we set $f = 0$.
3. Surrogate problems for linear vector optimization problems with an objective function $l(x)$ subject to $S^0 = \{x \in S \mid A(x) = a\}$, for which the feasible set S^0 is empty but S is nonempty.

Proceeding as in the paper of [337], we define the mapping $L : U \times X \times \mathcal{L}(U, \mathbb{R}^p) \times \mathcal{L}(V, \mathbb{R}^p) \to \mathbb{R}^p$ by

$$L(a, x, Y, Z) := l(x) + Y(a - A(x)) + Z(b - B(x)).$$

When three of the four variables $a \in U$, $x \in X$, $Y \in \mathcal{L}(U, \mathbb{R}^p)$, and $Z \in \mathcal{L}(X, \mathbb{R}^p)$ are fixed, then the corresponding partial mappings

$$L(\cdot, x, Y, Z), \ L(a, \cdot, Y, Z), \ L(a, x, \cdot, Z), \ L(a, x, Y, \cdot)$$

are affine. This property distinguishes our mapping L from the **Lagrangian** mapping usually associated with the problem (P(C)) (see, e.g., [362]).

In what follows we consider L as a function of two variables (a, x) and (Y, Z), and investigate **approximate saddle points** of L with respect to $(\mathcal{A} \times \mathcal{X}) \times (\mathcal{Y} \times \mathcal{Z})$, where

$$\mathcal{Y} := \partial f(0),$$
$$\mathcal{Z} := \{Z \in \mathcal{L}(V, \mathbb{R}^p) \mid Z[C_V] \subseteq C\}.$$

For short, we set $\mathcal{D} := (\mathcal{A} \times \mathcal{X}) \times (\mathcal{Y} \times \mathcal{Z})$.

Theorem 3.11.8. *Let y^* be a functional in $C^+ \backslash \{0\}$, let e be an element in C, and let (a_0, x_0, Y_0, Z_0) be an element in $\mathcal{D}$. Then (a_0, x_0, Y_0, Z_0) is a (y^*, e)-saddle point of L with respect to $\mathcal{D}$ if and only if the following conditions are satisfied:*

(i) $L(a_0, x_0, Y_0, Z_0) \in \text{Min}(\{L(a, x, Y_0, Z_0) \mid (a, x) \in \mathcal{A} \times \mathcal{X}\}, y^*, e)$,
(ii) $B(x_0) - b \in C_V$,
(iii) $y^*(Y_0(a_0 - A(x_0))) + y^*(Z_0(b - B(x_0))) \geq y^*(f(a_0 - A(x_0))) - y^*(e)$.

PROOF. *Necessity.* Condition (i) follows from (3.130) in the definition of a (y^*, e)-saddle point. In order to prove (ii), we suppose that $B(x_0) - b \notin C_V$. According to a strict separation theorem (see, e.g., Theorem 3.18 in [195]), there exists a functional $\mu \in C_{V+}$ such that $\mu(B(x_0) - b) < 0$. Let k be a point chosen from int C. With μ and k we define the mapping $Z : V \to \mathbb{R}^p$ by

$$Z(v) := \frac{\mu(v)}{\mu(b - B(x_0))}(e + k) + Z_0(v). \tag{3.132}$$

Obviously, Z belongs to $\mathcal{Z}$. Taking into account that $y^*(k) > 0$, we also see that

$$y^*((Z - Z_0)(b - B(x_0))) = y^*(e + k) > y^*(e).$$

This result implies that

$$\begin{aligned} y^*(L(a_0, x_0, Y_0, Z)) &= y^*(L(a_0, x_0, Y_0, Z_0)) + y^*((Z - Z_0)(b - B(x_0))) \\ &> y^*(L(a_0, x_0, Y_0, Z_0)) + y^*(e), \end{aligned}$$

which contradicts

$$L(a_0, x_0, Y_0, Z_0) \in \text{Max}(\{L(a_0, x_0, Y, Z); \ (Y, Z) \in \mathcal{Y} \times \mathcal{Z}\}, y^*, e). \tag{3.133}$$

Therefore, condition (ii) must be satisfied. Next we note that (3.133) implies

$$y^*(L(a_0, x_0, Y, Z)) \leq y^*(L(a_0, x_0, Y_0, Z_0)) + y^*(e) \tag{3.134}$$

for every $(Y, Z) \in \mathcal{Y} \times \mathcal{Z}$. By specializing (Y, Z) in (3.134), we obtain (iii). Indeed, from (3.134) it follows that for any mapping $Y \in \mathcal{Y}$ with the property $Y(a_0 - A(x_0)) = f(a_0 - A(x_0))$ and for $Z = 0$ the relation

$$y^*(f(a_0 - A(x_0))) \leq y^*(Y_0(a_0 - A(x_0))) + y^*(Z_0(b - B(x_0))) + y^*(e)$$

holds. This means that (iii) is true.

Sufficiency. (i) is equivalent to (3.130) in the definition of a (y^*, e)-saddle point. We have to prove that (3.131) also holds. Let (Y, Z) be an arbitrary pair in $\mathcal{Y} \times \mathcal{Z}$. Then we have from (3.127),

$$f(a_0 - A(x_0)) \in Y(a_0 - A(x_0)) + \mathbb{R}_+^p \subseteq Y(a_0 - A(x_0)) + C$$

and (ii)

$$Z(B(x_0) - b) \in C.$$

Since $y^* \in C^+$, we conclude that

$$y^*(Y(a_0 - A(x_0))) \leq y^*(f(a_0 - A(x_0)))$$

and

$$y^*(Z(b - B(x_0))) \leq 0.$$

These inequalities imply

$$y^*(Y(a_0 - A(x_0))) + y^*(Z(b - B(x_0))) \leq y^*(f(a_0 - A(x_0))).$$

Taking (iii) into consideration, it follows that

$$\begin{aligned} y^*(Y(a_0 - A(x_0)) &+ y^*(Z(b - B(x_0))) \\ &\leq y^*(Y_0(a_0 - A(x_0))) + y^*(Z_0(b - B(x_0))) + y^*(e). \end{aligned}$$

Consequently, we have

$$y^*(L(a_0, x_0, Y, Z)) \leq y^*(L(a_0, x_0, Y_0, Z_0)) + y^*(e).$$

Since (Y, Z) was arbitrarily chosen, the latter inequality holds for all $(Y, Z) \in \mathcal{Y} \times \mathcal{Z}$. So, (3.131) has been proved. $\qquad \square$

Corollary 3.11.9. *Let y^* be a functional in $C^+ \setminus \{0\}$, let e be an element in C, and let $(a_0, x_0, Y_0, Z_0) \in \mathcal{D}$ be a (y^*, e)-saddle point of L with respect to $\mathcal{D}$. Then the following properties are true:*

(j) $(a_0, x_0) \in S$,
(jj) $y^*(Y_0(a_0 - A(x_0))) \geq y^*(f(a_0 - A(x_0))) - y^*(e)$,
(jjj) $y^*(Z_0(b - B(x_0))) \geq -y^*(e)$.

PROOF. Obviously, (j) results from (ii) in Theorem 3.11.8. In order to prove (jj) and (jjj), we note that (3.134) implies

$$y^*(Y(a_0 - A(x_0))) + y^*(Z(b - B(x_0)))$$
$$\leq y^*(Y_0(a_0 - A(x_0))) + y^*(Z_0(b - B(x_0))) + y^*(e) \ \forall Y \in \mathcal{Y} \ \forall Z \in \mathcal{Z}. \tag{3.135}$$

By setting in (3.135) a $Y \in \mathcal{Y}$ with the property $Y(a_0 - A(x_0)) = f(a_0 - A(x_0))$ and $Z = Z_0$, we obtain (jj), while by setting $Y = Y_0$ and $Z = 0$ in (3.135), we obtain (jjj).   □

Remark 3.11.10. Item (jjj) in Corollary 3.11.9 can be interpreted as a condition of **approximate complementary slackness** for Z_0 and $b - B(x_0)$. Namely, we have

$$-y^*(e) \leq y^*(Z_0(b - B(x_0))) \leq 0.$$

Putting $e = 0$, this relation implies the well-known condition

$$y^*(Z_0(b - B(x_0))) = 0.$$

Theorem 3.11.11. *Let y^* be a functional in $C^+ \setminus \{0\}$, let e be an element in C, and let $(a_0, x_0, Y_0, Z_0) \in \mathcal{D}$ be a (y^*, e)-saddle point of L with respect to $\mathcal{D}$. Then (a_0, x_0) is a $(y^*, \bar{e})$-minimal element of $F[S]$, where*

$$\bar{e} := 2e$$

*is the **approximation error**.*

PROOF. According to property (j) in Corollary 3.11.9, we have $(a_0, x_0) \in S$. Let be (a, x) an arbitrary point in S. From (i) in Theorem 3.11.8, we have

$$y^*(L(a_0, x_0, Y_0, Z_0)) - y^*(e) \leq y^*(L(a, x, Y_0, Z_0)).$$

On the other hand, condition (iii) in Theorem 3.11.8 implies that

$$y^*(F(a_0, x_0)) - y^*(e) \leq y^*(L(a_0, x_0, Y_0, Z_0)).$$

Therefore, the following inequality is true:

$$y^*(F(a_0, x_0)) - y^*(\bar{e}) \leq y^*(L(a, x, Y_0, Z_0)), \tag{3.136}$$

with $\bar{e} := 2e$.

Taking now into account that

$$f(a - A(x)) \in Y_0(a - A(x)) + \mathbb{R}^p_+ \subseteq Y_0(a - A(x)) + C$$

and that

$$Z_0(B(x) - b) \in C,$$

we conclude that

$$y^*(Y_0(a - A(x))) \le y^*(f(a - A(x)))$$

and

$$y^*(Z_0(b - B(x))) \le 0.$$

These inequalities imply

$$y^*(L(a, x, Y_0, Z_0)) \le y^*(F(a, x)).$$

From this and (3.136), we obtain

$$y^*(F(a_0, x_0)) - y^*(\bar{e}) \le y^*(F(a, x)).$$

Since (a, x) was arbitrarily chosen the latter inequality holds for all $(a, x) \in S$. This means that (a_0, x_0) is a $(y^*, \bar{e})$-minimal element of $F[S]$. □

Theorem 3.11.12. *We assume the existence of a feasible point* $(\bar{a}, \bar{x}) \in S$ *with* $B(\bar{x}) - b \in \text{int } C_V$. *Let* y^* *be a functional in* $C^+ \setminus (-C^+)$, *let* e *be an element in* C, *and let* $(a_0, x_0) \in S$ *be a* (y^*, e)-*minimal element of* $F[S]$.

Then there exist operators $Y_0 \in \mathcal{Y}$ *and* $Z_0 \in \mathcal{Z}$, *such that* (a_0, x_0, Y_0, Z_0) *is a* (y^*, e)-*saddle point of* L *with respect to* $\mathcal{D}$.

PROOF. We consider the **scalarized Lagrangian** defined by

$$L^{y^*}(a, x, Y, v^*) := y^*(l(x)) + y^*(Y(a - A(x))) + v^*(b - B(x))$$

over $\mathcal{D}' := (A \times \mathcal{X}) \times (\mathcal{Y} \times C_{V^*}^+)$.

We show that for this Lagrangian the assumptions of Theorem 49.A in [391] are satisfied. Obviously, the assumptions (H1) and (H2) of this theorem are true. To show $(H3^*)$, we consider a sequence $\{(Y^n, v_n^*)\} \subset \mathcal{Y} \times C_{V^*}^+$ with $\|(Y^n, v_n^*)\| \to \infty$ if $n \to \infty$. Since $Y^n \in \partial f(0) \; \forall n$, there is a constant $\alpha > 0$ such that $\|Y^n\| \le \alpha \; \forall n$. Thus we have $\|v_n^*\| \to \infty$ if $n \to \infty$.

From $B(\bar{x}) - b \in \text{int } C_V$ follows the existence of a $\delta > 0$ such that

$$v^*(b - B(\bar{x})) \le -\delta \qquad \forall v^* \in \{v^* \in C_{V^*}^+; \; \|v^*\| = 1\}.$$

So, we have

$$
\begin{aligned}
L^{y^*}(\bar{a}, \bar{x}, Y^n, v_n^*) &= y^*(l(\bar{x})) + y^*(Y^n(\bar{a} - A(\bar{x}))) + v_n^*(b - B(\bar{x})) \\
&\le y^*(l(\bar{x})) + \alpha \|y^*\| \|\bar{a} - A(\bar{x})\| - \delta \|v_n^*\| \\
&\xrightarrow{n \to \infty} -\infty,
\end{aligned}
$$

which proves (v). Applying the mentioned theorem, there exist $Y_0 \in \mathcal{Y}$, $v_0^* \in C_{V^*}^+$ satisfying

$$\inf_{(a,x) \in A \times \mathcal{X}} L^{y^*}(a, x, Y_0, v_0^*) = \sup_{(Y, v^*) \in \mathcal{Y} \times C_{V^*}^+} \inf_{(a,x) \in A \times \mathcal{X}} L^{y^*}(a, x, Y, v^*),$$

and we have

$$\inf_{(a,x)\in\mathcal{A}\times\mathcal{X}} \sup_{(Y,v^*)\in\mathcal{Y}\times C_{V*}^+} L^{y^*}(a,x,Y,v^*)$$

$$= \sup_{(Y,v^*)\in\mathcal{Y}\times C_{V*}^+} \inf_{(a,x)\in\mathcal{A}\times\mathcal{X}} L^{y^*}(a,x,Y,v^*).$$

Regarding that $\sup_{(Y,v^*)\in\mathcal{Y}\times C_{V*}^+} L^{y^*}(a,x,Y,v^*) = y^*(F(a,x))$ whenever $(a,x)\in S$, we get

$$\sup_{(Y,v^*)\in\mathcal{Y}\times C_{V*}^+} L^{y^*}(a_0,x_0,Y,v^*)$$

$$= y^*(F(a_0,x_0))$$

$$\leq \inf_{(a,x)\in\mathcal{A}\times\mathcal{X}} y^*(F(a,x)) + y^*(e)$$

$$= \inf_{(a,x)\in\mathcal{A}\times\mathcal{X}} \sup_{(Y,v^*)\in\mathcal{Y}\times C_{V*}^+} L^{y^*}(a,x,Y,v^*) + y^*(e)$$

$$= \sup_{(Y,v^*)\in\mathcal{Y}\times C_{V*}^+} \inf_{(a,x)\in\mathcal{A}\times\mathcal{X}} L^{y^*}(a,x,Y,v^*) + y^*(e)$$

$$= \inf_{(a,x)\in\mathcal{A}\times\mathcal{X}} L^{y^*}(a,x,Y_0,v_0^*) + y^*(e). \tag{3.137}$$

From (3.137) it follows on the one hand that

$$L^{y^*}(a_0,x_0,Y,v^*) - y^*(e) \leq L^{y^*}(a_0,x_0,Y_0,v_0^*) \qquad \forall (Y,v^*) \in \mathcal{Y}\times C_{V*}^+, \tag{3.138}$$

and on the other hand that

$$L^{y^*}(a_0,x_0,Y_0,v_0^*) \leq L^{y^*}(a,x,Y_0,v_0^*) + y^*(e) \qquad \forall (a,x) \in \mathcal{A}\times\mathcal{X}. \tag{3.139}$$

Finally, we have to show that for $v_0^* \in C_{V*}^+$ and $y^* \in C^+ \setminus (-C^+)$ there is a mapping $Z \in \mathcal{Z}$ such that $v_0^*(v) = y^*(Z(v))$ for all $v \in V$. Let $k \in C$ be a point such that $y^*(k) > 0$. Define the mapping $Z_0 : V \to \mathbb{R}^p$ by

$$Z_0(v) := \frac{v_0^*(v)}{y^*(k)} k.$$

Then we have $Z_0 \in \mathcal{L}(V,\mathbb{R}^p)$. Since $v_0^*(v) \geq 0$ for all $v \in C_V$, we conclude that $Z_0[C_V] \subseteq C$, i.e., $Z \in \mathcal{Z}$. Furthermore, we get for all $v \in V$,

$$y^*(Z_0(v)) = \frac{v_0^*(v)}{y^*(k)} y^*(k) = v_0^*(v).$$

This yields together with (3.138) and (3.139) the desired assertion. $\qquad \square$

Theorem 3.11.13. *Let $\tilde{C}$ be a pointed convex cone in $\mathbb{R}^p$ satisfying $\tilde{C} \supseteq C$, let e be an element in C, and let (a_0,x_0,Y_0,Z_0) be an element in $\mathcal{D}$. Then (a_0,x_0,Y_0,Z_0) is an $(\tilde{C},e)$-saddle point of L with respect to $\mathcal{D}$ if and only if the following conditions are satisfied:*

(i) $L(a_0, x_0, Y_0, Z_0) \in \mathrm{Eff}_{\mathrm{Min}}(\{L(a, x, Y_0, Z_0) \mid (a, x) \in \mathcal{A} \times \mathcal{X}\}, \tilde{C}, e)$,

(ii) $B(x_0) - b \in C_V$,

(iii) $Y_0(a_0 - A(x_0)) + Z_0(b - B(x_0)) \notin f(a_0 - A(x_0)) - e - (\tilde{C} \setminus \{0\})$.

PROOF. *Necessity.* Condition (i) follows from (3.128) in the definition of a $(\tilde{C}, e)$-saddle point. In order to prove (ii) and (iii), we argue similarly as in the proof of necessity of Theorem 3.11.8. First, we suppose that $B(x_0) - b \notin C_V$. Then we apply the strict separation theorem and conclude that there is a functional $\mu \in C_{V*}^+$ such that $\mu(B(x_0) - b) < 0$. Let k be a point chosen from $C \setminus \{0\}$. By means of μ and k we define the mapping $Z : V \to \mathbb{R}^p$ by (3.132). Obviously, Z belongs to $\mathcal{Z}$. Taking in account that $C \subseteq \tilde{C}$, we also see that

$$(Z - Z_0)(b - B(x_0)) = e + k \in e + (C \setminus \{0\}) \subset e + (\tilde{C} \setminus \{0\}).$$

This result implies that

$$\begin{aligned} L(a_0, x_0, Y_0, Z) &= L(a_0, x_0, Y_0, Z_0) + (Z - Z_0)(b - B(x_0)) \\ &\subset L(a_0, x_0, Y_0, Z_0) + e + (\tilde{C} \setminus \{0\}), \end{aligned}$$

which contradicts

$$L(a_0, x_0, Y_0, Z_0) \in \mathrm{Eff}_{\mathrm{Max}}(\{L(a_0, x_0, Y, Z) \mid (Y, Z) \in \mathcal{Y} \times \mathcal{Z}\}, \tilde{C}, e). \quad (3.140)$$

Therefore, condition (ii) must be satisfied. Next we apply (3.140) again and conclude that

$$L(a_0, x_0, Y, Z) \notin L(a_0, x_0, Y_0, Z_0) + e + (\tilde{C} \setminus \{0\}) \quad (3.141)$$

for every $(Y, Z) \in \mathcal{Y} \times \mathcal{Z}$. By specializing (Y, Z) in (3.141), we obtain (iii). Indeed, from (3.141) it follows that for any mapping $Y \in \mathcal{Y}$ with the property $Y(a_0 - A(x_0)) = f(a_0 - A(x_0))$ and for $Z = 0$ the relation

$$f(a_0 - A(x_0)) \notin Y_0(a_0 - A(x_0)) + Z_0(b - B(x_0)) + e + (\tilde{C} \setminus \{0\})$$

holds. This means that (iii) is true.

Sufficiency. (i) is equivalent to (3.128) in the definition of a $(\tilde{C}, e)$-saddle point. We have to prove that (3.129) also holds. To this end we suppose that there is a pair $(Y, Z) \in \mathcal{Y} \times \mathcal{Z}$ such that

$$L(a_0, x_0, Y, Z) \in L(a_0, x_0, Y_0, Z_0) + e + (\tilde{C} \setminus \{0\}).$$

Then we have

$$\begin{aligned} Y(a_0 - A(x_0)) &+ Y(b - B(x_0)) \\ &\in Y_0(a_0 - A(x_0)) + Z_0(b - B(x_0)) + e + (\tilde{C} \setminus \{0\}), \end{aligned}$$

which implies that

$$Y(a_0 - A(x_0)) + Z(b - B(x_0)) + C$$
$$\subseteq Y_0(a_0 - A(x_0)) + Z_0(b - B(x_0)) + e + C + (\tilde{C} \setminus \{0\})$$
$$\subseteq Y_0(a_0 - A(x_0)) + Z_0(b - B(x_0)) + e + (\tilde{C} \setminus \{0\}).$$

But on the other hand, from (3.127),

$$f(a_0 - A(x_0)) \in Y(a_0 - A(x_0)) + \mathbb{R}_+^p \subseteq Y(a_0 - A(x_0)) + C,$$

and (ii)

$$Z(B(x_0) - b) \in C$$

it follows that

$$f(a_0 - A(x_0)) \in Y(a_0 - A(x_0)) + Z(b - B(x_0)) + Z(B(x_0) - b) + C$$
$$\subseteq Y(a_0 - A(x_0)) + Z(b - B(x_0)) + C + C$$
$$\subseteq Y(a_0 - A(x_0)) + Z(b - B(x_0)) + C.$$

Consequently, we have

$$f(a_0 - A(x_0)) \in Y_0(a_0 - A(x_0)) + Z_0(b - B(x_0)) + e + (\tilde{C} \setminus \{0\}),$$

which contradicts (iii). $\qquad\qquad\square$

Corollary 3.11.14. *Let $\tilde{C}$ be a pointed convex cone in $\mathbb{R}^p$ satisfying $\tilde{C} \supseteq C$, let e be an element in C, and let $(a_0, x_0, Y_0, Z_0) \in \mathcal{D}$ be a $(\tilde{C}, e)$-saddle point of L with respect to $\mathcal{D}$. Then the following properties are true:*

(j) $(a_0, x_0) \in S$,
(jj) $Y_0(a_0 - A(x_0)) \notin f(a_0 - A(x_0)) - e - (\tilde{C} \setminus \{0\})$,
(jjj) $Z_0(b - B(x_0)) \notin -e - (\tilde{C} \setminus \{0\})$.

PROOF. Obviously, (j) results from (ii) in Theorem 3.11.13. In order to prove (jj) and (jjj), we note that (3.141) implies

$$Y(a_0 - A(x_0)) + Z(b - B(x_0))$$
$$\leq Y_0(a_0 - A(x_0)) + Z_0(b - B(x_0)) + e + (\tilde{C} \setminus \{0\}) \ \forall Y \in \mathcal{Y} \ \forall Z \in \mathcal{Z}.$$
$$(3.142)$$

By setting in (3.142) a $Y \in \mathcal{Y}$ with the property $Y(a_0 - A(x_0)) = f(a_0 - A(x_0))$ and $Z = Z_0$, we obtain (jj), while by setting $Y = Y_0$ and $Z = 0$ in (3.142), we obtain (jjj). $\qquad\qquad\square$

Remark 3.11.15. Item (jjj) in Corollary 3.11.14 can be interpreted as a condition of **approximate complementary slackness** for Z_0 and $b - B(x_0)$. Namely, we have

$$Z_0(b - B(x_0)) \in -C,$$
$$Z_0(b - B(x_0)) \notin -e - (\tilde{C} \setminus \{0\}).$$

Hence, $Z_0(b - B(x_0))$ is contained in the set $-C \setminus [-e - (\tilde{C})] \cup \{e\}$, which is a bounded set if we claim that $\tilde{C} \supset C$. Putting $e = 0$, this relation implies the well-known condition

$$Z_0(b - B(x_0)) = 0.$$

Theorem 3.11.16. *Let $\tilde{C}$ be a pointed convex cone in $\mathbb{R}^p$ satisfying $\tilde{C} \supseteq C$, let e be an element in C, and let $(a_0, x_0, Y_0, Z_0) \in \mathcal{D}$ be a $(\tilde{C}, e)$-saddle point of L with respect to $\mathcal{D}$. Then (a_0, x_0) is a $(\tilde{C}, \bar{e})$-minimal element of $F[S]$, where*

$$\bar{e} := e + f(a_0 - A(x_0)) - Y_0(a_0 - A(x_0)) - Z_0(b - B(x_0))$$

*is the **approximation error**.*

PROOF. According to property (j) in Corollary 3.11.14, we have $(a_0, x_0) \in S$. We suppose that there is an $(a, x) \in S$ such that

$$F(a, x) \in F(a_0, x_0) - \bar{e} - (\tilde{C} \setminus \{0\}).$$

This means that

$$F(a, x) \in L(a_0, x_0, Y_0, Z_0) - e - (\tilde{C} \setminus \{0\}).$$

Hence we have

$$F(a, x) - C \subseteq L(a_0, x_0, Y_0, Z_0) - e - (\tilde{C} \setminus \{0\}).$$

But, on the other hand, in view of

$$f(a - A(x)) - Y_0(a - A(x)) - Z_0(b - B(x)) \in \mathbb{R}_+^p + C \subseteq C,$$

we have

$$L(a, x, Y_0, Z_0) = F(a, x) - [f(a, x) - Y_0(a - A(x)) - Z_0(b - B(x))]$$
$$\in F(a, x) - C.$$

Consequently,

$$L(a, x, Y_0, Z_0) \in L(a_0, x_0, Y_0, Z_0) - e - (\tilde{C} \setminus \{0\}),$$

which contradicts condition (i) in Theorem 3.11.13.   □

Theorem 3.11.17. *We assume the existence of a feasible point $(\bar{a}, \bar{x}) \in S$ with $B(\bar{x}) - b \in \text{int } C_V$. Let $\tilde{C}$ be a pointed convex cone in $\mathbb{R}^p$ satisfying $\text{int } (\tilde{C}) \cup \{0\} \supseteq \text{cl } C$, let e be an element in C, and let $(a_0, x_0) \in S$ be a $(\tilde{C}, e)$-minimal element of $F[S]$.*

Then there exist operators $Y_0 \in \mathcal{Y}$ and $Z_0 \in \mathcal{Z}$, such that (a_0, x_0, Y_0, Z_0) is a (C, e)-saddle point of L with respect to $\mathcal{D}$.

PROOF. Under the given assumptions there exists an element $y^* \in \text{int } C^+$ (cf. Theorem 5.11 in [195]) such that (a_0, x_0) belongs to $\text{Min}(F[S], y^*, e)$. Theorem 3.11.12 implies the existence of a pair $(Y_0, Z_0) \in \mathcal{Y} \times \mathcal{Z}$, such that (a_0, x_0, Y_0, Z_0) is a (y^*, e)-saddle point of L with respect to $\mathcal{D}$. From the strict C-monotonicity of y^*, we can conclude that (a_0, x_0, Y_0, Z_0) also is a (C, e)-saddle point of L with respect to $\mathcal{D}$.   □

4

Applications

4.1 Approximation Problems

4.1.1 General Approximation Problems

Location and approximation problems play an important role in optimization theory, and many practical problems can be described as location or
approximation problems. Besides problems with one objective function, several authors have even investigated vector-valued (synonymously vector or
multicriteria) location and approximation problems.

In this section we will consider a general vector control approximation
problem and derive necessary conditions for approximate solutions of this
problem.

In the whole section we assume

(A1) $(X, \| \cdot \|_X)$, $(Y, \| \cdot \|_Y)$ and $(Z, \| \cdot \|_Z)$ are real reflexive Banach spaces;
(A2) $C \subset Y$ is a pointed closed cone with $k^0 \in C \setminus (-C)$.

Moreover, we assume that C is a cone with $\operatorname{int} C \neq \emptyset$ having the Daniell
property, which means that every decreasing net (i.e., $i \leq j$ implies $x_j \leq x_i$)
having a lower bound converges to its infimum (see Section 2.1). Further, we
suppose that C has a weakly compact base (cf. Lemma 2.2.17).

In order to formulate our vector control approximation problem, we will
introduce a **vector-valued norm**:
$\| \cdot \| : Z \longrightarrow C$ is called a vector-valued norm if $\forall z, z_1, z_2 \in Z$, $\forall \lambda \in \mathbb{R}$,

1. $\|z\| = 0 \Longleftrightarrow z = 0$;
2. $\|\lambda z\| = |\lambda| \, \|z\|$;
3. $\|z_1 + z_2\| \in \|z_1\| + \|z_2\| - C$.

In the following we assume that $\| \cdot \|$ is continuous. The set of linear continuous
mappings from X to Y is denoted by $L(X, Y)$. Suppose that $f_1 \in L(X, Y)$,
$A_i \in L(X, Z)$, and $\alpha_i \geq 0$ $(i = 1, \ldots, n)$. Here A_i^* denotes the adjoint operator

to A_i. For brevity and clarity we sometimes emit parentheses in connection with y^* and A_i^*.

Then we consider for $x \in X$ and $a^i \in Z$ $(i = 1, \ldots, n)$ the vector-valued function

$$f(x) := f_1(x) + \sum_{i=1}^{n} \alpha_i \|A_i(x) - a^i\|.$$

Now we will introduce the following **vector control approximation problem**:

(P1):   Compute the set $\mathrm{Eff}(f[X], C)$.

Remark 4.1.1. The problem (P1) contains the following practically important special cases:

1. Vector-valued optimal control problems of the form (cf. Section 4.5)

$$\mathrm{Eff}(F_1[U], \mathbb{R}_+^2),$$

   with

$$F_1(u) := \begin{pmatrix} \|A(u) - a\|_1 \\ \|u\|_2 \end{pmatrix}, \qquad u \in U \subset X,$$

   where H_1 and H_2 are Hilbert spaces, $A \in L(H_1, H_2)$, $a \in H_2$, $U \subset H_1$ is a nonempty closed convex set, and $\mathbb{R}_+^2$ denotes the usual ordering cone in $\mathbb{R}^2$. Here u denotes the so-called control variable; the image $z = Au$ denotes the state variable.
2. Scalar location and approximation problems $(Y = R, f_1 \equiv 0)$:

$$\sum_{i=1}^{n} \alpha_i \|A_i(x) - a^i\| \longrightarrow \inf_{x \in X},$$

   where $\|\cdot\|$ is a norm in Z (cf. Section 3.7 and 4.2).
3. Vector approximation and location problems $(f_1 \equiv 0, n = 1$, cf. Jahn [195], Gerth and Pöhler [130], Henkel and Tammer [163, 164], Jahn [198], Tammer [342, 344], Wanka [366], Oettli [277]).
4. Linear vector optimization problems $(\alpha_i = 0$ for all $i = 1, \ldots, n)$.
5. Surrogate problems for linear vector optimization problems with an objective function $f(x) := f_1(x)$ subject to $x \in X$ and $A(x) = a$, for which the feasible set is empty.
6. Perturbed linear vector optimization problems.
7. Tychonoff regularization for linear vector optimization problems.

In the following theorem we will derive **necessary conditions** for ap**proximately efficient solutions** of $\mathrm{Eff}(f[X], C_{\varepsilon k^0})$ (cf. Section 3.1.1, Definition 3.1.1) using the concept of C-convexity introduced in Section 2.4 and the subdifferential of a C-convex function $f : X \to Y$ (see Definition 2.4.7). Taking into account the notation of this section we consider the **subdifferential** of the C-convex function $f : X \longrightarrow Y$ at $x_0 \in X$:

$$\partial^{\leq} f(x_0) := \{ \ M \in L(X,Y) \mid M\,(x - x_0) \in f(x) - f(x_0) - C \ \forall\, x \in X \ \}.$$

The **subdifferential of the vector-valued norm** $\|.\| : Z \longrightarrow Y$ has the following form (cf. Jahn [195]):

$$\partial^{\leq} \|\cdot\|(z_0) = \{M \in L(Z,Y) \mid M(z_0) = \|z_0\|, \ \|z\| - M(z) \in C \ \forall\, z \in Z\}. \quad (4.1)$$

The following result will be used in the proof of Theorem 4.1.3.

Lemma 4.1.2. (Jahn [195]) *Let S be a nonempty subset of a partially ordered reflexive Banach space Y with a pointed nontrivial ordering cone C. If the set $S+C$ is convex and has a nonempty topological interior, then for each efficient element $\bar{y} \in S$ of the set S there exists a linear functional $y^* \in C^+ \setminus \{0\}$ with the property*

$$y^*(\bar{y}) \leq y^*(y) \qquad for\ all \qquad y \in S.$$

Now we formulate the main result.

Theorem 4.1.3. *Under the assumptions of this section, for any $\varepsilon > 0$ and any approximately efficient element $f(x^0) \in \mathrm{Eff}(f[X], C_{\varepsilon k^0})$ there exist an element x_ε, a functional $y^* \in C^+ \setminus \{0\}$, and linear continuous mappings $M_{i\varepsilon} \in L(Z,Y)$ with*

$$M_{i\varepsilon}(A_i(x_\varepsilon) - a^i) = \|A_i(x_\varepsilon) - a^i\|,$$

$$\|z\| - M_{i\varepsilon}(z) \in C \qquad \forall\, z \in Z \quad (i = 1, \dots, n),$$

such that

$(\alpha) \quad f(x_\varepsilon) \in f(x^0) - \sqrt{\varepsilon}\|x^0 - x_\varepsilon\|_X k^0 - C, \ \ f(x_\varepsilon) \in \mathrm{Eff}(f[X], C_{\varepsilon k^0}),$

$(\beta) \quad \|x_\varepsilon - x^0\|_X \leq \sqrt{\varepsilon},$

$(\gamma) \quad \|y^* f_1 + \sum_{i=1}^{n} \alpha_i A_i^* y^* M_{i\varepsilon}\|_* \leq \sqrt{\varepsilon} y^*(k^0).$

PROOF. We assume $f(x^0) \in \mathrm{Eff}(f[X], C_{\varepsilon k^0})$. Under the given assumptions Corollary 3.10.14 implies the existence of an element x_ε with

$(\alpha') \quad f(x_\varepsilon) \in f(x^0) - \sqrt{\varepsilon}\|x^0 - x_\varepsilon\|_X k^0 - C,$

$(\beta') \quad \|x^0 - x_\varepsilon\|_X \leq \sqrt{\varepsilon},$

$(\gamma') \quad f_{\varepsilon k^0}(x_\varepsilon) \in \mathrm{Eff}(f_{\varepsilon k^0}[X], C),$

where $f_{\varepsilon k^0}(x) := f(x) + \sqrt{\varepsilon}\|x - x_\varepsilon\|_X k^0$. Because of the C-convexity of f and by assertion (γ') it is possible to conclude from Lemma 4.1.2 that there exists a functional $y^* \in C^+ \setminus \{0\}$ with

$$y^*(f(x_\varepsilon)) \leq \inf_{x \in X} \left(y^*(f(x)) + \sqrt{\varepsilon}\|x - x_\varepsilon\|_X y^*(k^0) \right).$$

This means that x_ε minimizes the function

$$x \longrightarrow y^*(f(x)) + \sqrt{\varepsilon}\,\|x - x_\varepsilon\|_X y^*(k^0).$$

Then the subdifferential calculus of convex functionals (cf. Aubin and Ekeland [12], Corollary 4.3.6) implies that

$$0 \in \partial f(x_\varepsilon) + \sqrt{\varepsilon} y^*(k^0) B^0,$$

where B^0 denotes the (closed) unit ball in Y. So it follows immediately that there is a linear continuous functional $l_\varepsilon : X \longrightarrow \mathbb{R}$ belonging to the subdifferential of the scalarized function $y^* \circ f$ at the point $x_\varepsilon \in X$ with

$$\|l_\varepsilon\|_* \leq \sqrt{\varepsilon} \, y^*(k^0). \tag{4.2}$$

Further, we have

$$\partial(y^* \circ f)(x_\varepsilon) = \partial\left(y^*(f_1(\cdot)) + \sum_{i=1}^{n} \alpha_i y^* \|A_i(\cdot) - a^i\| \right)(x_\varepsilon),$$

because y^* is a linear functional.

The rule of sums for subdifferentials yields the relation

$$\partial\left(y^*(f_1(\cdot)) + \sum_{i=1}^{n} \alpha_i y^* \|A_i(\cdot) - a^i\| \right)(x_\varepsilon)$$

$$= \partial y^*(f_1(\cdot))(x_\varepsilon) + \sum_{i=1}^{n} \alpha_i \partial y^* \|A_i(\cdot) - a^i\|(x_\varepsilon).$$

Moreover, from Lemma 2.4.8 and Corollary 4.3.6 in Aubin and Ekeland [12] we get the following equation:

$$\partial(y^* \circ f)(x) = \partial y^*(f_1(\cdot))(x) + \sum_{i=1}^{n} \alpha_i \partial y^* \|A_i(\cdot) - a^i\|(x)$$

$$= y^* f_1 + \sum_{i=1}^{n} \alpha_i A_i^* \partial(y^* \| \cdot \|)(A_i(x) - a^i)$$

$$= y^* f_1 + \sum_{i=1}^{n} \alpha_i A_i^* y^* \partial^{\leq} \| \cdot \|(A_i(x) - a^i) = y^* \circ \partial^{\leq} f(x), \tag{4.3}$$

where $f_1 \in \partial^{\leq} f_1(x)$.

Applying (4.1) statement (4.3) implies

$$\partial(y^* \circ f)(x) = y^* f_1 + \sum_{i=1}^{n} \alpha_i A_i^* y^* \partial(\| \cdot \|)(A_i(x) - a^i) = y^* f_1 + \sum_{i=1}^{n} \alpha_i A_i^* y^* M_{i\varepsilon}$$

$$\tag{4.4}$$

with

$$M_{i\varepsilon} \in L(Z, Y) , \quad M_{i\varepsilon}(A_i(x_\varepsilon) - a^i) = \|A_i(x_\varepsilon) - a^i\|$$

and

$$\|z\| - M_{i\varepsilon}(z) \in C \quad \forall z \in Z \quad (i = 1, \ldots, n).$$

From (4.2), (4.3), and (4.4) we get the desired inequality

$$\left\| y^* f_1 + \sum_{i=1}^{n} \alpha_i A_i^* y^* M_{i\varepsilon} \right\|_* \le \sqrt{\varepsilon}\, y^*(k^0). \qquad \square$$

Remark 4.1.4.   1. Obviously, if we use a scalarization of the approximation problem (P1) with linear continuous functionals $y^* \in C^+ \setminus \{0\}$ and Ekeland's original result [105], then we can show in the same way as in Theorem 4.1.3 that for any $\varepsilon > 0$ and any approximate solution $x^0 \in X$ with $y^*(f(x^0)) \le \inf_{x \in X} y^*(f(x)) + \varepsilon$, $y^* \in C^+ \setminus \{0\}$, there exists an element x_ε and linear continuous mappings $M_{i\varepsilon} \in L(Z, Y)$ with

$$M_{i\varepsilon}(A_i(x_\varepsilon) - a^i) = \|A_i(x_\varepsilon) - a^i\|,$$

$$\|z\| - M_{i\varepsilon}(z) \in C \qquad \forall z \in Z \quad (i = 1, \ldots, n),$$

such that
(α″) $y^*(f(x_\varepsilon)) \le y^*(f(x^0)) - \sqrt{\varepsilon}\|x^0 - x_\varepsilon\|_X,$
(β″) $\|x_\varepsilon - x^0\|_X \le \sqrt{\varepsilon},$
(γ″) $\|y^* f_1 + \sum_{i=1}^{n} \alpha_i A_i^* y^* M_{i\varepsilon}\|_* \le \sqrt{\varepsilon}.$
But the assertion (α) in Theorem 4.1.3 is a sharper result than (α″).
2. The assertion (γ) of Theorem 4.1.3 is a sharper result than the assertion in Theorem 2 of [163], because the mappings $M_{i\varepsilon}$ ($i = 1, \ldots, n$) in Theorem 4.1.3 do not depend on a direction $v \in X$.

As a direct consequence of Theorem 4.1.3 we get an extension of Theorem 5.4.3 of Aubin/Ekeland [12] for real-valued functions to vector-valued functions. Now we want to apply this result to the vector-valued approximation problem (P1).
Here we will use the following set:

$$\text{dom } f := \{x \in X \mid \exists\, y^* \in C^{\#} \quad \text{with} \quad y^*(f(x)) < \infty\}.$$

Corollary 4.1.5. *Under the assumptions of this section the set of points where f is subdifferentiable is dense in* dom f; *i.e., for each $\bar{x} \in$ dom f there exists a sequence $\{x_k\}, k \in \mathbb{N}$, with*

(α) $x_k \to \bar{x},$
(β) $y^*(f(x_k)) \to y^*(f(\bar{x}))$ *for an element* $y^* \in C^{\#},$
(γ) $\forall i = 1, \ldots n, \ \exists\, M_{ik} \in L(Z, Y)$ *with*

$$M_{ik}(A_i(x_k) - a^i) = \|A_i(x_k) - a^i\|, \ \|z\| - M_{ik}(z) \in C \ \forall z \in Z \quad and$$

$$f_1 + \sum_{i=1}^{n} \alpha_i A_i^* y^* M_{ik} \in \partial^{\le}(f(x_k)) \ne \emptyset \qquad \forall\, k.$$

Remark 4.1.6. The assumption in Theorem 4.1.3 and Corollary 4.1.5 that C is a convex cone with a weakly compact base may be replaced by the assumption that X is a Gâteaux differentiability space (compare [195], [35]).

In the following we will study some practically important **special cases** of the general approximation problem (P1).

Let us now assume, that the space Y is the space of real numbers $\mathbb{R}$. Suppose that $f_1 \in L(X, \mathbb{R})$, $A_i \in L(X, Z)$ and $\alpha_i \geq 0$ $(i = 1, \ldots, n)$. Then we consider for $x \in X$ and $a^i \in Z$ $(i = 1, \ldots, n)$ the following **real-valued approximation problem**:

(P2) $\tilde{f}_2(x) := f_1(x) + \sum_{i=1}^{n} \alpha_i \|A_i(x) - a^i\|_Z \longrightarrow \inf_{x \in X}$. If we put $f_1 = 0$ and $A_i = I$ $\forall i = 0, \ldots, n$, we get the special case of the **real-valued location problem**, i.e.,

(P3): $\hat{f}_3(x) = \sum_{i=1}^{n} \alpha_i \|x - a^i\| \longrightarrow \inf_{x \in X}$. In the following corollaries we will see that for the special approximation problems (P2) and (P3) the assertions of Theorem 4.1.3 get an easier form.

Corollary 4.1.7. *We consider the real-valued problem* (P2), *which is a special case of* (P1) *if we put* $k^0 = 1$ *and* $C = \{x \in \mathbb{R} \mid x \geq 0\}$.
Then for any $\varepsilon > 0$ *and any approximate solution* x_0 *with* $\tilde{f}_2(x^0) \leq \inf_{x \in X} \tilde{f}_2(x) + \varepsilon$ *there exist an element* x_ε *and linear continuous functionals* $l_{i\varepsilon} \in Z^*$ *with*

$$l_{i\varepsilon}(A_i(x_\varepsilon) - a^i) = \|A_i(x_\varepsilon) - a^i\|,$$

and

$$\|l_{i\varepsilon}\|_* = 1$$

such that
(α) $\tilde{f}_2(x_\varepsilon) \leq \tilde{f}_2(x_0) - \sqrt{\varepsilon}\|x^0 - x_\varepsilon\|_X$,
(β) $\|x_\varepsilon - x^0\|_X \leq \sqrt{\varepsilon}$,
(γ) $\|f_1 + \sum_{i=1}^{n} \alpha_i A_i^* l_{i\varepsilon}\|_* \leq \sqrt{\varepsilon}$.

Corollary 4.1.8. *We consider the real-valued location problem* (P3), *which is a special case of* (P1) *if we additionally put* $k^0 = 1$ *and* $C = \{x \in \mathbb{R} \mid x \geq 0\}$.
Then for any $\varepsilon > 0$, $\lambda > 0$ *and any approximate solution* x_0 *with* $\hat{f}_3(x^0) \leq \inf_{x \in X} \hat{f}_3(x) + \varepsilon$ *there exist an element* x_ε *and linear continuous functionals* $l_{i\varepsilon} \in Z^*$ *with*

$$l_{i\varepsilon}(x_\varepsilon - a^i) = \|x_\varepsilon - a^i\|,$$

and

$$\|l_{i\varepsilon}\|_* = 1$$

such that
(α) $\hat{f}_3(x_\varepsilon) \leq \hat{f}_3(x_0) - \sqrt{\varepsilon}\|x^0 - x_\varepsilon\|_X$,
(β) $\|x_\varepsilon - x^0\|_X \leq \sqrt{\varepsilon}$,
(γ) $\|\sum_{i=1}^{n} \alpha_i l_{i\varepsilon}\|_* \leq \sqrt{\varepsilon}$.

Corollary 4.1.9. *Consider the scalar optimization problem* (P2), *which is a special case of the problem* (P1) *if we additionally put* $k^0 = 1$ *and* $C = \{x \in \mathbb{R} \mid x \geq 0\}$. *Then the assertion* (γ) *of Corollary 4.1.5 takes the following form:*

(γ') $\exists\, l_{in} \in Z^*$ *with* $l_{in}(A_i(x_n) - a^i) = \|A_i(x_n) - a^i\|$, $\|l_{in}\|_* = 1$, $(i = 1, \ldots, n)$

with $\quad f_1 + \sum_{i=1}^{n} \alpha_i A_i^* l_{in} \in \partial^{\leq} \tilde{f}_2(x_n) \neq \emptyset.$

Corollary 4.1.10. *Consider the scalar location problem* (P3), *which is a special case of the problem* (P1) *if we additionally put* $k^0 = 1$ *and* $C = \{x \in \mathbb{R} \mid x \geq 0\}$. *Then the assertion* (γ) *of Corollary 4.1.5 takes the following form:*

(γ') $\quad \exists\, l_{in} \in Z^*$ *with* $l_{in}(x_n - a^i) = \|x_n - a^i\|$, $\|l_{in}\|_* = 1$, $(i = 1, \ldots, n)$

with
$$\sum_{i=1}^{n} \alpha_i l_{in} \in \partial^{\leq} \hat{f}_3(x_n) \neq \emptyset.$$

4.1.2 Finite-dimensional Approximation Problems

We consider a class of vector-valued control approximation problems in which each objective function is a sum of two terms, a linear function and a power of a norm of a linear vector function. Necessary conditions for approximate solutions of these problems were derived using a vector-valued variational principle, the subdifferential calculus, and the directional derivative, respectively.

In Section 4.1.1 we have derived necessary conditions for approximately efficient elements of a class of abstract approximation problems in which the objective function takes its values in a reflexive Banach space. However, these conditions are not easy to utilize, so that for more special problems it would be worthwhile to find necessary conditions that are easier to handle. The aim of this section is to derive necessary conditions for approximately efficient solutions for another class of approximation problems having a finite-dimensional image space of the objective function, which is more useful for control problems in comparison with the approximation problem in Section 4.1.1. In our proofs it is possible to use the special structure of the subdifferential of a power of the norm and a vector-valued variational principle.

We will introduce a general vector-valued control approximation problem (P4). Using the variational principle and the subdifferential calculus we derive necessary conditions for approximately efficient elements of the vector-valued control approximation problem (P4). Moreover, we show necessary conditions for approximately efficient elements of the vector-valued control approximation problem (P4) using the directional derivative of the objective function. For special cases of (P4), for instance, for the case of a real-valued control approximation problem, ε-variational inequalities will be presented.

We suppose that

(A) $(X, \|\cdot\|_X)$ and $(Y_i, \|\cdot\|_i)$ $(i = 1, \ldots, n)$ are real Banach spaces, $x \in X$, $a^i \in Y_i$, $\alpha_i \geq 0$, $\beta_i \geq 1$, $A_i \in L(X, Y_i)$, $(i = 1, \ldots, n)$, $f_1 \in L(X, \mathbb{R}^n)$,

(B) $D_j \subseteq X$ $(j = 1, \ldots, m)$ are closed and convex sets, and $D = \bigcap_{j=1}^m D_j$ is nonempty,

(C) $C \subset \mathbb{R}^n$ is a pointed closed convex cone with nonempty interior and $C + \mathbb{R}^n_+ \subseteq C$.

Moreover, $\|\cdot\|_*$ denotes the dual norm to $\|\cdot\|_X$, and $\|\cdot\|_{i*}$ the dual norm to $\|\cdot\|_i$. Let us recall that the dual norm $\|\cdot\|_*$ to $\|\cdot\|_X$ is defined by

$$\|p\|_* := \sup_{\|x\|_X = 1} |p(x)|.$$

Now we consider the following **vector control approximation problem**

(P4)   Compute the set $\mathrm{Eff}(f[D], C)$,

where

$$f(x) := f_1(x) + \begin{pmatrix} \alpha_1 \|A_1(x) - a^1\|_1^{\beta_1} \\ \cdots \\ \alpha_n \|A_n(x) - a^n\|_n^{\beta_n} \end{pmatrix}$$

is the objective vector function. Additionally, we assume that

(D) $f : X \longrightarrow \mathbb{R}^n$ is bounded from below on D, i.e., there exists some $z \in \mathbb{R}^n$ with $f[D] \subset z + C$.

It is well known that the set $\mathrm{Eff}(f[D], C)$ may be empty in the general non-compact case (compare Section 3.2).

We will derive necessary conditions for approximately efficient elements using a vector-valued variational principle and the subdifferential calculus. In order to apply the subdifferential calculus we have to show that the objective function in (P4) is C-convex (compare Section 2.4).

In order to prove our main result we need the following assertion of Aubin and Ekeland [12] concerning the subdifferential of norm terms:

Lemma 4.1.11. *If X is a Banach space then we have*

$$\partial \|x\| = \begin{cases} \{p \in L(X, \mathbb{R}) \mid p(x) = \|x\|, \ \|p\|_* = 1\} & \text{if } x \neq 0, \\ \{p \in L(X, \mathbb{R}) \mid \|p\|_* \leq 1\} & \text{if } x = 0, \end{cases}$$

and for $\beta > 1$,

$$\partial \left(\frac{1}{\beta} \|\cdot\|^\beta \right)(x) = \{p \in L(X, \mathbb{R}) \mid \|p\|_* = \|x\|^{\beta - 1}, \ p(x) = \|x\|^\beta\}.$$

In the next theorem we will derive necessary conditions for approximately efficient solutions of $\mathrm{Eff}(f[X], C_{\varepsilon k^0})$ using the subdifferential calculus. First, we state necessary conditions for approximately efficient elements of the vector-valued approximation problem (P4) with $D = X$.

Theorem 4.1.12. *Under the assumptions (A), (C), and (D) for any $\varepsilon > 0$ and any $x^0 \in X$ with $f(x^0) \in \mathrm{Eff}(f[X], C_{\varepsilon k^0})$ there exist an element $x_\varepsilon \in X$, a functional $y^* \in C^+ \setminus \{0\}$, and linear continuous mappings $M_{i\varepsilon} \in L(Y_i, \mathbb{R})$ with*

$$M_{i\varepsilon}(A_i(x_\varepsilon) - a^i) = \|A_i(x_\varepsilon) - a^i\|_i^{\beta_i}$$

and

$$\|M_{i\varepsilon}\|_{i*} \begin{cases} \leq 1 & \text{if } \beta_i = 1 \text{ and } A_i(x_\varepsilon) = a^i, \\ = \|A_i(x_\varepsilon) - a^i\|_i^{\beta_i - 1} & \text{otherwise,} \end{cases}$$

for all $i = 1, \ldots, n$, such that

(α) $\quad f(x_\varepsilon) \in f(x^0) - \sqrt{\varepsilon}\|x^0 - x_\varepsilon\|_X k^0 - C$,

(β) $\quad \|x_\varepsilon - x^0\|_X \leq \sqrt{\varepsilon}$,

(γ) $\quad \|y^* f_1 + \sum_{i=1}^n \alpha_i \beta_i A_i^* y_i^* M_{i\varepsilon}\|_* \leq \sqrt{\varepsilon} y^*(k^0)$.

PROOF. We assume $f(x^0) \in \mathrm{Eff}(f[X], C_{\varepsilon k^0})$. Under the given assumptions Corollary 3.10.14 implies the existence of an element $x_\varepsilon \in X$ with

(α') $\quad f(x_\varepsilon) \in f(x^0) - \sqrt{\varepsilon}\|x^0 - x_\varepsilon\|_X k^0 - C$,

(β') $\quad \|x^0 - x_\varepsilon\|_X \leq \sqrt{\varepsilon}$,

(γ') $\quad f_{\varepsilon k^0}(x_\varepsilon) \in \mathrm{Eff}(f_{\varepsilon k^0}[X], C)$,

where $f_{\varepsilon k^0}(x) := f(x) + \sqrt{\varepsilon}\|x - x_\varepsilon\|_X k^0$. Because of the C-convexity of f and by assertion (γ') we get that there exists a functional $y^* \in C^+ \setminus \{0\}$ with

$$y^*(f(x_\varepsilon)) \leq \inf_{x \in X} \left(y^*(f(x)) + \sqrt{\varepsilon}\|x - x_\varepsilon\|_X y^*(k^0)\right). \tag{4.5}$$

This means that x_ε minimizes the function

$$x \longrightarrow y^*(f(x)) + \sqrt{\varepsilon}\|x - x_\varepsilon\|_X y^*(k^0),$$

and then from the subdifferential calculus of convex functionals (cf. Aubin and Ekeland [12], Corollary 4.3.6) we get

$$0 \in \partial(y^* \circ f)(x_\varepsilon) + \sqrt{\varepsilon} y^*(k^0) B,$$

where B is the closed unit ball in X. It follows immediately that there is a linear continuous functional $l_\varepsilon : X \longrightarrow \mathbb{R}$ belonging to the subdifferential of the scalarized function $y^* \circ f$ at the point $x_\varepsilon \in X$ with

$$\|l_\varepsilon\|_* \leq \sqrt{\varepsilon} \, y^*(k^0).$$

Furthermore, we have

$$\partial(y^* \circ f)(x_\varepsilon) = \partial\left(y^*(f_1(\cdot)) + \sum_{i=1}^n \alpha_i y_i^* \|A_i(\cdot) - a^i\|_i^{\beta_i}\right)(x_\varepsilon),$$

because y^* is a linear functional.

Lemma 4.1.11 and the rule of sums for subdifferentials yield the relation

$$\partial\left(y^*(f_1(\cdot)) + \sum_{i=1}^{n} \alpha_i y_i^* \|A_i(\cdot) - a^i\|_i^{\beta_i}\right)(x_\varepsilon)$$

$$= y^* \partial^{\leq}(f_1(\cdot))(x_\varepsilon) + \sum_{i=1}^{n} \alpha_i y_i^* \partial \|A_i(\cdot) - a^i\|_i^{\beta_i}(x_\varepsilon). \qquad (4.6)$$

Applying Lemma 4.1.11, relation (4.6), implies

$$\partial(y^* \circ f)(x_\varepsilon) = y^* f_1 + \sum_{i=1}^{n} \alpha_i A_i^* y_i^* \partial(\|u\|_i^{\beta_i})\mid_{u=A_i(x_\varepsilon)-a^i}$$

$$= \left\{ y^* f_1 + \sum_{i=1}^{n} \alpha_i \beta_i A_i^* y_i^* M_{i\varepsilon} \mid M_{i\varepsilon} \in L(Y_i, \mathbb{R}),\ M_{i\varepsilon}(A_i(x_\varepsilon) - a^i) = \right.$$

$$\|A_i(x_\varepsilon) - a^i\|_i^{\beta_i},\ \|M_{i\varepsilon}\|_{i*} \leq 1 \text{ if } \beta_i = 1 \text{ and } A_i(x_\varepsilon) = a^i,$$

$$\left. \|M_{i\varepsilon}\|_{i*} = \|A_i(x_\varepsilon) - a^i\|_i^{\beta_i-1}\text{otherwise } (\forall\, i,\ 1 \leq i \leq n)\right\}.$$

Then we get the desired inequality

$$\left\| y^* f_1 + \sum_{i=1}^{n} \alpha_i \beta_i A_i^* y_i^* M_{i\varepsilon} \right\|_* \leq \sqrt{\varepsilon}\, y^*(k^0). \qquad \square$$

In the next theorem we derive necessary conditions for approximately efficient elements of the **vector approximation problem (P4) with restrictions**. In order to formulate necessary conditions we need the indicator functions χ_{D_j} of D_j defined by $\chi_{D_j}(x) = 0$ if $x \in D_j$, and $\chi_{D_j}(x) = +\infty$ otherwise. It is well known that the **subdifferential of the indicator function** χ_{D_j} at $x^0 \in X$ is the normal cone $N_{D_j}(x^0)$ to D_j at $x^0 \in X$, defined by

$$N_{D_j}(x^0) = \begin{cases} \{p \in L(X, \mathbb{R}):\ p(x' - x^0) \leq 0 \quad \forall\, x' \in D_j\} & \text{if } x^0 \in D_j, \\ \emptyset & \text{otherwise.} \end{cases}$$

Theorem 4.1.13. *Under the assumptions* (A), (B), (C), *and* (D) *for any* $\varepsilon > 0$ *and any* $x^0 \in D$ *with* $f(x^0) \in \mathrm{Eff}(f[D], C_{\varepsilon k^0})$ *there exist an element* $x_\varepsilon \in D$, *a functional* $y^* \in C^+ \setminus \{0\}$, *linear continuous mappings* $M_{i\varepsilon} \in L(Y_i, \mathbb{R})$ *with*

$$M_{i\varepsilon}(A_i(x_\varepsilon) - a^i) = \|A_i(x_\varepsilon) - a^i\|_i^{\beta_i},$$

$$\|M_{i\varepsilon}\|_{i*} \begin{cases} \leq 1 & \text{if } \beta_i = 1 \text{ and } A_i(x_\varepsilon) = a^i, \\ = \|A_i(x_\varepsilon) - a^i\|_i^{\beta_i-1} & \text{otherwise,} \end{cases}$$

for all $i = 1, \dots, n$, *and elements* $r_j \in N_{D_j}(x_\varepsilon)$ *for all* $j = 1, \dots, m$, *such that*
(α) $f(x_\varepsilon) \in f(x^0) - \sqrt{\varepsilon}\|x^0 - x_\varepsilon\|_X k^0 - C,$
(β) $\|x_\varepsilon - x^0\|_X \leq \sqrt{\varepsilon},$
(γ) $\left\| y^* f_1 + \sum_{i=1}^{n} \alpha_i \beta_i A_i^* y_i^* M_{i\varepsilon} + \sum_{j=1}^{m} r_j \right\|_* \leq \sqrt{\varepsilon} y^*(k^0).$

PROOF. We can follow the line of the proof of Theorem 4.1.12 taking into consideration that our problem has the feasible set D instead of X.

Then we get instead of inequality (4.5) the inequality

$$y^*(f(x_\varepsilon)) \le \inf_{x \in D} \left(y^*(f(x)) + \sqrt{\varepsilon}\|x - x_\varepsilon\|_X y^*(k^0) \right).$$

This problem is equivalent to the following unconstrained minimization problem:

$$F(x) := y^*(f(x)) + \sqrt{\varepsilon}\|x - x_\varepsilon\|_X y^*(k^0) + \sum_{j=1}^{m} \chi_{D_j}(x) \longrightarrow \min_{x \in X}.$$

Taking into account the fact that the subdifferential of the indicator function χ_{D_j} at x_ε is the normal cone to D_j at x_ε, we can conclude the statement of the theorem in the same way as in the proof of Theorem 4.1.12.   □

In the following we will study special cases of (P4).

Under the assumptions (A), (B), (C), and (D) we consider a special case of the vector-valued approximation problem (P4): $\alpha_i = 1$, $\beta_i = 1$ for all $i = 1, \ldots, n$, and $K = \mathbb{R}^n_+$, namely,

(P5)   Compute the set $\mathrm{Eff}(f[D], C)$

with

$$f(x) := f_1(x) + \begin{pmatrix} \|A_1(x) - a^1\|_1 \\ \|A_2(x) - a^2\|_2 \\ \ldots \\ \|A_n(x) - a^n\|_n \end{pmatrix}.$$

In order to derive necessary conditions for approximately efficient elements of (P5) we will use besides the subdifferential, the directional derivative.

Definition 4.1.14. *The **directional derivative** of the function $f : X \longrightarrow \mathbb{R}^n$ at $x \in D$ in the direction $v \in X$ is defined by*

$$f'(x)(v) := \lim_{t \to +0} \frac{f(x + tv) - f(x)}{t}.$$

Corollary 3.10.14 implies necessary conditions for approximately efficient elements. Such necessary conditions can be used in order to derive numerical algorithms. The following lemma is a direct consequence of Corollary 3.10.14 if we use the directional derivative of the norms $\|A_i(x) - a^i\|_i$, $i = 1, \ldots, n$ (see Jahn [195], Theorem 2.27). So we derive necessary conditions for approximately efficient solutions of the problem (P5). In order to formulate the next lemma we introduce a set of linear continuous mappings

$$\mathcal{L}_1(x_\varepsilon) := \{l = (l_1, \ldots, l_n) \mid l_i \in L(Y_i, \mathbb{R}), \ \|l_i\|_{i^*} = 1,$$
$$l_i(A_i(x_\varepsilon) - a^i) = \|A_i(x_\varepsilon) - a^i\|_i, \ i = 1, \ldots, n\}.$$

Theorem 4.1.15. *Under the assumptions given above, for any $\varepsilon > 0$ and any $x^0 \in D$ with*

$$f(x^0) \in \text{Eff}(f[D], C_{\varepsilon k^0}),$$

there exists some $x_\varepsilon \in D$ with

$$f(x_\varepsilon) \in f(x^0) - \sqrt{\varepsilon}\|x_\varepsilon - x^0\|_X k^0 - C, \qquad \|x_\varepsilon - x^0\|_X \leq \sqrt{\varepsilon},$$

such that for any feasible direction v at x_ε with respect to D having $\|v\|_X = 1$ there is a linear continuous mapping $l_\varepsilon \in \mathcal{L}_1(x_\varepsilon)$, with

$$\begin{pmatrix} l_{\varepsilon 1} A_1(v) \\ \cdots \\ l_{\varepsilon n} A_n(v) \end{pmatrix} \in \begin{pmatrix} l_1 A_1(v) \\ \cdots \\ l_n A_n(v) \end{pmatrix} + C \quad \text{for all} \quad l = (l_1, \dots, l_n) \in \mathcal{L}_1 \qquad (4.7)$$

and

$$f_1(v) + \begin{pmatrix} (A_1^* \, l_{\varepsilon 1})(v) \\ (A_2^* \, l_{\varepsilon 2})(v) \\ \cdots \\ (A_n^* \, l_{\varepsilon n})(v) \end{pmatrix} \notin -\sqrt{\varepsilon}\, k^0 - \text{int}\, C. \qquad (4.8)$$

Finally, we will study the special case of a real-valued approximation problem. We suppose that X and Y are real Banach spaces, $n = 1$, $\alpha_1 = 1$, $k^0 = 1$, $A(= A_1) \in L(X, Y)$, $f_1 \in L(X, \mathbb{R})$, and $a(= a^1) \in Y$. In this case we study the real-valued objective function

$$f(x) = f_1(x) + \|A(x) - a\|,$$

which is to be minimized over D.

According to Theorem 4.1.15 we get for $C = \mathbb{R}_+$, $k^0 = 1$ the following **ε-Kolmogorov condition**:

$$\max\{f_1(v) + A^* l(v) \mid l \in L(Y, \mathbb{R}), \, l(A(x_\varepsilon) - a) = \|A(x_\varepsilon) - a\|, \, \|l\|_* = 1\}$$
$$\geq -\sqrt{\varepsilon}. \qquad (4.9)$$

Here the maximality property (4.7) in Theorem 4.1.15 implies that we have to maximize the left-hand side of the inequality (4.9) with respect to $l \in L(Y, \mathbb{R})$ satisfying

$$l(A(x_\varepsilon) - a) = \|A(x_\varepsilon) - a\| \quad \text{and} \quad \|l\|_* = 1,$$

and the ε-variational inequality (4.8) implies the right-hand side of the inequality (4.9).

If we suppose that A is the identical operator and $f_1 = 0$, then we derive from (4.7) and (4.8) in Theorem 4.1.15 the following ε-variational inequality:

$$\max\{\, l(v) \mid l \in L(Y, \mathbb{R}), \, l(x_\varepsilon - a) = \|x_\varepsilon - a\|, \, \|l\|_* = 1\} \geq -\sqrt{\varepsilon}.$$

Furthermore, by putting $\varepsilon = 0$ in the last inequality we get a well-known necessary condition for solutions of the real-valued approximation problem (see Jahn [195]):

$$\max\{l(v) \mid l \in L(Y, \mathbb{R}),\ l(x_\varepsilon - a) = \|x_\varepsilon - a\|,\ \|l\|_* = 1\} \geq 0.$$

But we recall that the assertion of Theorem 4.1.15 is true only for $\varepsilon > 0$, since for $\varepsilon = 0$ the existence of efficient solutions is not guaranteed.

4.1.3 L_p-Approximation Problems

In approximation theory L_p-**approximation problems** play an important role. Now we will apply Theorem 4.1.15 in order to derive necessary conditions for approximately efficient elements of an L_p-approximation problem. Let us assume that $S \subset \mathbb{R}^n$ is a closed convex set, $Y = \mathbb{R}^m$, $C = \mathbb{R}^m_+$, $\varepsilon > 0$, $k^0 \in C \setminus \{0\}$. Suppose that $\Omega \subset \mathbb{R}^q$ is compact. For elements $f_j^i \in L_{p_i}(\Omega)$, $1 \leq p_i \leq \infty$, $[i = 1, \ldots, m]$, $[j = 0, \ldots, n]$, we define the vector-valued function

$$f(x) := \begin{pmatrix} \|\sum_{j=1}^n x_j f_j^1 - f_0^1\|_{p_1} \\ \cdots \\ \|\sum_{j=1}^n x_j f_j^m - f_0^m\|_{p_m} \end{pmatrix}.$$

In the following we study the vector optimization problem

(P6) $\text{Eff}(f[S], C).$

As a consequence of Theorem 4.1.15 we get necessary conditions for approximately efficient elements of problem (P6).

Corollary 4.1.16. *Under the assumptions given above for any $\varepsilon > 0$ and any approximately efficient element*

$$f(x^0) \in \text{Eff}(f[S], C_{\varepsilon k^0})$$

there exist elements $x_\varepsilon \in \mathbb{R}^n$ with

$$f(x_\varepsilon) \in \text{Eff}(f[S], C_{\varepsilon k^0}),$$

$$\|x^0 - x_\varepsilon\|_{\mathbb{R}^n} \leq \sqrt{\varepsilon},$$

such that for any feasible direction v at x_ε with respect to S having $\|v\|_X = 1$ there is a linear continuous mapping $l_\varepsilon \in \mathcal{L}_2$, where

$$\mathcal{L}_2 := \Big\{ l = (l_1, \ldots, l_m) \mid l_i \in L_{p_i^*},\ \|l_i\|_{p_i*} = 1,\ \forall\, i = 1, \ldots, m, \\ \int_\Omega l_i(t) \Big(\sum_{j=1}^n x_{\varepsilon j} f_j^i(t) - f_0^i(t) \Big) dt = \Big\| \sum_{j=1}^n x_{\varepsilon j} f_j^i - f_0^i \Big\|_{p_i} \Big\},$$

$$\text{with } p_i^* = \begin{cases} p_i/(p_i-1) & : & 1 < p_i < \infty \\ \infty & : & p_i = 1 \\ 1 & : & p_i = \infty, \end{cases}$$

$$\begin{pmatrix} l_{\varepsilon 1}(\sum_{j=1}^n v_j f_j^1) \\ \cdots \\ l_{\varepsilon m}(\sum_{j=1}^n v_j f_j^m) \end{pmatrix} \in \begin{pmatrix} l_1(\sum_{j=1}^n v_j f_j^1) \\ \cdots \\ l_m(\sum_{j=1}^n v_j f_j^m) \end{pmatrix} + C \quad \forall l = (l_1, \ldots, l_m) \in \mathcal{L}_2,$$

and

$$\begin{pmatrix} \sum_{j=1}^n v_j (\int_\Omega f_j^1(t) l_{\varepsilon i}(t) dt) \\ \cdots \\ \sum_{j=1}^n v_j (\int_\Omega f_j^m(t) l_{\varepsilon i}(t) dt) \end{pmatrix} \notin -\sqrt{\varepsilon} k^0 - \text{int } C,$$

respectively

$$\sum_{i=1}^m y_i^* \sum_{j=1}^n v_j \left(\int_\Omega f_j^i(t) l_{\varepsilon i}(t)\, dt \right) \geq -\sqrt{\varepsilon} y^*(k^0) \qquad \text{for a} \qquad y^* \in C^+ \setminus \{0\}.$$

Moreover, in the case that $S = \mathbb{R}^n$ and f is Gâteaux differentiable in a neighborhood of x^0, we have

$$\left\| \sum_{i=1}^m y_i^* \begin{pmatrix} \int_\Omega f_1^i(t) l_{\varepsilon i}(t) dt \\ \cdots \\ \int_\Omega f_n^i(t) l_{\varepsilon i}(t) dt \end{pmatrix} \right\|_* \leq \sqrt{\varepsilon} y^*(k^0).$$

4.1.4 Example: The Inverse Stefan Problem

We discuss our method using the example of the **inverse Stefan problem** following Reemtsen [307] and Jahn [195]. We show necessary conditions for approximative solutions of the inverse Stefan problem, which are important for numerical algorithms. We consider the problem of melting ice, where the temperature distribution $u(x,t)$ in the water at time t is described by the heat-flow equation $u_{xx}(x,t) - u_t(x,t) = 0$. We assume that the motion of the melting interface is known and some other boundary condition has to be determined; i.e., the ablating boundary $\delta(\cdot)$ is a known function of t and the heat input $g(t)$ along $x = 0$ is to be determined. Physically, the boundary condition has to be determined such that the melting interface moves in the prescribed way $x = \delta(t)$, $t \geq 0$.

Suppose that $\delta(t) \in C^1[0,T]$, $T > 0$, is a given function, $0 \leq t \leq T$, $0 \leq x \leq \delta(t)$, and $\delta(0) = 0$. Put

$$D(\delta) := \{(x,t) \in \mathbb{R}^2 \mid 0 < x < \delta(t), \ 0 < t \leq T\} \quad \text{for} \quad \delta \in C^1[0,T].$$

Now consider the parabolic boundary value problem

$$u_{xx}(x,t) - u_t(x,t) = 0, \ (x,t) \in D(\delta), \tag{4.10}$$

$$u_x(0,t) = g(t), \ 0 < t \leq T, \tag{4.11}$$

where $g \in C([0, T])$, $g(0) < 0$ is to be determined,

$$u(\delta(t), t) = 0, \quad \dot{\delta}(t) = -u_x(\delta(t), t), \quad 0 < t \leq T. \tag{4.12}$$

The inverse Stefan problem (4.10), (4.11), (4.12) was discussed by Crank [85]. In the following we apply the results of Section 4.1.3 for a characterization of approximate solutions of this problem in form of ε-Kolmogorov conditions using the settings

$$\bar{u}(x, t, a) = \sum_{i=0}^{l} a_i w_i(x, t), \quad l > 0 \text{ integer, fixed,}$$

with $w_i(x, t) = \sum_{k=0}^{\left[\frac{i}{2}\right]} \frac{i!}{(i - 2k)!k!} x^{i-2k} t^k, \quad i = 0, \ldots, l,$

and as an ansatz $g(t) = c_0 + c_1 t + c_2 t^2$, $c_0 \leq 0, c_1 \leq 0, c_2 \leq 0$.
So we get an objective function given by three error functions

$$\varphi_1(t, a, c) := \bar{u}(\delta(t), t, a) - 0, \quad \varphi_2(t, a, c) := \bar{u}_x(0, t, a) - g(t),$$
$$\varphi_3(t, a, c) := \bar{u}_x(\delta(t), t, a) - (-\dot{\delta}(t)),$$
$$\varphi(a, c) := \begin{pmatrix} \|\varphi_1(\cdot, a, c)\|_1 \\ \|\varphi_2(\cdot, a, c)\|_2 \\ \|\varphi_3(\cdot, a, c)\|_3 \end{pmatrix}.$$

Moreover, assume $S \subset \mathbb{R}^l \times \mathbb{R}^3$ and $S := \{s \in \mathbb{R}^l \times \mathbb{R}^3 \mid s_i \in \mathbb{R} \,\forall i = 1, \ldots, l + 3; \, s_i \leq 0 \,\forall i = l + 1, \ldots, l + 3\}$. Now we study the problem to determine the set $\mathrm{Eff}(\varphi[S], \mathbb{R}_+^3)$ in order to compute approximate solutions of the inverse Stefan problem. This is a special case of problem (P5) with

$$f(s) := \begin{pmatrix} \|A_1(s) - a^1\|_1 \\ \|A_2(s) - a^2\|_2 \\ \|A_3(s) - a^3\|_3 \end{pmatrix},$$

where $A_i \in L(\mathbb{R}^l \times \mathbb{R}^3, Y_i)$, Y_i are reflexive L_q-spaces, especially

$$A_1(t) = (w_1(\delta(t), t), w_2(\delta(t), t), \ldots, w_l(\delta(t), t), 0, 0, 0),$$
$$A_2(t) = (w_{1x}(0, t), w_{2x}(0, t), \ldots, w_{lx}(0, t), -1, -t, -t^2),$$
$$A_3(t) = (w_{1x}(\delta(t), t), w_{2x}(\delta(t), t), \ldots, w_{lx}(\delta(t), t), 0, 0, 0),$$
$$s^T = (a_1, a_2, \ldots, a_l, c_0, c_1, c_2),$$
$$a^1 = (0, \ldots, 0) \in Y_1, \quad a^2 = (0, \ldots, 0) \in Y_2, \quad a^3 = -\dot{\delta} \in Y_3 = L_q[0, T],$$

and $\| \cdot \|_i$ $(i = 1, 2, 3)$ denotes a norm in a reflexive L_q-space Y_i.
Now it is possible to apply Theorem 4.1.15:
Under the assumptions given above for any $\varepsilon > 0$ and any approximately efficient element $\varphi(s^0) \in \varepsilon k^0 - \mathrm{Eff}(\varphi[S], \mathbb{R}_+^3)$ there exists an element $s_\varepsilon \in S$

with $\varphi(s_\varepsilon) \in \varepsilon k^0 - \text{Eff}(\varphi[S], \mathbb{R}^3_+)$, $\|s_\varepsilon - s^0\|_{\mathbb{R}^{l+3}} \le \sqrt{\varepsilon}$, such that for any feasible direction v at s_ε with respect to S having $\|v\| = 1$ there is a linear continuous mapping $l_\varepsilon \in \mathcal{L}_1$, where $\mathcal{L}_1 := \{l = (l_1, l_2, l_3) \mid l_i \in L(Y_i, \mathbb{R}) :$ $\|l_i\|_{i*} = 1, \ l_i(A_i(s_\varepsilon) - a^i) = \|A_i(s_\varepsilon) - a^i\|_i \ \forall i = 1, 2, 3\}$, with

$$\begin{pmatrix} l_{\varepsilon 1} A_1(v) \\ l_{\varepsilon 2} A_2(v) \\ l_{\varepsilon 3} A_3(v) \end{pmatrix} \in \begin{pmatrix} l_1 A_1(v) \\ l_2 A_2(v) \\ l_3 A_3(v) \end{pmatrix} + \mathbb{R}^3_+ \quad \forall l = (l_1, l_2, l_3) \in \mathcal{L}_1 \quad \text{and}$$

$$\begin{pmatrix} (A_1^* l_{\varepsilon 1})(v) \\ (A_2^* l_{\varepsilon 2})(v) \\ (A_3^* l_{\varepsilon 3})(v) \end{pmatrix} \notin -\sqrt{\varepsilon} k^0 - \text{int } \mathbb{R}^3_+,$$

respectively

$$\sum_{i=1}^{3} y_i^* ((A_i^* l_{\varepsilon i})(v)) \ge -\sqrt{\varepsilon} y^* (k^0) \qquad \text{for a} \qquad y^* \in \mathbb{R}^3_+ \setminus \{0\}. \qquad (4.13)$$

Remark 4.1.17. For $\varepsilon = 0$ the condition (4.13) coincides with the well-known **Kolmogorov condition** (cf. Jahn [195]), which means that the directional derivative at the optimal point is greater than or equal to zero.

Moreover, necessary conditions for approximate solutions of the type (4.13) are important for numerical algorithms, especially for proximal-point algorithms (cf. Benker, Hamel, and Tammer [23]).

4.2 Solution Procedures

4.2.1 A Proximal-Point Algorithm for Real-Valued Control Approximation Problems

We consider the real-valued optimization problem

$$f(x) := c(x) + \sum_{i=1}^{n} \alpha_i \left\| A^i(x) - a^i \right\|^{\beta_i} \rightarrow \min_{x \in D},$$

which is an extension of a problem studied by Idrissi, Lefebvre, and Michelot [170, 172]. This class of problems contains many practically important special cases such as **approximation, location,** and **optimal control problems, perturbed linear programming problems,** and **surrogate problems for linear programming.** Necessary and sufficient optimality conditions are derived using the subdifferential calculus. A proximal-point algorithm is modified by the method of partial inverse (see Spingarn [328]) in order to solve the optimality conditions. For further references see [22, 23, 185]).

Many authors have studied generalizations of the well-known Fermat Weber problem from the theoretical and computational points of view (see

[53, 54, 55, 56, 57], [101], [102], [120], [170], [195], [218], [258], [373], [375], [376], [378]).

The aim of this section is to extend the results of Michelot and Lefebvre [258] and Idrissi, Lefebvre, and Michelot [170] to the following general **approximation problem** (compare [22, 23])

$$f(x) = (c \mid x) + \sum_{i=1}^{n} \alpha_i \left\| A^i(x) - a^i \right\|^{\beta_i} \to \min_{x \in D}, \qquad (4.14)$$

where $c \in H$, a^i, and $x \in H$, $\alpha_i \geq 0$, $\beta_i > 0$, $A^i \in L(H, H)$, $(i = 1, \ldots, n)$, ($L(H, H)$ denotes the space of linear continuous operators from H to H), H is a Hilbert space, $D_j \subset H$ ($j = 1, \ldots, m$), are closed and convex sets, $D_0 \subseteq H$ is a linear subspace, and $D = \bigcap_{j=0}^{m} D_j$. Furthermore, B^0 denotes the unit ball associated with the dual norm of the norm $\|.\|$, $P_D(e)$ denotes the projection of e onto a set $\mathcal{D}$.

We assume that the inverse operator $(A^{iT})^{-1}$ of the adjoint operator A^{iT} to A^i exists for all $i = 1, \ldots, n$. In order to derive a primal dual algorithm we assume in the following that any suitable constraint qualification (generalized Slater condition, stability, etc.) is satisfied (cf. [138]).

Furthermore, $N_{D_j}(x^0)$ denotes the **normal cone** to D_j at $x^0 \in H$ defined by

$$N_{D_j}(x^0) = \begin{cases} \{x \in H \mid \quad (x \mid x' - x^0) \leq 0 \quad \forall\, x' \in D_j\} & : \quad x^0 \in D_j, \\ \emptyset & : \quad \text{otherwise.} \end{cases}$$

Remark 4.2.1. (1) If a set D_j is bounded ($j = 1, \ldots, m$), then the existence of a solution of the problem (4.14) is ensured.

(2) The problem (4.14) contains the following practically important special cases:

- **linear programming** if $\alpha_i = 0 \ \forall\, i$;
- **surrogate problems for linear programming problems**

$$f(x) := (c \mid x) \longrightarrow \text{minimum}$$

subject to

$$x \in D \quad \text{and} \quad A^i(x) = a^i \quad (i = 1, \ldots, n),$$

for which the feasible set is empty but D is nonempty,
- **perturbed linear programming problems**,
- **approximation** and **location problems** if $c = 0$,
- **optimal control problems (optimal regulator problems)** of the form ($c = 0, n = 2, \beta_i = p$)

$$\|A(u) - r\|^p + \alpha \|u\|^p \to \min_{u \in U \subset H}.$$

We apply duality assertions in order to obtain lower bounds for the approximation problem (4.14). Furthermore, we derive optimality conditions for problem (4.14) using the subdifferential calculus. These optimality conditions are useful for an application of Spingarn's proximal-point algorithm (see [328], [170], and [258]) for solving problem (4.14). We present two proximal-point algorithms with different subspaces in the formulation of the algorithm. Moreover, we study two special cases of (4.14).

Duality theorems (compare Section 3.7) can be used in order to obtain lower bounds for the problem (4.14).
For the special case

$$f(x) = \sum_{i=1}^{n} \left\| A^i(x) - a^i \right\|^p + \alpha \left\| x \right\|^p \to \min_{x \in H}, \tag{4.15}$$

where $\alpha > 0$ and $p = 1$ or 2, we can apply the duality theorem of approximation theory (see [20]). Following [20] we obtain the following **lower bounds** for the problem (4.15).

Lemma 4.2.2. *For $p = 1$ we have the estimation*

$$\min_{x \in H} \left(\sum_{i=1}^{n} \left\| A^i(x) - a^i \right\| + \alpha \left\| x \right\| \right) \geq \frac{\sum\limits_{i=1}^{n} \left\| a^i \right\|}{\max \left\{ 1, \frac{1}{\alpha} \sum\limits_{i=1}^{n} \left\| A^i \right\| \right\}}, \tag{4.16}$$

and for $p = 2$,

$$\min_{x \in H} \left(\sum_{i=1}^{n} \left\| A^i(x) - a^i \right\|^2 + \alpha \left\| x \right\|^2 \right) \geq \frac{\alpha \left(\sum\limits_{i=1}^{n} \left\| a^i \right\|^2 \right)}{\alpha + \sum\limits_{i=1}^{n} \left\| A^i \right\|^2}. \tag{4.17}$$

PROOF. With $A(x) := (A^1(x), \ldots, A^n(x))$ we define a linear continuous operator $A \in L(H, H^n)$.

Now we can write the problem (4.15) in the form

$$\left\| A(x) - a \right\|^p_{H^n} + \alpha \left\| x \right\|^p_H \to \min_{x \in H},$$

where

$$a := (a^1, \ldots, a^n),$$

$$\left\| a \right\|^2_{H^n} = (a \mid a)_{H^n} := \sum_{i=1}^{n} (a^i \mid a^i)_H = \sum_{i=1}^{n} \left\| a^i \right\|^2_H \quad \text{for } p = 2,$$

and

$$\left\| a \right\|_{H^n} := \sum_{i=1}^{n} \left\| a^i \right\|_H \quad \text{for } p = 1.$$

For $p = 2$ we obtain

$$\|A\|^2 \le \sum_{i=1}^{n} \|A^i\|^2 .$$

In a similar way it follows that

$$\|A\| \le \max\left\{1, \frac{1}{\alpha}\sum_{i=1}^{n}\|A^i\|\right\} \qquad \text{for } p = 1.$$

Now Lemma 1 from [20] yields the lower bounds (4.16) and (4.17). □

Applying a general duality theorem, we can also calculate **lower bounds** for another special case of problem (4.14). Let us consider the problem

(P) $f(x) = (c\,|\,x) + \sum_{i=1}^{n} \alpha_i \|A^i(x) - a^i\| \longrightarrow \min_{x \in C_H, B(x) - b \in C_V},$

where additionally to the assumptions given above, V is a reflexive Banach space, $B \in L(H, V)$, $b \in V$, $C_H \subset H$ and $C_V \subset V$ are closed convex cones, and C_H^+ and C_V^+ the corresponding dual cones.

We have shown in Section 3.7.3, Example 3.7.1, that the following problem (D) can be considered as the dual problem to (P):

(D) $g(y, z) = \sum_{i=1}^{n} \alpha_i y_i(a^i) + z(b) \longrightarrow \max_{(y,z) \in \mathcal{D}},$

where

$$\mathcal{D} = \left\{(y, z) \mid y = (y_1, \dots, y_n), y_i \in H, \; \alpha_i\|y_i\|_* \le \alpha_i \; (i = 1, \dots, n), z \in C_V^+, \right.$$

$$\left. c - \sum_{i=1}^{n} \alpha_i A^{iT} y_i - B^T z \in C_H^+ \right\}.$$

Here $\|\cdot\|_*$ denotes the dual norm to $\|\cdot\|$.

Lemma 4.2.3. *Consider the problems (P) and (D).*

1. For any $(y, z) \in \mathcal{D}$ we have

$$\inf_{x \in K_H, B(x) - b \in K_V} \left((c\,|\,x) + \sum_{i=1}^{n} \alpha_i\|A^i(x) - a^i\|\right) \ge \sum_{i=1}^{n} \alpha_i y_i(a^i) + z(b).$$

2. Assume that there exist an element $\bar{x} \in C_H$ with $B(\bar{x}) - b \in \operatorname{int} C_V$ and an element $(\bar{y}, \bar{z}) \in \mathcal{D}$ with $c - \sum_{i=1}^{n} \alpha_i A^{iT}\bar{y}_i - B^T\bar{z} \in \operatorname{int} C_H^+$. Then the problems (P) and (D) have optimal solutions x^0 and (y^0, z^0), and

$$(c\,|\,x^0) + \sum_{i=1}^{n} \alpha_i\|A^i(x^0) - a^i\| = \sum_{i=1}^{n} \alpha_i y_i^0(a^i) + z^0(b).$$

PROOF. Analogous to the proof of duality assertions in Section 3.7.3 it is possible to derive the assertions from a general saddle point theorem (Theorem 47.1 in Zeidler [391]). □

The following approach is motivated by recent papers of Michelot and Lefebvre [258] and Idrissi, Lefebvre, and Michelot [170]. Using the *indicator functions* χ_{D_j} of D_j defined by $\chi_{D_j}(x) = 0$, if $x \in D_j$ and $\chi_{D_j}(x) = +\infty$ otherwise, the problem 4.14 is equivalent to the following *unconstrained minimization problem*:

$$F(x) = (c \,|\, x) + \sum_{i=1}^{n} \|A^i(x) - a^i\|^{\beta_i} + \sum_{j=0}^{m} \chi_{D_j}(x) \to \min_{x \in H}, \qquad (4.18)$$

where we put $\alpha_i = 1$ $(i = 1, \dots, n)$ without loss of generality.

The functional $F(x)$ is convex under the given assumptions. Therefore, the subdifferential condition

$$0 \in \partial F(x^0) \qquad (4.19)$$

is necessary and sufficient for the optimality of x^0. By calculating the subdifferential of the functional F from (4.18), the condition (4.19) is equivalent to the following optimality conditions (4.20), (4.21), (4.22):

$$q_i \in \partial(\|A^i(x^0) - a^i\|^{\beta_i}), \qquad i = 1, 2, \dots, n, \qquad (4.20)$$

$$r_j \in N_{D_j}(x^0), \qquad j = 1, 2, \dots, m, \qquad (4.21)$$

$$c + \sum_{i=1}^{n} q_i + \sum_{j=0}^{m} r_j \in D_0^\perp. \qquad (4.22)$$

In order to reformulate the optimality conditions (4.20)–(4.22) in a more practical way, we introduce a Hilbert space E, suitable subspaces $\mathcal{A}$, $\mathcal{B}$, and an operator T. This will be done in different ways. For the first algorithm we will use the following notation:

We introduce the Hilbert space

$$E := (H)^{n+m+1}$$

with the inner product

$$(u \,|\, v) := \sum_{i=1}^{n+m+1} (u_i \,|\, v_i)$$

for $u = (u_1, \dots, u_{n+m+1})$ and $v = (v_1, \dots, v_{n+m+1}) \in E$ and the subspaces

$$\mathcal{A} := \{y \in E \mid y = (y_1, \dots, y_{n+m+1}); \; y_j = x, \, j = 1, \dots, n+m+1; \, x \in D_0\},$$

$$\mathcal{B} := \left\{ p \in E \mid p = (p_1, \dots, p_{n+m+1}), \; \sum_{i=1}^{n+m+1} p_i \in D_0^\perp \right\},$$

and the operator T defined on E by

$$T(y) := \prod_{i=1}^{n+m+1} T^i(x) = (T^1(x), \ldots, T^{n+m+1}(x)) \qquad (4.23)$$

where

$$\begin{aligned}
T^i(x) &:= \partial(\|A^i(x) - a^i\|^{\beta_i}), \qquad i = 1, \ldots, n, \\
T^{n+j}(x) &:= N_{D_j}(x), \qquad j = 1, \ldots, m, \\
T^{n+m+1}(x) &:= c.
\end{aligned}$$

With this notation the optimality conditions (4.20)–(4.22) can be rewritten as

$$\text{Find} \quad (y^0, p^0) \in \mathcal{A} \times \mathcal{B} \quad \text{such that} \quad p^0 \in T(y^0). \qquad (4.24)$$

Remark 4.2.4. (1) Clearly, $\mathcal{A}$ and $\mathcal{B}$ are closed linear subspaces of E with $\mathcal{A} \perp \mathcal{B}$. We prove additionally that the subspaces $\mathcal{A}$ and $\mathcal{B}$ are complementary; i.e., $H = \mathcal{A} \oplus \mathcal{B}$:
We define the operator

$$P_\mathcal{A}(e) = \frac{1}{n+m+1} \left(P_{D_0}\left(\sum_{i=1}^{n+m+1} e_i \right), \ldots, P_{D_0}\left(\sum_{i=1}^{n+m+1} e_i \right) \right),$$

where $e = (e_1, \ldots, e_{n+m+1}) \in E$ is arbitrary.

Trivially, we have $\mathcal{B} = \mathrm{Ker}(P_\mathcal{A})$. Applying the necessary and sufficient condition

$$(e - P_\mathcal{A}(e) \,|\, a) = 0 \qquad \forall\, a \in \mathcal{A}$$

for an operator $P_\mathcal{A}$ to be a projection onto $\mathcal{A}$, we see that $P_\mathcal{A}$ is the projection of E onto $\mathcal{A}$; i.e., $R(P_\mathcal{A}) = \mathcal{A}$.

(2) It is easy to see that the operator T, defined in (4.23), realizes a maximal monotone multifunction, since T is composed of subdifferentials. For a numerical solution we need the condition (4.24) in the form $0 \in T(y^0) - p^0$, but the new multifunction $\tilde{T}(y, p) := T(y) - p$ is not maximal monotone. Therefore, we use in the following an idea of Spingarn [328].

Applying the results of Spingarn [328], the problem (4.24) is equivalent to

Find $(y^0, p^0) \in \mathcal{A} \times \mathcal{B}$ such that

$$0 \in T_\mathcal{A}(y^0 + p^0), \qquad (4.25)$$

where $T_\mathcal{A}$ is the *partial inverse* of T with respect to $\mathcal{A}$. The operator $T_\mathcal{A}$ is described by its graph in the following way:

$$\{(y_A + p_B, p_A + y_B) \mid p \in T(y),\ p = p_A + p_B,\ y = y_A + y_B\},$$

where v_A and v_B are the projections of v onto $\mathcal{A}$ and $\mathcal{B}$, respectively. We remark that T_A is maximal monotone iff T is maximal monotone. Setting

$$z = z_A + z_B$$

and

$$z_A = y, \qquad z_B = p,$$

then the problem (4.25) has the following form:

$$\text{Find} \qquad z^0 \in E \qquad \text{such that} \qquad 0 \in T_A(z^0). \tag{4.26}$$

Now we can apply the **proximal-point algorithm** for solving (4.26) (compare Michelot/Lefebvre [258] and Idrissi/Lefebvre/Michelot [170]). This algorithm has the form

$$z^{k+1} = (I + c_k T_A)^{-1}(z^k) \qquad k = 1, 2, \ldots, \tag{4.27}$$

where the starting point z^1 is arbitrary, and c_k is a sequence of positive real numbers. If (c_k) is bounded away from zero (we set $c_k = 1$), then either (z^k) converges weakly to a zero of T_A, or $\|z^k\| \to \infty$ and T_A has no zeros (see [308]).

The iteration (4.27) can be rewritten in the form

$$y^{k+1} + p^{k+1} = (I + T_A)^{-1}(y^k + p^k),$$

which is equivalent to

$$y^k - y^{k+1} + p^k - p^{k+1} \in T_A(y^{k+1} + p^{k+1}). \tag{4.28}$$

Now we will realize this iteration in terms of the multifunction T instead of T_A. Setting

$$\widetilde{p}^k = y^k - y^{k+1} + p^{k+1},$$
$$\widetilde{y}^k = p^k - p^{k+1} + y^{k+1},$$

which yields

$$\widetilde{y}^k + \widetilde{p}^k = y^k + p^k, \qquad y^{k+1} = (\widetilde{y}^k)_A, \qquad p^{k+1} = (\widetilde{p}^k)_B,$$

and applying the definition of the partial inverse T_A, we obtain the iteration (4.28) as

$$\widetilde{p}^k \in T(\widetilde{y}^k), \qquad \text{or} \qquad \widetilde{p}^k \in T(y^k + p^k - \widetilde{p}^k). \tag{4.29}$$

Further, considering the definition of the operator T, the iteration (4.29) is equivalent to

$$\widetilde{p}_i^k \in \partial(\|A^i(x^k + p_i^k - \widetilde{p}_i^k) - a^i\|^{\beta_i}) \qquad i = 1, \ldots, n, \tag{4.30}$$

$$\widetilde{p}_{n+j}^{k} \in N_{D_j}(x^k + p_{n+j}^k - \widetilde{p}_{n+j}^k), \qquad j = 1, \ldots, m, \tag{4.31}$$

$$\widetilde{p}_{n+m+1}^{k} = c. \tag{4.32}$$

For an implementation of the iteration algorithm (4.30)–(4.32) it remains to calculate the subdifferential from (4.30) and the normal cone from (4.31).

Setting

$$z_i^k := A^i(x^k + p_i^k - \widetilde{p}_i^k) - a^i,$$

we get for the subdifferential from (4.30),

$$\partial(\| z_i^k \|^{\beta_i}) = A^{iT} \beta_i \| z_i^k \|^{\beta_i - 1} \partial(\| z_i^k \|),$$

and the iteration condition (4.30) now has the following form:

$$(A^{iT})^{-1} \beta_i^{-1} \widetilde{p}_i^k \in \| z_i^k \|^{\beta_i - 1} \partial(\| z_i^k \|). \tag{4.33}$$

To apply this expression in practice, a computable form is needed. This is possible only for special cases.

Special case 1:

Let $\beta_i = 1$ for $i = 1, \ldots, n$. Then (4.33) has the form

$$(A^{iT})^{-1} \widetilde{p}_i^k \in \partial(\| z_i^k \|) = \{ z_i^* \in H \mid \| z_i^* \| = 1, \, (z_i^* \mid z_i^k) = \| z_i^k \| \}.$$

Hence

$$\left\| (A^{iT})^{-1} \widetilde{p}_i^k \right\| = 1$$

and

$$\| z_i^k \| = ((A^{iT})^{-1} \widetilde{p}_i^k \mid z_i^k).$$

Using the inequality

$$\| z_i^k \| \geq (z_i^* \mid z_i^k) \qquad \forall z_i^* \qquad \text{with} \qquad \| z_i^* \| = 1 \quad (\Leftrightarrow z_i^* \in B^0),$$

we get

$$((A^{iT})^{-1} \widetilde{p}_i^k \mid z_i^k) \geq (z_i^* \mid z_i^k).$$

It follows that

$$(z_i^* - (A^{iT})^{-1} \widetilde{p}_i^k \mid z_i^k) \leq 0 \qquad \forall z_i^* \in B^0.$$

Substituting the definition for z_i^k into the last inequality, we get

$$(z_i^* - (A^{iT})^{-1} \widetilde{p}_i^k \mid A^i(x^k + p_i^k - \widetilde{p}_i^k) - a^i) \leq 0 \qquad \forall z_i^* \in B^0, \tag{4.34}$$

which is equivalent to

$$z_i^k = A^i(x^k + p_i^k - \widetilde{p}_i^{*k}) - a^i \in N_{B^0}((A^{iT})^{-1}\widetilde{p}_i^{*k}).$$

Now we can transform the inequality (4.34) in the form

$$(\widetilde{p}_i^{*k} - x^k - p_i^k + (A^i)^{-1}a^i \mid A^{iT}z_i^* - \widetilde{p}_i^{*k}) \geq 0 \qquad \forall z_i^* \in B^0, \tag{4.35}$$

which is equivalent to

$$\widetilde{p}_i^{*k} \in P_M(x^k + p_i^k - (A^i)^{-1}a^i), \tag{4.36}$$

where P_M is the projection operator onto the set

$$M = \left\{ A^{iT}z_i^* \mid z_i^* \in B^0 \right\}.$$

Therefore, we have to solve the minimization problem

$$\|v - B(u)\|^2 \to \min_{u \in B^0},$$

where

$$v := x^k + p_i^k - (A^i)^{-1}a^i \qquad \text{and} \qquad B := A^{iT}.$$

We find the solution of this quadratic minimization problem as a solution of the operator equation

$$(B^T B + \lambda I)u = B^T v,$$

where λ is to be determined such that $\|u\| = 1$. This completes the calculation of the iteration (4.30). It remains to solve the iteration (4.31), which is equivalent to

$$y_{n+j}^k - y_{n+j}^{k+1} + p_{n+j}^{k+1} \in N_{D_j}(x^k + p_{n+j}^k - \widetilde{p}_{n+j}^{*k}), \tag{4.37}$$

where

$$y_{n+j}^k = x^k \qquad \text{and} \qquad y_{n+j}^{k+1} = x^{k+1}.$$

The last relation (4.37) can be transformed into the form

$$x^{k+1} + p_{n+j}^k - p_{n+j}^{k+1} = P_{D_j}(x^k + p_{n+j}^k). \tag{4.38}$$

The **algorithm for the special case 1** thus takes the following final form:

Algorithm I:

- Choose the starting points

$$x^1 \in D_0 \qquad \text{and} \qquad p^1 \in E \qquad \text{with} \sum_{i=1}^{n+m+1} p_i^1 \in D_0^\perp. \tag{4.39}$$

• Compute x^{k+1} and p^{k+1} from

$$x^k - x^{k+1} + p_i^{k+1} = P_M(x^k + p_i^k - (A^i)^{-1} a^i), \quad i = 1, \ldots, n, \quad (4.40)$$

$$x^{k+1} + p_{n+j}^k - p_{n+j}^{k+1} = P_{D_j}(x^k + p_{n+j}^k), \qquad j = 1, \ldots, m, \quad (4.41)$$

$$x^k - x^{k+1} + p_{n+m+1}^{k+1} = c, \quad (4.42)$$

with

$$\sum_{i=1}^{n+m+1} p_i^{k+1} \in D_0^\perp \qquad \text{and} \qquad x^{k+1} \in D_0. \quad (4.43)$$

This means that we must first calculate the projections on M and D_j in the right-hand side of the equations (4.40) and (4.41). Then we have to solve the linear equations (4.40)–(4.42) such that the relations (4.43) are satisfied. The solution of the linear equations (4.40)–(4.42) is very simple, since we can eliminate the variables in the following way:

Setting for $1 \le i \le n-1$ and $1 \le j \le m$,

$$b_i^k = P_M\left(x^k + p_1^k - (A^i)^{-1} a^1\right) - P_M\left(x^k + p_{i+1}^k - (A^{i+1})^{-1} a^{i+1}\right),$$

$$b_{n+j-1}^k = P_M\left(x^k + p_1^k - (A^i)^{-1} a^1\right) + P_{D_j}\left(x^k + p_{n+j}^k\right) - x^k + p_{n+j}^k,$$

$$b_{n+m}^k = P_M\left(x^k + p_1^k - (A^i)^{-1} a^1\right) - c,$$

we obtain immediately the solutions

$$p_i^{k+1} = p_1^{k+1} - b_{i-1}^k, \qquad i = 2, \ldots, n+m+1, \quad (4.44)$$

where p_1^{k+1} is determined by the relations (4.43):

$$(n+m+1)\, p_1^{k+1} - \sum_{i=1}^{n+m} b_i^k \in D_0^\perp \quad \text{and} \quad (x^{k+1} =)\, x^k - c + p_1^{k+1} - b_{n+m}^k \in D_0. \quad (4.45)$$

Remark 4.2.5. (1) The determination of p_1^{k+1} from the relations (4.45) is possible only if the subspace D_0 is concretely given; for example, if $D_0 = H$, then from (4.45) follows

$$p_1^{k+1} = \frac{1}{n+m+1} \sum_{i=1}^{n+m} b_i^k \quad \text{and} \quad x^{k+1} = x^k - c + p_1^{k+1} - b_{n+m}^k. \quad (4.46)$$

(2) Another possibility for the solution of the equations (4.40)–(4.42) is given in the following way:

Using the definitions given above,

$$\bar{p}^k = y^k - y^{k+1} + p^{k+1},$$

$$\tilde{y}^k = p^k - p^{k+1} + y^{k+1},$$

which yields

$$\tilde{y}^k + \tilde{p}^k = y^k + p^k, \qquad y^{k+1} = (\tilde{y}^k)_{\mathcal{A}}, \qquad p^{k+1} = (\tilde{p}^k)_{\mathcal{B}},$$

we obtain

$$\tilde{p}_1^k, \ldots, \tilde{p}_n^k$$

from (4.40),

$$\tilde{y}_{n+1}^k, \ldots, \tilde{y}_{n+m+1}^k$$

from (4.41), (4.42), and the remaining components from

$$\tilde{p}_{n+j}^k = x^k + p_{n+j}^k - \tilde{y}_{n+j}^k \ (1 \le j \le m) \text{ and } \tilde{y}_i^k = x^k + p_i^k - \tilde{p}_i^k \ (1 \le i \le n).$$

Now we obtain the solution vectors

$$p^{k+1} = (p_1^{k+1}, \ldots, p_{n+m+1}^{k+1}) \text{ and } y^{k+1} = (x^{k+1}, \ldots, x^{k+1})$$

by calculating the projections $(\tilde{p}^k)_{\mathcal{B}}$ and $(\tilde{y}^k)_{\mathcal{A}}$ from $\tilde{p}^k$ and $\tilde{y}^k$ onto the subspaces $\mathcal{B}$ and $\mathcal{A}$, respectively.

Special case 2:

Let be $\beta_i = \beta > 0$ and $c = 0$; i.e., we consider the problem

$$f(x) = \sum_{i=1}^n \left\| A^i(x) - a^i \right\|^\beta \to \min_{x \in D = \bigcap_{j=0}^m D_j} . \tag{4.47}$$

Defining an operator $A \in L(H \to H^n)$ by

$$A(x) := (A^1(x), A^2(x), \ldots, A^n(x))$$

and a norm on $H^n = H \times H \times \cdots \times H$ by

$$\|u\|_{H^n} := \left(\sum_{i=1}^n \|u^i\|^\beta \right)^{1/\beta}, \qquad \text{where} \qquad u = (u^1, u^2, \ldots, u^n),$$

and using the indicator functions χ_{D_j} of D_j, the problem (4.47) is equivalent to the unconstrained minimization problem

$$F(x) = \|A(x) - a\|_{H^n} + \sum_{j=0}^n \chi_{D_j}(x) \to \min_{x \in H}, \tag{4.48}$$

where $a = (a^1, a^2, \ldots, a^n)$. Using the optimality condition $0 \in \partial F(x^0)$, we obtain analogously to problem (4.14) the following necessary and sufficient conditions for the problem (4.48):

$$q \in \partial(\|A(x^0) - a\|_{H^n}), \tag{4.49}$$

$$r_j \in N_{D_j}(x^0), \qquad j = 1, \ldots, m, \tag{4.50}$$

$$q + \sum_{j=1}^{m} r_j \in D_0^\perp. \tag{4.51}$$

It remains to verify condition (4.49), since the other conditions (4.50) and (4.51) are the same as in (4.21) and (4.22).

So we introduce the following subspaces:

$$\mathcal{A}_1 := \{ y \in E \mid y = (y_1, \ldots, y_{m+1}) \,;\, y_j = x \,, \, j = 1, \ldots, m+1 \,;\, x \in D_0 \},$$

and

$$\mathcal{B}_1 \quad := \quad \left\{ \; p \in E \mid p = (p_1, \ldots, p_{m+1}), \quad \sum_{i=1}^{m+1} p_i \in D_0^\perp \; \right\},$$

where $E = H^{m+1}$.

Using these subspaces $\mathcal{A}_1$ and $\mathcal{B}_1$, we can compute the subdifferential (4.49) in a similar manner as for the special case 1 and obtain

$$\tilde{p}_1^k = P_M(x^k + p_1^k - A^{-1}a),$$

where P_M is the projection operator onto the set

$$M = \{ A^T z^* \mid z^* \in B^0 \},$$

where

$$B^0 = \{ z^* \in H^n \mid \|z^*\|_{H^n} = 1 \}$$

and

$$A^T z^* = (A^{1T} z_1^* + \ldots + A^{nT} z_n^*) \in H.$$

Now the given algorithm for the special case 1 can be applied.

However, **Algorithm I** has some disadvantages. For example, it needs the operators A^i to be regular and invertible, and some operator equation has to be solved if $A^i \neq I$ (I = identity). If we define the subspaces $\mathcal{A}$ and $\mathcal{B}$ in a different way as above, we get another specification of Spingarn's algorithm under weaker assumptions. The new idea for deriving a proximal-point algorithm is to put the operators A^i in the definition of the subspaces $\mathcal{A}$ and $\mathcal{B}$.

In the following we derive a proximal-point algorithm for control approximation problems (**Algorithm II**, cf. [347]):

$$f(x) = c(x) + \sum_{i=1}^{n} \alpha_i \|A^i(x) - a^i\|_{(i)}^{\beta_i} \to \min_{x \in D},$$

$$\| \cdot \|_{(i)} \; : \; \text{norm in } \mathbb{R}^k, \; (i = 1, \ldots, n)$$

$$x, c \in \mathbb{R}^k, \; a^i \in \mathbb{R}^k, \alpha_i \geq 0, \; \beta_i \geq 1, \; A^i \in L(\mathbb{R}^k, \mathbb{R}^k),$$

$$D = \bigcap_1^m D_j, \; D_j \subset \mathbb{R}^k \; \text{closed and convex,}$$

$$\text{(4.52)}$$

where $L(\mathbb{R}^k \to \mathbb{R}^k)$ denotes the set of linear continuous mappings from $\mathbb{R}^k$ to $\mathbb{R}^k$. Furthermore, suppose int $D \neq \emptyset$.

In contrast to problem (4.14) at the beginning of Section 4.2.1, we consider now a control approximation problem (4.52) with different norms in the objective function, in finite-dimensional spaces and without assuming that the operators A^i are regular.

The space E is defined by

$$E := \mathbb{R}^{k_1} \times \mathbb{R}^{k_2} \times \cdots \times \mathbb{R}^{k_{n+m+1}}, \tag{4.53}$$

where $k_1 = \cdots = k_{n+m+1} = k$. Define as a shortcut the operator $S : E \to \mathbb{R}^k$ by

$$S(e) := \sum_{i=1}^{n} A^{iT} e_i + \sum_{j=1}^{m+1} e_{n+j}, \quad e = (e_1, \ldots, e_{n+m+1}) \in E.$$

Then the subspaces have the form

$$\mathcal{A}_2 := \{y \in E \mid y = \underbrace{(A^1 x, A^2 x, \ldots, A^n x, x, \ldots, x)}_{n+m+1}, x \in R^s\}, \tag{4.54}$$

$$\mathcal{B}_2 := \{p \in E : S(p) = 0\}. \tag{4.55}$$

It is easy to see that the subspace $\mathcal{B}_2$ is the orthogonal complement of $\mathcal{A}_2$: With $v \in E$ and $a \in \mathcal{A}_2$, it yields

$$(a \mid v) = \sum_{i=1}^{n} (A^i x \mid v_i) + \sum_{j=1}^{m+1} (x \mid v_{n+j}) = (x \mid S(v)). \tag{4.56}$$

If $v \in \mathcal{B}_2$, then the right-hand side is zero, and therefore (4.56) yields $\mathcal{B}_2 \subset \mathcal{A}_2^{\perp}$. If $v \in \mathcal{A}_2^{\perp}$, then (4.56) has to be zero for all $a \in \mathcal{A}_2$ and therefore for all $x \in \mathbb{R}^k$. This means that the right part in the inner product has to be zero, so we have $v \in \mathcal{B}_2$ and $\mathcal{A}_2^{\perp} \subset \mathcal{B}_2 \subset \mathcal{A}_2^{\perp}$, $\mathcal{A}_2^{\perp} = \mathcal{B}_2$. With the closedness of $\mathcal{A}_2$ and $\mathcal{B}_2$ follows

$$E = \mathcal{A}_2 \oplus \mathcal{B}_2. \tag{4.57}$$

The operator $T : E \to 2^E$ has again the form

$$\tilde{p} \in T(\tilde{y}) \ \Leftrightarrow \ \tilde{p}_i \in T_i(\tilde{y}_i) \quad i = 1, \ldots, n + m + 1, \quad \tilde{p}, \tilde{y} \in E \ .$$

$$T_i(y_i) := \partial(\alpha_i \| y_i - a^i \|_{(i)}^{\beta_i}), \ i = 1, \ldots, n,$$

$$T_{n+j}(x) := N_{D_j}(x) \qquad j = 1, \ldots, m,$$

$$T_{n+m+1} := c \ .$$

The problem (4.20)–(4.22) is equivalent to the problem (4.24) with this choice of A_2, B_2 and T with $q_i = A^{iT} p_i^0$, $r_j = p_{n+j}^0$, and $y_{n+m+1}^0 = (A^1 x^0, \ldots, A^n x^0, x^0, \ldots, x^0)$. This means that if one of the problems has a solution, then the other one also has a solution, and the solution of (4.24) can be transferred into a solution of (4.20)–(4.22).

The projection onto the subspaces A_2 and B_2 is an important step for Spingarn's algorithm.

Let $v \in E$ be an arbitrary element,

$$v = v_{A_2} + v_{B_2},$$

$$v_{A_2} = (y_1, \ldots, y_{n+m+1}) \in A_2,$$

$$v_{A_2} = (A^1 x, \ldots, A^n x, x, \ldots, x) \in A_2.$$

Using the operator S $(S(e) = \sum_{i=1}^{n} A^{iT} e_i + \sum_{j=1}^{m+1} e_{n+j})$ on both sides, and using $S(p) = 0$ for $p \in B_2$, one gets

$$S(v) = S(v_{A_2}) \ ,$$

$$S(v) = \sum_{i=1}^{n} A^{iT} A^i x + (m+1)x$$

$$\sum_{i=1}^{n} A^{iT} v_i + \sum_{j=n+1}^{n+m+1} v_j = \sum_{i=1}^{n} A^{iT} A^i x + (m+1)x.$$

Setting

$$u = \sum_{i=1}^{n} A^{iT} v_i + \sum_{j=n+1}^{n+m+1} v_j$$

and

$$B = \sum_{i=1}^{n} A^{iT} A^i + (m+1)I \ ,$$

x can be computed by solving the equation

$$Bx = u \ .$$

This B is regular for all finite-dimensional linear operators $A^i : \mathbb{R}^k \to \mathbb{R}^k$, because all eigenvalues are greater than or equal to 1. So the inverse operator of B exists.

The projection of $v \in E$ onto $\mathcal{A}_2$ has the following form:

$$x := B^{-1} \left(\sum_{i=1}^{n} A^{iT} v_i + \sum_{j=n+1}^{n+m+1} v_j \right),$$

$$y_i := A^i x \quad i = 1(1)n,$$

$$y_{n+j} := x \quad j = 1(1)\,m+1,$$

$$v_{\mathcal{A}_2} = (y_1, \ldots, y_{n+m+1}).$$

Regarding $E = \mathcal{A}_2 \oplus \mathcal{B}_2$ (cf. (4.57)), the projection onto $\mathcal{B}_2$ is $v_{\mathcal{B}_2} = v - v_{\mathcal{A}_2}$.

The points $\tilde{p}^k$ and $\tilde{y}^k$ are connected via $\tilde{p}^k + \tilde{y}^k = p^k + y^k$ ($y^k \in \mathcal{A}_2, p^k \in \mathcal{B}_2$). The projection onto a linear subspace is additive, so this can be used to calculate one projection from the other:

$$\tilde{y}^k = y^k + p^k - \tilde{p}^k,$$

$$P_{\mathcal{A}_2}(\tilde{y}^k) = P_{\mathcal{A}_2}(y^k) + P_{\mathcal{A}_2}(p^k) - P_{\mathcal{A}_2}(\tilde{p}^k),$$

$$y^{k+1} - y^k = -P_{\mathcal{A}_2}(\tilde{p}^k).$$

Thus all steps of Spingarn's method of the partial inverse are solved for this problem. So it is possible to give another formulation of the proximal-point algorithm (with the sum $p^k + y^k$ as one variable, and $B_0^i = \{p : \|p\|_{i*} \le 1\}$).

The following algorithm is a procedure for solving the control approximation problem (4.52):

Algorithm II

1. Initialization

Choose $x^1 \in \mathbb{R}^k$ and $p^1 \in \mathcal{B}_2$, set $(p^1 + y^1)_i = p_i^1 + A^i x^1$ and $p_{n+j}^1 + y_{n+j}^1 = p_{n+j}^1 + x^1$.
Calculate $B^{-1} := \left((m+1)I + \sum_{i=1}^{n} A^{iT} A^i \right)^{-1}$.

2. Proximal Step

For $i = 1(1)n$: Set $b_i := y_i^k + p_i^k - a^i$.
If $\beta_i = 1$, then set

$$\tilde{p}_i^k = \begin{cases} b_i \text{ if } \|b_i\| < \alpha_i, \\ \alpha_i P_{B_0^i}(b_i/\alpha_i) \text{ otherwise.} \end{cases}$$

If $\beta_i > 1$ and $\| \cdot \|_{(i)}$ the sum, maximum, or Euclidean norm, then a special construction of $\tilde{p}_i^k$ is given in [347].

For $j = 1(1)m$: Set $\tilde{p}_{n+j}^k = p_{n+j}^k + y_{n+j}^k - P_{D_j}(p_{n+j}^k + y_{n+j}^k)$.
Set $\tilde{p}_{n+m+1} = c$.

3. Projection Step

Calculate:
$$\bar{p}^k := B^{-1}\underbrace{\left(\sum_{i=1}^{n} A^{iT}\tilde{p}_i^k + \sum_{j=1}^{m+1} \tilde{p}_{n+j} \right)}_{=:o_1}.$$

$x^{k+1} := x^k - \bar{p}^k.$

$p^{k+1} + y^{k+1} := \tilde{p}^k + (A^1(x^k - 2\bar{p}^k), \ldots, A^n(x^k - 2\bar{p}^k), x^k - 2\bar{p}^k, \ldots, x^k - 2\bar{p}^k).$

Stop

if $\|o_1\| + \|\bar{p}^k\| = \|S(\tilde{p})\| + \|\bar{p}^k\| < \varepsilon$ and $\|(p^{k+1} + y^{k+1}) - (p^k + y^k)\| < \varepsilon$ for a given value $\varepsilon > 0$.

Otherwise,

set $k = k + 1$ and return to the Proximal Step (step 2).

Special Case:

For the special case where all operators satisfy $A^i = I$, no B^{-1} is needed, and the projection step simplifies to

$$\bar{p}^k := \left(\sum_{i=1}^{n+m+1} \tilde{p}_i^k \right) / (n + m + 1) .$$

$x^{k+1} := x^k - \bar{p}^k.$

$(p^{k+1} + y^{k+1})_i := \tilde{p}_i^k + x^k - 2\bar{p}^k.$

Finally, we remark that the general **convergence results** for Spingarn's algorithm (compare [328]) also hold for our application of the proximal-point algorithm.

4.2.2 Computer programs for the application of the proximal-point algorithm

In [347] an application of Algorithm II for solving location problems in urban development (cf. (4.14) in Section 4.2.1 with $c = 0$) is given. The computer program for the proximal-point algorithm written in C^{++} is combined with a geographical information system (GIS) such that it is possible for the decision-maker to select on a map of his choice the given facilities a^i, the weights $\alpha_i \geq 0$, and the parameters $\beta_i \geq 1$ for $i = 1, \ldots, n$. Approximate solutions of the location problem computed by the proximal-point algorithm are visualized again on the map.

Figure 4.2.1 illustrates the use of this computer program.

Furthermore, we give an outline of a Mathematica program concerning the **proximal-point algorithm** for solving the problem (4.14) from Section

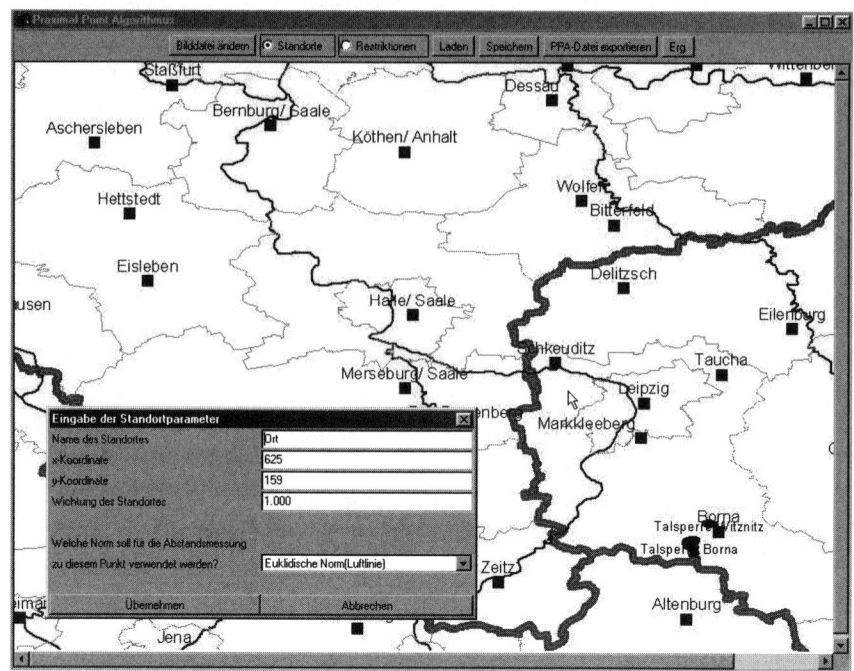

Figure 4.2.1. Application of Algorithm II in a C^{++} computer program.

4.2.1 with $c = 0$, $A^i = I$, and $\beta_i = 1$ for all $i = 1, \ldots, n$ (cf. Idrissi, Lefeb-vre, Michelot [171, 172] and Benker, Hamel, Spitzner, Tammer [21]). First, the decision-maker has to fix the locations a_i (points of the $\mathbb{R}^2$) and the corresponding weights α_i $(i = 1, \ldots, n)$ (real numbers greater than or equal to zero). These data are fed into the computer as a matrix:

$$\mathrm{loc} = \big\{ \{a_1, \alpha_1\}, \ldots, \{a_n, \alpha_n\} \big\}.$$

The number n of given points is not restricted. For each point we can choose a norm N_i $(i = 1, \ldots, n)$:

$$\mathrm{norm} = \{N_1, \ldots, N_n\}.$$

We take 1 for the maximum, 2 for the Euclidean, and 3 for the Lebesgue norm. Our algorithm for solving $(P(\lambda))$ is derived for location problems with a convex restriction set D. We use an intersection of half-spaces and circles. The half-spaces are given by

$$\xi_j \, x + \eta_j \, y + \varrho_j \leq 0,$$

and the circles by

$$(x - \xi_j)^2 + (y - \eta_j)^2 \leq \varrho_j^2,$$

where $(x, y) \in \mathbb{R}^2$.

We use $\varphi_j = 1$ for a half-plane and $\varphi_j = 2$ for a circle. Then we have to take

$$\mathrm{restr} = \big\{\{\xi_1, \eta_1, \varrho_1, \varphi_1\}, \ldots, \{\xi_n, \eta_n, \varrho_n, \varphi_n\}\big\}.$$

Finally, we need a limit of exactness, for example $\varepsilon = 10^{-10}$:

$$\mathrm{eps} = 10^{\char94}(-10).$$

Then we can start the program by

$$s[\mathrm{loc}, \mathrm{norm}, \mathrm{restr}, \mathrm{eps}].$$

The program gives as result three numbers. The first is the number of iterations needed to get the coordinates of the requested location, which are the other two numbers.

```
Needs[...];

parfo[ger_] := Module[...]; projki[ny_, nrn_] := Module[...];
projzj[pkt_, gk_] := Module[...]; ppa[parama_, paraml_, nor_,
xyk_, dgk_, n_, m_] := Module[...];

s[param_, nor_, dgk_, eps_] :=
Module[
 {i,par,n,dg,m,parama,paraml,ppv1,ppv2,l,differ,xyks,erg},
 ...

 xyks = {{0,0}}; ppv1 = xyks;
 Do[ ppv1=BlockMatrix[{{ppv1},{xyks}}], {i,1,n+m} ];
 ppv2 = ppa[parama,paraml,nor,ppv1,dgk,n,m];
 differ = Max[Abs[ppv2 - ppv1]];
 l = 1;
 While[
 differ > eps,
 ppv1 = ppv2;
 ppv2 = ppa[parama,paraml,nor,ppv1,dgk,n,m];
 l = l + 1;
 differ = Max[Abs[ppv2 - ppv1]];
 ];
 erg = Prepend[ppv2[[1]], l];
 Return[erg];
 ];
```

First, the data are transformed into a usable form. The routine **ppa** is our turnover of the proximal-point algorithm. In the subroutine **projzj** we project points on half-spaces or circles and in the subroutine `projkj` we project a vector on the unit ball of the used norm.

Let us calculate the same example as in the section with the geometric algorithm using as a limit of exactness $\varepsilon = 10^{-10}$. If we take the same weights for all locations and the maximum norm for all norms, that is,

$$\text{loc} = \{ \{-1.5, 3.5, 1\}, \{1, 3, 1\}, \{1, 0, 1\}, \{-3, -2, 1\},$$
$$\{3.5, -1.5, 1\}, \{2, 2, 1\}, \{-2, 2, 1\}, \{4, 1, 1\}, \{-3, 2, 1\}\},$$

we get as optimal solution $x_0^1 = (0.5, 1.5)$. But let us take different weights. For

$$\text{loc} = \{ \{-1.5, 3.5, 1\}, \{1, 3, 2\}, \{1, 0, 1\}, \{-3, -2, 4\},$$
$$\{3.5, -1.5, 3\}, \{2, 2, 1\}, \{-2, 2, 2\}, \{4, 1, 1\}, \{-3, 2, 1\}\},$$

we get $x_0^2(1.37944, 0.62056)$, and for

$$\text{loc} = \{ \{-1.5, 3.5, 1\}, \{1, 3, 3\}, \{1, 0, 3\}, \{-3, -2, 1\},$$
$$\{3.5, -1.5, 6\}, \{2, 2, 5\}, \{-2, 2, 1\}, \{4, 1, 1\}, \{-3, 2, 1\}\},$$

we get $x_0^3(0.289759, 0.710241)$.

These locations are plotted in Figure 4.2.2 by small circles.

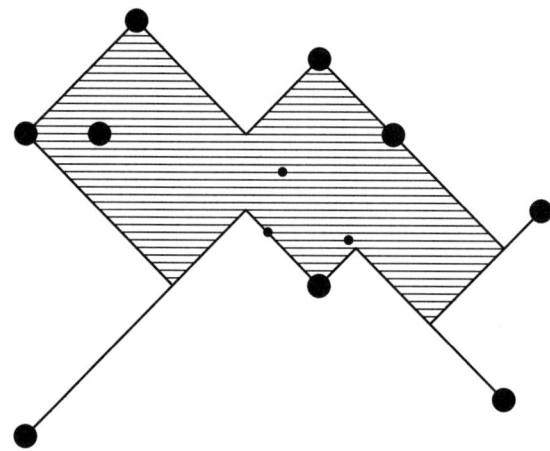

Figure 4.2.2. Solutions x_0^1, x_0^2, and x_0^3 of the location problem generated by the proximal-point algorithm choosing different weights α_i $(i = 1, \ldots, n)$.

4.2.3 An Interactive Algorithm for the Vector Control Approximation Problem

In order to formulate this problem we suppose that

(A) a^i, $x \in X$, $\alpha_i \geq 0$, $\beta_i \geq 1$, $f_1 \in L(X, \mathbb{R}^n)$, $A^i \in L(X, Y_i)$ $(i = 1, \ldots, n)$; ($L(X, Y)$ denotes the space of linear continuous operators from X to Y);

(B) $D_j \subset X$ $(j = 1, \ldots, m)$ are closed and convex sets, $D_0 \subseteq X$ is a closed linear subspace, and $D = \bigcap_{j=0}^{m} D_j$ is nonempty and bounded. Furthermore, we assume that a suitable constraint qualification is satisfied.

(C) One of the following conditions (C1) or (C2) is satisfied:
  (C1) H is a Hilbert space and $X = Y_i = H$ for all $i = 1, \ldots, n$;
  (C2) $X = \mathbb{R}^k$ and $Y_i = \mathbb{R}^k$ for all $i = 1, \ldots, n$.

(D) $C \subset \mathbb{R}^n$ is a convex cone with $\operatorname{cl} C + (\mathbb{R}_+^n \setminus \{0\}) \subset \operatorname{int} C$.

Now we consider the following **vector control approximation problem:**

(P)   Compute the set $\operatorname{Eff}(f[D], C)$,

where

$$f(x) := f_1(x) + \begin{pmatrix} \alpha_1 \|A^1(x) - a^1\|^{\beta_1} \\ \cdots \\ \alpha_n \|A^n(x) - a^n\|^{\beta_n} \end{pmatrix}. \tag{4.58}$$

We introduce a suitable scalarization of the vector control approximation (P), and we derive an interactive algorithm for the vector control approximation problem using a surrogate parametric optimization problem and taking into account stability results of this special parametric optimization problem.

Under the given assumptions it is easy to see that the vector-valued objective function $f : X \longrightarrow \mathbb{R}^n$ in (4.58) is ($\mathbb{R}_+^n$)-**convex**; i.e., for all x_1, $x_2 \in X$, $\mu \in [0, 1]$, we have $\mu f(x_1) + (1 - \mu) f(x_2) \in f(\mu x_1 + (1 - \mu) x_2) + \mathbb{R}_+^n$ (compare Section 2.4).

Then we can show (compare with [195]) that for each element $f(x^0) \in \operatorname{Eff}(f[D], C)$ there exists a parameter $\lambda \in \operatorname{int} C^+$ such that x^0 solves the real-valued optimization problem

(P(λ)) $f(x, \lambda) := \sum_{i=1}^{n} \lambda_i \big((f_1)_i(x) + \alpha_i \|A^i(x) - a^i\|^{\beta_i} \big) \longrightarrow \min_{x \in D}$.

In the following, for the case $\alpha_i > 0$ for all $i = 1, \ldots, n$, without loss of generality we replace (P(λ)) by

(P$'_s$) $\hat{f}_1(x) + \sum_{i=1}^{n} \|A^i(x) - a^i\|^{\beta_i} \longrightarrow \min_{x \in D}$,

where $\hat{f}_1 \in X^*$. Using the *indicator functions* χ_{D_j} of D_j defined by $\chi_{D_j}(x) = 0$, if $x \in D_j$ and $\chi_{D_j}(x) = +\infty$ otherwise, the problem (P$'_s$) is equivalent to the following unconstrained minimization problem:

$$(\mathbf{P}_s) \quad F(x) = \hat{f}_1(x) + \sum_{i=1}^{n} \left\| A^i(x) - a^i \right\|^{\beta_i} + \sum_{j=0}^{m} \chi_{D_j}(x) \to \min_{x \in X}.$$

First, we consider the case of ((A),(B),(C1)). Obviously, under the given assumptions the functional F is convex. In order to solve $(\mathbf{P}_s)$ it is possible to use **Algorithm I**.

Remark 4.2.6. Under the assumptions ((A), (B), (C2)) it is possible to apply **Algorithm II** (see Section 4.2.1) in the same way. In this case the weak convergence of the sequence $\{z^k\}$ implies even the norm convergence.

Now we present an interactive algorithm in which we have to solve the special parametric optimization problem under the assumptions ((A), (B), (C1)) or ((A), (B), (C2)):

$$(\mathbf{P}(\lambda)) \quad f(x, \lambda) = \textstyle\sum_{i=1}^{n} \lambda_i((f_1)_i(x) + \alpha_i \| A^i(x) - a^i \|^{\beta_i}) \longrightarrow \min_{x \in D},$$

where $\lambda \in \Lambda \subset \operatorname{int} C^+$.

Stability results for parametric optimization problems are important for an effective **interactive algorithm**. In other words, we need various types of continuity of the **optimal-value function**

$$\varphi(\lambda) := \inf\{f(x, \lambda) \mid x \in D\},$$

of the **optimal set mapping**

$$\psi(\lambda) := \{x \in D \mid f(x, \lambda) = \varphi(\lambda)\},$$

or of the ε-**optimal set mappings**

$$\psi_\varepsilon(\lambda) := \{x \in D \mid f(x, \lambda) < \varphi(\lambda) + \varepsilon\}$$

and

$$\bar{\psi}(\lambda, \varepsilon) := \{x \in D \mid f(x, \lambda) \leq \varphi(\lambda) + \varepsilon\}.$$

By stability of the mappings φ, ψ, ψ_ε, and $\bar{\psi}$ we mean in particular certain continuity attributes of these mappings. We use the concept of upper and lower continuity introduced in Definition 2.5.1 (a) and (b) in the following formulation:

Definition 4.2.7. *A multifunction* $\Gamma : \Lambda \rightrightarrows 2^X$, *where* (X, d_X) *and* (Λ, d_Λ) *are metric spaces, is called:*

1. **upper continuous** *at a point* λ^0 *if for each open set* Ω *containing* $\Gamma(\lambda^0)$ *there exists a* δ-*neighborhood* $V_\delta\{\lambda^0\}$ *of* λ^0 *such that*

$$\Gamma(\lambda) \subset \Omega \qquad \text{for all} \qquad \lambda \in V_\delta\{\lambda^0\};$$

2. **lower continuous** *at a point* λ^0 *if for each open set* Ω *satisfying* $\Omega \cap \Gamma(\lambda^0) \neq \emptyset$ *there exists a* δ-*neighborhood* $V_\delta\{\lambda^0\}$ *of* λ^0 *such that*

$$\Gamma(\lambda) \cap \Omega \neq \emptyset \qquad \text{for all} \qquad \lambda \in V_\delta\{\lambda^0\}.$$

The existence of continuous selection functions is closely related to the condition that ψ is lower continuous and all optimal sets are nonempty and convex. If one considers ε-optimal solutions, then the strong condition of lower continuity of ψ may be avoided.

Theorem 4.2.8. *Suppose the assumptions* (A), (B), (C1) *are satisfied. Then for each* $\varepsilon > 0$ *the* ε-**optimal set mapping** ψ_ε *is lower continuous in the sense of Definition 4.2.7.*

PROOF. The assumptions of Theorem 4.2.4 in [147] are satisfied, since f is continuous on $X \times \Lambda$, D is closed and does not depend on the parameter λ, and φ is continuous regarding the continuity of f on the set D. Theorem 4.2.4 in [147] yields that for each $\varepsilon > 0$ the ε-optimal set mapping ψ_ε is lower continuous. □

In the finite-dimensional case we can derive some additional results:

Theorem 4.2.9. *We consider the problem* $(P(\lambda))$ *subject to the assumptions* (A), (B) *and* (C2). *Suppose that* $\psi(\lambda^0)$ *is nonempty and bounded. Then*

(i) *the the optimal-value function* φ *is continuous at* λ^0;
(ii) *the optimal set mapping* ψ *is upper continuous in the sense of Definition 4.2.7 at* λ^0;
(iii) *the* ε-*optimal set mapping* ψ_ε *is lower continuous in the sense of Definition 4.2.7 at* λ^0 *for each* $\varepsilon > 0$;
(iv) *the mapping* $\bar{\psi}$ *defined by*

$$\bar{\psi}(\lambda, \varepsilon) := \{x \in M \mid f(x, \lambda) \leq \varphi(\lambda) + \varepsilon\}, \quad \lambda \in \Lambda, \ \varepsilon \geq 0,$$

is upper continuous in the sense of Definition 4.2.7 at $(\lambda^0, 0)$.

PROOF. The results follow immediately from Theorem 9 in Hogan [167]. □

Remark 4.2.10. The last property is of interest if the problems corresponding to the parameters λ^t, $t = 1, 2, \ldots$, $\lambda^t \longrightarrow \lambda^0$, constitute certain substitute problems with respect to the problem to be solved, $(P(\lambda^0))$, and the problems $(P(\lambda^t))$ are solved with increasing accuracy $\varepsilon_t \longrightarrow 0$. Then each sequence $\{x^t\}$ of ε_t-optimal solutions of $(P(\lambda^t))$ possesses an accumulation point, and each of its accumulation points is contained in the solution set $\psi(\lambda^0)$.

Moreover, under the assumptions of Theorem 4.2.8 a continuous selection theorem of Michael [255] can be used if we additionally assume the compactness of Λ.

Theorem 4.2.11. *Assume that* (A), (B), (C) *hold, and that Λ is a compact set. Then there exists a function $g \in C(\Lambda, X)$ such that*

$$g(\lambda) \in \operatorname{cl} \psi_\varepsilon(\lambda) \quad \forall \lambda \in \Lambda.$$

PROOF. We can conclude from Theorems 4.2.8 and 4.2.9 that ψ_ε is lower continuous in the sense of Definition 4.2.7. Moreover, the image sets $\psi_\varepsilon(\lambda)$ are nonempty and convex for all $\lambda \in \Lambda$. Then we get the desired result from the continuous selection theorem of Michael [255].

Using these stability statements we can derive the following **interactive algorithm** for the vector control approximation problem (P). At least under ((A), (B), (C2)) it is possible to seek elements of a neighborhood of the set of proper efficient points that corresponds to the individual interest of the decision-maker (compare [23]).

In the following interactive procedure for solving the vector control approximation problem we can use **Algorithm I** or **Algorithm II** from Section 4.2.1.

Step 1: Choose $\bar{\lambda} \in \Lambda$. Compute an approximate solution (x^0, p^0) with the primal–dual algorithm **Algorithm I** (or **II**). If (x^0, p^0) is accepted by the decision-maker, then **stop**. Otherwise, go to Step 2.

Step 2: Put $k = 0$, $t_0 = 0$. Choose $\hat{\lambda} \in \Lambda$, $\hat{\lambda} \neq \bar{\lambda}$. Go to Step 3.

Step 3: Choose t_{k+1} with $t_k < t_{k+1} \leq 1$ and compute an approximate solution (x^{k+1}, p^{k+1}) of

$$\mathrm{P}(t_{k+1}, \bar{\lambda}, \hat{\lambda}) \quad \min_{x \in D} \left\{ \sum_{i=1}^{n} (\bar{\lambda}_i + t_{k+1}(\hat{\lambda}_i - \bar{\lambda}_i))((f_1)_i(x) + \alpha_i \| A^i(x) - a^i \|^{\beta_i}) \right\}$$

with the **Algorithm I** (or **II**) and use (x^k, p^k) as starting point. If an approximate solution of $\mathrm{P}(t, \bar{\lambda}, \hat{\lambda})$ cannot be found for $t > t_k$, then go to Step 1. Otherwise, go to Step 4.

Step 4: The point (x^{k+1}, p^{k+1}) is to be evaluated by the decision-maker. If it is accepted by the decision-maker, then **stop**. Otherwise, go to Step 5.

Step 5: If $t_{k+1} \geq 1$, then go to Step 1. Otherwise, set $k = k + 1$ and go to Step 3.

Remark 4.2.12. Under the assumptions (A), (B) and (C2) of Theorem 4.2.9, a sufficiently good approximation of a solution of $(\mathrm{P}(t_{k+1}, \bar{\lambda}, \hat{\lambda})$ can be generated if we use an approximate solution of $(\mathrm{P}(t_k, \bar{\lambda}, \hat{\lambda}))$ as starting point.

4.2.4 Proximal Algorithms for Vector Equilibrium Problems

The **proximal-point method** was introduced by Martinet (see [252] and [253]) as a regularization method in the context of convex optimization in

Hilbert spaces. It has since been studied by several authors for monotone inclusion problems and variational inequalities (see [308] for a survey). The purpose of this section is to present generalized proximal-point methods with Bregman functions applied to vector equilibrium problems and discuss the convergence properties of these algorithms. Our method consists, in the first place, in scalarizing the vectorial problem, then adapting the real proximal-point algorithm, as proposed by [263] and [118], to the **weak vector equilibrium problem** (WVEP).

During this section, we restrict attention to a finite-dimensional space $X = \mathbb{R}^m$, even if much of what will be said carries over to the reflexive Banach setting. Suppose Y is a Hausdorff topological vector space, and $P \subset Y$ a pointed closed convex cone with nonempty interior int P.

Let $M \subset X$ and $f : M \times M \longrightarrow Y$ be a vector-valued function and consider the problem

(WVEP)   find $\bar{x} \in M$ such that $f(\bar{x}, y) \notin -\operatorname{int} P$ for all $y \in M$.

Associated with this problem is the **scalar equilibrium problem**

(SEP)   find $\bar{x} \in M$ such that $F(\bar{x}, y) \geq 0$ for all $y \in M$.

Here $F : M \times M \longrightarrow \mathbb{R}$ is defined by $F(x, y) = \varphi(f(x, y))$, where φ is a suitable functional (see Section 2.3) defined on Y by $\varphi(z) := \inf\{t \in \mathbb{R} \mid z \in tu^0 - P\}$ and u_0 is a fixed arbitrary element in int P.

We will make use of the following scalarization properties: see Theorem 2.3.1 and Corollary 2.3.5: φ is continuous, sublinear, P-monotone ($z_2 - z_1 \in P \Rightarrow \varphi(z_1) \leq \varphi(z_2)$ and $z_2 - z_1 \in \operatorname{int} P \Rightarrow \varphi(z_1) < \varphi(z_2)$), and for every $\lambda \in \mathbb{R}$ the sublevel and strict sublevel sets of φ of height λ are given by $\operatorname{lev}_\varphi(\lambda) = \{z \in Y \mid \varphi(z) \leq \lambda\} = \lambda u^0 - P$ and $\operatorname{lev}_\varphi^<(\lambda) = \{z \in Y \mid \varphi(z) < \lambda\} = \lambda u^0 - \operatorname{int} P$. Moreover, for every $\lambda \in \mathbb{R}$ and $z \in Y$ one has $\varphi(z + \lambda u^0) = \varphi(z) + \lambda$.

Note that from the above characterizations of the sublevel sets of φ, one can also obtain

$$\varphi(z) > \lambda \Leftrightarrow z - \lambda u^0 \notin -P,$$
$$\varphi(z) \geq \lambda \Leftrightarrow z - \lambda u^0 \notin -\operatorname{int} P,$$
$$\varphi(z) = \lambda \Leftrightarrow z - \lambda u^0 \in -\operatorname{bd} P.$$

Thus the problems (WVEP) and (SEP) are equivalent in the sense that the solution set of each of the two problems coincides with each other.

Bregman functions.

For a given real function h defined on a nonempty closed convex subset S of X, with int $S \neq \emptyset$, we let $D_h(x, y) := h(x) - h(y) - \langle x - y, \nabla h(y) \rangle$, for each $x \in S, y \in \operatorname{int} S$. The function D_h is called the **Bregman distance**. Let us notice that the term "distance" is misleading: Neither is D_h symmetric nor does it satisfy the triangle inequality. However, the function D_h has some good distance features if the function h is nice enough; see Lemmas 4.2.20 and 4.2.21.

Definition 4.2.13. *A function $h : S \longrightarrow \mathbb{R}$ is called a **Bregman function** with zone* int S, *for short B-function, if*

(B1) h *is continuous and strictly convex on S;*
(B2) h *is continuously differentiable on* int S;
(B3) *for every $t \in \mathbb{R}, x \in S$ and $y \in$ int S, the level set $L(x,t) := \{y \in$ int $S \mid D_h(x,y) \le t\}$ is bounded;*
(B4) *if $y_n \longrightarrow y_0$ in S, then $D_h(y_0, y_n) \longrightarrow 0$;*
(B5) *if $\{y_n\} \subset$ int S converges to y_0 in S, $\{x_n\}$ is bounded in S, and $D_h(x_n, y_n) \longrightarrow 0$, then $x_n \longrightarrow y_0$.*

Note that letting $h(x) = +\infty$ for $x \notin S$, we have that $h : X \to \mathbb{R} \cup \{+\infty\}$ is a proper convex lower semicontinuous function on X, with domain dom $h = S$. For the function D_h, the above conditions ensure that the natural domain of D_h is $S \times$ int S.

Remark 4.2.14. When all $\{x_n\}, \{y_n\}$, and y_0 are in int S, conditions (B4) and (B5) hold automatically, as a consequence of (B1)–(B3). So (B4) and (B5) need to be checked only at points on the boundary bd $S := $ cl $S \setminus$ int S of S.

Remark 4.2.15. When $S = \mathbb{R}^m$, a sufficient condition for a strictly convex and differentiable function h to be a B-function is a coercivity condition, i.e., $\lim_{\|x\|\to\infty} h(x)\|x\|^{-1} = +\infty$.

Let us cite some **examples of B-functions**; other examples can be found in [59], [74], [104], [118], [187], [186].

Example 4.2.16. Let $h(x) = \frac{1}{2}\|x\|^2$ for which $D_h(x, y) = \frac{1}{2}\|x - y\|^2$ for every $x, y \in X$. Then all the properties (B1)–(B5) are satisfied for $S = X$; i.e., h is a B-function.

Example 4.2.17. Suppose that $S = \mathbb{R}^m$ and for $x = (x_1, \ldots, x_m) \in S$, let $h(x) = \frac{1}{2}\sum_{i=1}^m |x_i|^p$. Then h is a B-function with zone $\mathbb{R}^m$.
    On could also consider $S = \mathbb{R}^m_+$ and the function $h(x) = \sum_{i=1}^m x_i \log x_i$ on $\mathbb{R}^m_{++} := $ int S and $h(x) = 0$ if $x \in$ bd S. In this case,

$$D_h(x, y) = \sum_{i=1}^m \left(x_i \log \frac{x_i}{y_i} + y_i - x_i \right).$$

Example 4.2.18. Suppose that $S = [-1, +1]^m$ and for $x = (x_1, \ldots, x_m) \in S$, let $h(x) = -\sum_{i=1}^m \sqrt{1 - x_i^2}$. Then h is a B-function with zone $]-1, +1[^m$.

Example 4.2.19. Suppose that $S = \mathbb{R}^m_+$ and for $x = (x_1, \ldots, x_m) \in S$, let $h(x) = \sum_{i=1}^m (x_i^\alpha - x_i^\beta)$, with $\alpha \ge 1$ and $0 < \beta < 1$. Then

- for $\alpha = 2$ and $\beta = \frac{1}{2}$, we get $D_h(x, y) = \|x - y\|^2 + \sum_{i=1}^{m} \frac{1}{2\sqrt{y_i}} (\sqrt{x_i} - \sqrt{y_i})^2$;
- and for $\alpha = 1$ and $\beta = \frac{1}{2}$, we get $D_h(x, y) = \sum_{i=1}^{m} \frac{1}{2\sqrt{y_i}} (\sqrt{x_i} - \sqrt{y_i})^2$.

Lemma 4.2.20. *For every $x \in S$ and every $y \in$ int S, we have $D_h(x, y) \geq 0$, and $D_h(x, y) = 0$ if and only if $x = y$.*

PROOF. This lemma follows from the definition of D_h and the property (B1). $\qquad \square$

Lemma 4.2.21. Three points Lemma. *For every $x, y, z \in$ int S, we have*

$$D_h(x, z) = D_h(x, y) + D_h(y, z) + \langle x - y, \nabla h(y) - \nabla h(z) \rangle. \qquad (4.59)$$

PROOF. This follows from a straightforward substitution of the definition of D_h. $\qquad \square$

In the rest of this section, let us consider the following assumptions concerning the data of the studied vector equilibrium problem:

(A0) M is a nonempty closed convex subset of X.

(A1) $f : M \times M \to Y$ satisfies $\forall x, y \in M$ $\forall \lambda \in \mathbb{R}$: $f(x, y) \notin \lambda u^0 - $ int P implies $f(y, x) \in -\lambda u^0 - P$, and $f(x, x) = 0$.

(A2) In the second argument f is P-convex and P-lower semicontinuous on M.

(A3) In the first argument f is P-upper semicontinuous on each line segment in M.

(B6) $h : X \longrightarrow \bar{\mathbb{R}}$ is a B-function with a zone S whose interior contains M.

(B7) h is strongly convex on int S with modulus $\alpha > 0$; i.e., for all $x, y \in$ int S one has

$$h(x) - h(y) \geq \langle x - y, \nabla h(y) \rangle + \frac{\alpha}{2} \|x - y\|^2.$$

(B8) The gradient ∇h of h is Lipschitz continuous on int S with modulus K; i.e., for all $x, y \in$ int S one has

$$\|\nabla h(x) - \nabla h(y)\| \leq K\|x - y\|.$$

Equivalently, in terms of subdifferential operators, h is strongly convex iff its subdifferential operator is strongly monotone, that is to say, $\forall x, y \in$ int S

$$\langle y - x, \nabla h(y) - \nabla h(x) \rangle \geq \alpha \|x - y\|^2. \qquad (4.60)$$

One remarks also that if h is strongly convex, then h is strictly convex and has bounded level sets.

The Lipschitz continuity of ∇h implies the following general condition: $\forall x, y \in$ int S

$$\langle y - x, \nabla h(y) - \nabla h(x) \rangle \leq K\|x - y\|^2, \qquad (4.61)$$

The relations (4.60) and (4.61) lead also to

$$\frac{\alpha}{2} \|x - y\|^2 \leq D_h(x, y) \leq K\|x - y\|^2.$$

Remark 4.2.22. (1) Let us notice that (A1) is equivalent to $\forall x, y \in M$, $\forall \lambda \in \mathbb{R}$, $F(x,y) \geq \lambda \Rightarrow F(y,x) \leq -\lambda$. This condition is also satisfied when we suppose that $\forall x, y \in M$, $F(x,y) + F(y,x) \leq 0$.

(2) Condition (A2) implies that F is convex and lower semicontinuous in the second argument.

PROOF. To prove the first statement it suffices to use the sublinearity of φ and $\mathrm{lev}_\varphi(0) := \{z \in Y \mid \varphi(z) \leq 0\} = -P$. For the second statement, we remark that for $x \in M$,

$$
\begin{aligned}
\mathrm{lev}_{F(x,\cdot)}(\lambda) &:= \{y \in M \mid F(x,y) \leq \lambda\} \\
&= \{y \in M \mid f(x,y) \in \lambda u^0 - P\} =: \mathrm{lev}_{f(x,\cdot)}(\lambda u^0).
\end{aligned}
$$

Thus, P-lower semicontinuity (resp. P-quasi-convexity) of $f(x, \cdot)$ implies lower semicontinuity (resp. quasi-convexity) of $F(x, \cdot)$.

To prove convexity of $F(x, \cdot)$, we use only the sublinearity and P-monotonicity of φ.   □

One-phase proximal algorithms.

We start with the following approximate iteration scheme (VPA1) (called the **one-phase vector proximal algorithm**) for solving (WVEP):

Algorithm (VPA1).

Consider sequences $\{\alpha_k\}$ and $\{\varepsilon_k\}$ of nonnegative real numbers.

1. Start at an arbitrary $x^0 \in M$ and $\xi_0 \in B^*$.
2. If (x^k, ξ_k) is in the current iterate, the next iterate $(x^{k+1}, \xi_{k+1}) \in M \times B^*$ is a solution of

$$
\alpha_k \langle f(x^{k+1}, x), \xi_{k+1} \rangle + h(x) - h(x^{k+1}) - \langle x - x^{k+1}, \nabla h(x^k) \rangle + \varepsilon_k \geq 0 \,\forall\, x \in M. \tag{4.62}
$$

The idea underlying the vector proximal algorithm for solving the problem (WVEP) for a vector-valued mapping is basically the same as the scalar one for solving the problem (SPE). More precisely, if x^k is the current point, the next term x^{k+1} of the iteration (4.62) produced by (VPA1) is a solution of

$$
\alpha_k F(x^{k+1}, x) - D_h(x^{k+1}, x^k) + D_h(x, x^k) + \varepsilon_k \geq 0 \quad \forall\, x \in M. \tag{4.63}
$$

Lemma 4.2.23. *Suppose that (A2) and (B6) are satisfied. Then for $\varepsilon_k = 0$ a solution of iteration (4.63) is a solution of (4.64), and conversely,*

$$
\alpha_k F(x^{k+1}, x) + \langle x^{k+1} - x, \nabla h(x^k) - \nabla h(x^{k+1}) \rangle \geq 0 \quad \forall\, x \in M. \tag{4.64}
$$

PROOF. Suppose that x^{k+1} is a solution of (4.63). By way of the sub-differential operator of the proper convex lower semicontinuous functions $\phi(x) := \alpha_k F(x^{k+1}, x) + D_h(x, x^k) + \delta_M(x)$, the relation (4.63) is equivalent to

$$0 \in \partial(\alpha_k F(x^{k+1}, \cdot) + D_h(\cdot, x^k) + \delta_M(\cdot))(x^{k+1}). \qquad (4.65)$$

Applying the subdifferential calculus to the sum of convex functions, and using the domain qualification $\mathrm{rint\,dom}(F(x^{k+1}, \cdot)) \cap \mathrm{rint\,dom}(D_h(\cdot, x^k)) \cap \mathrm{rint}\,M = \mathrm{rint}\,M \neq \emptyset$, we have [1]

$$0 \in \partial(\alpha_k F(x^{k+1}, \cdot) + \delta_M)(x^{k+1}) + \partial D_h(\cdot, x^k)(x^{k+1}).$$

Let $y_1 \in \partial(\alpha_k F(x^{k+1}, \cdot) + \delta_M)(x^{k+1})$ and $y_2 \in \partial D_h(\cdot, x^k)(x^{k+1})$ be such that $0 = y_1 + y_2$; then

$$-y_2 = \nabla h(x^k) - \nabla h(x^{k+1}) \in \partial(\alpha_k F(x^{k+1}, \cdot) + \delta_M)(x^{k+1}).$$

In other words,

$$0 \le \alpha_k F(x^{k+1}, x^{k+1}) \le \alpha_k F(x^{k+1}, x) + \langle x^{k+1} - x, \nabla h(x^k) - \nabla h(x^{k+1}) \rangle \; \forall x \in M.$$

We then obtain the relation (4.64).

For the converse, it suffices to remark that $\nabla h(x^{k+1})$ is a subgradient of the convex function h at x^{k+1}. $\qquad \square$

Let us remark that for Lemma 4.2.23 we base our argument on the sum formula of subdifferentials that is now a part of the folklore of scalar convex analysis; see Rockafellar [308].

Theorem 4.2.24. (Convergence result of VPA1) *Assume that M, f, h, and P satisfy assumptions (A1)–(A3) and (B1)–(B6). Suppose in addition that*

$$\varepsilon_k = 0 \text{ and } 0 < \lambda \le \alpha_k \le \Lambda < +\infty \text{ for each } k \in \mathbb{N}. \qquad (4.66)$$

Then the sequence $\{x_k\}$ generated by (VPA1) converges to a solution $\bar{x}$ of (WVEP), and for each solution x^* one has $f(\bar{x}, x^*), f(x^*, \bar{x}) \in - \mathrm{bd}\,P$.

PROOF. By way of Lemma 4.2.23, we have

$$0 \le \alpha_k F(x^{k+1}, x) + \langle x^{k+1} - x, \nabla h(x^k) - \nabla h(x^{k+1}) \rangle \; \forall x \in M. \qquad (4.67)$$

Choose $x = x^*$ in the solution set of (WVEP). Then combining (A1) and Remark 4.2.22 yields $\alpha_k F(x^{k+1}, x^*) \le 0$ for each $k \in \mathbb{N}$, and thus

$$\langle x^{k+1} - x^*, \nabla h(x^k) - \nabla h(x^{k+1}) \rangle \ge 0. \qquad (4.68)$$

Applying Lemma 4.2.21, we obtain

---

[1] $\mathrm{rint}\,M$ is the relative interior of M, i.e., the interior of M relative to $\mathrm{aff}\,M$, the intersection of all linear manifolds containing M.

$$D_h(x^*, x^k) \geq D_h(x^*, x^{k+1}) + D_h(x^{k+1}, x^k). \tag{4.69}$$

By Lemma 4.2.20, we deduce that the sequence $\{D_h(x^*, x^k)\}$ is nonnegative and decreasing. Thus $\{D_h(x^*, x^k)\}$ is convergent. On account of boundedness of the level set $L(x^*, D_h(x^*, x^0)) := \{x \in \text{int } S \mid D_h(x^*, x) \leq D_h(x^*, x^0)\}$, we can confirm that $\{x^k\}$ is bounded.

Let $\bar{x} \in M$ be an accumulation point of $\{x^k\}$. Then for some subsequence $\{x^{k_j}\}$ we have $\lim_{j \to \infty} x^{k_j} = \bar{x}$. Taking for $j \in \mathbb{N}$, $u_j = x^{k_j}$ and $v_j = x^{k_j+1}$, we have $\lim_{j \to \infty} D_h(v_j, u_j) = 0$. Using (B5), we deduce that $\{u_j\}$ and $\{v_j\}$ have the same limit $\bar{x}$.

Fix some $x \in M$; we obtain from (4.67) and (B2) that

$$\liminf_{k \to +\infty} F(x^{k_j+1}, x) \geq \liminf_{k \to +\infty} \frac{1}{\alpha_k} \langle x^{k_j+1} - x, \nabla h(x^{k_j+1}) - \nabla h(x^{k_j}) \rangle = 0. \tag{4.70}$$

By assumptions (A1), Remark 4.2.22, and (A2) we have

$$F(x, \bar{x}) \leq \liminf_{j \to \infty} F(x, x^{k_j+1}) \leq 0.$$

Thus $\bar{x}$ is a solution of $F(x, \bar{x}) \leq 0$ for all $x \in M$.

The assertion of this theorem is not yet proved. For this, let $y \in K$ and consider, for $t \in]0, 1[$, $x_t = ty + (1 - t)\bar{x}$. Since M is convex, then for each $t \in]0, 1[$, $F(y_t, \bar{x}) \leq 0$.

From (A2) (see Remark 4.2.22) it follows that for every $t \in]0, 1[$,

$$0 = F(x_t, x_t) \leq tF(x_t, y) + (1 - t)F(x_t, \bar{x}) \leq tF(x_t, y).$$

Letting $t \searrow 0$, from the assumption (A3), upper semicontinuity on $[\bar{x}, y]$ for the first argument of F yields $F(\bar{x}, y) \geq 0$; it follows that $\bar{x}$ is a solution of (WVEP).

Let us prove that the whole sequence $\{x^k\}$ converges to $\bar{x}$. Since $\bar{x}$ is a solution of (WVEP), we obtain from (4.69) that $\{D_h(\bar{x}, x^k)\}$ converges to some limit $r \geq 0$. Using the Bregman assumption (B4), we have $\lim_{j \to \infty} D_h(\bar{x}, x^{k_j}) = 0$. Therefore, $r = 0$.

Let us consider any convergent subsequence $\{v_j = x^{k_j}\}$ of $\{x^k\}$ and $\bar{x}'$ the corresponding limit point. Then, for $u_j = \bar{x} \ \forall j \in \mathbb{N}$, we have $\lim_{j \to \infty} D_h(u_j, v_j) = 0$, and thus (B5) gives $\bar{x} = \bar{x}'$. This implies that the whole sequence $\{x^k\}$ converges to $\bar{x}$.

Let x^* be an arbitrary solution of (WVEP); one deduces that $F(\bar{x}, x^*) \geq 0$ and $F(x^*, \bar{x}) \geq 0$. Using (A1), it follows that $F(\bar{x}, x^*) = F(x^*, \bar{x}) = 0$; this completes the proof. $\qquad \square$

Remark 4.2.25. In the algorithm (VPA1) the iteration x^{k+1} is obtained through the optimal solution of the equilibrium problem (4.64), which is difficult or even impossible to solve exactly. In practice, we may expect to compute an approximation to the optimal solution x^{k+1} of Problem (4.64), i.e., $\varepsilon_k > 0$. Thus we need to take into account that $\{\varepsilon_k\}$ is a sequence of decreasing positive numbers such that $\sum_{k=0}^{\infty} \sqrt{\varepsilon_k} < +\infty$.

We shall leave it to the reader to verify this assertion; we refer to [118].

Remark 4.2.26. Replacing (A1) by $\forall x, y \in M \ \forall \lambda \in \mathbb{R}$ one has that $x \neq y$ and $f(x, y) \notin \lambda u^0 - \text{int } P$ imply $f(y, x) \in -\lambda u^0 - \text{int } P$; we can assert that the solution set of (WVEP) is reduced to the unique solution $\bar{x}$.

Indeed, consider x^* another solution of (WVEP). One deduces that $f(\bar{x}, x^*)$ and $f(x^*, \bar{x})$ are not in $-\text{int } P$. Using the proposed assumptions with $\lambda = 0$, it follows that $f(\bar{x}, x^*) \in -\text{int } P$, a contradiction. $\qquad\square$

Remark 4.2.27. When we suppose that X is a real reflexive Banach space, we need assumptions (B7) and (B8) on the B-function h to realize weak convergence of a subsequence of $\{x^k\}$ to a solution of (WVEP). If f satisfies instead of (A1) the condition in Remark 4.2.26, we conclude that the whole sequence $\{x^k\}$ weakly converges to the unique solution of (WVEP).

PROOF. As in the proof of the previous theorem, we could justify that $\{x^k\}$ is bounded and $D_h(x^{k+1}, x^k) \leq D_h(x^*, x^k) - D_h(x^*, x^{k+1})$. Taking into account (B7), we have $\frac{\alpha}{2}\|x^k - x^{k+1}\|^2 \leq D_h(x^{k+1}, x^k)$, and $\{D_h(x^*, x^k)\}$ is decreasing to some nonnegative limit; thus

$$\lim_{k \to +\infty} \|x^k - x^{k+1}\| = 0.$$

Coming back to (4.70), and using (A2) and (B2), which ensures the weak lower semicontinuity of $F(x, \cdot)$, we deduce from (B8) that for all $x \in M$,

$$
\begin{aligned}
F(x, \bar{x}) &\leq \liminf_{j \to \infty} F(x, x^{k_j + 1}) \\
&\leq \frac{1}{\lambda} \limsup_{j \to \infty} \langle x^{k_j + 1} - x, \nabla h(x^{k_j}) - \nabla h(x^{k_j + 1}) \rangle \\
&\leq \frac{K}{\lambda} \limsup_{j \to \infty} \|x^{k_j} - x^{k_j + 1}\| \cdot \|x^{k_j + 1} - x\| \\
&\leq 0.
\end{aligned}
$$

This being true for all $x \in M$, and following the proof of the previous theorem, we may conclude that $\bar{x}$ is a solution of (WVEP).

Since the solution of (WVEP) is unique and equal to $\bar{x}$, by considering any weakly converging subsequence of $\{x^k\}$, the corresponding limit point must be equal to $\bar{x}$. This implies that the whole sequence $\{x^k\}$ converges to $\bar{x}$. $\quad\square$

Remark 4.2.28. When we suppose, instead of the hypothesis (B7), that F is strongly monotone, i.e., $\exists \delta > 0$ such that $F(x, y) + F(y, x) \leq -\delta\|x - y\|^2 \ \forall x, y \in M$, then the weak convergence of $\{x^k\}$ becomes strong convergence: $\|x^k - \bar{x}\| \longrightarrow 0$. If we suppose $\delta > K/\lambda$, we obtain the rate of convergence

$$\|x^{k+1} - \bar{x}\| \leq \left(\frac{K}{\lambda\delta}\right)^k \|x^1 - x^0\|.$$

PROOF. Setting $x = \bar{x}$ in (4.67), we have

$$0 \leq F(x^{k+1}, \bar{x}) + \frac{1}{\alpha_k} \langle x^{k+1} - \bar{x}, \nabla h(x^k) - \nabla h(x^{k+1}) \rangle$$

$$\leq -F(\bar{x}, x^{k+1}) - \delta \|\bar{x} - x^{k+1}\|^2 + \frac{1}{\lambda} \|\nabla h(x^k) - \nabla h(x^{k+1})\| \cdot \|x^{k+1} - \bar{x}\|$$

$$\leq -\delta \|\bar{x} - x^{k+1}\|^2 + \frac{K}{\lambda} \|x^k - x^{k+1}\| \cdot \|x^{k+1} - \bar{x}\|.$$

Hence $\|x^{k+1} - \bar{x}\| \leq \frac{K}{\lambda \delta} \|x^k - x^{k+1}\|$; and since $\lim_{k \to +\infty} \|x^k - x^{k+1}\| = 0$, we deduce that the sequence $\{x^k\}$ strongly converges to $\bar{x}$.

By induction we obtain the estimate $\|x^{k+1} - \bar{x}\| \leq \left(\frac{K}{\lambda \delta}\right)^k \|x^0 - x^1\|$, and the convergence is ensured by $\frac{K}{\lambda \delta} < 1$.   □

For some similar results for the convergence of the one-phase proximal algorithm one can refer to [118] and [263].

Two-phase proximal algorithms

In this paragraph, instead of taking at each iteration an equilibrium point, which can be interpreted as a fixed point of a solution set mapping of optimization problems (i.e., $x^{k+1} \in U_k(x^{k+1}) := \operatorname{argmin}\{\alpha_k F(x^{k+1}, x) + D_h(x, x^k) \mid x \in M\}$ the solution set of $\alpha_k F(x^{k+1}, \cdot) + D_h(\cdot, x^k)$ on M), we choose a simultaneous optimization method.

Algorithm (VPA2).

1. Start at an arbitrary x^0.
2. If x^k is the current iterate, the next iterate x^{k+1} is found from the following two-phase procedure:

phase 1: $x^{k+\frac{1}{2}} \hookleftarrow x^k$ is a solution of $\forall x \in M$,

$$\alpha_k \left(F(x^k, x^{k+\frac{1}{2}}) - F(x^k, x)\right) \leq h(x) - h(x^{k+\frac{1}{2}}) - \langle x - x^{k+\frac{1}{2}}, \nabla h(x^k) \rangle + \varepsilon_k;$$

phase 2: $x^{k+1} \hookleftarrow x^{k+\frac{1}{2}}$ is a solution of $\forall x \in M$,

$$\alpha_k \left(F(x^{k+\frac{1}{2}}, x^{k+1}) - F(x^{k+\frac{1}{2}}, x)\right) \leq h(x) - h(x^{k+1}) - \langle x - x^{k+1}, \nabla h(x^k) \rangle + \varepsilon'_k.$$

Remark 4.2.29. Note that the iterations in the **two-phase algorithm** (VPA2) are equivalent to the following proximal-point method with Bregman distances:

1: $x^{k+\frac{1}{2}} \in \varepsilon_k - \operatorname{argmin}\{\alpha_k F(x^k, x) + D_h(x, x^k) : x \in M\}$;
2: $x^{k+1} \in \varepsilon'_k - \operatorname{argmin}\{\alpha_k F(x^{k+\frac{1}{2}}, x) + D_h(x, x^k) : x \in M\}$.

We now address an important question relating to algorithms (VPA1) and (VPA2): Given any start point $x^0 \in M$, does the sequence $\{x^k\}$ generated by these algorithms exist?

For the algorithm (VPA1) we need conditions that ensure the existence of vector equilibria, so that similar assumptions as in Theorem 3.8.30 are imposed.

For the algorithm (VPA2) of this subsection, we need only conditions that guarantee the existence of ε_k-optimal solutions $x^{k+\frac{1}{2}}$ and x^{k+1} to the problems of minimizing the scalar functions $\phi_1(x) := \alpha_k F(x^k, x) + D_h(x, x^k)$ and $\phi_2(x) := \alpha_k F(x^{k+\frac{1}{2}}, x) + D_h(x, x^k)$ over the set-constraints M.

When $\varepsilon_k > 0$, these optimal solutions $x^{k+\frac{1}{2}}$ and x^{k+1} always exist.

When $\varepsilon_k = 0$, the existence of $x^{k+\frac{1}{2}}$ and x^{k+1} is guaranteed under conditions of existence of the minimum of convex lower semicontinuous functions, for instance when ϕ_i, $i = 1, 2$, are coercive, or M is compact.

This two-phase proximal algorithm for the equilibrium problems was suggested by Antipin in a set of papers (see [6, 7, 8]) and mainly [118].

The following theorem establishes a **global convergence** of the two-phase algorithm (VPA2).

Theorem 4.2.30. *(Convergence result of (VPA2))* Assume, in addition to the hypotheses of Theorem 4.2.24, that M, f, h satisfy that there exists $\gamma > 0$ such that $0 \leq \Lambda\gamma < 1$ and for each $x_1, x_2, y_1, y_2 \in M$ we have

$$\varphi(f(x_1, y_1)) - \varphi(f(x_1, y_2)) - \varphi(f(x_2, y_1)) + \varphi(f(x_2, y_2))$$
$$\geq -\gamma \left(D_h(x_1, x_2) + D_h(y_1, y_2) \right). \qquad (4.71)$$

Then the conclusion of Theorem 4.2.24 remains true.

PROOF. By a similar argument as in the proof of Lemma 4.2.23, we have for each x in M,

$$\alpha_k \left(F(x^k, x^{k+\frac{1}{2}}) - F(x^k, x) \right) \leq \langle x^{k+\frac{1}{2}} - x, \nabla h(x^k) - \nabla h(x^{k+\frac{1}{2}}) \rangle \qquad (4.72)$$

and

$$\alpha_k \left(F(x^{k+\frac{1}{2}}, x^{k+1}) - F(x^{k+\frac{1}{2}}, x) \right) \leq \langle x^{k+1} - x, \nabla h(x^k) - \nabla h(x^{k+1}) \rangle. \qquad (4.73)$$

Setting $x = x^{k+1}$ in (4.72) and $x = x^*$ a solution of the problem (WVEP) in (4.73), by adding these two last inequalities and using the condition (A1), we may obtain

$$\alpha_k \big(F(x^k, x^{k+\frac{1}{2}}) - F(x^k, x^{k+1}) - F(x^{k+\frac{1}{2}}, x^{k+\frac{1}{2}}) + F(x^{k+\frac{1}{2}}, x^{k+1}) \big)$$
$$\leq \alpha_k \big(F(x^k, x^{k+\frac{1}{2}}) - F(x^k, x^{k+1}) - F(x^{k+\frac{1}{2}}, x^*) + F(x^{k+\frac{1}{2}}, x^{k+1}) \big)$$
$$\leq \langle x^{k+\frac{1}{2}} - x^*, \nabla h(x^k) \rangle + \langle x^{k+1} - x^{k+\frac{1}{2}}, \nabla h(x^{k+\frac{1}{2}}) \rangle$$
$$+ \langle x^* - x^{k+1}, \nabla h(x^{k+1}) \rangle$$
$$= D_h(x^*, x^k) - D_h(x^*, x^{k+1}) - D_h(x^{k+1}, x^{k+\frac{1}{2}}) - D_h(x^{k+\frac{1}{2}}, x^k). \qquad (4.74)$$

The last equality follows by expanding the definition of D_h and direct algebra.
    Using assumption (4.71), it follows that

$$F(x^k, x^{k+\frac{1}{2}}) - F(x^k, x^{k+1}) - F(x^{k+\frac{1}{2}}, x^{k+\frac{1}{2}}) + F(x^{k+\frac{1}{2}}, x^{k+1})$$
$$\geq -\alpha_k \gamma \left(D_h(x^{k+\frac{1}{2}}, x^k) + D_h(x^{k+1}, x^{k+\frac{1}{2}}) \right).$$

Now invoking the condition $0 \leq \Lambda\gamma < 1$, we get

$$D_h(x^*, x^k) \geq D_h(x^*, x^{k+1}) + D_h(x^{k+1}, x^{k+\frac{1}{2}}) + D_h(x^{k+\frac{1}{2}}, x^k)$$
$$-\alpha_k \gamma \left(D_h(x^{k+\frac{1}{2}}, x^k) + D_h(x^{k+1}, x^{k+\frac{1}{2}}) \right)$$
$$\geq D_h(x^*, x^{k+1}).$$

We obtain that $\{D_h(x^*, x^k)\}$ is a nonincreasing and nonnegative sequence;
thus $\lim_{k\to\infty} D_h(x^*, x^k)$ exists and

$$\lim_{k\to\infty} D_h(x^{k+\frac{1}{2}}, x^k) = \lim_{k\to\infty} D_h(x^{k+1}, x^{k+\frac{1}{2}}) = 0.$$

As a straight adaptation of the proof of Theorem 4.2.24, we can confirm
that the sequences $\{x^k\}, \{x^{k+\frac{1}{2}}\}$, and $\{x^{k+1}\}$ converge to some solution $\bar{x}$ of
(WVEP). □

Remark 4.2.31. The condition (4.71) simplifies when $h(x) = \frac{1}{2}\|x\|^2$ to
$\forall x_1, x_2, y_1, y_2 \in M$,

$$F(x_1, y_1) - F(x_1, y_2) - F(x_2, y_1) + F(x_2, y_2) \geq -\frac{\gamma}{2}\left(\|x_1 - x_2\|^2 + \|y_1 - y_2\|^2\right),$$

which holds when f satisfies the following vector P-Hölder condition:

$$f(x_1, y_1) - f(x_1, y_2) + \frac{\gamma}{2}\|y_1 - y_2\|^2 u^0 \in P,$$
$$f(x_1, y_1) - f(x_2, y_1) + \frac{\gamma}{2}\|x_1 - x_2\|^2 u^0 \in P. \qquad (4.75)$$

PROOF. When $x_1, x_2, y_1, y_2 \in M$, the P-Hölder condition (4.75) yields

$$F(x_1, y_1) - F(x_1, y_2) \geq -\frac{\gamma}{2}\|y_1 - y_2\|^2, \quad F(x_1, y_1) - F(x_2, y_1) \geq -\frac{\gamma}{2}\|x_1 - x_2\|^2,$$

so that

$$F(x_1, y_1) - F(x_1, y_2) - F(x_2, y_1) + F(x_2, y_2)$$
$$= \frac{1}{2}\left(F(x_1, y_1) - F(x_1, y_2)\right) + \frac{1}{2}\left(F(x_1, y_1) - F(x_2, y_1)\right)$$
$$+ \frac{1}{2}\left(F(x_2, y_2) - F(x_2, y_1)\right) + \frac{1}{2}\left(F(x_2, y_2) - F(x_1, y_2)\right)$$
$$\geq -\frac{\gamma}{2}\left(D_h(y_1, y_2) + D_h(x_1, x_2) + D_h(y_2, y_1) + D_h(x_2, x_1)\right)$$
$$= -\frac{\gamma}{2}\left(\|x_1 - x_2\|^2 + \|y_1 - y_2\|^2\right).$$

4.2.5 Relaxation and Penalization

In this section we are interested in analyzing the **perturbation of the vector equilibrium problem**. More precisely, we give conditions under which a relaxation of the domain of feasible decisions M and the penalization of the vector criterion mapping f do not change the set of solutions of the considered problems.

Assume that we are given a closed convex cone P with nonempty interior in the space $Y = \mathbb{R}^m$, a closed convex subset M of $X = \mathbb{R}^n$, and $f : X \times X \longrightarrow Y$.

Consider the vector equilibrium problem

$(\mathcal{P}_0)$ find $\overline{x} \in M$ such that $f(\overline{x}, y) \notin -\operatorname{int} P$ for all $y \in M$.

Associated with this problem, let us consider the family of equilibrium problems

$(\mathcal{P}_\lambda)$ find $\overline{x} \in D$ such that $f(\overline{x}, y) + \lambda \Phi(\overline{x}, y) \notin -\operatorname{int} P$ for all $y \in D$.

Here the replacement of M by D represents a **relaxation** of the constraints domain, and $f + \lambda \Phi$ represents the **penalization** of the objective function f.

We are interested in seeing under what conditions the problems $(\mathcal{P}_0)$ and $(\mathcal{P}_\lambda)$ are equivalent in the sense that the solution sets $S(\mathcal{P}_0)$ and $S(\mathcal{P}_\lambda)$ of the two problems coincide.

Lemma 4.2.32. *Let P and P_0 be two cones in Y such that P_0 is closed and $\emptyset \neq P_0 \setminus \{0\} \subset \operatorname{int} P$. Then for each $\rho > 0$, there exists a real $\delta_0 > 0$ such that for every $\delta > \delta_0$,*

$$\delta P_0 \cap U + B(0, \rho) \subset \operatorname{int} P,$$

where $U := \{x \in Y : \|x\| = 1\}$ and $B(0, \rho) := \{x \in Y : \|x\| \leq \rho\}$.

PROOF. Suppose the assertion of the lemma is false. Then we can find some $\rho_0 > 0$ and sequences $\{k_n\}$ (of positive integers) $\{u_n\}$ and $\{v_n\}$ such that for each $n \in \mathbb{N}^*$,

$$k_n \geq n, \|v_n\| \leq \rho_0, u_n \in P_0 \cap U \text{ and } k_n u_n + v_n \notin \operatorname{int} P.$$

By the compactness of $P_0 \cap U$ in Y, one can find a convergent subsequence of $\{u_n\}$, also denoted by $\{u_n\}$, to some $\overline{u} \in P_0 \cap U$.
We conclude from "int P is a cone" that $u_n + \frac{1}{k_n} v_n \notin \operatorname{int} P$, hence that $\overline{u} \notin \operatorname{int} P$ when $n \to +\infty$, and finally that $\overline{u} \in P_0 \setminus \operatorname{int} P$. This contradicts our assumption "$P_0 \setminus \{0\} \subset \operatorname{int} P$." $\square$

We are now ready to provide conditions under which a relaxation of the domain and a penalization of the objective vector mapping do not change the set of solutions of (VEP).

Theorem 4.2.33. *Let M and D be two nonempty compact subsets of X with $M \subset D$. We denote by $\Pi_M(x)$ the metric projection of x on M. Suppose that $f, \Phi : D \times D \to Y$ satisfy*

(A1) there exist $L > 0, \alpha > 0$ such that $\|f(x,y)\| \leq L\|x - y\|^\alpha \quad \forall x \in D \setminus M, \forall y \in \Pi_M(x);$
(A2) Φ is continuous on $M \times D;$
(A3) $\Phi(x,y) \in -P \quad \forall x, y \in M;$
(A4) there exists a closed cone P_0 such that $\emptyset \neq P_0 \setminus \{0\} \subset -\operatorname{int} P$ and

$$\Phi(x,y) \in P_0 \setminus \{0\} \quad \forall x \in D \setminus M, \forall y \in \Pi_M(x);$$

(A5) for each $z \in M$, there exists a neighborhood $V(z)$ and $\varepsilon(z) > 0$ such that

$$\|\Phi(x,y)\| \geq \varepsilon(z)\|x - y\|^\alpha \quad \forall x \in V(z) \cap (D \setminus M), \forall y \in \Pi_M(x).$$

Then there exists some $\mu_0 > 0$ such that for each $\mu > \mu_0$, each solution of the problem $(\mathcal{P}_\mu)$ is a solution of $(\mathcal{P}_0)$; i.e., $S(\mathcal{P}_\mu) \subset S(\mathcal{P}_0)$.

PROOF. Taking into account (A3), it suffices to find some $\mu_0 > 0$ such that for every $\mu > \mu_0$, each solution of $(\mathcal{P}_\mu)$ is contained in M.

Step 1. Since M is compact and covered by the family $\{V(z) : z \in M\}$, we conclude that there exists a finite subset $\{z_1, \ldots, z_k\}$ of M such that $M \subset \cup_{i=1}^k V(z_i) =: V$.

Setting $\rho := \max_{1 \leq i \leq k} \frac{L}{\varepsilon(z_i)}$, which is positive, and using (A1) and (A5), then for every $x \in V \cap (D \setminus M)$ and $y \in \Pi_M(x)$ one has

$$\frac{1}{\|\Phi(x,y)\|} f(x,y) \in B(0,\rho). \tag{4.76}$$

On the other hand, (A4) implies that for every $x \in D \setminus M$ and $y \in \Pi_M(x)$ one has

$$\frac{1}{\|\Phi(x,y)\|} \Phi(x,y) \in P_0 \cap U. \tag{4.77}$$

Using Lemma 4.2.32, we obtain the existence of some $\eta_0 > 0$ such that $\forall \mu > \eta_0, \forall x \in V \cap (D \setminus M)$ and $\forall y \in \Pi_M(x)$,

$$\frac{1}{\|\Phi(x,y)\|} (f(x,y) + \mu\Phi(x,y)) \in -\operatorname{int} P. \tag{4.78}$$

Multiplying by $\|\Phi(x,y)\|$, since $-\operatorname{int} P$ is a cone, we conclude that for every $\mu > \eta_0$ the solution set of the problem $(\mathcal{P}_\mu)$ does not intersect the subset $V \cap (D \setminus M)$.

Step 2. Set $D_0 = D \setminus V$, and consider the marginal function α defined on D_0 by $\alpha(x) := \inf_{y \in \Pi_M(x)} \|\Phi(y,x)\|$.

Since D_0 and M are compact sets, one can confirm, via the closedness of the graph of the multifunction Π_M, that Π_M is upper continuous on D_0.

Also, Φ is continuous on $M \times D_0$. Then using Berge's maximum theorem (see [25, p. 123], [12, Chap. III, Corollary 9 and Proposition 21]), it follows that the function α is lower semicontinuous on D_0. Since D_0 is compact, we deduce that α admits a minimum point $\overline{x}$ on D_0; i.e., $\alpha(\overline{x}) = \min_{x \in D_0} \alpha(x)$.

Set $M_\Phi := \inf_{x \in D_0, y \in \Pi_M(x)} \|\Phi(x,y)\| = \inf_{x \in D_0} \alpha(x)$. Then taking into account compactness of $\Pi_M(\overline{x})$, (A2), and (A4), we have for some $\overline{y} \in \Pi_M(\overline{x})$,

$$M_\Phi = \alpha(\overline{x}) = \min_{y \in \Pi_M(\overline{x})} \|\Phi(\overline{x},y)\| = \|\Phi(\overline{x},\overline{y})\| > 0.$$

Return now to (A1). We have for every $x \in D_0, y \in \Pi_M(x)$,

$$\|f(x,y)\| \leq L\|y-x\|^\alpha \leq \sup_{x \in D_0, y \in M} L\|y-x\|^\alpha =: M_f. \qquad (4.79)$$

We have $M_f < \infty$, since M and D_0 are bounded.

Taking $\rho := \frac{M_f}{M_\Phi} y$, which is positive, and using Lemma 4.2.32, we can find some $\eta_1 > 0$ such that for every $\mu > \eta_1$ one has

$$\frac{1}{\|\Phi(x,y)\|}\left(f(x,y) + \mu\Phi(x,y)\right) \in -\operatorname{int} P \quad \forall x \in D_0, \ \forall y \in \Pi_M(x). \qquad (4.80)$$

Hence, for every $\mu > \eta_1$, each solution of $(\mathcal{P}_\mu)$ does not belong to D_0.

Step 3. Choose $\mu_0 := \max(\eta_0, \eta_1)$. We confirm that each solution of $(\mathcal{P}_\mu)$ does not belong to $D \setminus M$, and the proof is complete. $\qquad \square$

Remark 4.2.34. By making use of boundedness of f on $\{x\} \times D$ for every $x \in M$, it is possible to replace (A2) and (A4) by

(A2') for every $x \in M$ the mapping $\Phi(x, \cdot)$ is continuous on D;
(A4') for every $x \in M$, there exists $y \in D \setminus M$ such that $\Phi(y,x) \in P_0 \setminus \{0\}$.

PROOF. Indeed, let us follow lines of the proof of the above theorem and pass directly to the second step. Fix $z_0 \in M$ and use (A2'); we obtain $\min_{x \in D_0} \|\Phi(z_0, x)\| =: M_\Phi > 0$.

The boundedness of f implies that $M_f := \sup_{x \in D_0} \|f(z_0, x)\|$ is finite. If we set $\rho := M_f/M_\Phi$, it follows from (A4') that $\rho > 0$. Then, following the proof of the above theorem, one can prove the result. $\qquad \square$

Remark 4.2.35. When the relaxation of the domain D is too large and we have doubts about the control in the assumption (A1), we can restrict our control to a neighborhood of M. But in this case we must suppose the boundedness of f on the whole relaxed domain $D \times M$. So instead of (A1) one can suppose

(A1') f is bounded on $D \times M$ and there exist $L > 0$, $\alpha > 0$ and an open subset Ω of X such that $M \subset \Omega$ and

$$\|f(x,y)\| \leq L\|y-x\|^\alpha \qquad \forall x \in (D \setminus M) \cap \Omega, \forall y \in \Pi_M(x).$$

Then the conclusion of Theorem 4.2.33 remains true.

To prove this conclusion it suffices to remark that M_f is finite in (4.79) can be justified by using only boundedness of f on $D_0 \times M$, which is included in $D \times M$.   □

The next result deals with the inverse inclusion of the solution sets: $S(\mathcal{P}_0) \subset S(\mathcal{P}_\mu)$.

Theorem 4.2.36. *Let M and D be two nonempty compact subsets of X with $M \subset D$. Suppose that in addition to (A2) and (A3), the following assumptions are satisfied:*

(A6) *there exist $L > 0$ and $\alpha > 0$ and an open subset Ω of X such that $M \subset \Omega$ and*

$$\|f(x,y) - f(x,z)\| \leq L\|z - y\|^\alpha \ \forall x \in M, \forall y \in (\Omega \cup D) \setminus M, \ \forall z \in \Pi_M(y);$$

(A7) *$\forall y \in D, \Phi(\cdot, y)$ is constant on M;*
(A8) *there exists a closed cone P_1 such that $\emptyset \neq P_1 \setminus \{0\} \subset \operatorname{int} P$ and*

$$\Phi(z, y) \in P_1 \setminus \{0\} \ \forall y \in D \setminus M, \forall z \in \Pi_M(y);$$

(A9) *for each $x \in M$, there exists a neighborhood $V(x)$ in Ω and $\varepsilon(x) > 0$ such that*

$$\|\Phi(z,y)\| \geq \varepsilon(x)\|z - y\|^\alpha \ \forall y \in V(x) \cap (D \setminus M), \forall z \in \Pi_M(y).$$

Then there exists some $\mu_1 > 0$ such that for each $\mu > \mu_1$, $S(\mathcal{P}_0) \subset S(\mathcal{P}_\mu)$.

PROOF. Let $\bar{x} \in M$ be a solution of $(\mathcal{P}_0)$. Then from (A3) we immediately have $\forall \mu \in \mathbb{R}$,

$$f(\bar{x}, y) + \mu\Phi(\bar{x}, y) \notin -\operatorname{int} P \qquad \forall y \in M. \tag{4.81}$$

Let us verify that (4.81) holds for $y \in D \setminus M$. Following the lines of Step 1 in the proof of Theorem 4.2.33 and using (A6), (A8), (A9), we can find an open set $V \subset \Omega$ and $\eta_2 > 0$ such that $\forall \mu > \eta_2$,

$$f(\bar{x}, y) - f(\bar{x}, z) + \mu\Phi(z, y) \in \operatorname{int} P \ \forall y \in V \cap (D \setminus M), \forall z \in \Pi_M(y). \tag{4.82}$$

Let us fix some $y \in D \setminus V$, which is a compact set. By continuity of Φ on $M \times D$, (A7), and (A8) we have

$$M_\Phi := \inf_{x \in M, y \in D \setminus V} \|\Phi(x, y)\| = \inf_{y \in D \setminus V} \inf_{z \in \Pi_M(y)} \|\Phi(z, y)\| > 0.$$

Since $D \setminus V$ and M are compact and f satisfies (A6), we have $\forall y \in D \setminus V, \forall z \in \Pi_M(y)$,

$$\|f(\bar{x}, y) - f(\bar{x}, z)\| \leq L\|y - z\|^\alpha \leq \sup_{z \in M, \ y \in D \setminus V} \|y - z\|^\alpha := M_f < +\infty.$$

Set $\rho := \frac{M_f}{M_\Phi} > 0$, we have $\frac{1}{\|\Phi(z,y)\|} \left(f(\overline{x}, y) - f(\overline{x}, z) \right) \in B(0, \rho) \quad \forall y \in D \setminus V, \forall z \in \Pi_M(y)$.

Apply Lemma 4.2.32, (A7), and (A8). There exists $\eta_3 > 0$ such that $\forall \mu > \eta_3$,

$$f(\overline{x}, y) - f(\overline{x}, z) + \mu\Phi(z, y) \in \operatorname{int} P \quad \forall y \in D \setminus V, \forall z \in \Pi_M(y). \qquad (4.83)$$

Using (A7), (4.82), and (4.83) we have $\forall \mu > \mu_1 = \max(\eta_2, \eta_3)$,

$$f(\overline{x}, y) - f(\overline{x}, z) + \mu\Phi(\overline{x}, y) \in \operatorname{int} P \quad \forall y \in D \setminus M, \forall z \in \Pi_M(y). \qquad (4.84)$$

Let us recall that $\overline{x}$ is a solution of $(\mathcal{P}_0)$. Then since $z \in \Pi_M(y) \subset M$,

$$f(\overline{x}, z) \notin -\operatorname{int} P. \qquad (4.85)$$

Combining (4.84) and (4.85), and using $Y \setminus \operatorname{int} P - \operatorname{int} P \subset Y \setminus P \subset Y \setminus \operatorname{int} P$, we deduce that for $\mu > \eta_2$,

$$f(\overline{x}, y) + \mu\Phi(\overline{x}, y) \notin -\operatorname{int} P \quad \forall y \in D \setminus M. \qquad (4.86)$$

If we put $\mu_1 = \max(\eta_2, \eta_3)$, we get (4.81) for every $y \in D$; therefore $\overline{x}$ is a solution of $(\mathcal{P}\mu)$ for every $\mu > \mu_1$. $\qquad \square$

Discrete equilibrium problems.

Since in Theorems 4.2.33 and 4.2.36, we suppose M to be only compact (without convexity), one can treat the **discrete case**.

Consider $X = \mathbb{R}^n$ and $Y = \mathbb{R}^p$ endowed with the sum norms (which are equivalent to the usual Euclidean structure) and the ordering defined by $P = \mathbb{R}_+^p$. The domain is supposed to be discrete, since $M = A^n$, where $A = \{0, 1\}$.

Consider the relaxation of the domain M as

$$D = \left\{ x = (x_1, \dots, x_n) \in X \mid 0 \leq x_i \leq 1 \; \forall i \in I = \{1, \dots, n\} \right\}.$$

The penalty term we consider is defined for $(x, y) \in D \times D$ by

$$\Phi(x, y) := \begin{pmatrix} \langle e - x, x \rangle - \langle e - y, y \rangle \\ \vdots \\ \langle e - x, x \rangle - \langle e - y, y \rangle \end{pmatrix} \in \mathbb{R}^p, \quad \text{where } e = \begin{pmatrix} 1 \\ \vdots \\ 1 \end{pmatrix} \in \mathbb{R}^n.$$

Theorem 4.2.37. *Suppose that $f : D \times D \to Y$ satisfies:*

(1) *there exist $L > 0$ and an open subset Ω such that $M \subset \Omega$ and*

$$\|f(x, y) - f(x, z)\| \leq L\|z - y\| \quad \forall x \in D, \forall y \in (\Omega \cup D) \setminus M, \forall z \in \Pi_M(y);$$

(2) *$f(x, x) = 0 \; \forall x \in \Omega \cap D$.*

Then there exists $\mu_2 > 0$ such that for each $\mu > \mu_2$, the solution sets of the problems $(\mathcal{P}_0)$ and $(\mathcal{P}\mu)$ coincide; i.e., $S(\mathcal{P}_0) = S(\mathcal{P}\mu)$.

PROOF. To prove this result we will have to verify all assumptions of Theorem 4.2.33 and Theorem 4.2.36. Firstly, we remark that for $\alpha = 1$, the assumptions (A1) and (A6) are satisfied. Suppose we are given the cones $P_0 = \{u \in Y : u_1 = \cdots = u_p \leq 0\}$ and $P_1 = -P_0$. Then assumptions (A2)–(A4) and (A7)–(A8) are obviously satisfied.

In order to prove (A5) and (A9), let us for every fixed $z \in M$ take the neighborhood $V(z) = B(z, \frac{1}{2}) \cap \Omega$, where $B(z, \frac{1}{2})$ denotes the open ball around z with radius $\frac{1}{2}$. Let $x \in U(z) := V(z) \cap (D \setminus M)$. Then the subset $\{y \in \Pi_M(x) : x \in U(z)\}$ is reduced to $\{z\}$. Thus, for $\alpha = 1$, (A5) and (A9) jointly become

$$\|\Phi(x, z)\| = \|\Phi(z, x)\| \geq \varepsilon(z)\|x - z\| \quad \forall x \in V(z) \cap (D \setminus M). \qquad (4.87)$$

Take an arbitrary point $x \in U(z)$. Then by setting $I_1 = \{i \in I : z_i = 0\}$ and $I_2 = \{i \in I : z_i = 1\}$, we have $I = I_1 \cup I_2$ and

$$\|\Phi(x, z)\| - \frac{1}{2}\|x - z\| = p\sum_{i=1}^{n} x_i(1 - x_i) - \frac{1}{2}\left[\sum_{i \in I_1} x_i + \sum_{i \in I_2}(1 - x_i)\right]$$

$$\geq \sum_{i=1}^{n} x_i(1 - x_i) - \frac{1}{2}\sum_{i \in I_1} x_i - \frac{1}{2}\sum_{i \in I_2}(1 - x_i)$$

$$\geq \sum_{i \in I_1} x_i\left(\frac{1}{2} - x_i\right) - \frac{1}{2}\sum_{i \in I_2}(1 - x_i)\left(x_i - \frac{1}{2}\right) > 0.$$

The last inequality is valid because, when $i \in I_1$, $0 \leq x_i \leq \frac{1}{2}$, and when $i \in I_2$, $\frac{1}{2} \leq x_i \leq 1$.

We conclude that all assumptions (A1)–(A9) are satisfied. Thus for $\mu > \mu_2 = \max(\mu_0, \mu_1)$, the solution sets $S(\mathcal{P}_0)$ and $S(\mathcal{P}\mu)$ coincide. $\qquad \square$

4.3 Location Problems

4.3.1 Formulation of the Problem

Urban development is connected with conflicting requirements of areas for dwelling, traffic, disposal of waste, recovery, trade, and others. Conflicting potential consists in the problem that on the one hand municipalities require an economical use of urban areas, and on the other hand, demand of urban areas is increasing, even if the population is constant or is decreasing (compare [335]). Up to now this problem has been solved in the following way:

- to build more compactly (but this may be connected with a reduction of quality of life);

- to build much more into the urban surroundings or into the natural areas (but this is problematic for ecological reasons).

Consequently, it is necessary to use the available urban areas in an optimal sense. Sustainable oriented **town planning** has to solve not only the problem of which institutions or establishments are necessary, but also at which location they are needed, and in each case in dependence on the given supply and inventory.

Using methods of location theory may be one way supporting urban planning to determine the best location for a special new construction, establishment, or for equipments.

The area of a town can be thought of as a mosaic in which the whole is made of smaller units. Different kinds of units represent different kinds of use areas. In real towns, however, the elements of the mosaic continually change position and shape or disappear to be replaced by other or new elements. The job of the planner is to recognize the opportunities and constraints, to consider acute deficits as well as to present needs produced by this shifting mosaic, and he has to propose a development plan for the town.

In our investigations we have to consider the special situation in East German towns. One of the actual main problems of town planning is the traffic problem, due to the extremely high increase of motorized individual traffic in the last years. The lack of parking space is a part of the traffic problem. This is typical for many newly built residential areas in East Germany. Such a residential area is Halle-Silberhöhe, which was built at the beginning of the 1980s. In this district 5-, 6- and 11-story buildings dominate. In our example we consider two residential sections that contain about 9300 inhabitants. This area has a size of 800 m $\times 1000$ m. There exist 1750 parking facilities, a deficit of 1950. The impact is that many inhabitants park their cars on green areas.

One way to solve this problem of "missing parking facilities" is to build multistory parking garages. Now the problem is to find the **best location** for such a multistory garage.

Moreover, we study an example where we apply our methods in order to determine a location for a children's playground in a newly built residential area.

The aim of this section is to consider a location problem in town planning, to formulate a multicriteria location problem, to derive optimality conditions, and to present several **algorithms for solving multicriteria location problems with the corresponding computer programs**. It is well known that the set of solutions in vector optimization (set of efficient elements) may be large, and so finally, we will carry out a comparison of alternatives by using a graphical representation.

Location and approximation problems have been studied by many authors from the theoretical as well as the computational point of view (Chalmet, Francis, and Kolen [63], Kuhn [218], Idrissi, Lefebvre, and Michelot [171, 172, 173], Idrissi, Loridan, and Michelot [174], Gerth (Tammer), and Pöhler [130],

Pelegrin and Fernández [287], Tammer, and Tammer [346], Wendell, Hurter, and Lowe [375], Hamacher and Nickel [155], and many others). An interesting overview on algorithms for planar location problems is given in the book of Hamacher [153].

It would be possible to formulate our problem as a real-valued location problem (**Fermat–Weber problem**), which is the problem to determine a location x of a new facility such that the weighted sum of distances between p given facilities a^i $(i = 1, \ldots, p)$ and x is minimal. In its first and simplest form such a problem was posed by the jurist and mathematician Fermat in 1629. He asked for the point realizing the minimal sum of distances from three given points. In 1909 this problem appeared, in a slightly generalized form, in the pioneering work "Über den Standort der Industrien" of Weber [371]. Later, Föppl [119] introduced the notation "Vial-Zentrum" for the optimal point. In the following decades, this location problem has influenced a great number of useful generalizations (cf. [218], [298], [103] [79], [299, 300, 301]) and applications (cf. [258]). The problem in town planning above introduced leads to a location problem of determining the minimal weighted sum of distances between the given facilities and the unknown Vial x. Using this approach it is very difficult to say how the weights λ_i $(i = 1, \ldots, p)$ should be chosen. Another difficulty may arise if the solution of the corresponding optimal location is practically not useful. Then we need new weights, and again we don't know how to choose the weights.

So the following approach is of interest. Here we formulate the problem as a **multicriteria location problem**

(P) $\mathrm{Eff}_{\mathrm{Min}}(f[\mathbb{R}^2], \mathbb{R}^n_+),$

where

$$f(x) := \begin{pmatrix} \|x - a^1\|_{\max} \\ \|x - a^2\|_{\max} \\ \ldots \\ \|x - a^n\|_{\max} \end{pmatrix},$$

$x, a^i \in \mathbb{R}^2$, $(i = 1, \ldots, n)$,

$$\|x\|_{\max} = \max\{|x_1|, |x_2|\}.$$

Remark 4.3.1. For applications in town planning it is important that we can choose **different norms** in the formulation of (P). The decision which of the norms will be used depends on the course of the roads in the city or in the district.

This section is organized as follows:

In Section 4.3.2 we derive a **geometric algorithm** based on duality assertions (cf. [130], [346]), and in Section 4.3.3 we present a **Mathematica program** for solving a multicriteria location problem. The algorithm in Section 4.3.2 generates the whole set of efficient elements. Using the method of comparison of alternatives and the corresponding visualization in Section 4.3.4, the decision-maker can choose the alternatives in the set of efficient solutions that he prefers. Moreover, in Section 4.3.5 we study applications to a problem of town planning.

4.3.2 An Algorithm for the Multicriteria Location Problem

Using duality assertions we will present an algorithm for solving (P) (compare Chalmet, Francis, and Kolen [63], Gerth (Tammer), and Pöhler [130]). In Section 3.7.3 we derived the following dual problem for (P):

(D) $\mathrm{Eff}_{\mathrm{Max}}(f^*[\mathcal{B}], \mathbb{R}^n_+),$

where $f^*(Y) := \begin{pmatrix} Y^1(a^1) \\ \cdots \\ Y^n(a^n) \end{pmatrix}$ and

$$\mathcal{B} = \{Y = (Y^1, \ldots, Y^n), Y^i \in L(\mathbb{R}^2, \mathbb{R}) : \exists \lambda^* \in \mathrm{int}\,\mathbb{R}^n_+$$

with

$$\sum_{i=1}^n \lambda_i^* Y^i = 0, \quad \text{and} \quad \| Y^i \|_* \leq 1 \quad (i = 1, \ldots, n)\}.$$

Here $\| \cdot \|_*$ denotes the Lebesgue norm.

We can use the conditions $\sum_{i=1}^n \lambda_i^* Y^i = 0$, and $\|Y^i\|_* \leq 1$ $(i = 1, \ldots, n)$ in order to derive an algorithm (cf. [130]). Consider the following sets with respect to the given facilities $a^i \in \mathbb{R}^2$ $(i = 1, \ldots, n)$, which are related to the structure of the subdifferential of the maximum norm:

$$\begin{aligned}
s_1(a^i) &= \{x \in \mathbb{R}^2 \mid a_1^i - x_1 = a_2^i - x_2 \geq 0\}, \\
s_2(a^i) &= \{x \in \mathbb{R}^2 \mid a_1^i - x_1 = a_2^i - x_2 \leq 0\}, \\
s_3(a^i) &= \{x \in \mathbb{R}^2 \mid a_1^i - x_1 = x_2 - a_2^i \geq 0\}, \\
s_4(a^i) &= \{x \in \mathbb{R}^2 \mid a_1^i - x_1 = x_2 - a_2^i \leq 0\}, \\
s_5(a^i) &= \{x \in \mathbb{R}^2 \mid a_2^i - x_2 > |a_1^i - x_1|\}, \\
s_6(a^i) &= \{x \in \mathbb{R}^2 \mid x_2 - a_2^i > |a_1^i - x_1|\}, \\
s_7(a^i) &= \{x \in \mathbb{R}^2 \mid a_1^i - x_1 > |a_2^i - x_2|\}, \\
s_8(a^i) &= \{x \in \mathbb{R}^2 \mid x_1 - a_1^i > |a_2^i - x_2|\}.
\end{aligned}$$

Moreover, we introduce the sets

$$\mathcal{S}_r := \{x \in \mathcal{N} \mid \exists i \in \{1, \ldots, n\} \quad \text{and} \quad x \in s_r(a^i)\}$$

$(r = 5, 6, 7, 8)$, where $\mathcal{N}$ denotes the smallest level set of the dual norm to the maximum norm (Lebesgue norm) containing the points a^i $(i = 1, \ldots, n)$.

Now we are able to describe the following algorithm for solving the multi-criteria location problem (see Gerth (Tammer) and Pöhler [130]):

$$\mathcal{X}_{\text{Eff}} = \{(\operatorname{cl}\mathcal{S}_5 \cap \operatorname{cl}\mathcal{S}_6) \cup [(\mathcal{N}\setminus\mathcal{S}_5)\cap(\mathcal{N}\setminus\mathcal{S}_6)]\} \cap$$
$$\{(\operatorname{cl}\mathcal{S}_7 \cap \operatorname{cl}\mathcal{S}_8) \cup [(\mathcal{N}\setminus\mathcal{S}_7)\cap(\mathcal{N}\setminus\mathcal{S}_8)]\}.$$

4.3.3 A Mathematica Program for Solving the Multicriteria Location Problem

In this section we explain the essential elements of a Mathematica program (foundations of Mathematica are presented in Kofler [213]) that uses the theory of Section 4.3.2. At the beginning, the decision-maker has to transform the real locations a^i ($i = 1,\ldots,n$) into mathematical coordinates $\{xi, yi\}$. This set of points is to be fed into the computer in the form of a matrix:

```
points = {{x1,y1}, {x2,y2}, ... , {xn,yn}}.
```

It is possible to choose between the maximum norm

```
no = 1
```

and the Lebesgue norm

```
no = 2
```

to represent the efficient set. (This set is plotted in red, and the given points are plotted in green.) The calculation will be started by

```
project[points,no];
```

In a first step we choose coordinates independent of the selected norm and construct a lattice that contains the given points. Next we define the sets $\mathcal{S}_r$ and $\operatorname{cl}\mathcal{S}_r$, $r = 5,6,7,8$, introduced in [130]. The module `efficient[...]` determines the set of efficient points which we show in a last step depending on the selected norm. This gives the following outline of our program:

```
Needs[...];

c[points_] := Module[... ]; lattice[c_] := Module[... ];
snset[lattice_, nr_] := Module[... ]; efficient[lattice_] :=
Module[... ]; showno1[c1_, c2_, c3_, c4_, gralt_] :=
Module[... ]; showno2[c1_, c2_, c3_, c4_, gralt_] :=
Module[... ]; picture[points_, c_, no_, eff_] :=
Module[... ];

project[points_, no_] := Module[ {cs,git,effic,final}, cs =
If[no==1, c[points], points]; git = lattice[cs]; effic =
efficient[git]; final = picture[points,cs,no,effic];
Show[final, DisplayFunction->$DisplayFunction,
AspectRatio->Automatic, Axes->False, Frame->True] ];
```

In the following we demonstrate the use of the program with an example. First, we have to fix the given points:

```
points = {{-1.5,3.5},{1,3},{1,0},{-3,-2},{3.5,-1.5},
{2,2},{-2,2},{4,1},{-3,2}}.
```

If the decision-maker prefers the maximum norm, he chooses the parameter 1:

```
project[points,1].
```

This gives Figure 4.3.3.

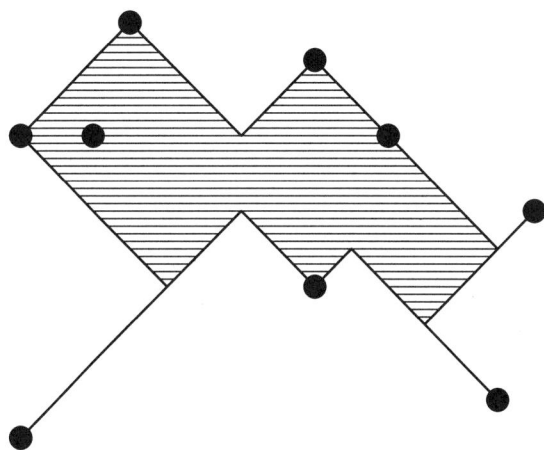

Figure 4.3.3. The set of efficient elements of the multicriteria location problem (P) with the maximum norm.

But if the decision-maker prefers the Lebesgue norm, he has to take the parameter 2:

```
project[points,2].
```

In this case the algorithm generates the solution set represented by Figure 4.3.4.

Note that it is possible that the program will not be able to render the graphic if the number of given points is too large. (This is a problem of finite memory in the computer.) In this case one should reduce the number of given points or try to locate a combination of some facilities at one point.

4.3.4 Comparison of Alternatives

It is well known that the set $\mathcal{X}_{\mathrm{Eff}}$ of efficient points in vector optimization may be large, and all its elements may not be comparable a priori. So we

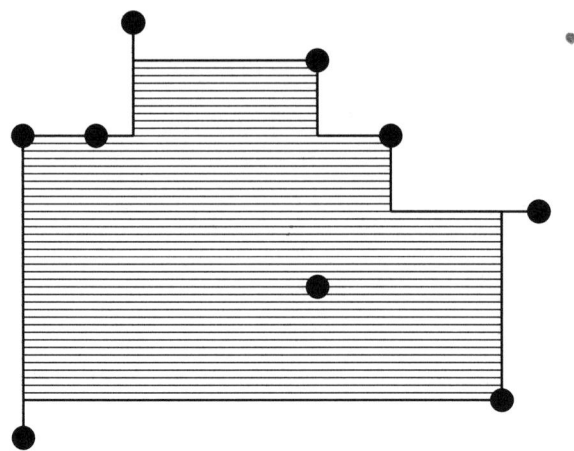

Figure 4.3.4. The set of efficient points of the multicriteria location problem (P) with the Lebesgue norm instead of the maximum norm.

have the problem to choose a solution in $\mathcal{X}_{\mathrm{Eff}}$ that mostly corresponds the preferences of the decision-maker. In Section 4.3.5 we will use several criteria for the location of a children's playground and carry out a comparison of alternatives.

In order to study in addition to the minimization of the distances some other criteria $C_1, \ldots, C_l$ and to compare the alternatives, we realize the following procedure.

In a **first step** we choose some **representatives** $x^1,\ x^2, \ldots, x^m$ in the set $\mathcal{X}_{\mathrm{Eff}}$ of efficient points of the multicriteria location problem (P).
Next, in a **second step** we **evaluate** the alternatives with respect to the criteria.
Further, in a **third step** we use a graphical representation of the alternatives by a so-called **radar chart**. In the radar chart representation, each criterion is represented by an axis emanating from the center of the circle. Data are scaled so that the worst value of each criterion corresponds to the center of the circle and the best value to a point on the periphery. So it is possible to compare the alternatives.
Finally, in a **fourth step** we **decide** on an alternative.

Remark 4.3.2. We assume that the decision-maker accepts the **Pareto principle**, which says that the alternative x^i is better than or equal to x^j (x^i dominates or is equal to x^j) iff $C_r(x^i) \geq C_r(x^j)$ for all $r = 1, \ldots, l$.

Our Mathematica program works with three representatives x_1, x_2, x_3. Then all of these representatives are evaluated with respect to each criterion by an arbitrary number. Every representative x_i and every criteria c_j gets an evaluation aij. These data are put in into the computer as a matrix:

```
rate = {{ a11, a21 ,a31},...,{a11,a21,a31}}.
```

One gets the geometric representation by

```
pareto[rate].
```

The representation is red, green, and blue for the first, second, and third locations, respectively.

In the following we give a short outline of our program:

```
pn[crit_, m_] := Module[... ]; pareto[points_] :=
Module[...picture = Show[gr0,gr01, gr11,gr12, gr21,gr22,
gr31,gr32, AspectRatio->Automatic, Axes->False,
DisplayFunction->$DisplayFunction]; ];
```

Let us consider an example. We choose in the set of efficient elements generated by the Mathematica program in Section 4.3.3 three alternatives: x^1 (strong line), x^2 (middle line), and x^3 (weak line). Taking

```
rate = {{4,4,5}, {5,3,1}, {4,3,3}, {3,3,3}, {2,5,2}};
```

we get the following geometric representation using the Mathematica-program (Figure 4.3.5).

Other methods of visualization are presented by Vetschera [364], [363].

4.3.5 Application to a Problem of Town Planning

In this section we will apply the algorithms from Sections 4.3.2 and 4.3.4 in order to determine a **location for a children's playground in a newly built residential area**.

In a neighborhood of a small lake in a new part of Halle there are apartment blocks located at $a^1 = (3.5, 3.5)$, $a^2 = (4.5, 4)$, $a^3 = (4, 2.5)$, $a^4 = (5, 1.5)$, $a^5 = (6, 1)$, $a^6 = (6.5, 1.5)$, $a^7 = (6, 4)$, $a^8 = (7, 3.5)$, $a^9 = (8, 3)$, $a^{10} = (9.5, 3)$, $a^{11} = (10.5, 3.5)$, $a^{12} = (10.5, 5)$, $a^{13} = (6, 5.5)$, $a^{14} = (6.5, 6.5)$, $a^{15} = (11, 6.5)$, $a^{16} = (10, 7.5)$, $a^{17} = (9, 8)$, schools located at $a^{18} = (2, 4.5)$, $a^{19} = (2, 6)$, day nurseries located at $a^{20} = (3, 4.5)$, $a^{21} = (6, 3)$ and railway stations located at $a^{22} = (8, 6.5)$, $a^{23} = (3, 13.5)$.

We note that in the blocks located at a^3 and a^{17} there live a great number of children aged of 5 to 12.

So we study the problem of determining the set of efficient points of

$$
(P_L) \quad \begin{pmatrix} \|x - a^1\|_{\max} \\ \|x - a^2\|_{\max} \\ \dots \\ \|x - a^{23}\|_{\max} \end{pmatrix} \longrightarrow v - \min_{x \in \mathbb{R}^2}.
$$

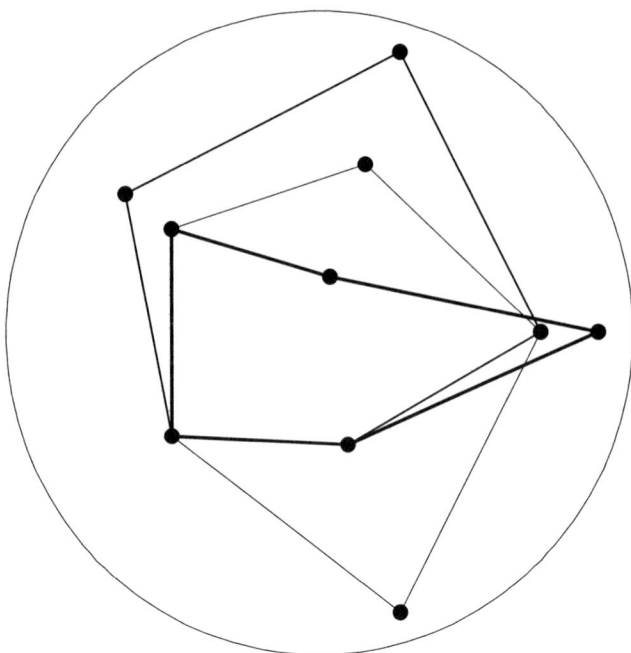

Figure 4.3.5. Comparison of the alternatives x^1, x^2, x^3 with respect to the criteria C_1, C_2, C_3, C_4, C_5.

Applying the algorithm from Section 4.3.2 we determine the set of efficient points of (P_L) by

$$\mathcal{X}_{\mathrm{Eff}} = \{ (\mathrm{cl}\,\mathcal{S}_5 \cap \mathrm{cl}\,\mathcal{S}_6) \cup [(\mathcal{N} \setminus \mathcal{S}_5) \cap (\mathcal{N} \setminus \mathcal{S}_6)] \} \cap$$
$$\{ (\mathrm{cl}\,\mathcal{S}_7 \cap \mathrm{cl}\,\mathcal{S}_8) \cup [(\mathcal{N} \setminus \mathcal{S}_7) \cap (\mathcal{N} \setminus \mathcal{S}_8)] \}.$$

Figure 4.3.6 represents the set of efficient points of the multicriteria location problem (P_L).

Remark 4.3.3. It is also possible to formulate the problem (P_L) with the Lebesgue norm instead of the maximum norm. The decision as to which of the norms will be used depends on the course of the roads in the city or in the district.

Then we can carry out a corresponding algorithm (compare [63], [130]) with sets $S_r(a^i)$ related to the structure of the subdifferential of the Lebesgue norm.

Figure 4.3.7 shows the set of efficient points $\mathcal{X}_{\mathrm{Eff}}^{\mathrm{Leb}}$ of (P_L) with the Lebesgue norm instead of the maximum norm (compare Chalmet, Francis, and Kolen [63] and Gerth and Pöhler [130]).

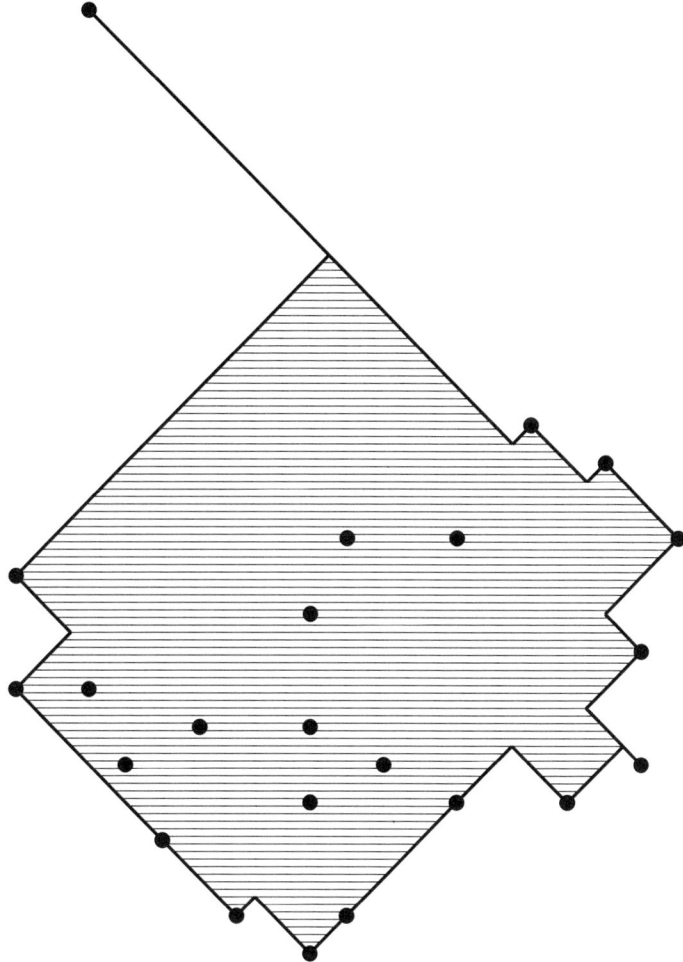

Figure 4.3.6. The set of efficient points of the multicriteria location problem (P_L).

The approach in the multicriteria location problem (P_L) considers only the distances between the apartment blocks, schools, day nurseries, and railway stations located at a^i $(i = 1, \ldots, 23)$ and the location of the children's playground. So we get in the first place a view of all efficient solutions of the multicriteria location problem. In order to choose from this set of efficient solutions $\mathcal{X}_{\text{Eff}}$ an element that mostly corresponds to the preferences of the decision-maker, we look at several other criteria and carry out a comparison of alternatives. We will use a graphical representation of the alternatives with

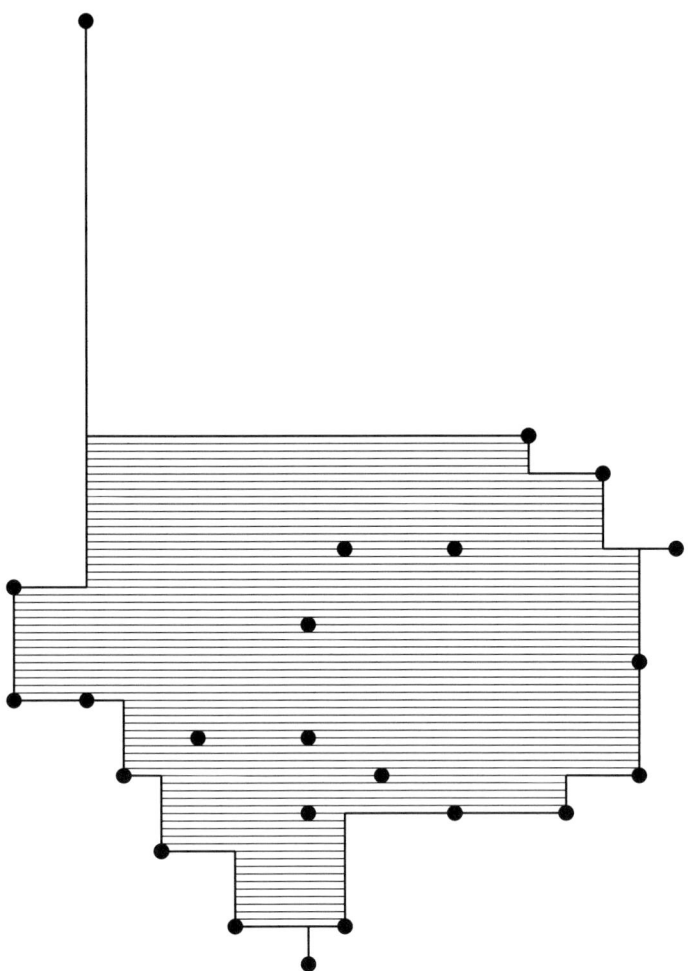

Figure 4.3.7. The set of efficient points $\mathcal{X}_{\mathrm{Eff}}^{\mathrm{Leb}}$ of (P_L) with the Lebesgue norm instead of the maximum norm.

respect to relevant criteria (see Section 4.3.4). Such an approach supports the user in formulating and evolving preferences toward decision alternatives.

For the problem to determine a **location for a children's playground** we will consider the following criteria in order to choose an element in the set of efficient points $\mathcal{X}_{\mathrm{Eff}}$:

(C_1) Without risk to reach on foot.
(C_2) Relations to already existing or projected parks.

(C_3) Connection with already existing centers for schools or sports grounds.
(C_4) Distance to apartment blocks where a great number of children (5 to 12 years old) live.

Now, in order to solve our problem of finding a location for a children's playground we recall that in the apartment blocks a^3 and a^{17} a great number of children aged of 5 to 12 live. Then we have the following procedure:

Step 1: Choose three representatives, for example, $x^1 = (5;7)$, $x^2 = (9;6)$, $x^3 = (6;3)$ with $x^1, x^2, x^3 \in \mathcal{X}_{\text{Eff}}$.

Step 2: Evaluation of the alternatives with respect to the criteria.

Criteria	x^1	x^2	x^3
C_1	4	4	5
C_2	5	3	1
C_3	4	3	3
C_4	3	3	3

Step 3: Graphical representation and comparison of alternatives:

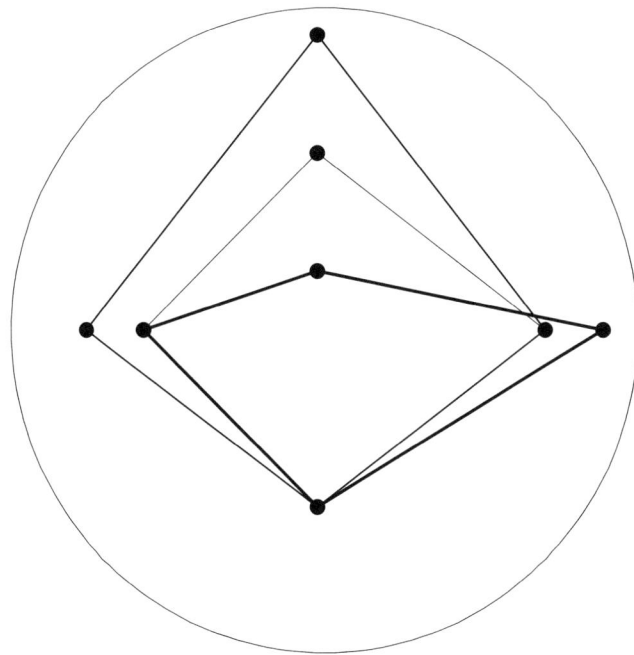

Figure 4.3.8. Radar chart for the alternatives x^1, x^2, and x^3.

The radar chart (Figure 4.3.8) shows that alternative x^1 dominates x^2. So the alternatives x^1 and x^3 are non dominated with respect to the Pareto relation.

Step 4: Decision. If for the decision-maker the criteria C_2 and C_3 are more important than the other criteria then the decision-maker decides for alternative x^1. In the case that the most important criterion is C_1 then he chooses alternative x^3.

Of course, it is possible to include additional criteria or to replace some criteria by others. Moreover, if the results do not coincide with the preferences of the user, he can choose other representatives $x^i \in X_{\mathrm{Eff}}$ $(i = 1, \ldots, m)$ and start again with Step 1.

4.4 Multicriteria Fractional Programming

4.4.1 Solution Concepts

Many aims in real decision problems can be expressed by a fractional objective function (cf. [316]); thus the field of fractional optimization has real-world application.

Example 4.4.1. The problem in economics to minimize a cost functional $f_1(x) = a^T x$ and to maximize profitability $f_2(x) = \frac{b^T x}{c^T x}$, where $x, a, b, c \in \mathbb{R}^n$, can be formulated as the problem to determine the set of efficient elements of a **multicriteria fractional programming problem**

$$\mathrm{Eff}(f[X], \mathbb{R}_+^2),$$

with $f(x) = (f_1(x), f_2(x))$ and $X \subseteq \mathbb{R}^n$.

For the case of optimization problems with only one fractional objective function, Dinkelbach [97] has proposed a parametric solution approach. This approach is based on the relation to a special parametric problem, which is described without the original ratios. However, it requires, additionally, the generation of that unknown parameter value, which makes the problem equivalent to our original problem. Many other authors have published results to generalize Dinkelbach's idea to efficient and properly efficient solutions of vectorial optimization problems with m fractional objective functions (cf. Bector and Chandra [16], Kaul and Lyall [206], Weir [374], Tammer, Tammer, and Ohlendorf [349]). But some of those results are not entirely correct. Moreover, up to now, for $m > 1$ a general convergent algorithm to generate the required parameter vector does not exist. However, in the case $m = 1$ such an algorithm exists.

The aim of this section is to extend the results of Dinkelbach and other authors to different sets of approximate efficient and properly efficient solutions of multicriteria fractional optimization problems. As a by-product we obtain the corrected formulations of corresponding results for the exact solutions.

The main part of this section is devoted to the mentioned relations between the (approximate) solutions of the original multicriteria fractional problem and the transformed one. Moreover, we discuss possibilities for solving the transformed problem by a three-level interactive approach following ideas of the book [147].

In this section we assume that $\mathbf{F}$ and $\mathbf{B}$ are nonempty subsets of the m-dimensional Euclidean space $\mathbb{R}^m$. As usual (compare Section 3.1.1, Definition 3.1.1), we define the set $\mathrm{Eff}(\mathbf{F}, \mathbf{B})$ of efficient elements of the set $\mathbf{F}$ with respect to the set $\mathbf{B}$ in the form

$$\mathrm{Eff}(\mathbf{F}, \mathbf{B}) = \{\bar{y} \in \mathbf{F} \mid \mathbf{F} \cap (\bar{y} - (\mathbf{B} \setminus \{\mathbf{0}\})) = \emptyset\},$$

and we study the vector optimization problem $(\mathcal{P})$ to determine $\mathrm{Eff}(\mathbf{F}, \mathbf{B})$.

As the most frequently considered special case of $(\mathcal{P})$ we mention the situation that

$$\mathbf{F} = f(\mathbf{X}) = \{(f_1(x), \ldots, f_m(x))^T \mid x \in \mathbf{X}\}, \ \mathbf{X} \subseteq \mathbb{R}^n, \mathbf{B} = \mathbb{R}^m_+. \quad (4.88)$$

Besides the set $\mathrm{Eff}(\mathbf{F}, \mathbb{R}^m_+)$ we also want to consider the set $\mathrm{PrEff}(\mathbf{F}, \mathcal{Z})$ of properly efficient elements of $\mathbf{F}$ in the sense of [129] with respect to a family $\mathcal{Z}$ of sets $\mathbf{Q} \subseteq \mathbb{R}^m$ satisfying

$$\mathrm{cl}\,\mathbf{Q} + (\mathbb{R}^m_+ \setminus \{0\}) \subseteq \mathrm{int}\,\mathbf{Q} \ \text{ and } \ 0 \in \mathrm{cl}\,\mathbf{Q} \quad (4.89)$$

defined in the form

$$\mathrm{PrEff}(\mathbf{F}, \mathcal{Z}) \ = \ \bigcup_{\mathbf{Q} \in \mathcal{Z}} \mathrm{Eff}(\mathbf{F}, \mathbf{Q}),$$

as well as the set

$$\mathrm{GEff}(\mathbf{F}) = \{\bar{y} \in \mathbf{F} \mid \exists c > 0, \ \forall y \in \mathbf{F}, \ \forall i = 1, \ldots, m \text{ with } y_i < \bar{y}_i \\ \exists j \in \{1, \ldots, m\} \setminus \{i\} \ \text{ with } \ c(y_j - \bar{y}_j) \geq \bar{y}_i - y_i\}$$

of properly efficient elements of $\mathbf{F}$ in the sense of Geoffrion [125].

Some relations between the given concepts that are important for multi-criteria fractional programming problems are summarized in the following proposition (compare Section 3.1.1).

Proposition 4.4.2. *1. We have $\mathrm{PrEff}(\mathbf{F}, \mathcal{Z}) \subseteq \mathrm{Eff}(\mathbf{F}, \mathbb{R}^m_+)$.*
2. If $\mathbf{F} \subseteq \mathbb{R}^m$ is convex or if $\mathbf{F}$ is defined according to (4.88) with $\mathbf{X}$ convex and $f_i, i = 1, \ldots, m$, convex on $\mathbf{X}$, then
   $\mathrm{GEff}(\mathbf{F}) = \{\bar{y} \in \mathbf{F} \mid \exists \mu \in \mathrm{int}\,\mathbb{R}^m_+ \text{ with } \mu^T \bar{y} = \min_{y \in \mathbf{F}} \mu^T y\}$.
3. For the case $\mathbf{B} = \mathbb{R}^m_+$, $\mathcal{Z} = \mathcal{Z}_0$ with $\mathcal{Z}_0 = \{\mathbf{Q}^l, l = 1, 2, \ldots\}$,
   $\mathbf{Q}^l = \bigcup_{i=1,\ldots,m} \mathbf{Q}^l_i$, $\mathbf{Q}^l_i = \{y \in \mathbb{R}^m / y_i > 0, \ y_i + \alpha^l y_j > 0, \forall j \neq i\}$, where
   $\{\alpha^l\}$ is an arbitrary positive sequence with $\alpha^l \to \infty$, we have $\mathrm{GEff}(\mathbf{F}) = \mathrm{PrEff}(\mathbf{F}, \mathcal{Z}_0)$.

PROOF. Statement 1 follows from $\mathbb{R}_+^m \setminus \{0\} \subseteq \operatorname{int} \mathbf{Q}$, which is a consequence of (4.89). The first part of statement 2 was given in [35], the second one in [125].

To prove statement 3 let us assume $\bar{y} \in \mathbf{F}$. The condition $\bar{y} \notin \operatorname{GEff}(\mathbf{F})$ means that for each $c > 0$ there is a point $y(c) \in \mathbf{F}$ and at least one index $i \in \{1, \ldots, m\}$ with $y_i(c) < \bar{y}_i$ such that for all $j \neq i$ we have $c(y_j(c) - \bar{y}_j) < \bar{y}_i - y_i(c)$. Now let $\{\alpha^l\}$ be an arbitrary positive and divergent sequence. Choosing $c = \alpha^l, l = 1, 2, \ldots$, we see that $y^l = y(\alpha^l)$ satisfies $\bar{y} - y^l \in \mathbf{Q}^l$, and hence $\bar{y} \notin \operatorname{PrEff}(\mathbf{F}, \mathcal{Z}_0)$.

If, on the other hand, $\bar{y} \notin \operatorname{PrEff}(\mathbf{F}, \mathcal{Z}_0)$, then because of the divergence of the sequence $\{\alpha^l\}$, for every $c > 0$ there is a number l with $\alpha^l \geq c$ such that $\bar{y} - y^l \in \mathbf{Q}^l$ implies $c(y_j^l - \bar{y}_j) < \bar{y}_i - y_i^l$ for a certain index i with $y_i^l < \bar{y}_i$ and all $j \neq i$, and hence $\bar{y} \notin \operatorname{GEff}(\mathbf{F})$.   □

Note that the set $\operatorname{PrEff}(\mathbf{F}, \mathcal{Z}_0)$ in statement 3 does not depend on the concrete choice of the sequence $\{\alpha^l\}$.

A very similar result to that of statement 3 was given in Section 5.2 of [373].

Different concepts for approximate solutions of vector optimization problems have been introduced in [94], [110], [127], [241], [271], [302], [329], [337], [340], [362] (cf. Section 3.1.1). We follow here Definitions 3.1.1 and 3.1.2 and introduce three different types of approximate solutions for the special case that $\mathbf{B} = \mathbb{R}_+^m$, where $k^0 \in \operatorname{int} \mathbb{R}_+^m$ and $\varepsilon \geq 0$.

According to Definition 3.1.1 the set of ε-efficient elements of $\mathbf{F}$ with respect to k^0 is given by

$$\varepsilon - \operatorname{Eff}(\mathbf{F}, k^0) = \operatorname{Eff}(\mathbf{F}, \mathbb{R}_+^m + \varepsilon k^0),$$

and the set of properly ε-efficient elements of $\mathbf{F}$ with respect to $\mathcal{Z}$ and k^0 is given by

$$\varepsilon \operatorname{PrEff}(\mathbf{F}, \mathcal{Z}, k^0) = \bigcup_{\mathbf{Q} \in \mathcal{Z}} \operatorname{Eff}(\mathbf{F}, \mathbf{Q} + \varepsilon k^0),$$

where $\mathcal{Z}$ is defined as in (4.89).

Under the assumption that $z : \mathbb{R}^p \to \mathbb{R}$ is monotone with respect to $\mathbb{R}_+^p$ ($y^1 \leq y^2 \Rightarrow z(y^1) \leq z(y^2)$), the set of ε-efficient elements of $\mathbf{F}$ with respect to z according to Definition 3.1.2 is defined by

$$\varepsilon - \operatorname{Eff}(\mathbf{F}, z) = \{\bar{y} \in \mathbf{F} \mid y \leq \bar{y} \Longrightarrow z(\bar{y}) \leq z(y) + \varepsilon\}.$$

Proposition 4.4.3. *For every $k^0 \in \operatorname{int} \mathbb{R}_+^p$ and $\varepsilon \geq 0$ we have:*

1. $\operatorname{Eff}(\mathbf{F}) \subseteq \varepsilon - \operatorname{Eff}(\mathbf{F}, k^0)$.
2. $\operatorname{PrEff}(\mathbf{F}, \mathcal{Z}) \subseteq \varepsilon \operatorname{PrEff}(\mathbf{F}, \mathcal{Z}, k^0)$.
3. $\varepsilon \operatorname{PrEff}(\mathbf{F}, \mathcal{Z}, k^0) \subseteq \varepsilon - \operatorname{Eff}(\mathbf{F}, k^0)$.

4. *For every $\bar{y} \in \varepsilon - \text{Eff}(\mathbf{F}, k^0)$ the functional $\hat{z}$ given by $\hat{z}(y) = z_0(y - \bar{y})$,
where*

$$z_0(y) = \inf\{t \in \mathbb{R} \mid y \in -\operatorname{cl} B + tk^0\} \tag{4.90}$$

for $B = \mathbb{R}_+^m$, is strictly monotone with respect to $\operatorname{int} \mathbb{R}_+^m$, and $\bar{y} \in \varepsilon - \text{Eff}(\mathbf{F}, \hat{z})$.

5. *Let the functional z_0 defined in (4.90) with $\operatorname{cl} B + \operatorname{int} \mathbb{R}_+^m \subseteq \operatorname{int} B$ be strictly monotone with respect to $\mathbb{R}_+^m$, subadditive, and continuous. Then there exists an open set $\mathbf{Q} \subseteq \mathbb{R}^m$ with $\mathbb{R}_+^m \setminus \{0\} \subseteq \mathbf{Q}$, $0 \in \operatorname{cl} \mathbf{Q} \setminus \mathbf{Q}$, $\operatorname{cl} \mathbf{Q} + (\mathbb{R}_+^m \setminus \{0\}) \subseteq \mathbf{Q}$ such that $\varepsilon - \text{Eff}(\mathbf{F}, z_0) \subseteq \varepsilon - \text{Eff}(\mathbf{F}, \mathbf{Q}, k^0)$.*

PROOF. Statements 1 and 2 follow from the fact that according to our assumptions, $\mathbb{R}_+^m + \varepsilon k^0 \subseteq \operatorname{int} \mathbb{R}_+^m$. In the same way as statement 1 from Proposition 4.3.1, we see that statement 3 of Proposition 4.3.2 is a consequence of $\mathbb{R}_+^m \setminus \{0\} \subseteq \operatorname{int} \mathbf{Q}$, which implies $\mathbb{R}_+^m + \varepsilon k^0 \setminus \{0\} \subseteq \operatorname{int}(\mathbf{Q} + \varepsilon k^0)$. Statements 4 and 5 were proved in [340]. □

4.4.2 Generalized Dinkelbach Transformation

Consider as a special case of $(\mathcal{P})$ a vectorial fractional optimization problem

$$(\mathcal{P}_f) \qquad\qquad \text{Eff}(f(\mathbf{X}), \mathbb{R}_+^m),$$

where $f(x) = \frac{g(x)}{h(x)}$, $\frac{g(x)}{h(x)} := (\frac{g_1(x)}{h_1(x)}, \ldots, \frac{g_m(x)}{h_m(x)})^T$ and $h_i(x) > 0 \; \forall x \in \mathbf{X}$, $i = 1, \ldots, m$. We show that $(\mathcal{P}_f)$ is closely related to a multiparametric vector optimization problem $\mathcal{P}(\lambda)$, which we call the corresponding **Dinkelbach-transformed problem**, namely,

$$\mathcal{P}(\lambda) \qquad\qquad \text{Eff}(H(\mathbf{X}, \lambda), \mathbb{R}_+^m),$$

where $H_i(x, \lambda) = g_i(x) - \lambda_i h_i(x)$, $i = 1, \ldots, m$, and $\lambda \in \mathbb{R}^m$ is a parameter that must be chosen in a suitable way.

The original result of Dinkelbach [97] from 1967 (and also the earlier result of Jagannathan [189] from 1966 for linear fractional problems) concerns the case $m = 1$ with only one objective function and says that a given point $\bar{x}$ is optimal for $(\mathcal{P}_f)$ iff it is optimal for $\mathcal{P}(\bar{\lambda})$ with $\bar{\lambda} = \frac{g(\bar{x})}{h(\bar{x})}$.

Corresponding results for the sets of efficient and properly efficient solutions, respectively, of both problems in the case $m \geq 1$ were given by Bector and Chandra [16], Kaul and Lyall [206], Weir [374], and others. Note that the formulation as well as the proof of the corresponding Lemma 1 in [206] and Theorem 4 in [374] are not entirely correct in the given form. Above all, the authors disregarded the fact that in the case of proper efficiency it is essential to assume, additionally, that all ratios $\frac{h_i}{h_j}$ are bounded below by positive bounds (and not only by zero).

In the following two theorems we formulate the relations between the sets of approximate solutions of $(\mathcal{P}_f)$ and $\mathcal{P}(\lambda)$. Proposition 4.4.3 gives a possibility to extend the results to the set of approximate solutions in the sense of [94].

Theorem 4.4.4. *Let $k^0 \in \operatorname{int} \mathbb{R}_+^m, \varepsilon \geq 0$, and $\bar{x} \in \mathbf{X}$. Then we have*
$$f(\bar{x}) \in \varepsilon - \operatorname{Eff}(f(\mathbf{X}), k^0) \iff H(\bar{x}, \bar{\lambda}) \in \varepsilon - \operatorname{Eff}(H(\mathbf{X}, \bar{\lambda}), \bar{k}) \text{ for}$$

$$\bar{\lambda}_i = \frac{g_i(\bar{x})}{h_i(\bar{x})} - \varepsilon k_i^0 \qquad and \qquad \bar{k}_i = k_i^0 h_i(\bar{x}), \qquad \forall i = 1, \dots, m. \quad (4.91)$$

PROOF. The relation $H(\bar{x}, \bar{\lambda}) \notin \varepsilon - \operatorname{Eff}(H(\mathbf{X}, \bar{\lambda}), \bar{k})$ is equivalent to the existence of an element $x^1 \in \mathbf{X}$ with

$$g_i(x^1) - \bar{\lambda}_i h_i(x^1) \leq g_i(\bar{x}) - \bar{\lambda}_i h_i(\bar{x}) - \varepsilon \bar{k}_i \qquad \forall i = 1, \dots, m,$$

where for at least one index i the corresponding inequality must be strict. Dividing these inequalities by $h_i(x^1)$ and taking relation (4.91) into account, we get the equivalent inequalities

$$\frac{g_i(x^1)}{h_i(x^1)} \leq \frac{g_i(\bar{x})}{h_i(\bar{x})} - \varepsilon k_i^0 \qquad \forall i = 1, \dots, m,$$

where again for at least one index i the corresponding inequality must be strict. But this is equivalent to $f(\bar{x}) \notin \varepsilon - \operatorname{Eff}(f(\mathbf{X}), k^0)$. $\square$

For the special case $\varepsilon = 0$ we get the already mentioned result of [16] and [206] in the corrected form (namely, including the essential condition (4.91) for $\varepsilon = 0$, which actually was used there in the proofs but had been forgotten in the formulation of the statement).

Corollary 4.4.5. $f(\bar{x}) \in \operatorname{Eff}(f(\mathbf{X}), \mathbb{R}_+^m) \iff H(\bar{x}, \bar{\lambda}) \in \operatorname{Eff}(H(\mathbf{X}, \bar{\lambda}))$ *for $\bar{\lambda}$ according to (4.91) with $\varepsilon = 0$.*

Theorem 4.4.6. *Let $k^0 \in \operatorname{int} \mathbb{R}_+^m, \varepsilon \geq 0$, and $\bar{x} \in \mathbf{X}$, and assume that there is a positive number γ such that for all $i, j = 1, \dots, m$ and all $x \in \mathbf{X}$, $\frac{h_i(x)}{h_j(x)} \geq \gamma$. Then we have*

$$f(\bar{x}) \in \varepsilon \operatorname{PrEff}(f(\mathbf{X}), \mathcal{Z}_0, k^0) \iff H(\bar{x}, \bar{\lambda}) \in \varepsilon \operatorname{PrEff}(H(\mathbf{X}, \bar{\lambda}), \mathcal{Z}_0, \bar{k})$$

for $\bar{\lambda}$ and $\bar{k}$ according to (4.91).

PROOF. The relation $H(\bar{x}, \bar{\lambda}) \notin \varepsilon \, \mathrm{PrEff}(H(\mathbf{X}, \bar{\lambda}), \mathcal{Z}_0, \bar{k})$ implies that for every $l = 1, 2, \ldots$ there is a point $x^l \in \mathbf{X}$ with $H(\bar{x}, \bar{\lambda}) - H(x^l, \bar{\lambda}) - \varepsilon \bar{k} \in \mathbf{Q}^l$. This means that there is an index i with $(H_i(\bar{x}, \bar{\lambda}) - H_i(x^l, \bar{\lambda}) - \varepsilon \bar{k}_i) > 0$ such that $\forall j \neq i$, and for a certain positive and divergent sequence $\{\alpha^l\}$ we have

$$(H_i(\bar{x}, \bar{\lambda}) - H_i(x^l, \bar{\lambda}) - \varepsilon \bar{k}_i) \; + \; \alpha^l(H_j(\bar{x}, \bar{\lambda}) - H_j(x^l, \bar{\lambda}) - \varepsilon \bar{k}_j) \; > 0.$$

Taking into account the definition of H and of $\bar{\lambda}$ and $\bar{k}$ in 4.91, these inequalities can be written in the form

$$\left(\frac{g_i(\bar{x})}{h_i(\bar{x})} h_i(x^l) - g_i(x^l) - \varepsilon \bar{k}_i \right) \; + \; \alpha^l \left(\frac{g_j(\bar{x})}{h_j(\bar{x})} h_j(x^l) - g_j(x^l) - \varepsilon \bar{k}_j \right) > 0,$$

or, equivalently, in the form

$$(f_i(\bar{x}) - f_i(x^l) - \varepsilon k_i^0) \; + \; \frac{h_j(x^l)}{h_i(x^l)} \alpha^l (f_j(\bar{x}) - f_j(x^l) - \varepsilon k_j^0) \; > 0.$$

Together with $(f_i(\bar{x}) - f_i(x^l) - \varepsilon k_i^0) > 0$ this implies

$$f_i(\bar{x}) - f_i(x^l) - \varepsilon k_i^0 \; + \; \beta^l (f_j(\bar{x}) - f_j(x^l) - \varepsilon k_j^0) > 0,$$

where $\beta^l = \gamma \alpha^l \to \infty$. But this means that $f(\bar{x}) \notin \varepsilon \, \mathrm{PrEff}(f(\mathbf{X}), \mathcal{Z}_0, k^0)$.

In the same way, the assumption $f(\bar{x}) \notin \varepsilon \, \mathrm{PrEff}(f(\mathbf{X}), \mathcal{Z}_0, k^0)$ implies the existence of points $x^l \in \mathbf{X}$ for $l = 1, 2, \ldots$ satisfying for at least one index i the relation $(H_i(\bar{x}, \bar{\lambda}) - H_i(x^l, \bar{\lambda}) - \varepsilon \bar{k}_i) > 0$ and for all $j \neq i$,

$$(H_i(\bar{x}, \bar{\lambda}) - H_i(x^l, \bar{\lambda}) - \varepsilon \bar{k}_i) \; + \; \beta^l (H_j(\bar{x}, \bar{\lambda}) - H_j(x^l, \bar{\lambda}) - \varepsilon \bar{k}_j) \; > 0,$$

where $\beta^l = \gamma \alpha^l \to \infty$ such that $H(\bar{x}, \bar{\lambda}) \notin \varepsilon \, \mathrm{PrEff}(H(\mathbf{X}, \bar{\lambda}), \mathcal{Z}_0, \bar{k})$. $\qquad \square$

Note that the assertion of Theorem 4.4.6 does not remain true if we only assume (as was done in [206] and [374]) $h_i(x) > 0$ on $\mathbf{X}$ for $i = 1, \ldots, m$, since then $\inf\{ \frac{h_i(x)}{h_j(x)} / x \in \mathbf{X} \} = 0$ is not excluded.

This can be seen by the following small example. Let be $n = 1, m = 2, \mathbf{X} = \{x \in \mathbb{R}/x \geq 0\}, g_1(x) = e^{-x}$, $h_1 = 1$, $g_2(x) = x^2 + x + 1$, $h_2(x) = x^2 + 1$. Then, for instance, $\bar{x} = 0$ yields a properly efficient element (in the sense of Geoffrion) $H(\bar{x}, \bar{\lambda}) = (0, 0)^T$ for $\mathcal{P}(\bar{\lambda})$ with $\bar{\lambda} = (1, 1)^T$, but $(\frac{g_1(\bar{x})}{h_1(\bar{x})}, \frac{g_2(\bar{x})}{h_2(\bar{x})})^T = (1, 1)^T$ is not properly efficient in the sense of Geoffrion for $(\mathcal{P}_f)$. Similar examples can also be constructed for the other direction of Theorem 4.4.6.

Of course, if all functions h_i are equal or if there are positive lower and upper bounds for all functions h_i on $\mathbf{X}$, the required boundedness of all ratios $\frac{h_i}{h_j}$ by positive bounds is satisfied.

For the special case $\varepsilon = 0$ we get the corrected formulation of the inexact results in [206] and [374].

Corollary 4.4.7. *If there is a positive number γ such that $\frac{h_i(x)}{h_j(x)} \geq \gamma$ for all $i,j = 1, \ldots, m$ and all $x \in \mathbf{X}$, then we have*

$$f(\bar{x}) \in \mathrm{GEff}(f(\mathbf{X})) \iff H(\bar{x}, \bar{\lambda}) \in \mathrm{GEff}(H(\mathbf{X}, \bar{\lambda}))$$

for $\bar{\lambda}$ according to (4.91).

4.4.3 Possibilities for a Solution Approach

The reason for using models of vector optimization for solving concrete decision problems is the fact that very often it is impossible to formulate the interests of the decision-maker a priori by only one objective function. As a natural consequence of such an incomplete knowledge about the underlying decision problem we can observe the phenomenon that in vector optimization we get a great number of "solutions" enjoying a priori the same rights. Of course, in practical decision problems the final aim must be to find a feasible decision that corresponds to the decision-maker's interests in a certain "optimal" way.

As already described in [147], this can often be realized by organizing a learning process in the form of an interactive procedure in which one can compute and compare as many solutions as necessary to help the decision-maker to express his individual interests more precisely. Such an interactive procedure is usually a certain kind of two-level algorithm and needs essentially a suitable parametric surrogate optimization problem related to the underlying vector optimization problem.

Theoretically, all these ideas can also be applied directly to the fractional vector optimization problem $(\mathcal{P}_f)$ studied in the previous subsection. However, there may be computational difficulties in handling problems with complicated fractional objective functions. Moreover, there are also theoretical difficulties in ensuring convexity properties of the surrogate problem that are to be solved in such an interactive procedure. Note that even linear fractionals are not convex but only pseudoconvex and that for sums of fractionals even generalized convexity properties no longer hold. Hence, for instance, statement 2 of Proposition 4.4.2 cannot be used.

For this reason we want to discuss here possibilities to apply an interactive procedure not directly to the original fractional problem $(\mathcal{P}_f)$ but to the corresponding Dinkelbach-transformed problem $\mathcal{P}(\lambda)$. However, in such an approach we have to overcome another difficulty, namely, the generation of the essential parameter value $\bar{\lambda}$ satisfying (4.91). Hence, in contrast to interactive procedures in the usual case, for our considered case of fractional vector optimization problems we propose a **three-level interactive procedure**.

Let us explain our ideas for the most frequently used set $\mathrm{Eff}(\mathbf{F}, \mathbb{R}_+^m)$ of efficient solutions ($\varepsilon = 0$) of $(\mathcal{P}_f)$ and the most frequently used surrogate problem, in which the artificial objective function is the weighted sum of the original objective functions. Applied to $\mathcal{P}(\lambda)$ our parametric surrogate problem has the form

$$\mathcal{P}(\lambda, \mu) \qquad F(x, \lambda, \mu) \rightarrow \min \qquad \text{subject to} \quad x \in \mathbf{X},$$

where $\mu > 0$ and

$$F(x, \lambda, \mu) = \sum_{i=1}^{m} \mu_i \big(g_i(x) - \lambda_i h_i(x)\big).$$

The already mentioned three levels of an interactive procedure for $(\mathcal{P}_f)$ may be characterized in the following way:

Level 1: Compare all stored results and decide whether to stop the procedure. If not, choose a new parameter vector μ.

Level 2: Find for the value of μ given from Level 1 a vector λ such that there exists a solution x of $\mathcal{P}(\lambda, \mu)$ satisfying

$$H(x, \lambda) = 0. \tag{4.92}$$

Level 3: Find for the values μ and λ given in the Levels 1 and 2 a solution x of $\mathcal{P}(\lambda, \mu)$ satisfying $H(x, \lambda) = 0$ and store x together with additional information on x (especially the vector $\frac{g(x)}{h(x)}$). Go to Level 1.

Level 1 is the purely interactive part in which we have to generate a new parameter value μ as long as we are not satisfied with the generated results. Level 3 can often be realized successfully by path-following methods of parametric optimization. Because of the fact that possibilities of realizing Levels 1 and 3 have already been described extensively in earlier papers (cf. [147]), we concentrate our considerations here on the second level. The typical difficulty in this level is the fact that the essential equation (4.92) is given only implicitly, since the solution x of the third level is unknown at the time we have to solve the second level.

Let us study Level 2 under the following additional assumption (A4). Here we take the symbol $\mathbf{PC}^r$ $(r \geq 1)$ to denote the class of those $\mathbf{C}^{r-1}$-functions for which the derivations of order $r-1$ are piecewise $\mathbf{C}^1$ (cf. [317], [348]). Moreover, we use the concept of strong stability of stationary points described by Kummer ([223], Section 5), which generalizes corresponding concepts already known for problems described by $\mathbf{C}^2$-functions (cf. Kojima [214]).

(A4) $\mathbf{X} = \{x \in \mathbb{R}^n \mid q_l(x) \leq 0, \ l = 1, \ldots, p\}$, $g_i, h_i, q_l \in \mathbf{PC}^2$, and there exists a strongly stable stationary point (x^*, u^*) of $\mathcal{P}(\lambda^*, \mu^*)$ satisfying condition (4.92).

Proposition 4.4.8. *Let us assume (A4). Then we have:*

1. *There are neighborhoods $\mathbf{U}$ of λ^*, $\mathbf{V}$ of μ^*, and $\mathbf{W}$ of (x^*, u^*) such that for each $(\lambda, \mu) \in \mathbf{U} \times \mathbf{V}$ problem $\mathcal{P}(\lambda, \mu)$ has a unique stationary point $(\bar{x}(\lambda, \mu), \bar{u}(\lambda, \mu))$ in $\mathbf{W}$.*
2. *The vector function $(\bar{x}, \bar{u})$ belongs to the class $\mathbf{PC}^1$ on $\mathbf{U}$.*

3. *The vector function G, defined on* **U** *by*

$$G_i(\lambda) = g_i(\bar{x}(\lambda, \mu^*)) - \lambda_i h_i(\bar{x}(\lambda, \mu^*)), \quad i = 1, \ldots, m,$$

belongs to the class **PC**^{1}.

PROOF. Statements 1 and 2 follow from an implicit function theorem of [222]. Statement 3 is a consequence of statement 2 and a chain rule given in [317] and [348].   □

Obviously, under (A4) condition (4.92) can be reformulated in the form

$$G(\lambda) = 0, \qquad \lambda \in \mathbf{U}. \tag{4.93}$$

To solve (4.93) we can apply suitable generalizations of Newton's method for nonsmooth equations using generalized derivatives. In the papers [317] and [348] one can find possibilities to generate the generalized Jacobian of the vector functions $\bar{x}$ and G. To ensure convergence to the (of course unknown) point λ^* from assumption (A4), usually one needs a suitable initial point λ^0 in a sufficiently small neighborhood of λ^*.

Among the great number of contributions concerning generalizations of Newton's method to nonsmooth equations, we refer here to the rather general results given in [223] and [306]. Useful ideas to guarantee convergence in the second level even in the case that only approximate solutions of the third level may be generated (which may often be the case) can be found in [348].

Applying Lemma 2.1 in [200], the function $\bar{\varphi}(\lambda, \mu) = F(\bar{x}(\lambda, \mu), \lambda, \mu)$ belongs to the class **PC**^{2}, and for each $(\lambda, \mu) \in \mathbf{U} \times \mathbf{V}$, we have $\frac{\partial}{\partial \lambda_i} \bar{\varphi}(\lambda, \mu) = -\mu_i h_i(\bar{x}(\lambda, \mu))$, and hence

$$\sum_{j=1}^{m} \mu_j^* \frac{\partial}{\partial \lambda_i} G_j(\lambda) = -\mu_i^* h_i(\bar{x}(\lambda)) \tag{4.94}$$

with $\bar{x}(\lambda) = \bar{x}(\lambda, \mu^*)$. Hence, for the special case $m = 1$ (in which we can put without loss of generality $\mu = 1$) the function G belongs even to the class **PC**^{2} with $\nabla G(\lambda) = -h(\bar{x}(\lambda))$. In this way the iteration rule of Newton's method has the very simple form

$$\lambda^{s+1} = \lambda^s - \frac{g(\bar{x}(\lambda^s)) - \lambda^s h(\bar{x}(\lambda^s))}{-h(\bar{x}(\lambda^s))} = f(\bar{x}(\lambda^s)),$$

which is simply the iteration rule of Dinkelbach [97], who used this rule with assumptions other than those given in (A4), since convergence results are very much easier to obtain in the one-dimensional case.

Unfortunately, for $m > 1$ a formula of the type $\nabla G(\lambda) = -h(\bar{x}(\lambda))$ does not follow from (4.94).

4.5 Multicriteria Control Problems

4.5.1 The Formulation of the Problem

In the present section we study suboptimal controls of a class of multicriteria optimal control problems.

In control theory one often has the problem to minimize more than one objective function, for instance, a cost functional as well as the distance between the final state and a given point.

To realize this task one usually takes as objective function a weighted sum of the different objectives. However, the more natural way would be to study the set of efficient points of a vector optimization problem with the given objective functions. It is well known that the weighted sum is only a special surrogate problem to that of finding efficient points, which has the disadvantage that in the nonconvex case one cannot find all efficient elements in this way.

Necessary conditions for solutions of multiobjective dynamic programming or control problems have been derived by several authors; see Klötzler [211], Benker and Kossert [24], Breckner [45], Gorochowik and Kirillowa [145], Tammer [345], and Salukvadze [312]. It is difficult to show the existence of an optimal control (see Klötzler [211]), whereas suboptimal controls exist under very weak assumptions. So it is important to derive some assertions for suboptimal controls.

Ekeland-type variational principles (compare Section 3.10, [105], [106], [12]) are very useful results in optimization theory; they say that there exists an exact solution of a slightly perturbed optimization problem in a neighborhood of an approximate solution of the original problem. The aim of this section is to derive an ε-minimum principle for suboptimal controls of multicriteria control problems from the vector-valued variational principle.

We formulate a multicriteria control problem with an objective function that takes its values in the m-dimensional Euclidean space $\mathbb{R}^m$.

Let us assume:

(A1): (V, d) is a complete metric space,
         $C \subset \mathbb{R}^m$ is a pointed closed cone with $k^0 \in C \setminus \{0\}$.

(A2):   Let $F : V \longrightarrow R^m$ and $x_0 \in V$; for every $r \in R$ the set
         $\{x \in V \mid F(x) \leq_C F(x_0) + rk^0\}$ is closed.

The following theorem follows immediately from Corollary 3.10.14 regarding the fact that under the given assumptions there always exists an approximately efficient element $v_0 \in V$ with

$$F(v_0) \in \mathrm{Eff}(F[V], C_{\varepsilon k^0}).$$

Corollary 4.5.1. *Assume (A1) and (A2). Then for every $\varepsilon > 0$ there exists some point $v_\varepsilon \in V$ such that*

1. $F[V] \cap (F(v_\varepsilon) - \varepsilon k^0 - (C \setminus \{0\})) = \emptyset$,
2. $F_{\varepsilon k^0}[V] \cap (F_{\varepsilon k^0}(v_\varepsilon) - (C \setminus \{0\})) = \emptyset$, where $F_{\varepsilon k^0}(v) := F(v) + d(v, v_\varepsilon)\varepsilon k^0$.

Remark 4.5.2. Corollary 3.10.14 is slightly stronger than Corollary 4.5.1. The main difference concerns condition (3.119) in Corollary 3.10.14, which gives the whereabouts of point x_ε in V.

Remark 4.5.3. The main result of Corollary 4.5.1 says that v_ε is an efficient solution of a slightly perturbed vector optimization problem. This statement can be used to derive necessary conditions for approximately efficient elements. In the next section we will use Corollary 4.5.1 in order to derive an **ε-minimum principle in the sense of Pontryagin** for suboptimal solutions of multicriteria control problems.

4.5.2 An ε-Minimum Principle for Multicriteria Optimal Control Problems

In this section we will give an application of the multicriteria variational principle in Corollary 4.5.1 to control problems.

Consider the system of differential equations

$$\left. \begin{array}{rcl} \frac{dx}{dt}(t) & = & \varphi(t, x(t), u(t)), \\ x(0) & = & x_0 \in \mathbb{R}^n \end{array} \right\} \tag{4.95}$$

and the control restriction

$$u(t) \in U,$$

which must hold almost everywhere on $[0, T]$ with $T > 0$.

We assume that
(C1) $\varphi : [0, T] \times \mathbb{R}^n \times U \longrightarrow \mathbb{R}^n$ is continuous and U is compact.

The vector $x(t)$ describes the state of the system, $u(t)$ is the control at time t and belongs to the set U.

Furthermore, suppose that
(C2) $\frac{\partial \varphi}{\partial x_i}$, i=1,...,n, are continuous on $[0, T] \times \mathbb{R}^n \times U$,
(C3) $(x, \varphi(t, x, u)) \leq c(1 + \|x\|^2)$ for some $c > 0$.

Remark 4.5.4. Let $u : [0, T] \longrightarrow U$ be a measurable control. Condition (C2) and the continuity of φ ensure that there exists a unique solution x of the differential equation (4.95) on $[0, \tau]$ for a sufficiently small $\tau > 0$. By using Gronwall's inequality, condition (C3) implies

$$\|x(t)\|^2 \leq (\|x_0\|^2 + 2cT)e^{2cT},$$

and hence ensures the existence of the solution on the whole time interval $[0, T]$. Moreover, the last inequality yields

$$\left\|\frac{dx(t)}{dt}\right\| \leq \max\{\varphi(t,x,u) \mid (t,x,u) \in [0,T] \times B^0 \times U\},$$

where B^0 denotes the ball of radius $(\|x_0\|^2 + 2cT)^{\frac{1}{2}} e^{cT}$. Applying Ascoli's theorem, we see that the family of all trajectories x of the control system (4.95) is equicontinuous and bounded, and hence relatively compact in the uniform topology (compare Ekeland [105]).

In order to formulate the multicriteria control problem we introduce the objective function $f : \mathbb{R}^n \longrightarrow \mathbb{R}^m$ and suppose that

(C4) f is a differentiable vector-valued function,
(C5) $C \subset \mathbb{R}^m$ is a pointed closed cone, $k^0 \in C \setminus (-\bar{C})$.

Now we formulate the **multicriteria optimal control problem** under the assumptions (C1)–(C5)

(P): Find some measurable control $\bar{u}$ such that the corresponding trajectory $\bar{x}$ satisfies

$$f(x(T)) \notin f(\bar{x}(T)) - (C \setminus \{0\})$$

for all solutions x of (4.95).

It is well known that it is difficult to show the existence of optimal (or efficient) controls of (P), whereas suboptimal controls exist under very weak conditions. So it is important to derive some assertions for suboptimal controls.

An application of a variational principle for vector optimization problems yields an ε-minimum principle for (P), which is closely related to Pontryagin's minimum principle (for $\varepsilon = 0$).

Now we apply Corollary 4.5.1 in order to derive an ε-minimum principle for the multicriteria optimal control problem (P).

We introduce the space V of controls, defined as the set of all measurable functions $u : [0,T] \longrightarrow U$ with the metric

$$d(u_1, u_2) = \text{meas}\{t \in [0,T] : u_1(t) \neq u_2(t)\}.$$

In order to prove our main result we need the following lemmas:

Lemma 4.5.5. (Ekeland [105]) (V,d) is a complete metric space.

Lemma 4.5.6. (Ekeland [105]) The function $F : u \longrightarrow f(x(T))$ is continuous on V, where $x(\cdot)$ is the solution of (4.95) depending on $u \in V$.

Theorem 4.5.7. Consider the multicriteria control problem under the assumptions (C1)–(C5). Then for every $\varepsilon > 0$, there exists a measurable control u_ε with the corresponding admissible trajectory x_ε such that

1. $f(x(T)) \notin f(x_\varepsilon(T)) - \varepsilon k^0 - (C \setminus \{0\})$ *for all solutions x of (4.95),*
2.

$$\begin{pmatrix} (\varphi(t, x_\varepsilon(t), u(t)), p_\varepsilon^1(t)) \\ \cdots \\ (\varphi(t, x_\varepsilon(t), u(t)), p_\varepsilon^m(t)) \end{pmatrix} \notin \begin{pmatrix} (\varphi(t, x_\varepsilon(t), u_\varepsilon(t)), p_\varepsilon^1(t)) \\ \cdots \\ (\varphi(t, x_\varepsilon(t), u_\varepsilon(t)), p_\varepsilon^m(t)) \end{pmatrix} - \varepsilon k^0 - \operatorname{int} C$$

for every $u \in U$ and almost all $t \in [0, T]$, where $p_\varepsilon(\cdot) = (p_\varepsilon^1(\cdot), \ldots, p_\varepsilon^m(\cdot))$ is the solution of the linear differential system

$$\left.\begin{aligned} \frac{dp_{\varepsilon i}^s}{dt}(t) &= -\sum_{j=1}^n \frac{\partial \varphi_j}{\partial x_i}(t, x_\varepsilon(t), u_\varepsilon(t)) p_{\varepsilon j}^s \quad \text{for } i = 1, \ldots, n; \\ p_\varepsilon^s(T) &= f_s'(x_\varepsilon(T)) \quad \text{for } s = 1, \ldots, m. \end{aligned}\right\} \quad (4.96)$$

PROOF. Suppose that (C1)–(C5) are satisfied.

We consider the vector-valued function $F : u \longrightarrow f(x(T))$ and the space V of measurable controls $u : [0, T] \longrightarrow U$. From Lemma 4.5.5 we get that (V, d) is a complete metric space. Lemma 4.5.6 and Remark 4.5.3 yield that the assumptions of Corollary 4.5.1 are satisfied, and we can apply Corollary 4.5.1 to the vector-valued function F. This yields a measurable control $u_\varepsilon \in V$ such that

(i) $F(u_\varepsilon) \in \operatorname{Eff}(F[V], C_{\varepsilon k^0})$,
(ii) $F(u_\varepsilon) \in \operatorname{Eff}(F_{\varepsilon k^0}[V], C)$,

where

$$F_{\varepsilon k^0}(u) := F(u) + \varepsilon k^0 d(u, u_\varepsilon).$$

The corresponding admissible trajectory x_ε to u_ε satisfies

$$\frac{dx_\varepsilon(t)}{dt} = \varphi(t, x_\varepsilon(t), u_\varepsilon(t)) \qquad (4.97)$$

for almost all $t \in [0, T]$ and

$$x_\varepsilon(0) = x_0.$$

So we derive from (i)

$$f(x(T)) \notin f(x_\varepsilon(T)) - \varepsilon k^0 - (C \setminus \{0\})$$

for all solutions x of (4.95); i.e., statement 1 is satisfied.

In order to prove statement 2, we take $t_0 \in (0, T)$, where equality (4.97) holds, $u_0 \in U$, and define $v_\tau \in V$ for $\tau \geq 0$ by

$$v_\tau(t) := \begin{cases} u_0 & : t \in [0, T] \cap (t_0 - \tau, t_0), \\ u_\varepsilon(t) & : t \notin [0, T] \cap (t_0 - \tau, t_0). \end{cases}$$

For a sufficiently small τ,

$$d(u_\varepsilon, v_\tau) = \operatorname{meas}\{t \mid v_\tau(t) \neq u_\varepsilon(t)\} \leq \tau.$$

Furthermore, for the corresponding admissible trajectory x_τ (compare Pallu de la Barrière [284]),

$$\frac{d}{d\tau} f_s(x_\tau(T))\Big|_{\tau=0} = (\varphi(t_0, x_\varepsilon(t_0), u_0) - \varphi(t_0, x_\varepsilon(t_0), u_\varepsilon(t_0)) \, , \, p_\varepsilon^s(t_0)), \quad (4.98)$$

$s = 1, \ldots, m$, where $p_\varepsilon = (p_\varepsilon^1, \ldots, p_\varepsilon^m)$ satisfies (4.96).

Now we can conclude from (ii) for $u = v_\tau$ that

$$F(u_\tau) + \varepsilon k^0 d(u_\tau, u_\varepsilon) \notin F(u_\varepsilon) - (C \setminus \{0\})$$

and

$$f(x_\tau(T)) \notin f(x_\varepsilon(T)) - \varepsilon k^0 d(u_\tau, u_\varepsilon) - (C \setminus \{0\}).$$

For sufficiently small τ we get

$$f(x_\tau(T)) \notin f(x_\varepsilon(T)) - \varepsilon \tau k^0 - (C \setminus \{0\}), \qquad \tau > 0.$$

This implies

$$\frac{f(x_\tau(T)) - f(x_\varepsilon(T))}{\tau} \notin -\varepsilon k^0 - (C \setminus \{0\}), \qquad \tau > 0,$$

and

$$\lim_{\tau \to +0} \frac{f(x_\tau(T)) - f(x_\varepsilon(T))}{\tau} \notin -\varepsilon k^0 - (C \setminus \{0\}), \qquad \tau > 0,$$

i.e.,

$$\frac{d}{d\tau} f(x_\tau(T))\Big|_{\tau=0} \notin -\varepsilon k^0 - \operatorname{int} C.$$

Finally, (4.98) yields

$$\begin{pmatrix} (\varphi(t_0, x_\varepsilon(t_0), u_0) - (\varphi(t_0, x_\varepsilon(t_0), u_\varepsilon(t_0)) \, , \, p_\varepsilon^1(t_0)) \\ \cdots \\ (\varphi(t_0, x_\varepsilon(t_0), u_0) - (\varphi(t_0, x_\varepsilon(t_0), u_\varepsilon(t_0)) \, , \, p_\varepsilon^m(t_0)) \end{pmatrix} \notin -\varepsilon k^0 - \operatorname{int} C$$

for an arbitrary $u_0 \in U$ and almost all $t_0 \in [0, T]$. Hence,

$$\begin{pmatrix} (\varphi(t, x_\varepsilon(t), u(t)) \, , \, p_\varepsilon^1(t)) \\ \cdots \\ (\varphi(t, x_\varepsilon(t), u(t)) \, , \, p_\varepsilon^m(t)) \end{pmatrix} \notin \begin{pmatrix} (\varphi(t, x_\varepsilon(t), u_\varepsilon(t)) \, , \, p_\varepsilon^1(t)) \\ \cdots \\ (\varphi(t, x_\varepsilon(t), u_\varepsilon(t)) \, , \, p_\varepsilon^m(t)) \end{pmatrix} - \varepsilon k^0 - \operatorname{int} C$$

for an arbitrary $u \in U$ and almost all $t \in [0, T]$. $\qquad\square$

Remark 4.5.8. If we put $\varepsilon = 0$, then Theorem 4.5.7 coincides with the following assertion: Whenever there is a measurable control u_ε and the corresponding admissible trajectory x_ε with

1. $f(x(T)) \notin f(x_\varepsilon(T)) - (C \setminus \{0\})$

for all solutions x of (4.95), then

1.

$$\begin{pmatrix} (\varphi(t, x_\varepsilon(t), u(t)) \, , \, p^1_\varepsilon(t)) \\ \cdots \\ (\varphi(t, x_\varepsilon(t), u(t)) \, , \, p^m_\varepsilon(t)) \end{pmatrix} \notin \begin{pmatrix} (\varphi(t, x_\varepsilon(t), u_\varepsilon(t)) \, , \, p^1_\varepsilon(t)) \\ \cdots \\ (\varphi(t, x_\varepsilon(t), u_\varepsilon(t)) \, , \, p^m_\varepsilon(t)) \end{pmatrix} - \operatorname{int} C$$

for every $u \in U$ and almost all $t \in [0, T]$, where $p_\varepsilon(\cdot)$ is the solution of the linear differential system (4.96).

This means that u_ε satisfies a minimum principle for multicriteria control problems in the sense of Pontryagin (compare Gorochowik and Kirillowa [145]).

Remark 4.5.9. Now let us study the special case $Y = \mathbb{R}^1$ and put $\varepsilon = 0$. Then Theorem 4.5.7 coincides with the following assertion: Whenever

(i) $f(x_\varepsilon(T)) \leq \inf \ f(x(T))$ holds,

then

(ii) $(\varphi(t, x_\varepsilon(t), u_\varepsilon(t)) \, , \, p_\varepsilon(t)) \leq \min_{u \in U}(\varphi(t, x_\varepsilon(t), u(t)) \, , \, p_\varepsilon(t))$ almost everywhere on [0,T], where $p_\varepsilon(\cdot)$ is the solution of (4.96).

This is the statement of **Pontryagin's minimum principle**. However, Theorem 4.5.7 holds even if optimal solutions do not exist.

In Theorem 4.5.7 we have proved an ε-minimum principle without any scalarization of the multicriteria optimal control problem. Finally, we derive an ε-minimum principle using a suitable scalarization with a functional $z :$ $\mathbb{R}^m \longrightarrow \mathbb{R}^1$ defined for $y \in \mathbb{R}^m$ by

$$z(y) := \inf\{t \in \mathbb{R} \mid y \in - \operatorname{cl} C + tk^0\}. \tag{4.99}$$

In Section 2.3, Theorem 2.3.1, we have shown that the functional z in (4.99) is continuous, sublinear, and strictly (int C)-monotone.

Theorem 4.5.10. *Consider the multicriteria control problem under the assumptions (C1)–(C5). Then for every $\varepsilon > 0$, there exists a measurable control u_ε with the corresponding admissible trajectory x_ε such that*

1. $f(x(T)) \notin f(x_\varepsilon(T)) - \varepsilon k^0 - (C \setminus \{0\})$

   for all solutions x of (4.95),

2.

$$z\left(\begin{pmatrix} (\varphi(t, x_\varepsilon(t), u(t)) - \varphi(t, x_\varepsilon(t), u_\varepsilon(t)) \, , \, p^1_\varepsilon(t)) \\ \cdots \\ (\varphi(t, x_\varepsilon(t), u(t)) - \varphi(t, x_\varepsilon(t), u_\varepsilon(t)) \, , \, p^m_\varepsilon(t)) \end{pmatrix} \right) \geq -\varepsilon,$$

for every $u \in U$ and almost all $t \in [0, T]$, where $z : \mathbb{R}^m \longrightarrow \mathbb{R}^1$ is the continuous, sublinear, strictly (int C)-monotone functional defined by (4.99), and $p_\varepsilon = (p^1_\varepsilon, \ldots, p^m_\varepsilon)$ is the solution of the linear differential system (4.96).

PROOF. Under the given assumptions we can conclude from the second statement of Theorem 4.5.7 that

$$
\begin{pmatrix}
(\varphi(t, x_\varepsilon(t), u(t)) - \varphi(t, x_\varepsilon(t), u_\varepsilon(t)) \ , \ p_\varepsilon^1(t)) \\
\cdots \\
(\varphi(t, x_\varepsilon(t), u(t)) - \varphi(t, x_\varepsilon(t), u_\varepsilon(t)) \ , \ p_\varepsilon^m(t))
\end{pmatrix}
\notin -\varepsilon k^0 - \operatorname{int} C \quad (4.100)
$$

for every $u \in U$ and almost all $t \in [0, T]$, where $p_\varepsilon = (p_\varepsilon^1, \ldots, p_\varepsilon^m)$ is the solution of (4.96).

The functional z in (4.99) has for $y \in \mathbb{R}^m$ and $t \in \mathbb{R}$ the property (compare Section 2.3, Theorem 2.3.1)

$$
z(y) \geq t \iff y \notin - \operatorname{int} C + t k^0,
$$

so we can conclude from (4.100) that

$$
z\left(
\begin{pmatrix}
(\varphi(t, x_\varepsilon(t), u(t)) - \varphi(t, x_\varepsilon(t), u_\varepsilon(t)) \ , \ p_\varepsilon^1(t)) \\
\cdots \\
(\varphi(t, x_\varepsilon(t), u(t)) - \varphi(t, x_\varepsilon(t), u_\varepsilon(t)) \ , \ p_\varepsilon^m(t))
\end{pmatrix}
\right) \geq -\varepsilon,
$$

for every $u \in U$ and almost all $t \in [0, T]$, where $p_\varepsilon(\cdot)$ is the solution of (4.96). □

4.5.3 A Multicriteria Stochastic Control Problem

In this section we consider a multicriteria stochastic control problem and derive necessary conditions for approximate solutions of the control problem using a multicriteria variational principle of Ekeland's type.

The restrictions in the multicriteria stochastic control problem are formulated by dynamical equations. The solution of these dynamical equations can be obtained by applying the Girsanov measure transformation. Furthermore, the objective functions are terminal cost $g_i(x(1))$, for which we consider the expected value of control u, i.e., $E_u[g_i(x(1))] = F_i(u)$ $(i = 1, \ldots, l)$, where E_u denotes the expectation constructed from control u.

We introduce the following model of a partially observed stochastic control problem:

Let $\{B_t\}_{t \in [0,1]}$ be a Brownian motion on a probability space $(\Omega, \mathcal{A}, \mu)$ taking values in $\mathbb{R}^m$ and let $\mathcal{C}$ be the space of continuous functions from $[0, 1]$ to $\mathbb{R}^m$ endowed with the usual filtering $\mathcal{F}_t = \sigma(x(s) : s \leq t)$ of measurable states up to time t.

The σ-algebra of measurable states depending on t is denoted by $\mathcal{F}_t = \sigma(x_s \mid s \leq t)$.

In the following we assume

(C1) σ is an $m \times m$ matrix-valued mapping $\sigma = (\sigma_{ij})$ defined on $[0,1] \times \mathcal{C}$ with $\mathcal{C} = \mathcal{C}[0,1]$; $\sigma(t,x)$ is nonsingular; for $1 \leq i, \ j \leq m$, $\sigma_{ij}(t,x)$ is $\mathcal{F}_t$-measurable in its second argument and Lebesgue measurable in its first; each σ_{ij} satisfies a uniform Lipschitz condition in x,

$$\|x\|_s = \sup_{0 \leq t \leq s} |x(t)|;$$

there is a constant $k_0 < \infty$ such that

$$\sum \int_0^1 \sigma_{ij}^2 dt \leq k_0 \quad \text{a.s.P.}$$

(a.s.P. means: almost surely with respect to P).

Consider the stochastic differential equation

$$\left. \begin{aligned} dx(t) &= f(t,x,u)dt + \sigma(t,x)dB_t, \ 0 \leq t \leq 1, \\ x(0) &= x_0 \in \mathbb{R}^m, \end{aligned} \right\} \tag{4.101}$$

where $x(t)$ splits into an observed component $y(t) \in \mathbb{R}^n$ and an unobserved component $z(t) \in \mathbb{R}^{m-n}$.
Furthermore, consider the observation σ-algebra $\mathcal{Y}_t$ generated by $\{y(s) : s \leq t\}$.

Definition 4.5.11. *An admissible partially observable feedback control is defined to be a $\mathcal{Y}_t$-predictable mapping $u : [0,1] \times \mathcal{C} \longrightarrow U$, where U is a Borel subset of $\mathbb{R}$, such that $E|u(\tau,.)| < \infty$.*

The set of such controls is denoted by V. We define on V, for $u_1, u_2 \in V$,

$$d(u_1, u_2) = \tilde{P}(\{(t,x) \mid |u_1(t,x) - u_2(t,x)| > 0\}), \tag{4.102}$$

where $\tilde{P}$ is the product measure of λ and P (λ is the Lebesgue measure on $[0,1]$ and P is the probability measure on $(\mathcal{C}, \mathcal{F}_1)$ induced by the solution $\bar{x}$ of

$$\left. \begin{aligned} d\bar{x}(t) &= \sigma(t,\bar{x})dB_t, \\ \bar{x}(0) &= x_0 \in R^m, \end{aligned} \right\} \tag{4.103}$$

i.e. $P(A) = \mu\{\omega \in \Omega : \bar{x}(\omega) \in A\}$ for $A \in \mathcal{F}_1$).

We will see that (V,d) is a complete metric space (cf. Elliott and Kohlmann [108]).

Moreover, we suppose that

(C2) $\varphi : [0,1] \times \mathcal{C} \times U \longrightarrow \mathbb{R}^m$ is measurable, causal, and continuous in its third component.

We recall that a mapping $\Phi : [0,1] \times \mathcal{C} \longrightarrow \mathbb{R}^m$ is called causal if Φ is optional and

$$|\sigma^{-1}(t,x)\Phi(t,x)| \le M(1 + \|x\|_t).$$

Under these assumptions we can apply *Girsanov's theorem* in order to construct a probability measure P_u on $(\mathcal{C}, \mathcal{F}_1)$ that is absolutely continuous with respect to P and a Wiener process $\{w_t^u\}_{t\in[0,1]}$ on $(\mathcal{C}, \mathcal{F}_1, P_u)$ such that $(\mathcal{C}, \mathcal{F}_1, P_u, \{w_t^u\}_{t\in[0,1]}, \xi)$, where ξ is the canonical process on $\mathcal{C}$ (i.e. $\xi(t,x) = x(t)$), is a weak solution of (4.101) for each $u \in V$.

In fact $\{w_t\}_{t\in[0,1]}$ defined by

$$w_t(x) = \int_0^t \sigma^{-1}(s,x)dx(s)$$

is a Wiener process on $(\mathcal{C}, \mathcal{F}_1, P)$ and by Girsanov's theorem $\{w_t^u\}_{t\in[0,1]}$ defined by

$$\begin{aligned} dw_t^u(x) &= dw_t(x) - \sigma^{-1}(t,x)f(t,x,u(t,x))dt \\ &= \sigma^{-1}(t,x)\left[dx(t) - f(t,x,u(t,x))dt\right] \end{aligned}$$

is a Wiener process on $(\mathcal{C}, \mathcal{F}_1, P_u)$ where P_u is the probability measure on $(\mathcal{C}, \mathcal{F}_1)$ defined by

$$\begin{aligned} P_u(A) &= \int_A \exp\bigg\{ \int_0^1 f^T(t,x,u(t,x))(\sigma(t,x)\sigma^T(t,x))^{-1}dx(t) \\ &\quad -\frac{1}{2}\int_0^1 f^T(t,x,u(t,x))(\sigma(t,x)\sigma^T(t,x))^{-1}f(t,x,u(t,x))dt \bigg\}P(dx) \\ &=: \int_A p_0^1(u)P(dx) \end{aligned}$$

for $A \in \mathcal{F}_1$ and we have

$$dx(t) = f(t,x,u(t,x))dt + \sigma(t,x)dw_t^u(x).$$

In order to formulate the multicriteria stochastic control problem we introduce the multiobjective function

$$J(u) := \begin{pmatrix} E_u[g_1(x(1))] \\ \cdots \\ E_u[g_l(x(1))] \end{pmatrix},$$

where E_u denotes the expectation (constructed from the control u) with respect to P_u; g_i are bounded $\mathcal{F}_1$-measurable functions.

We suppose

(C3) $C \subset \mathbb{R}^l$ is a pointed closed cone, $k^0 \in C \setminus (-\bar{C})$.

Now we formulate the **multicriteria stochastic control problem** under the assumptions (C1)–(C3):

(P_C): Compute a feasible control $\bar{u}$ such that

$$J(u) \notin J(\bar{u}) - (C \setminus \{0\})$$

for all admissible controls u.

In this way we study an extension of the stochastic control problem introduced by Elliott and Kohlmann [108], [109].

Consider the space V of all partially observable admissible controls and the distance d on V introduced by (4.102).

Lemma 4.5.12. (cf. Elliott and Kohlmann [108]) (V, d) *is a complete metric space.*

Furthermore, we introduce a vector-valued mapping F associated with the multicriteria control problem (P_C) for which the assumptions of the variational principle in Corollary 4.5.1 are satisfied.

Lemma 4.5.13. (cf. Elliott and Kohlmann [108]) *Suppose* (C1)–(C3) *hold. Then the mapping* $F : (V, d) \longrightarrow (\mathbb{R}^l, \|.\|_{\mathbb{R}^l})$ *defined by*

$$F(u) := \begin{pmatrix} E_u[g_1(x(1))] \\ \cdots \\ E_u[g_l(x(1))] \end{pmatrix}$$

is continuous.

Remark 4.5.14. Lemmas 4.5.12 and 4.5.13 show together with assumption (C3) that the assumptions (A1), (A2) of Corollary 4.5.1 are satisfied for the multicriteria stochastic control problem (P_C).

Theorem 4.5.15. *Assume that* (C1)–(C3) *hold. Then for every* $\varepsilon > 0$ *there exists an* $\mathcal{F}_t$-*predictable process* γ_ε *such that for every* $t \in [0, 1]$, *every* $A \in \mathcal{Y}_t$, *and every admissible control* $u \in V$ *the following statements are true:*

1. *For the control* $u_\varepsilon \in V$,

$$\begin{pmatrix} E_u[g_1(x(1))] \\ \cdots \\ E_u[g_l(x(1))] \end{pmatrix} \notin \begin{pmatrix} E_{u_\varepsilon}[g_1(x(1))] \\ \cdots \\ E_{u_\varepsilon}[g_l(x(1))] \end{pmatrix} - \varepsilon k^0 - (C \setminus \{0\}).$$

2. *For* $\tau > 0$ *we get the following assertion:*

$$\begin{pmatrix} \int_t^{t+\tau} \int_A p_0^t(u_\varepsilon) p_t^{t+\tau}(u) \gamma_{1\varepsilon} \sigma^{-1}(f_s^u - f_s^{u_\varepsilon}) dP ds \\ \cdots \\ \int_t^{t+\tau} \int_A p_0^t(u_\varepsilon) p_t^{t+\tau}(u) \gamma_{l\varepsilon} \sigma^{-1}(f_s^u - f_s^{u_\varepsilon}) dP ds \end{pmatrix} \notin -\varepsilon k^0 \tau P(A) - (C \setminus \{0\}),$$

where $f_s^u := f(s, x, u(s, x))$.

PROOF. We consider the vector-valued function $F : u \longrightarrow J(u)$ and the space V of admissible controls $u : [0, T] \to U$.

From Lemma 4.5.12 we get that (V, d) is a complete metric space. Lemma 4.5.13 yields that F is lower semicontinuous with respect to k^0 and B and bounded from below on V. So we can conclude that the assumptions of Corollary 4.5.1 are satisfied, and we can apply Corollary 4.5.1 to the vector-valued function F. This yields an admissible control $u_\varepsilon \in V$ such that

(i) $F(u_\varepsilon) \in \mathrm{Eff}(F[V], C_{\varepsilon k^0})$,

(ii) $F_{\varepsilon k^0}(u_\varepsilon) \in \mathrm{Eff}(F_{\varepsilon k^0}[V], C)$,

where

$$F_{\varepsilon k^0}(u) := F(u) + \varepsilon k^0 d(u, u_\varepsilon).$$

So we derive from (i)

$$J(u) \notin J(u_\varepsilon) - \varepsilon k^0 - (C \setminus \{0\})$$

for all feasible controls u; i.e., statement 1 is satisfied.

Furthermore, the families of conditional expectations

$$G_i^t = E_{u_\varepsilon}[g_i(x(1)) \mid \mathcal{F}_t], \quad i = 1, \ldots, l,$$

are martingales and thus have the representation

$$G_i^t = F_i(u_\varepsilon) + \int_0^t \gamma_{i\varepsilon} dw_\varepsilon, \quad i = 1, \ldots, l,$$

where w_ε is the process defined by

$$dw_\varepsilon = \sigma^{-1}(dx - f^{u_\varepsilon} dt),$$

so that we can conclude from Girsanov's theorem that w_ε is a Brownian motion under the measure P_{u_ε}.

In order to prove statement 2 we take $t \in [0, 1]$, $A \in \mathcal{Y}_t$, and $u \in V$ and define $v_\tau \in V$ for $\tau > 0$ by

$$v_\tau(s, x) := \begin{cases} u(s, x) : & (s, x) \in (t, t + \tau] \times A, \\ u_\varepsilon(s, x) : & (s, x) \in [0, t] \times C \cup (t, t + \tau] \times A' \cup (t + \tau, 1] \times C, \end{cases}$$

where $A' = \Omega \setminus A$.

The indicator function of $B = (t, t + \tau] \times A$, denoted by χ_B, is a $\mathcal{Y}_t$-predictable map. Regarding that v_τ can be written as $\chi_B u + \chi_{B'} u_\varepsilon$, it follows that v_τ is predictable and an admissible control in V.

Now we apply the martingale representations given above for $t = 1$ and $i = 1, \ldots, l$:

$$g_i(x(1)) = F_i(u_\varepsilon) + \int_0^1 \gamma_{i\varepsilon} dw_\varepsilon.$$

So we get

$$E_{v_\tau}[g_i] = F_i(v_\tau) = F_i(u_\varepsilon) + E_{v_\tau}\left[I_A \int_t^{t+\tau} \gamma_{i\varepsilon}\sigma^{-1}(f^u - f^{u_\varepsilon})ds\right].$$

Then we may conclude from statement (ii) in Corollary 4.5.1,

$$F_{\varepsilon k^0}(u_\varepsilon) \in \text{Eff}(F_{\varepsilon k^0}[V], C),$$

such that

$$F(v_\tau) + \varepsilon k^0 d(u_\tau, u_\varepsilon) \notin F(u_\varepsilon) - (C \setminus \{0\}).$$

Regarding

$$d(v_\tau, u_\varepsilon) \leq \tau P(A),$$

it follows that

$$F(v_\tau) + \varepsilon k^0 \tau P(A) \notin F(u_\varepsilon) - (C \setminus \{0\}).$$

Together with the definition of P_{u_τ} and the properties given above we derive

$$\left(\begin{array}{c} \int_t^{t+\tau} \int_A p_0^t(u_\varepsilon)p_t^{t+\tau}(u)\gamma_{1\varepsilon}\sigma^{-1}(f_s^u - f_s^{u_\varepsilon})dP\ ds \\ \cdots \\ \int_t^{t+\tau} \int_A p_0^t(u_\varepsilon)p_t^{t+\tau}(u)\gamma_{l\varepsilon}\sigma^{-1}(f_s^u - f_s^{u_\varepsilon})dP\ ds \end{array}\right) \notin -\varepsilon k^0\ \tau\ P(A) - (C\setminus\{0\}).$$

$\square$

Using the martingale representation results given above we derive the following necessary condition that the approximate solution u_ε must satisfy. This is a condition of the following kind: u_ε must be an εk^0-weakly efficient element of the conditional expectation of a certain Hamiltonian.

Theorem 4.5.16. *Consider the stochastic control problem (P_C) under the assumptions* (C1)–(C3). *Then there exists for each $\varepsilon > 0$ an admissible control u_ε, such that*

1.
$$\left(\begin{array}{c} E_u[g_1(x(1))] \\ \cdots \\ E_u[g_l(x(1))] \end{array}\right) \notin \left(\begin{array}{c} E_{u_\varepsilon}[g_1(x(1))] \\ \cdots \\ E_{u_\varepsilon}[g_l(x(1))] \end{array}\right) - \varepsilon k^0 - (C \setminus \{0\})$$

for all feasible controls $u \in V$,

2.
$$\left(\begin{array}{c} E[p_{1\varepsilon}\sigma^{-1}f_t^u \mid \mathcal{Y}_t] \\ \cdots \\ E[p_{l\varepsilon}\sigma^{-1}f_t^u \mid \mathcal{Y}_t] \end{array}\right) \notin \left(\begin{array}{c} E[p_{1\varepsilon}\sigma^{-1}f_t^{u_\varepsilon} \mid \mathcal{Y}_t] \\ \cdots \\ E[p_{l\varepsilon}\sigma^{-1}f_t^{u_\varepsilon} \mid \mathcal{Y}_t] \end{array}\right) - \varepsilon k^0 - \text{int}\ C$$

for all $u \in V$ and the $\mathcal{F}_t$-predictable process

$$p_\varepsilon = \left(\begin{array}{c} p_0^t(u_\varepsilon)\gamma_{1\varepsilon} \\ \cdots \\ p_0^t(u_\varepsilon)\gamma_{l\varepsilon} \end{array}\right).$$

PROOF. We will differentiate the left-hand side in the vector-valued inequality of the last theorem. From the second condition in the assertions of Corollary 4.5.1 we derive for $(\tau > 0)$

$$\frac{1}{\tau} \begin{pmatrix} \int_t^{t+\tau} \int_A p_0^t(u_\varepsilon) p_t^{t+\tau}(u) \gamma_{1\varepsilon} \sigma^{-1} (f_s^u - f_s^{u_\varepsilon}) dP\, ds \\ \cdots \\ \int_t^{t+\tau} \int_A p_0^t(u_\varepsilon) p_t^{t+\tau}(u) \gamma_{l\varepsilon} \sigma^{-1} (f_s^u - f_s^{u_\varepsilon}) dP\, ds \end{pmatrix} \notin -\varepsilon k^0\, P(A) - \frac{1}{\tau}(C \backslash \{0\}).$$

This yields, regarding that C is a cone,

$$\lim_{\tau \to 0} \frac{1}{\tau} \begin{pmatrix} \int_t^{t+\tau} \int_A p_0^t(u_\varepsilon) p_t^{t+\tau}(u) \gamma_{1\varepsilon} \sigma^{-1} (f_s^u - f_s^{u_\varepsilon}) dP\, ds \\ \cdots \\ \int_t^{t+\tau} \int_A p_0^t(u_\varepsilon) p_t^{t+\tau}(u) \gamma_{l\varepsilon} \sigma^{-1} (f_s^u - f_s^{u_\varepsilon}) dP\, ds \end{pmatrix} \notin -\varepsilon k^0\, P(A) - \text{int}\, C.$$

Now we compute the left-hand side of this variational inequality. We observe that $\mathcal{Y}_t$ is countably generated for any rational number r, $0 \le r \le 1$, by sets $\{A_{nr}\}$, $n = 1, 2, \ldots$, since the trajectories are continuous, almost surely. Furthermore, u_{nr} can be considered as an admissible control over the time interval $(t, t+\tau]$ for $t \ge r$, and we can consider a perturbation of u_ε by u_{nr} for $t \ge r$ and $x \in A \in \mathcal{Y}_t$, as in the above section. Under the given assumptions the following limit exists, and

$$\lim_{\tau \to 0} \frac{1}{\tau} \begin{pmatrix} \int_t^{t+\tau} \int_{A_{nr}} p_0^{t+\tau}(u_\varepsilon) \gamma_{1\varepsilon} \sigma^{-1}(f(s,x,u_{mr}) - f(s,x,u_\varepsilon)) dP\, ds \\ \cdots \\ \int_t^{t+\tau} \int_{A_{nr}} p_0^{t+\tau}(u_\varepsilon) \gamma_{l\varepsilon} \sigma^{-1}(f(s,x,u_{mr}) - f(s,x,u_\varepsilon)) dP\, ds \end{pmatrix}$$

$$= \begin{pmatrix} \int_{A_{nr}} p_0^t(u_\varepsilon) \gamma_{1\varepsilon} \sigma^{-1}(f(s,x,u_{mr}) - f(s,x,u_\varepsilon)) dP \\ \cdots \\ \int_{A_{nr}} p_0^t(u_\varepsilon) \gamma_{l\varepsilon} \sigma^{-1}(f(s,x,u_{mr}) - f(s,x,u_\varepsilon)) dP \end{pmatrix}$$

for almost all $t \in [0,1]$; i.e., there is a set $T_1 \subset [0,1]$ of zero measure such that the equation given above is true for $t \notin T_1$ and all n, m, r. Moreover, there is a set $T_2 \subset [0,1]$ of zero measure such that if $t \notin T_2$,

$$\lim_{\tau \to 0} \frac{1}{\tau} \begin{pmatrix} \int_t^{t+\tau} E_{u_\varepsilon}(\gamma_{1\varepsilon}^2) ds \\ \cdots \\ \int_t^{t+\tau} E_{u_\varepsilon}(\gamma_{l\varepsilon}^2) ds \end{pmatrix} = \begin{pmatrix} E_{u_\varepsilon}(\gamma_{1u_\varepsilon}^2) ds \\ \cdots \\ E_{u_\varepsilon}(\gamma_{lu_\varepsilon}^2) ds \end{pmatrix}.$$

Then we can conclude by applying Lemma 5.1 in Elliott and Kohlmann [108]:

$$\lim_{\tau \to 0} \frac{1}{\tau} \begin{pmatrix} \int_t^{t+\tau} \int_{A_{nr}} p_0^t(u_\varepsilon) p_t^{t+\tau}(u_{mr}) \gamma_{1\varepsilon} \sigma^{-1}(f(s,x,u_{mr}) - f(s,x,u_\varepsilon)) dP\, ds \\ \cdots \\ \int_t^{t+\tau} \int_{A_{nr}} p_0^t(u_\varepsilon) p_t^{t+\tau}(u_{mr}) \gamma_{l\varepsilon} \sigma^{-1}(f(s,x,u_{mr}) - f(s,x,u_\varepsilon)) dP\, ds \end{pmatrix}$$

$$= \begin{pmatrix} \int_{A_{nr}} p_0^t(u_\varepsilon) \gamma_{1\varepsilon} \sigma^{-1}(f(s,x,u_{mr}) - f(s,x,u_\varepsilon)) dP \\ \cdots \\ \int_{A_{nr}} p_0^t(u_\varepsilon) \gamma_{l\varepsilon} \sigma^{-1}(f(s,x,u_{mr}) - f(s,x,u_\varepsilon)) dP \end{pmatrix}$$

for $t \notin T_1 \cup T_2$, all $r \le t$, and all n, m. Finally, this implies that

$$
\begin{pmatrix} E[p_{1\varepsilon}\sigma^{-1}f_t^u \mid \mathcal{Y}_t] \\ \cdots \\ E[p_{l\varepsilon}\sigma^{-1}f_t^u \mid \mathcal{Y}_t] \end{pmatrix} \notin \begin{pmatrix} E[p_{1\varepsilon}\sigma^{-1}f_t^{u_\varepsilon} \mid \mathcal{Y}_t] \\ \cdots \\ E[p_{l\varepsilon}\sigma^{-1}f_t^{u_\varepsilon} \mid \mathcal{Y}_t] \end{pmatrix} - \varepsilon k^0 - \operatorname{int} C
$$

for all $u \in V$ and a $\mathcal{F}_t$-predictable process

$$
p_\varepsilon = \begin{pmatrix} p_0^t(u_\varepsilon)\gamma_{1\varepsilon} \\ \cdots \\ p_0^t(u_\varepsilon)\gamma_{l\varepsilon} \end{pmatrix}.
$$

The proof is complete. □

Remark 4.5.17. Using martingale representation results and a variational principle for multicriteria optimization problems we obtain a necessary condition that u_ε must satisfy. In fact, u_ε is an ε-weakly minimal solution a.s.$\tilde{P}$. in the sense of multicriteria optimization for the conditional expectation of a certain Hamiltonian of the stochastic system. Here the expectation is taken with respect to the observed σ-field.

Remark 4.5.18. Problems in using the necessary condition presented in Theorem 4.5.16 for the development of a numerical algorithm are discussed by Heyde, Grecksch, Tammer [165] and Grecksch, Heyde, Isac, Tammer [146].

4.6 Stochastic Efficiency in a Set

In this section we will mention an example for a binary relation in abstract spaces: stochastic dominance. For more details see papers by Ogryczak and Ruszczynski [282], [283].

Comparison of random variables is usually related to the problem of choice among risky alternatives in a given attainable set. For example, in the well-known problem of portfolio selection (Markowitz [249], [250], [251]) the feasible set of random variables is defined as all convex combinations of a given collection of investment opportunities (securities).

Decision problems with real-valued outcomes, such as return, net profit, or number of lives saved, are studied by Ogryczak [281], Ogryczak and Ruszczynski [283], Michalowski and Ogryczak [256] and [257]. An important example, originating from financial planning, is the problem of choice among investment opportunities or portfolios having uncertain returns.

General problems of comparing real-valued random variables (distributions) are considered by Ogryczak and Ruszczynski [282]. In the stochastic dominance approach (cf. [282] and [283]) random variables are compared by pointwise comparison of some performance functions constructed from their distribution functions.

Let X be a random variable with the probability measure P_X. The first performance function $F_X^{(1)}$ is defined as the right-continuous cumulative distribution function itself:

$$F_X^{(1)}(\eta) = F_X(\eta) = P\{X \leq \eta\} \quad \text{for} \quad \eta \in \mathbb{R}.$$

The weak relation of the **first-degree stochastic dominance** (FSD) (cf. [282]) is defined as follows:

$$X \succeq_{FSD} Y \iff F_X(\eta) \leq F_Y(\eta) \quad \text{for all} \quad \eta \in \mathbb{R}.$$

The second performance function $F_X^{(2)}$ is given by areas below the distribution function F_X,

$$F_X^{(2)}(\eta) = \int_{-\infty}^{\eta} F_X(\xi) d\xi \quad \text{for} \quad \eta \in R,$$

and defines the weak relation of the **second-degree stochastic dominance** (SSD),

$$X \succeq_{SSD} Y \iff F_X^{(2)}(\eta) \leq F_Y^{(2)}(\eta) \quad \text{for all} \quad \eta \in \mathbb{R}.$$

The corresponding strict dominance relations $\succ_{FSD}$ and $\succ_{SSD}$ are defined by the standard rule

$$X \succ Y \iff X \succeq Y \quad \text{and} \quad Y \not\succeq X.$$

Thus, one says that X dominates Y under the FSD rules ($X \succ_{FSD} Y$), if $F_X(\eta) \leq F_Y(\eta)$ for all $\eta \in \mathbb{R}$, where at least one strict inequality holds. Similarly, we say that X dominates Y under the SSD rules ($X \succ_{SSD} Y$), if $F_X^{(2)}(\eta) \leq F_Y^{(2)}(\eta)$ for all $\eta \in \mathbb{R}$, with at least one strict inequality. Compare [51], [282], and [283] for stochastic efficiency concepts.

Note that $F_X(\eta)$ expresses the probability of underachievement for a given target value η. Thus the first stochastic dominance is based on the multidimensional (continuum-dimensional) objective defined by the probabilities of underachievement for all target values. The FSD is the most general relation. If $X \succ_{FSD} Y$, then X is preferred to Y within all models preferring larger outcomes, no matter how risk-averse or risk-seeking they are.

For decision-making under risk, most important is the second-degree stochastic dominance relation, associated with the function $F_X^{(2)}$. If $X \succ_{SSD} Y$, then X is preferred to Y within all risk-averse preference models that prefer larger outcomes. It is therefore a matter of primary importance that an approach to the comparison of random outcomes be consistent with the second-degree stochastic dominance relation.

Mean-risk approaches (cf. [283]) are based on comparing two scalar characteristics (summary statistics), the first of which, denoted by μ, represents the expected outcome (reward), and the second, denoted by r, is some measure of risk. The weak relation of mean-risk dominance is defined as follows:

$$X \succeq_{\mu/r} Y \Longleftrightarrow \mu_X \geq \mu_Y \quad \text{and} \quad r_X \leq r_Y.$$

The corresponding strict dominance relation $\succ_{\frac{\mu}{r}}$ is defined in the standard way. We say that X dominates Y under the μ/r rules ($X \succ_{\mu/r} Y$), if $\mu_X \geq \mu_Y$ and $r_X \leq r_Y$, and at least one of these inequalities is strict.

An important advantage of mean-risk approaches is the possibility of a pictorial trade-off analysis. Having assumed a trade-off coefficient λ between the risk and the mean, one may directly compare real values of $\mu_X - \lambda r_X$ and $\mu_Y - \lambda r_Y$. Indeed, the following implication holds:

$$X \succeq_{\mu/r} Y \Longrightarrow \mu_X - \lambda r_X \geq \mu_Y - \lambda r_Y \quad \text{for all} \quad \lambda > 0.$$

In [282] is discussed that the trade-off approach is consistent with the mean-risk dominance. Suppose that the mean-risk model is consistent with the SSD model by the implication

$$X \succeq_{\text{SSD}} Y \Longrightarrow X \succeq_{\mu/r} Y.$$

The mean-risk and trade-off approaches lead to guaranteed results:

$$X \succ_{\mu/r} Y \Longrightarrow Y \nsucceq_{\text{SSD}} X$$

and

$$\mu_X - \lambda r_X > \mu_Y - \lambda r_Y \quad \text{for some} \quad \lambda > 0 \Longrightarrow Y \nsucceq_{\text{SSD}} X.$$

In other words, they cannot strictly prefer an inferior decision.

Example 4.6.1. Consider a **portfolio optimization problem** following the original Markowitz formulation (cf. [249], [250], [251]) that is based on a single-period model of investment.

At the beginning of a period, an investor allocates capital among various securities. Assuming that each security is represented by a variable, this is equivalent to assigning a nonnegative weight to each of the variables. During the investment period, a security generates a certain (random) rate of return. The change of capital invested observed at the end of the period is measured by the weighted average of the individual rates of return. This leads to a decision problem assuming that larger outcomes are preferred. A feasible random variable $X \in Q$ is called efficient under the relation $\succeq$ if there is no $Y \in Q$ such that $Y \succ X$.

List of Abbreviations

aint $A = A^i$	—	algebraic interior (or core) of the set A
A_∞	—	asymptotic cone of the nonempty set A
$\mathcal{A} \perp \mathcal{B}$	—	orthogonal closed linear subspaces.
$H = \mathcal{A} \oplus \mathcal{B}$	—	complementary subspaces, see under line (4.24)
a.c.	—	asymptotically compact
A^*	—	adjoint operator to A
A^T	—	adjoint operator in Hilbert spaces
$(A^T)^{-1}$	—	inverse operator of the adjoint operator A^T to A
lin A	—	linear hull of a nonempty set A
aff A	—	affine hull of a nonempty set A
conv A	—	convex hull of a nonempty set A
$A + B$	—	addition of sets, see Section 2.1
$[A]_C$	—	see full set, Definition 2.1.21
A^c	—	$Y \setminus A$, where $A \subset Y$
A^r	—	see Theorem 3.10.12
A_x	—	$\{y \in Y \mid (x,y) \in A\}$, where $A \subset X \times Y$ is nonempty
a.s.P.	—	almost surely w.r.t. probability P
$\alpha(y)$	—	sup-support function
$\mathcal{B}(x)$	—	neighborhood base of x
B^0	—	unit ball
$B(x,r)$	—	(closed) ball of center x and radius r
BEff$(D;C)$	—	the set of Benson-efficient points of D
B-function	—	Bregman function
cl B or $\overline{B}$	—	closure of the set B
τ- cl B	—	closure of the set B w.r.t. the topology τ
$\Delta \circ \Gamma$	—	composition of multifunctions Δ and Γ
∂B	—	$B \setminus$ int B, see Section 3.3
bd B	—	usual (topological) boundary of B
$\beta(y)$	—	inf-support function
C_ε	—	Henig dilating cone
C-l.c.	—	C-lower continuous
C-l.s.c.	—	C-lower semicontinuous
C^+	—	continuous dual cone of the cone C
$\inf_C A$	—	infimum of A w.r.t. C
$C^\#$	—	quasi interior of the cone C^+
$C[0,1]$	—	see Example 2.1.12
$C^1[0,T]$	—	see inverse Stefan problem
$\leq_C$	—	see Section 3.2.3
$\preceq_{k^0}$	—	a preorder, see Section 3.10 and Theorem 3.10.7
C-seq-b-regular	—	see under Corollary 3.10.16
(CP)	—	containment property
$\mathcal{C} : X \rightrightarrows Y$	—	multifunction, whose values are pointed convex cones
$\mathcal{C}^c$	—	see Section 3.5.2

CP	—	complementarity problem
$\|P\|$	—	cardinality of P
$D^{\perp}$	—	orthogonal space to a linear subspace D
$\partial^{\leq} f(x_0)$	—	subdifferential of f at $x_0 \in \text{dom } f$
$\partial^{\leq c} f(x_0)$	—	subdifferential of f at $x_0 \in \text{dom } f$
dim	—	dimension
$D\Gamma(x, y)(u)$	—	Dini upper derivative of Γ at (x, y) in the direction u
$\underline{D}\Gamma(x, y)(u)$	—	Dini lower derivative of Γ at (x, y) in the direction u
$S\Gamma(\overline{x}, \overline{y})(u)$	—	derivative in the sense of Shi in the direction u
$f'(x)(v)$	—	directional derivative of f at x in the direction v
$\text{dist}(x, A), d(x, A)$	—	$\inf_{a \in A} d(x, a)$, where $A \subset X$ and d a metric on X
$\text{dom } f$	—	domain of f
(DP)	—	domination property
$\varphi_{\text{cl } B, k^0}(y)$	—	scalarization functional, see Section 2.3
$\text{Eff}(M, B_{\varepsilon k^0})$	—	the set of approximately efficient points
$\varepsilon\text{-Eff}^{\varphi}(M, C)$	—	see Definition 3.1.2
$\text{Eff}_{\text{Min}}(P, C)$	—	see Remark 2.1.3 or Section 3.7
$\text{Eff}_{\text{Max}}(D, C)$	—	see Remark 2.1.3 or Section 3.7
$\text{Eff}_{\text{Min}}(F, B, e)$	—	(B, e)-minimal elements; see Section 3.11.2
$\text{Eff}_{\text{Max}}(F, B, e)$	—	the set of all (B, e)-maximal elements of F
$\emptyset$	—	empty set
EP	—	equilibrium problem
EVP	—	Ekeland's variational principle
epi f	—	epigraph of f
$e(A, B)$	—	excess of A over B
y_{ε}	—	see Section 3.1.1
E	—	expectation operator
$\mathcal{F}_t$	—	σ-algebra
$\mathcal{F}(M)$	—	the family of all nonempty finite subsets of M
$\mathcal{F}(M, x)$	—	the family of all finite subsets of M containing x
$\nabla_x f(x)$	—	Fréchet derivative of f w.r.t. x
$\text{gr } \Gamma$	—	graph (of a multifunction Γ)
Γ_C	—	see Section 2.4
$\Gamma_{f,C}$	—	see Section 2.4
Γ^P	—	see Theorem 3.5.12
$\text{GEff}(\mathbf{F})$	—	the set of properly efficient points in the sense of Geoffrion
GVEP	—	generalized vector equilibrium problem
GVVI	—	generalized vector variational inequality
$\text{HEff}(D; C)$	—	the set of Henig-proper efficient points of D w.r.t. C
H-C-u.c.	—	Hausdorff C-upper continuous (w.r.t. cone C)
H.l.c.s.	—	Hausdorff locally convex space
H-l.c.	—	Hausdorff lower continuous
H.t.v.s.	—	Hausdorff topological vector space
H-u.c.	—	Hausdorff upper continuous
HVIS	—	hemivariational inequalities system; see line (3.101)

$\operatorname{Im}\Gamma$	—	image (of a multifunction Γ)
χ_D	—	indicator function of D
$\operatorname{int} C$	—	interior of the set C
$\Gamma^{+1}(B)$	—	inverse image; see Section 2.4
$\Gamma^{-1}(B)$	—	inverse image; see Section 2.4
∞	—	element adjoined to Y to get $Y^\bullet$
$\operatorname{Ker}A$	—	kernel of an operator A
KKM-lemma	—	Knaster, Kuratowski, and Mazurkiewicz lemma
(K,L)-monotonicity	—	see Definition 3.8.5
$\Phi(x,z^*)$	—	Lagrangian (see Remark 3.11.5)
Λ^α for $\alpha\in\,]0,1[$	—	see under Definition 2.4.1
ℓ^∞, ℓ^p	—	see Example 2.2.3
L^P or L_p	—	see Example 2.2.3
$\liminf_{x\to x_0}\Gamma(x)$	—	limit inferior of Γ at x_0
$L(X,Y)$, $\mathcal{L}(X,Y)$	—	the set of linear continuous mappings from X in Y
$\mathcal{L}_1(x_\varepsilon)$, $\mathcal{L}_2(x_\varepsilon)$	—	see Section 4.1.2
$\operatorname{lev}_\Gamma(y)$	—	sublevel set; see Section 2.4
$\operatorname{lev}_\Gamma^<(y)$	—	strict sublevel set; see Section 2.4
l.c.	—	lower continuous
l.c.s.	—	locally convex space
l.s.c.	—	lower semicontinuous
$(X,\mathcal{P})$	—	l.c.s. with a family $\mathcal{P}$ of seminorms
ℓ_2^+	—	the common ordering cone in the Hilbert space ℓ_2
$\operatorname{meas}A$	—	measure of a set A
$\Gamma:X\rightrightarrows Y$	—	multifunction
$\operatorname{Max}(M_0,\mathcal{R})$	—	the class of maximal elements of M_0 with respect to $\mathcal{R}$
$\operatorname{Min}(M_0,\mathcal{R})$	—	the class of minimal elements of M_0 with respect to $\mathcal{R}$
$\operatorname{HMax}(Y;C)$	—	see Section 3.2.6
$\operatorname{PrMax}(Y;C)$	—	see Section 3.2.6
$(x_i)_{i\in I}$	—	net or generalized sequence
$\mathcal{N}_\tau(x)$	—	the class of all neighborhoods of x w.r.t. a topology τ
$\mathbb{N}$	—	the set of nonnegative integers
$\mathbb{N}^*$	—	the set of positive integers
$\|\cdot\|_*$	—	dual norm to $\|\cdot\|$
$\|\|\cdot\|\|$	—	vector-valued norm
$N_D(x^0)$	—	normal cone to D at $x^0\in X$
$(X,\|\cdot\|)$	—	normed space X
n.v.s.	—	normed vector space
(a,b)	—	open interval
$[a,b)$	—	half open interval
$\mathcal{P}(X)$	—	the class of subsets of X
$p\text{-Eff}(\mathbf{F},\mathcal{Z})$	—	the set of properly efficient points w.r.t. a family $\mathcal{Z}$
P-pseudomonotonicity	—	see Lemma 3.8.29
$\operatorname{PrEff}(Y;C)$	—	the set of properly efficient points of Y w.r.t. C
$\mathbf{PC}^r$ $(r\geq 1)$	—	see Section 4.4.3

p_A	—	Minkowski functional of the set A
U^0	—	polar of the set U
P_X	—	projection of $X \times Y$ onto X
$P_{\mathcal{D}}(e)$	—	projection of e onto a set $\mathcal{D}$
P_u	—	probability measure; see Section 4.5.3
$\overline{\mathbb{R}}$	—	$\mathbb{R} \cup \{+\infty\} \cup \{-\infty\}$
$\mathbb{R}^{\bullet}$	—	$\mathbb{R} \cup \{+\infty\}$
$R(P_{\mathcal{A}})$	—	range of the operator $P_{\mathcal{A}}$
raint	—	relative algebraic interior (of a set)
$r : X \times X \to P$	—	a P-valued metric, P a cone
$\mathbb{R}^n$	—	the real n-dimensional vector space
$\mathbb{R}_+ := [0, \infty[$	—	the set of nonnegative real numbers
$\mathbb{R}^n_+$	—	usual ordering cone of $\mathbb{R}^n$
$\mathcal{R}^{-1}$	—	inverse of the relation (or multifunction) $\mathcal{R}$
$Y_t^+(x)$	—	upper section of $Y \subset X$ with respect to t and $x \in X$
$Y_t^-(x)$	—	lower section of $Y \subset X$ with respect to t and $x \in X$
$\mathcal{S} \circ \mathcal{R}$	—	composition of two relations $\mathcal{R}$ and $\mathcal{S}$
$\succeq$	—	compare "directed set"
$X \succeq_{FSD} Y$	—	see Section 4.6
$X \succeq_{SSD} Y$	—	see Section 4.6
SEP	—	scalar equilibrium problem
SGVEP	—	strong generalized vector equilibrium problem
S(f,P)	—	the set of solutions of a GVEP
$(c^n) \to c$	—	sequence (c^n) converging to c
$M \times M$	—	set of ordered pairs of elements of M
$(M, \mathcal{R})$	—	set M with order structure $\mathcal{R}$
Min(F, y^*, e)	—	the set of all (y^*, e)-minimal elements of F
Max(F, y^*, e)	—	the set of all (y^*, e)-maximal elements of F
s.t.	—	subject to
SPEff($A; C$)	—	class of strong proper efficient points of set A w.r.t. C
SVCP	—	strong vector complementarity problem
$T_{\mathcal{A}}$	—	partial inverse of T with respect to $\mathcal{A}$
$T_B(A; a)$	—	Bouligand tangent cone (or contingent cone) of A at a
$T_U(A; a)$	—	Ursescu tangent cone (or adjacent cone) of A at a
t.l.s. (or t.v.s.)	—	topological linear (or topological vector) space
$(x_i) \xrightarrow{\tau} x$	—	convergence (w.r.t. topology τ)
$\tau_1 \times \tau_2$	—	product topology
a^T	—	transposed vector to $a \in \mathbb{R}^n$
$U^{\bullet}$	—	$U \setminus \{x_0\}$, where U is a neighborhood of x_0
u.c.	—	upper continuous
VOP	—	vector optimization problem
VSP	—	vector saddle point problem
VVI	—	vector variational inequality
w-pointedness	—	see Section 3.8.1
$w := \sigma(X, X^*)$	—	weak topology of X

$w^* := \sigma(X^*, X)$ —   weak* topology of X^*

w-normal   —   weakly normal

$x_i \xrightarrow{w} x$ —   weakly convergent net

w.r.t.   —   with respect to

WVCP   —   weak vector complementarity problem

WVEP   —   weak vector equilibrium problem

WVSP   —   weak vector saddle point problem

WGVEP   —   weak generalized vector equilibrium problem.

$w\,\mathrm{Eff}(A; C)$ —   set of weakly efficient points

X^* —   continuous dual space to the space X

(X, ρ) —   metric space with metric ρ

(X, τ) —   topological space X equipped with topology τ

τ-a.c. set   —   asymptotically compact set w.r.t. topology τ

$x^*(x) = \langle x, x^* \rangle$ —   value of linear continuous functional x^* at x

$(x \mid y)$ —   inner product in a Hilbert space

$Y^\bullet$ — $Y \cup \{\infty\}$, see Section 2.4

(Y, C_Y) —   linear space Y ordered by a cone C_Y

References

1. Amahroq, T., Taa, A.: On Lagrange–Kuhn–Tucker multipliers for multiobjective optimization problems. Optimization, **41**, 159–172 (1997)
2. Ansari, Q.H.: On generalized vector variational-like inequalities. Ann. Sci. Math. Québec, **19**, 131–137 (1995)
3. Ansari, Q.H.: Vector equilibrium problems and vector variational inequalities. In: Giannessi, F. (ed.) Vector Variational Inequalities and Vector Equilibria. Mathematical Theories. Nonconvex Optimization and its Applications. Kluwer, Dordrecht, pp. 1–15 (2000)
4. Ansari, Q.H., Siddiqi, A.H.: A generalized vector variational-like inequality and optimization over an efficient set. In: Brokate, M., Siddiqi, A.H. (eds.) Functional Analysis with Current Applications in Science, Technology and Industry. Pitman Research Notes in Mathematics Series, No. 377, Longman, Essex, pp. 177–191 (1998)
5. Ansari, Q.H., Siddiqi, A.H., Yao, J.C.: Generalized vector variational like inequalities and their scalarization. In: Giannessi, F. (ed.) Vector Variational Inequalities and Vector Equilibria. Mathematical Theories. Nonconvex Optimization and its Applications. Kluwer, Dordrecht, pp. 18–37 (2000)
6. Antipin, A.S.: Controlled proximal differential systems for saddle problems. Differential Equations, **28**, 1498–1510 (1992)
7. Antipin, A.S.: Feed-back controlled saddle gradient processes, Automat. Remote Control, **55**, 311–320 (1994)
8. Antipin, A.S.: Convergence and estimates of the rate of convergence of proximal methods to fixed points of extremal mappings. Zh. Vychisl. Mat. Mat. Fiz., **35**, 688–704 (1995)
9. Arrow, K.J., Barankin, E.W., Blackwell, D.: Admissible points of convex sets. In: Kuhn, H.W., Tucker, A.W. (eds.) Contributions to the Theory of Games **2**. Ann. of Math. Studies **28**, Princeton Univ. Press, Princeton, pp. 87–91 (1953)
10. Attouch, H., Riahi, H.: Stability results for Ekeland's ε-variational principle and cone extremal solutions. Math. Oper. Res., **18**, 173–201 (1993)
11. Aubin, J.-P.: Optima and Equilibria. An Introduction to Nonlinear Analysis. Springer, Berlin Heidelberg (1993)
12. Aubin, J.-P., Ekeland, I.: Applied Nonlinear Analysis. John Wiley and Sons, New York (1984)
13. Aubin, J.-P., Frankowska, H.: Set-Valued Analysis. Birkhäuser, Boston (1990)

14. Baiocchi, C., Capelo, A.: Variational and Quasivariational Inequalities, Applications to Free Boundary Problems. John Wiley and Sons, New York (1984)
15. Ballestero, E., Romero, C.: Multiple Criteria Decision Making and its Applications to Economic Problems. Kluwer, Boston (1998)
16. Bector, C.R., Chandra, S.: Multiobjective fractional programming duality: a parametric approach. Research Report, Faculty of Management, University of Manitoba (1987)
17. Bednarczuk, E.: Some stability results for vector optimization problems in partially topological vector spaces. In: Proceedings of the First World Congress of Nonlinear Analysts (Tampa, Florida, 1992), pp. 2371–2382 (1992)
18. Bednarczuk, E.: An approach to well-posedness in vector optimization: consequences to stability. Control Cybernet., **23**, 107–121 (1994)
19. Bednarczuk, E.: Berge-type theorems for vector optimization problems. Optimization, **32**, 373–384 (1995)
20. Benker, H.: Upper and lower bounds for minimal norm problems under linear constraints. Banach Center Publ., **14**, 35–45 (1985)
21. Benker, H., Hamel, A., Spitzner, J., Tammer, Chr.: A proximal point algorithm for location problems. In: Göpfert, A. et al. (eds.) Methods of Multicriteria Decision Theory. Deutsche Hochschulschriften 2398, Hänsel-Hohenhausen-Verlag, pp. 203–211 (1997)
22. Benker, H., Hamel, A., Tammer, Chr.: A proximal point algorithm for control approximation problems. Z. Oper. Res., **43**, 261–280 (1996)
23. Benker, H., Hamel, A., Tammer, Chr.: An algorithm for vectorial control approximation problems. In: Fandel, G., Gal, T. (eds.) Multiple Criteria Decision Making. Springer, Berlin, pp. 3–12 (1997)
24. Benker, H., Kossert, S.: Remarks on quadratic optimal control problems in Hilbert spaces. Z. Anal. Anwendungen, **1** (3), 13–21 (1982)
25. Berge, C.: Espaces Topologiques et Fonctions Multivoques. Dunod, Paris (1959)
26. Berge, C.: Topological Spaces. Macmillan Co., New York (1963)
27. Bernau, H.: Interactive methods for vector optimization. In: Brosowski, B., Martensen, E. (eds.) Optimization in Mathematical Physics. Peter Lang, Frankfurt am Main, pp. 21–36 (1987)
28. Bianchi, M., Hadjisavvas, N., Schaible, S.: Vector equilibrium problems with generalized monotone bifunctions. J. Optim. Theory Appl., **92**, 527–542 (1997)
29. Bianchi, M., Schaible, S.: Generalized monotone bifunctions and equilibrium problems. J. Optim. Theory Appl., **90**, 31–43 (1996)
30. Bigi, G., Pappalardo, M.: Regularity conditions in vector optimization. J. Optim. Theory Appl., **102**, 83–96 (1999)
31. Bitran, G.R.: Duality for nonlinear multiple-criteria optimization problems. J. Optim. Theory Appl., **35**, 367–401 (1981)
32. Bitran, G.R., Magnanti, T.L.: The structure of admissible points with respect to cone dominance. J. Optim. Theory Appl., **29**, 573–614 (1979)
33. Blum, E., Oettli, W.: Variational principles for equilibrium problems. In: Guddat, J., et al. (eds.) Parametric Optimization and Related Topics III. Peter Lang, Frankfurt am Main, pp. 79–88 (1993)
34. Blum, E., Oettli, W.: From optimization and variational inequalities to equilibrium problems. The Mathematics Students, **63**, 123–145 (1994)
35. Borwein, J.M.: Proper efficient points for maximizations with respect to cones. SIAM J. Control Optim., **15**, 57–63 (1977)

36. Borwein, J.M.: A Lagrange multiplier theorem and a sandwich theorem for convex relations. Math. Scand., **48**, 198–204 (1981)
37. Borwein, J.M.: Convex relations in analysis and optimization. In: Schaible, S., Ziemba, W.T. (eds.) Generalized Concavity in Optimization and Economics. Academic Press, New York, pp. 335–377 (1981)
38. Borwein, J.M.: Continuity and differentiability properties of convex operators. Proc. London Math. Soc., **44**, 420–444 (1982)
39. Borwein, J.M.: On the existence of Pareto efficient points. Math. Oper. Res., **8**, 64–73 (1983)
40. Borwein, J.M.: Convex cones, minimality notions and consequences. In: Jahn, J., Krabs, W. (eds.) Recent Advances and Hystorical Development of Vector Optimization. Lecture Notes in Econom. and Math. Systems, 294, Springer, Heidelberg, pp. 64–73 (1987)
41. Borwein, J.M., Preiss, D.: A smooth variational principle with applications to subdifferentiability and to differentiability of convex functions. Trans. Amer. Math. Soc., **303**, 517–527 (1987)
42. Borwein, J.M., Zhuang, D.: Super efficiency in vector optimization. Trans. Amer. Math. Soc., **338**, 105–122 (1993)
43. Breckner, W.W.: Dualität bei Optimierungsaufgaben in halbgeordneten topologischen Vektorräumen (I). Rev. Anal. Numér. Théor. Approx., **1**, 5–35 (1972)
44. Breckner, W.W.: Dualität bei Optimierungsaufgaben in halbgeordneten topologischen Vektorräumen (II). Rev. Anal. Numér. Théor. Approx., **2**, 27–35 (1973)
45. Breckner, W.W.: Derived sets for weak multiobjective optimization problems with state and control variables. J. Optim. Theory Appl., **93**, 73–102 (1997)
46. Brezis, H., Browder, F.E.: A general principle on ordered sets in nonlinear functional analysis. Adv. Math., **21**, 355–364 (1976)
47. Brezis, H., Nirenberg, L., Stampacchia, G.: A remark on Ky Fan's minimax principle. Boll. Un. Mat. Ital., Sez.A, **4**(6), 293–300 (1972)
48. Brøndsted, A.: On a lemma of Bishop and Phelps. Pacific J. Math., **55**, 335–341 (1974)
49. Brosowski, B., Conci, A.: On vector optimization and parametric programming. Segundas Jornados Latino Americans de Matematica Aplicada, **2**, 483–495 (1983)
50. Brumelle, S.: Duality for multiple objective convex programs. Math. Oper. Res., **6**, 159–172 (1981)
51. Caballero, R., Cerda Tena, E., Munoz, M., Rey, L.: Relations among several efficiency concepts in stochastic multiple objective programming. In: Haimes, Y.Y., Steuer, R.E. (eds.) Research and Practice in Multiple Criteria Decision Making. Lecture Notes in Econom. and Math. Systems, 487, Springer, Berlin Heidelberg, pp. 57–69 (2000)
52. Cârjă, O.: Elements of Nonlinear Functional Analysis. (Roumanian) Editura Universităţii "Al.I.Cuza" Iaşi, Iaşi (1998)
53. Carrizosa, E., Conde, E., Fernandez, R., Puerto, J.: Efficiency in Euclidean constrained location problems. Oper. Res. Lett., **14**, 291–295 (1993)
54. Carrizosa, E., Fernandez, R.: A polygonal upper bound for the efficient set for single-location problems with mixed norms. Top. Soc. Estad. Investig. Oper., Madrid, **1**, 107–116 (1993)

55. Carrizosa, E., Fernandez, R., Puerto, J.: Determination of a pseudoefficient set for single-location problems with mixed polyhedral norms. In: Orban, F., Rasson, J.P. (eds.) Proceedings of the Fifth Meeting of the EURO Working Group on Locational Analysis, FUNDP, Namur, Belgium, pp. 27–39 (1990)

56. Carrizosa, E., Fernandez, R., Puerto, J.: An axiomatic approach to location criteria. In Orban, F., Rasson, J.P. (eds.) Proceedings of the Fifth Meeting of the EURO Working Group on Locational Analysis, FUNDP, Namur, Belgium, pp. 40–53 (1990)

57. Carrizosa, E., Plastria, F.: A characterization of efficient points in constrained location problems with regional demand. Working paper, BEIF/53, Vrije Universiteit Brussel, Brussels, Belgium (1993)

58. Castellani, M., Mastroeni, G., Pappalardo, M.: On regularity for generalized systems and applications. In: Di Pillo, G., Giannessi, F. (eds.) Nonlinear Optimization and Applications. Plenum Press, New York, pp. 13–26 (1996)

59. Censor, Y., Zenios, S.: The proximal minimization algorithm with D-functions. J. Optim. Theory Appl., **73**, 451–464 (1992)

60. Cesari, L., Suryanarayana, M.B.: Existence theorems for Pareto optimization; multivalued and Banach space valued functionals. Trans. Amer. Math. Soc., **244**, 37–65 (1978)

61. Chadli, O., Chbani, Z., Riahi, H.: Equilibrium problems with generalized monotone bifunctions and applications to variational inequalities. J. Optim. Theory Appl., **105**, 299–323 (2000)

62. Chadli, O., Riahi, H.: On generalized equilibrium problems. J. Global Optim., **16**, 59–75 (2000)

63. Chalmet, L.G., Francis, R.L., Kolen, A.: Finding efficient solutions for rectilinear distance location problem efficiently. Eur. J. Oper. Res., **6**, 117–124 (1981)

64. Chen, G.Y.: Existence of solutions for a vector variational inequality: an extension of Hartman–Stampacchia theorem. J. Optim. Theory Appl., **74**, 445–456 (1992)

65. Chen, G.Y., Chen, G.M.: Vector variational inequality and vector optimization. In: Lecture Notes in Econom. and Math. Systems, 285, Springer, New York, pp. 408–416 (1987)

66. Chen, G.Y., Craven, B.D.: Approximate dual and approximate vector variational inequality for multiobjective optimization. J. Austral. Math. Soc. Ser. A, **47**, 418–423 (1989)

67. Chen, G.Y., Craven, B.D.: A vector variational inequality and optimization over an efficient set. ZOR, **3**, 1–12 (1990)

68. Chen, G.Y., Goh, C.J., Yang, X.Q.: On gap functions for vector variational inequalities. In: Giannessi, F. (ed.) Vector Variational Inequalities and Vector Equilibria. Mathematical Theories. Nonconvex Optimization and its Applications. Kluwer, Dordrecht, pp. 55–72 (2000)

69. Chen, G.Y., Goh, C.J., Yang, X.Q.: Existence of solution for generalized vector variational inequalities. Optimization, **50**, 1–5 (2001)

70. Chen, G.Y., Hou, S.H.: Existence of solutions for vector variational inequalities. In: Giannessi, F. (ed.) Vector Variational Inequalities and Vector Equilibria. Mathematical Theories. Nonconvex Optimization and its Applications. Kluwer, Dordrecht, pp. 73–86 (2000)

71. Chen, G.Y., Huang, X.X.: A unified approach to the existing three types of variational principles for vector valued functions. Math. Meth. Oper. Res., **48**, 349–357 (1998)
72. Chen, G.Y., Huang, X.X.: Ekeland's ε-variational principle for set-valued mappings. Math. Meth. Oper. Res., **48**, 181–186 (1998)
73. Chen, G.Y., Huang, X.X., Lee, G.M.: Equivalents of an approximate variational principle for vector-valued functions and applications. Math. Meth. Oper. Res., **49**, 125–136 (1999)
74. Chen, G.Y., Teboulle, M.: Convergence analysis of a proximal-like minimization algorithm using Bregman functions. SIAM J. Optim., **3**, 538–543 (1993)
75. Chen, G.Y., Yang, X.Q.: Vector complementarity problem and its equivalences with weak minimal element in ordered spaces. J. Math. Anal. Appl., **153**, 136–158 (1990)
76. Chen, G.Y., Yang, X.Q.: On the existence of solutions to vector complementarity problems. In: Giannessi, F. (ed.) Vector Variational Inequalities and Vector Equilibria. Mathematical Theories. Nonconvex oOptimization and its Applications. Kluwer, Dordrecht, pp. 87–95 (2000)
77. Chew, K.L.: Maximal points with respect to cone dominance in Banach spaces and their existence. J. Optim. Theory Appl., **44**, 1–53 (1984)
78. Chowdhury, M.S.R., Tan, K.K.: Generalized variational inequalities for quasi-monotone operators and applications. Bull. Polish Acad. Sci. Math., **45**, 25–54 (1997)
79. Cieslik, D.: The Fermat–Steiner–Weber problem in Minkowski spaces. Optimization, **19**, 485–489 (1988)
80. Clarke, F.H.: Optimization and Nonsmooth Analysis. John Wiley and Sons, New York (1983)
81. Clarke, F.H., Ledyaev, Y.S., Stern, R.J., Wolenski, P.R.: Nonsmooth Analysis and Control Theory. Springer, New York (1998)
82. Corley, H.W.: An existence result for maximizations with respect to cones. J. Optim. Theory Appl., **31**, 277–281 (1980)
83. Corley, H.W.: Duality for maximizations with respect to cones. J. Math. Anal. Appl., **84**, 560–568 (1981)
84. Corley, H.W.: Existence and Lagrangian duality for maximizations of set-valued maps. J. Optim. Theory Appl., **54**, 489–501 (1987)
85. Crank, J.: Free and Moving Boundary Problems. Clarendon Press, Oxford (1984)
86. Dafermos, S.: Traffic equilibrium and variational inequalities. Transportation Sc., **14**, 42–54 (1980)
87. Daneš, J.: A geometric theorem useful in nonlinear functional analysis. Boll. Un. Mat. Ital., **6**, 369–375 (1972)
88. Daniele, P., Maugeri, A.: Vector variational inequalities and modelling of a continuum traffic equilibrium problem. In: Giannessi, F. (ed.) Vector Variational Inequalities and Vector Equilibria. Mathematical Theories. Nonconvex Optimization and its Applications. Kluwer, Dordrecht, pp. 97–111 (2000)
89. Daniilidis, A.: Arrow–Barankin–Blackwell theorems and related results in cone duality: A survey. In: Nguyen, V.H., Strodiot, J.-J., Tossings, P. (eds.) Optimization. Lecture Notes in Econom. and Math. Systems, 481, Springer, Berlin Heidelberg, pp. 119–131 (2000)
90. Daniilidis, A., Hadjisavvas, N.: Existence theorems for vector variational inequalities. Bull. Austral. Math. Soc., **54**, 473–481 (1996)

91. Dauer, J.P., Stadler, W.: A survey of vector optimization in infinite-dimensional spaces. Part II. J. Optim. Theory Appl., **51**, 205–241 (1986)
92. Dedieu, J.-P.: Critères de fermeture pour l'image d'un fermé non convexe par une multiapplication. C.R. Acad. Sci., Paris Ser. I, **287**, 941–943 (1978)
93. Dentcheva, D., Helbig, S.: On several concepts for ε-efficiency. OR Spektrum, **16**, 179–186 (1994)
94. Dentcheva, D., Helbig, S.: On variational principles, level sets, well-posedness, and ε-solutions in vector optimization. J. Optim. Theory Appl., **89**, 325–349 (1996)
95. Ding, X.P., Tarafdar, E.: Generalized vector variational-like inequalities. In: Giannessi, F. (ed.) Vector Variational Inequalities and Vector Equilibria. Mathematical Theories. Nonconvex Optimization and its Applications. Kluwer, Dordrecht, pp. 113–124 (2000)
96. Ding, X.P., Tarafdar, E.: Generalized vector variational-like inequalities with c_x-η-pseudomonotone set-valued mappings. In: Giannessi, F. (ed.) Vector Variational Inequalities and Vector Equilibria. Mathematical Theories. Nonconvex Optimization and its Applications. Kluwer, Dordrecht, pp. 125–140 (2000)
97. Dinkelbach, W.: On nonlinear fractional programming. Management Sci., **13**, 492–498 (1967)
98. Dolecki, S., Malivert, C.: Polarities and stability in vector optimization. In: Jahn, J., Krabs, W. (eds.) Recent Advances and Historical Development of Vector Optimization (Darmstadt, 1986). Springer, Berlin, pp. 96–113 (1987)
99. Dolecki, S., Malivert, C.: General duality in vector optimization. Optimization, **27**, 97–119 (1993)
100. Dunford, N., Schwartz, J.T.: Linear Operators. Part I: General Theory. John Wiley and Sons, New York (1988)
101. Durier, R.: On Pareto optima, the Fermat–Weber problem and polyhedral gauges. Math. Programming, **47**, 65–79 (1990)
102. Durier, R., Michelot, C.: Geometrical properties of the Fermat–Weber problem. Eur. J. Oper. Res., **20**, 332–343 (1985)
103. Durier, R., Michelot, C.: Sets of efficient points in a normed space. J. Math. Anal. Appl., **117**, 506–528 (1986)
104. Eckstein, J.: Nonlinear proximal point algorithms using Bregman functions with applications to convex programming. Math. Oper. Res., **18**, 202–226 (1993)
105. Ekeland, I.: On the variational principle. J. Math. Anal. Appl., **47**, 324–353 (1974)
106. Ekeland, I.: Nonconvex minimization problems. Bull. Amer. Math. Soc., **1**, 443–474 (1979)
107. El Abdouni, B., Thibault, L.: Lagrange multipliers for Pareto nonsmooth programming problems in Banach spaces. Optimization, **26**, 277–285 (1992)
108. Elliott, R.-J., Kohlmann, M.: The variational principle and stochastic optimal control. Stochastics, **3**, 229–241 (1980)
109. Elliott, R.-J., Kohlmann, M.: The variational principle for optimal control of diffusions with partial information. Systems Control Lett., **12**, 63–69 (1989)
110. Ester, J.: Systemanalyse und Mehrkriterielle Entscheidung. Verlag Technik, Berlin (1987)
111. Fan, K.: A Generalization of Tychonoff's fixed point theorem. Math. Ann., **142**, 305–310 (1961)

112. Fan, K.: A minimax inequality and applications, In: Shisha, O. (ed.) Inequalities, III. Academic Press, New York, pp. 103–113 (1972)

113. Fan, K.: A survey of some results closely related the Knaster–Kuratowski–Mazuriewicz theorem, game theory and applications, economic theory, econometrics and mathematical economics. Academic Press, New York, pp. 358–370 (1990)

114. Ferro, F.: An optimization result for set-valued mappings and a stability property in vector problems with constraints. J. Optim. Theory Appl., **90**, 63–77 (1996)

115. Ferro, F.: Optimization and stability results through cone lower semicontinuity. Set-Valued Anal., **5**, 365–375 (1997)

116. Ferro, F.: A new ABB theorem in Banach spaces. Optimization, **46**, 353–362 (1999)

117. de Figueiredo, D.G.: Lectures on the Ekeland's Variational Principle with Applications and Detours. Tata Res. Inst., Bombay, Springer, Berlin (1989)

118. Flåm, S.D., Antipin, A.S.: Equilibrium programming using proximal-like algorithms. Math. Programming, **78**, 29–41 (1997)

119. Föppl, A.: Vorlesungen über Technische Mechanik. Band 7. Teubner-Verlag, Leipzig (1923)

120. Francis, R.L., White, J.A.: Facility Layout and Location: an Analytical Approach. Prentice-Hall, Englewood Cliffs, New Jersey (1974)

121. Friedrichs, K.O.: Ein Verfahren der Variationsrechnung, das Minimum eines Integrals als das Maximum eines anderen Ausdruckes darzustellen. Nachr. der Gesellsch. der Wiss. zu Göttingen, 13–20 (1929)

122. Fu, J.: Simultaneous vector variational inequalities and vector implicit complementarity problem. J. Optim. Theory Appl., **93**, 141–151 (1997)

123. Gajek, L., Zagrodny, D.: Countably orderable sets and their applications in optimization. Optimization, **26**, 287–301 (1992)

124. Gajek, L., Zagrodny, D.: Geometric variational principle. Dissertationes Math., **340**, 55–71 (1995)

125. Geoffrion, A.M.: Proper efficiency and the theory of vector maximization. J. Math. Anal. Appl., **22**, 618–630 (1968)

126. Georgiev, P.G.: The strong Ekeland variational principle, the strong drop theorem and applications. J. Math. Anal. Appl., **131**, 1–21 (1988)

127. Gerstewitz (Tammer), Chr.: Nichtkonvexe Dualität in der Vektoroptimierung. Wiss. Zeitschr. TH Leuna-Merseburg, **25**, 357–364 (1983)

128. Gerstewitz (Tammer), Chr., Göpfert, A.: Zur Dualität in der Vektoroptimierung. Seminarberichte der Sektion Mathematik der Humboldt-Universität zu Berlin, **39**, 67–84 (1981)

129. Gerstewitz (Tammer), Chr., Iwanow, E.: Dualität für nichtkonvexe Vektoroptimierungsprobleme. Wiss. Z. Tech. Hochsch. Ilmenau, **2**, 61–81 (1985)

130. Gerth (Tammer), Chr., Pöhler, K.: Dualität und algorithmische Anwendung beim vektoriellen Standortproblem. Optimization, **19**, 491–512 (1988)

131. Gerth (Tammer), Chr., Weidner, P.: Nonconvex separation theorems and some applications in vector optimization. J. Optim. Theory Appl., **67**, 297–320 (1990)

132. Giannessi, F.: Theorems of alternative, quadratic programs and complementarity problems. In: Cottle, R. W. et.al. (eds.) Variational Inequalities and Complementarity Problems. John Wiley and Sons, New York, pp. 151–186 (1980)

133. Giannessi, F. (ed.): Vector Variational Inequalities and Vector Equilibria. Mathematical Theories. Kluwer Academic Publishers, Dordrecht Boston London (1999)

134. Giannessi, F., Mastroeni, G., Pellegrini, L.: On the theory of vector optimization and variational inequalities. Image space analysis and separation. In: Giannessi, F. (ed.) Vector Variational Inequalities and Vector Equilibria. Mathematical Theories. Nonconvex Optimization and its Applications. Kluwer, Dordrecht, pp. 153–215 (2000)

135. Goh, C.J., Yang, X.Q.: Scalarization methods for vector variational inequality. In: Giannessi, F. (ed.) Vector Variational Inequalities and Vector Equilibria. Mathematical Theories. Nonconvex Optimization and its Applications. Kluwer, Dordrecht, pp. 217–232 (2000)

136. Gong, X.H.: Connectedness of efficient solution sets for set-valued maps in normed spaces. J. Optim. Theory Appl., **83**, 83–96 (1994)

137. Gong, X.H., Fu, W.T., Liu, W.: Super efficiency for a vector equilibrium in locally convex topological vector spaces. In: Giannessi, F. (ed.) Vector Variational Inequalities and Vector Equilibria. Mathematical Theories. Nonconvex Optimization and its Applications. Kluwer, Dordrecht, pp. 233–252 (2000)

138. Göpfert, A., Gerth (Tammer), Chr.: Über die Skalarisierung und Dualisierung von Vektoroptimierungsproblemen. Z. Anal. Anwendungen, **5**, 377–384 (1986)

139. Göpfert, A., Nehse, R.: Vektoroptimierung. Theorie, Verfahren und Anwendungen. BSB B. G. Teubner Verlagsgesellschaft (1990)

140. Göpfert, A., Tammer, Chr.: ε-approximate solutions and conical support points. A new maximal point theorem. ZAMM, **75**, 595–596 (1995)

141. Göpfert, A., Tammer, Chr.: A new maximal point theorem. Z. Anal. Anwendungen, **14**, 379–390 (1995)

142. Göpfert, A., Tammer, Chr.: Maximal point theorems in product spaces and applications for multicriteria approximation problems. In: Haimes, Y.Y., Steuer, R.E. (eds.) Research and Practice in Multiple Criteria Decision Making. Lecture Notes in Econom. and Math. Systems, 487, Springer, Berlin Heidelberg, pp. 93–104 (1999)

143. Göpfert, A., Tammer, Chr., Zălinescu, C.: A new minimal point theorem in product spaces. Z. Anal. Anwendungen, **18**, 767–770 (1999)

144. Göpfert, A., Tammer, Chr., Zălinescu, C.: On the vectorial Ekeland's variational principle and minimal points in product spaces. Nonlinear Analysis, **39**, 909–922 (2000)

145. Gorochowik, B.B., Kirillowa, F.M.: About scalarization of vector optimization problems. (Russian) Dokl. Akad. Nauk Soviet Union, **19**, 588–591 (1975)

146. Greeksch, W., Heyde, F., Isac, G., Tammer, Chr.: A characterization of approximate solutions of multiobjective stochastic optimal control problems. Optimization, **52**, 153–170 (2003)

147. Guddat, J., Guerra Vasquez, F., Tammer, K., Wendler, K.: Multiobjective and Stochastic Optimization Based on Parametric Optimization. Akademie-Verlag, Berlin (1985)

148. Ha, T.X.D.: On the existence of efficient points in locally convex spaces. J. Global Optim., **4**, 265–278 (1994)

149. Ha, T.X.D.: A note on a class of cones ensuring the existence of efficient points in bounded complete sets. Optimization, **31**, 141–152 (1994)

150. Hadjisavvas, N., Schaible, S.: Quasimonotone variational inequalities in Banach spaces. J. Optim. Theory Appl., **90**, 95–111 (1996)

151. Hadjisavvas, N., Schaible, S.: From scalar to vector equilibrium problems in the quasimonotone case. J. Optim. Theory Appl., **96**, 297–309 (1998)

152. Hamacher, H.W.: Mathematische Lösungsverfahren für Planare Standortprobleme. Vieweg Lehrbuch, Braunschweig/Wiesbaden (1995)

153. Hamacher, W., Nickel, S.: Multicriterial planar location problems. Preprint 243, Fachbereich Mathematik, Universität Kaiserslautern, Germany (1993)

154. Hamel, A.: Phelps' lemma, Danes' drop theorem and Ekeland's principle in locally convex topological vector spaces. Proc. Amer. Math. Soc. (accepted) (2003)

155. Hansen, P., Perreur, J., Thisse, J.F.: Location theory, dominance and convexity: some further results. Oper. Res., **28**, 1241–1250 (1980)

156. Hansen, P., Thisse, J.F.: Recent advances in continuous location theory. Sistemi Urbani, **1**, 33–54 (1983)

157. Hanson, M.A.: On sufficiency of the Kuhn–Tucker conditions. J. Math. Anal. Appl., **80**, 545–450 (1982)

158. Hartley, R.: On cone-efficiency, cone-convexity and cone-compactness. SIAM J. Appl. Math., **34**, 211–222 (1978)

159. Hartman, P., Stampacchia, G.: On some nonlinear elliptic differential–functional equations. Acta Math., **115**, 271–310 (1966)

160. Hazen, G.B., Morrin, T.L.: Optimality conditions in nonconical multiple-objective programming. J. Optim. Theory Appl., **40**, 25–60 (1983)

161. Henig, M.I.: Proper efficiency with respect to cones. J. Optim. Theory Appl., **36**, 387–407 (1982)

162. Henig, M.I.: Existence and characterization of efficient decisions with respect to cones. Math. Programming, **23**, 111–116 (1982)

163. Henkel, E.-C., Tammer, Chr.: ε-variational inequalities in partially ordered spaces. Optimization, **36**, 105–118 (1996)

164. Henkel, E.-C., Tammer, Chr.: ε-inequalities for vector approximation problems. Optimization, **38**, 11–21 (1996)

165. Heyde, F., Grecksch, W., Tammer, Chr.: Exploitation of necessary and sufficient conditions for suboptimal solutions of multiobjective stochastic control problems. ZOR - Math. Meth. Oper. Res., **54**, 425–438 (2001)

166. Hirche, J.: Zur Lösung von Optimierungsproblemen mit monoton-linear zusammengesetzten Zielfunktionen. Beiträge zur Num. Math., **9**, 87–94 (1981)

167. Hogan, W.W.: The continuity of the perturbation function of a convex program. Oper. Res., **21**, 351–352 (1973)

168. Holmes, R.B.: Geometric Functional Analysis and its Applications. Springer, New York, Graduate Texts in Mathematics, No. 24 (1975)

169. Huang, X.X.: Equivalents of a general approximate variational principle for set-valued maps and application to efficiency. Math. Meth. Oper. Res., **51**, 433–442 (2000)

170. Idrissi, F., Lefebvre, O., Michelot, C.: A primal dual algorithm for a constrained Fermat–Weber problem involving mixed gauges. Revue d'Automatique d'Informatique et de Recherche Operationelle / Operations Research, **22**, 313–330 (1988)

171. Idrissi, H.F., Lefebvre, O., Michelot, C.: Duality for constrained multifacility location problems with mixed norms and applications. Ann. Oper. Res., **18**, 71–92 (1989)

172. Idrissi, H.F., Lefebvre, O., Michelot, C.: Applications and numerical convergence of the partial inverse method. In: Optimization, Fifth French–German conference, Castel Novel 1988, Lecture Notes Math., 1405, Springer, Berlin, pp. 39–54 (1989)

173. Idrissi, H.F., Lefebvre, O., Michelot, C.: Solving constrained multifacility minimax location problems. Working Paper, Centre de Recherches de Mathématiques Statistiques et Economie Mathématique, Université de Paris 1, Panthéon-Sorbonne, Paris, France (1991)

174. Idrissi, H.F., Loridan, P., Michelot, C.: Approximation of solutions for location problems. J. Optim. Theory Appl., **56**, 127–143 (1988)

175. Isac, G.: Sur l'existence de l'optimum de Pareto. Riv. Mat. Univ. Parma, IV. Ser., **9**, 303–325 (1983)

176. Isac, G.: Supernormal cones and absolute summability. Libertas Math., **5**, 17–31 (1985)

177. Isac, G.: Supernormal cones and fixed point theory. Rocky Mt. J. Math., **17**, 219–226 (1987)

178. Isac, G.: Pareto optimization in infinite dimensional spaces: the importance of nuclear cones. J. Math. Anal. Appl., **182**, 393–404 (1994)

179. Isac, G.: The Ekeland's principle and the Pareto ε-efficiency. In: Tamiz, M. (ed.) Multi-Objective Programming and Goal Programming: Theories and Applications. Lecture Notes in Econom. and Math. Systems, 432, Springer, Berlin, pp. 148–163 (1996)

180. Isac, G.: Ekeland's principle and nuclear cones: A geometrical aspect. Math. Comput. Modelling, **26** (11), 111–116 (1997)

181. Isac, G.: Full nuclear cones associated to a normal cone. Application to Pareto efficiency. Applied Math. Letters, **15**, 633–639 (2002)

182. Isac, G., Bulavsky, A.V., Kalashnikov, V.V.: Complementarity, Equilibrium, Efficiency and Economics. Kluwer Academic Publishers, Boston Dordrecht London (2002)

183. Isac, G., Postolica, V.: The Best Approximation and Optimization in Locally Convex Spaces. Peter Lang, Frankfurt (1993)

184. Isac, G., Yuan, G.X.Z.: The existence of essentially connected components of solutions for variational inequalities. In: Giannessi, F. (ed.) Vector Variational Inequalities and Vector Equilibria. Mathematical Theories. Nonconvex Optimization and its Applications. Kluwer, Dordrecht, pp. 253–265 (2000)

185. Iusem, A.N.: Some properties of generalized proximal point methods for quadratic and linear programming. J. Optim. Theory Appl., **85**, 593–612 (1995)

186. Iusem, A.N.: On some properties of generalized proximal point methods for variational inequalities. J. Optim. Theory Appl., **96**, 337–362 (1998)

187. Iusem, A.N., Svaiter, B.F., Teboulle, M.: Entropy-like proximal point methods in convex programming. Math. Oper. Res., **19**, 790–814 (1994)

188. Iwanow, E., Nehse, R.: Some results on dual vector optimization problems. Math. Operationsforsch. Statist. Ser. Optim., **16**, 505–517 (1985)

189. Jagannathan, R.: On some properties of programming problems in parametric form pertaining to fractional programming. Management. Sci., **12**, 609–615 (1966)

190. Jahn, J.: Duality in vector optimization. Math. Programming, **25**, 343–353 (1983)

191. Jahn, J.: Scalarization in vector optimization. Math. Programming, **29**, 203–218 (1984)

192. Jahn, J.: A characterization of properly minimal elements of a set. SIAM J. Control Optim., **23**, 649–656 (1985)
193. Jahn, J.: Duality in partially ordered sets. Lecture Notes in Econom. and Math. Systems, 294, Springer, Berlin, pp. 160–172 (1986)
194. Jahn, J.: Existence theorems in vector optimization. J. Optim. Theory Appl., **50**, 397–406 (1986)
195. Jahn, J.: Mathematical Vector Optimization in Partially Ordered Spaces. Peter Lang, Frankfurt Bern New York (1986)
196. Jahn, J.: Parametric approximation problems arising in vector optimization. J. Optim. Theory Appl., **54**, 503–516 (1987)
197. Jahn, J.: A generalization of a theorem of Arrow, Barankin and Blackwell. SIAM J. Control Optim., **26**, 999–1005 (1988)
198. Jahn, J., Krabs, W.: Applications of multicriteria optimization in approximation theory. In: Stadler, W. (ed.) Multicriteria Optimization in Engineering and the Sciences. Plenum Press, New York, pp. 49–75 (1988)
199. Jameson, G.: Ordered Linear Spaces. Lecture Notes Math., 141, Springer, Berlin (1970)
200. Jongen, H.Th., Möbert, Th., Tammer, K.: On iterated minimization in non-convex optimization. Math. Oper. Res., **11**, 679–691 (1986)
201. Kalmoun, E.M.: Résultats d'éxistence pour les problèmes d'équilibre vectoriels et leurs applications. Ph.D. thesis, University Cadi Ayyad of Marrakech (2001)
202. Kalmoun, E.M., Riahi, H.: Topological KKM theorems and generalized vector equilibria on G-convex spaces with applications. Proc. Amer. Math. Soc., **129**, 1335–1348 (2001)
203. Kalmoun, E.M., Riahi, H.: Generalized vector equilibrium problems and applications to variational and hemivariational inequalities. To appear in J. Optim. Theory Appl.
204. Kalmoun, E.M., Riahi, H., Tanaka, T.: On vector equilibrium problems: remarks on a general existence theorem and applications. Nihonkai Math. J., **12**, 149–164 (2001)
205. Kalmoun, E.M., Riahi, H., Tanaka, T.: Remarks on a new existence theorem for generalized vector equilibrium problems and its applications. To appear in RIMS Kokyuroku (Kyoto University Press) edited by W. Takahashi (2001)
206. Kaul, R.N., Lyall, V.: A note on nonlinear fractional vector maximization. Opsearch, **26**(2), 108–121 (1989)
207. Kawasaki, H.: Conjugate relations and weak subdifferentials. Math. Oper. Res., **6**, 593–607 (1981)
208. Kawasaki, H.: A duality theorem in multiobjective nonlinear programming. Math. Oper. Res., **7**, 95–110 (1982)
209. Kazmi, K.R.: Existence of solutions for vector saddle-point problems. In: Giannessi, F. (ed.) Vector Variational Inequalities and Vector Equilibria. Mathematical Theories. Nonconvex Optimization and its Applications. Kluwer, Dordrecht, pp. 267–275 (2000)
210. Khanh, P.Q.: On Caristi–Kirk's theorem and Ekeland's variational principle for Pareto extrema. Institute of Mathematics, Polish Academy of Sciences, Preprint 357 (1986)
211. Klötzler, R.: On a general conception of duality in optimal control. Lecture Notes Math., 703, pp. 183–196 (1979)
212. Klose, J.: Sensitivity analysis using the tangent derivative. Numer. Funct. Anal. Optim., **13**, 143–153 (1992)

213. Kofler, M.: Mathematica. Addison-Wesley, Bonn (1992)
214. Kojima, M.: Strongly stable stationary solutions in nonlinear programs. In: Robinson S.M. (ed.) Analysis and Computation of Fixed Points. Academic Press, New York, pp. 93–138 (1980)
215. Konnov, I.V., Yao, J.C.: On the generalized vector variational inequality problem. J. Math. Anal. Appl., **206**, 42–58 (1997)
216. Kosmol, P.: Optimierung und Approximation. Walter de Gruyter, Berlin New York (1991)
217. Krasnosel'skij, M.A.: Positive Solutions of Operator Equations. (Russian) Fizmatgiz, Moskow (1962)
218. Kuhn, H.W.: A note on Fermat's problem. Math. Programming, **4**, 98–107 (1973)
219. Kuhpfahl, I., Patz, R., Tammer, Chr.: Location problems in town planning. In: Schweigert, D. (ed.) Methods of Multicriteria Decision Theory. University of Kaiserslautern, pp. 101–112 (1996)
220. Kuk, H., Tanino, T., Tanaka, M.: Sensitivity analysis in vector optimization. J. Optim. Theory Appl., **89**, 713–730 (1996)
221. Kuk, H., Tanino, T., Tanaka, M.: Sensitivity analysis in parametrized convex vector optimization. J. Math. Anal. Appl., **202**, 511–522 (1996)
222. Kummer, B.: Lipschitzian inverse functions, directional derivatives and applications in $C^{1,1}$-optimization. J. Optim. Theory Appl., **70**, 561–582 (1991)
223. Kummer, B.: Newton's method based on generalized derivatives for nonsmooth functions: Convergence analysis. In: Oettli, W., Pallaschke, D. (eds.) Advances in Optimization, Proceedings, Lambrecht, 1991. Springer, Berlin, pp. 171–194 (1992)
224. Kuratowski, K.: Topology. Academic Press, New York and Polish Scientific Publishers, Warsaw (1966)
225. Lai, T.C., Yao, J.C.: Existence results for VVIP. Appl. Math. Lett., **9**, 17–19 (1996)
226. Larson, T., Partrisson, M.: Equilibrium characterizations of solutions to side constraint asymmetric traffic assignment models. Le Matematiche, **49**, 249–280 (1994)
227. Lee, G.M., Kim, D.S., Lee, B.S.: Generalized vector variational inequality. Appl. Math. Lett., **9**, 39–42 (1996)
228. Lee, G.M., Kim, D.S., Lee, B.S.: On noncooperative vector equilibrium. Indian J. Pure Appl. Math., **278**, 735–739 (1996)
229. Lee, G.M., Kim, D.S., Lee, B.S., Cho, S.J.: Generalized vector variational inequality and fuzzy extension. Appl. Math. Lett., **6**, 47–51 (1993)
230. Lee, G.M., Kim, D.S., Lee, B.S., Yen, N.D.: Vector variational inequality as a tool for studying vector optimization problems. Nonlinear Anal., **34**, 745–765 (1998)
231. Lee, G.M., Kim, D.S., Lee, B.S., Yen, N.D.: Vector variational inequality as a tool for studying vector optimization problems. In: Giannessi, F. (ed.) Vector Variational Inequalities and Vector Equilibria. Mathematical Theories. Nonconvex Optimization and its Applications. Kluwer, Dordrecht, pp. 277–305 (2000)
232. Lee, G.M., Kum, S.: On implicit vector variational inequalities. J. Optim. Theory Appl., **104**, 409–425 (2000)

233. Lee, G.M., Kum, S.: Vector Variational inequalities in a haussdorff topological vector space. In: Giannessi, F. (ed.) Vector Variational Inequalities and Vector Equilibria. Mathematical Theories. Nonconvex Optimization and its Applications. Kluwer, Dordrecht, pp. 307–320 (2000)

234. Lee, B.S., Lee, G.M., Kim, D.S.: Generalized vector-valued variational inequalities and fuzzy extension. J. Korean Math. Soc., **33**, 609–624 (1996)

235. Lee, B.S., Lee, G.M., Kim, D.S.: Generalized vector variational-like inequalities in locally convex Hausdorff topological vector spaces. Indian J. Pure Appl. Math., **28**, 33–41 (1997)

236. Levy, H.: Stochastic dominance and expected utility: survey and analysis. Management Sci., **38**, 555–593 (1992)

237. Li, Z.F., Wang, S.Y.: Lagrange multipliers and saddle points in multiobjective programming. J. Optim. Theory Appl., **83**, 63–81 (1994)

238. Li, S.J., Yang, X.Q., Chen, G.Y.: Vector Ekeland variational principle. In: Giannessi, F. (ed.) Vector Variational Inequalities and Vector Equilibria. Mathematical Theories. Nonconvex Optimization and its Applications. Kluwer, Dordrecht, pp. 321–333 (2000)

239. Lin, K.L., Yang, D.P., Yao, J.C.: Generalized vector variational inequalities. J. Optim. Theory Appl., **92**, 117–125 (1997)

240. Loridan, P.: ε-solutions in vector optimization problems. J. Optim. Theory Appl., **43**, 265–276 (1984)

241. Loridan, P.: A dual approach to the generalized Weber problem under locational uncertainty. Cahiers du C.E.R.O., **20**, 241–253 (1984)

242. Loridan, P., Morgan, J., Raucci, R.: Convergence of minimal and approximate minimal elements of sets in partially ordered vector spaces. J. Math. Anal. Appl., **239**, 427–439 (1999)

243. Luc, D.T.: An existence theorem in vector optimization. Math. Oper. Res., **14**, 693–699 (1989)

244. Luc, D.T.: Theory of Vector Optimization. Lecture Notes in Econom. and Math. Systems, 319, Springer, Berlin (1989)

245. Luc, D.T., Lucchetti, R., Malivert, C.: Convergence of efficient sets, Set-Valued Anal., **2**, 207–218 (1994)

246. Luc, D.T., Vargas, C.: A saddle point theorem for set-valued maps, J. Nonlinear Analysis: Theory, Methods and Appl., **18**, 1–7 (1992)

247. Makarov, E.K., Rachkowski, N.N.: Density theorems for generalized Henig proper efficiency. J. Optim. Theory Appl., **91**, 419–437 (1996)

248. Malivert, C.: Fenchel duality in vector optimization. In: Advances in Optimization (Lambrecht, 1991), pp. 420–438. Lecture Notes in Econom. and Math. Systems, 382. Springer, Berlin (1992)

249. Markowitz, H.M.: Portfolio selection. J. Finance, **7**, 77–91 (1952)

250. Markowitz, H.M.: Portfolio Selection. John Wiley and Sons, New York (1959)

251. Markowitz, H.M.: Mean-Variance Analysis in Portfolio Choice and Capital Markets. Blackwell, Oxford (1987)

252. Martinet, B.: Régularisation d'inéquations variationnelles par approximations successives. Rev. Française Inf. Rech. Opér., pp. 154–159 (1970)

253. Martinet, B.: Algorithmes pour la résolution de problèmes d'optimisation et de minimax. Thèse d'Etat, Université de Grenoble (1972)

254. Maugeri, A.: Optimization problems with side constraints and generalized equilibrium principles. Le Mathematiche, **49**, 305–312 (1994)

255. Michael, E.: Continuous selections I. Ann. of Math., **63**, 361–382 (1956)
256. Michalowski, W.; Ogryczak, W.: A recursive procedure for selecting optimal portfolio according to the MAD model. Control Cybernet., **28**, 725–738 (1999)
257. Michalowski, W.; Ogryczak, W.: Extending the MAD portfolio optimization model to incorporate downside risk aversion. Naval Res. Logist., **48**, 185–200 (2001)
258. Michelot, C., Lefebvre, O.: A primal–dual algorithm for the Fermat–Weber problem involving mixed gauges. Math. Programming, **39**, 319–335 (1987)
259. Miettinen, K.: Nonlinear Multiobjective Optimization. Kluwer, Boston (1999)
260. Miettinen, K., Mäkelä, M.M.: Tangent and normal cones in nonconvex multiobjective optimization. In: Haimes, Y.Y., Steuer, R.E. (eds.) Research and Practice in Multiple Criteria Decision Making. Lecture Notes in Econom. and Math. Systems, 487, Springer, Berlin Heidelberg, pp. 114–124 (2000)
261. Minami, M.: Weak Pareto-optimal necessary conditions in a nondifferentiable multiobjective program on a Banach space. J. Optim. Theory Appl., **41**, 451–461 (1983)
262. Mosco, U.: A remark on a theorem of F.E. Browder. J. Math. Anal. Appl., **20**, 90–93 (1967)
263. Moudafi, A.: From optimization and variational inequalities to equilibrium problems: an algorithmic point of view. To appear in J. of Nat. Geometry
264. Nagurney, N., Zhao, L.: Disequilibrium and variational inequalities. J. Comput. Appl. Math., **33**, 181–198 (1990)
265. Nakayama, H.: Geometric consideration of duality in vector optimization. J. Optim. Theory Appl., **44**, 625–655 (1984)
266. Nakayama, H.: Duality theory in vector optimization: an overview. In: Lecture Notes in Econom. and Math. Systems, 242, pp. 109–125 (1985)
267. Naniewicz, Z., Panagiotopoulos, P.D.: Mathematical Theory of Hemivariational Inequalities and Applications. Marcel Dekker Inc., New York (1995)
268. Nehse, R.: Duale Vektoroptimierungsprobleme vom Wolfe-Typ. Seminarberichte der Sektion Mathematik der Humboldt-Universität zu Berlin, **37**, 55–60 (1981)
269. Nemeth, A.B.: The nonconvex minimization principle in ordered regular Banach spaces. Mathematica (Cluj), **23**, 43–48 (1981)
270. Nemeth, A.B.: A nonconvex vector minimization problem. Nonlinear Anal., **10**, 669–678 (1986)
271. Nemeth, A.B.: Between Pareto efficiency and Pareto ε-efficiency. Optimization, **20**, 615–637 (1989)
272. Ng, K.F., Zheng, X.Y.: Existence of efficient points in vector optimization. Preprint, The Chinese University of Hong Kong, Department of Mathematics (2000)
273. Nieuwenhuis, J.W.: Supremal points and generalized duality. Math. Operationsforsch. Statist., Ser. Optimization, **11**, 41–59 (1980)
274. Nieuwenhuis, J.W.: Some minimax theorems in vector-valued functions. J. Optim. Theory Appl., **40**, 463–475 (1983)
275. Nikodem, K.: Continuity of K-convex set-valued functions. Bull. Pol. Acad. Sci., Math., **34**, 393–400 (1986)
276. Nikodem, K.: On midpoint convex set-valued functions. Aequationes Math., **33**, 46–56 (1987)

277. Oettli, W.: Approximate solutions of variational inequalities. In: Albach, H. et al. (eds.) Quantitative Wirtschaftsforschung. J.C.B. Mohr, Tübingen, pp. 535–538 (1977)

278. Oettli, W.: A remark on vector-valued equilibria and generalized monotonicity. Acta Math. Vietnam., **22**, 213–221 (1997)

279. Oettli, W., Schläger, D.: Existence of equilibria for monotone multivalued mappings. Set-valued optimization. Math. Methods Oper. Res., **48**, 219–228 (1998)

280. Oettli, W., Schläger, D.: Generalized vectorial equilibria and generalized monotonicity. In: Functional Analysis with Current Applications in Science, Technology and Industry (Aligarh, 1996). Pitman Res. Notes Math. Ser., 377, Longman, Harlow, pp. 145–154 (1998)

281. Ogryczak, W.: Multiple criteria linear programming model for portfolio selection. Ann. Oper. Res., **97**, 143–162 (2000)

282. Ogryczak, W.; Ruszczynski, A.: On consistency of stochastic dominance and mean-semideviation models. Math. Programming, **89**, 217–232 (2001)

283. Ogryczak, W.; Ruszczynski,: Dual stochastic dominance and related mean-risk models. SIAM J. Optim., **13**, 60–78 A. (2002)

284. Pallu de la Barrière, R.: Cours d'Automatique Théorique. Dunod, Paris (1966)

285. Panagiotopoulos, P.D.: Inequality Problems in Mechanics and Applications. Birkhäuser, Basel (1985)

286. Pascoletti, A., Serafini, P.: Scalarizing vector optimization problems. J. Optim. Theory Appl., **42**, 499–524 (1984)

287. Pelegrin, B., Fernández, F.R.: Determination of efficient points in multiple objective location problems. Naval Res. Logist., **35**, 697–705 (1988)

288. Penot, J.-P.: L'optimisation à la Pareto: dcux ou trois choses que je sais d'elle. In: Structures Economiques et Econométrie (Lyon, 1978), pp. 4.14–4.33 (1978)

289. Penot, J.-P.: Differentiability of relations and differential stability of perturbed optimization problems. SIAM J. Control Optim., **22**, 529–551 (1984)

290. Penot, J.-P.: The drop theorem, the petal theorem and Ekeland's variational principle. Nonlinear Anal., **10**, 813–822 (1986)

291. Penot, J.-P., Sterna-Karwat, A.: Parametrized multicriteria optimization: continuity and closedness of optimal multifunctions. J. Math. Anal. Appl., **120**, 150–168 (1986)

292. Penot, J.-P., Sterna-Karwat, A.: Parametrized multicriteria optimization; order continuity of the marginal multifunctions. J. Math. Anal. Appl., **144**, 1–15 (1989)

293. Penot, J.-P., Théra, M.: Semicontinuous mappings in general topology. Arch. Math. (Basel), **38**, 158–166 (1982)

294. Peressini, A. L.: Ordered Topological Vector Spaces. Harper and Row Publishers, New York (1967)

295. Petchke, M.: On a theorem of Arrow, Barankin, and Blackwell. SIAM J. Control Optim., **28**, 395–401 (1990)

296. Phelps, R.R.: Convex Functions, Monotone Operators and Differentiability. Lecture Notes Math., 1364, Springer, Berlin (1989)

297. Phelps, R.R.: Convex Functions, Monotone Operators and Differentiability (2nd ed.). Lect. Notes Math., 1364, Springer, Berlin (1993)

298. Planchart, A., Hurter, A.P.: An efficient algorithm for the solution of the Weber problem with mixed norms. SIAM J. Control, **13**, 650–665 (1975)

299. Plastria, F.: Localization in single facility location. Eur. J. Oper. Res., **18**, 215–219 (1984)

300. Plastria, F.: Continuous location anno 92: a progress report. Isolde VI survey papers. Studies in Locational Analysis, **5**, 85–127 (1993)
301. Plastria, F.: Continuous location problems: research, results and questions. Vrije Universiteit Brussel, Report BEIF/70 (1994)
302. Podhinovski, V.V., Nogin, V.D.: Pareto-Optimal Solutions of Multicriterial Problems. (Russian) Nauka, Moscow (1982)
303. Postolică, V.: Existence conditions of efficient points for multifunctions with values in locally convex spaces ordered by supernormal cones. Stud. Cercet. Mat., **41**, 325–339 (1989)
304. Postolică, V.: New existence results for efficient points in locally convex spaces. J. Global Optim., **3**, 233–242 (1993)
305. Puerto, F.P., Fernandez, F.R.: Multicriteria decisions in location. Studies in Locational Analysis, **7**, 185–199 (1994)
306. Ralph, D., Scholtes, S.: Sensitivity analysis and Newton's method for composite piecewise smooth equations. Manuscript (1995)
307. Reemtsen, R.: On level sets and an approximationproblem for the numerical solution of a free boundary problem. Computing, **27**, 27–35 (1981)
308. Rockafellar, R.T.: Augmented Lagrangians and applications of the proximal point algorithm in convex programming. Math. Oper. Res., **1**, 96–116 (1976)
309. Rockafellar, R.T., Wets, R.J.B.: Variational Analysis. Springer, Berlin Heidelberg (1998)
310. Rolewicz, S.: On drop property. Studia Math., **85**, 27–35 (1987)
311. Rosinger, E.E.: Multiobjective duality without convexity. J. Math. Anal. Appl., **66**, 442–450 (1978)
312. Salukvadze, M.E.: Vector optimization problems in control theory. Mecniereba, Tbilisi (1975)
313. Samuelson, P.A.: A spatial price equilibrium and linear programming. Amer. Econom. Rev., **42**, 283–303 (1952)
314. Sawaragi, Y., Nakayama, H., Tanino, T.: Theory of Multiobjective Optimization. Academic Press, Orlando, Florida (1985)
315. Schaible, S.: A note on the sum of a linear and linear-fractional function. Naval Res. Logist. Quart., **24**, 691–693 (1977)
316. Schaible, S., Ibaraki, T.: Fractional programming. Europ. J. Oper. Res., **12**, 325–338 (1983)
317. Schandl, B., Klamroth, K., Wiecek, M.M.: Using block norms in bicriteria optimization. In: Haimes, Y.Y., Steuer, R.E. (eds.) Research and Practice in Multiple Criteria Decision Making. Lecture Notes in Econom. and Math. Systems, 487, Springer, Berlin Heidelberg, pp. 149–160 (2000)
318. Schönfeld, P.: Some duality theorems for the nonlinear vector maximum problem. Unternehmensforsch., **14**, 51–63 (1970)
319. Shi, D.S.: Contingent derivatives of the perturbation in multiobjective optimization. J. Optim. Theory Appl., **70**, 351–362 (1991)
320. Shi, D.S.: Sensitivity analysis in convex vector optimization. J. Optim. Theory Appl., **77**, 145–159 (1993)
321. Siddiqi, A.H., Ansari, Q.H., Ahmad, R.: On vector variational-like inequalities. Indian J. Pure Appl. Math., **28**, 1009–1016 (1997)
322. Siddiqi, A.H., Ansari, Q.H., Khaliq, A.: On vector variational inequalities. J. Optim. Theory Appl., **84**, 171–180 (1995)
323. Siddiqi, A.H., Ansari, Q.H., Khan, M.F.: Variational-like inequalities for multivalued maps. Indian J. Pure Appl. Math., **3**, 161–166 (1999)

324. Smithson, R.E.: Subcontinuity for multifunctions. Pacific J. Math., **61**, 283–288 (1975)
325. Song, W.: Connectivity of efficient solution sets in vector optimization of set-valued mappings. Optimization, **39**, 1–11 (1997)
326. Song, W.: Generalized vector variational inequalities. In: Giannessi, F. (ed.) Vector Variational Inequalities and Vector Equilibria. Mathematical Theories. Nonconvex Optimization and its Applications. Kluwer, Dordrecht, pp. 381–401 (2000)
327. Sonntag, Y., Zălinescu, C.: Comparison of existence results for efficient points. J. Optim. Theory Appl., **105**, 161–188 (2000)
328. Spingarn, J.E.: Partial inverse of a monotone operator. Appl. Math. Optim., **10**, 247–265 (1983)
329. Staib, T.: On generalizations of Pareto minimality. J. Optim. Theory Appl., **59**, 289–306 (1988)
330. Sterna-Karwat, A.: On existence of cone maximal points in real topological linear spaces. Israel J. Math., **54**, 33–41 (1986)
331. Sterna-Karwat, A.: Continuous dependence of solutions on a parameter in a scalarization method, J. Optim. Theory Appl., **55**, 417–434 (1987)
332. Sterna-Karwat, A.: Remarks on convex cones. J. Optim. Theory Appl., **59**, 335–340 (1988)
333. Sterna-Karwat, A.: Convexity of the optimal multifunctions and its consequences to vector optimization. Optimization, **20**, 809–817 (1989)
334. Sterna-Karwat, A.: Approximating families of cones and proper efficiency of vector optimization. Optimization, **20**, 799–808 (1989)
335. Sukopp, H., Wittig, R.: Stadtökologie. Fischer-Verlag, Stuttgart (1993)
336. Takahashi, W.: Existence theorems generalizing fixed point theorems for multivalued mappings. In: Baillon, J.B., Thera, M. (eds.) Fixed Point Theory and Applications. Pitman Research Notes in Math. 252, Longman, Harlow, pp. 397–406 (1991)
337. Tammer, Chr.: A generalization of Ekeland's variational principle. Optimization, **25**(2–3), 129–141 (1992)
338. Tammer, Chr.: Existence results and necessary conditions for ε-efficient elements. In: Brosowski, B. et al. (eds.) Multicriteria Decision; Proceedings of the 14th Meeting of the German Working Group "Mehrkriterielle Entscheidung", Peter Lang Verlag, Frankfurt Bern, pp. 97–110 (1993)
339. Tammer, Chr.: Erweiterungen und Anwendungen des Variationsprinzips von Ekeland. ZAMM, **73**, 832–826 (1993)
340. Tammer, Chr.: Stability results for approximately efficient solutions. OR Spektrum, **16**, 47–52 (1994)
341. Tammer, Chr.: A variational principle and a fixed point problem. In: Henry, J., Yvon, J.-P. (eds.) System Modelling and Optimization. Lecture Notes in Control and Inform. Sci., 197, Springer, London, pp. 248–257 (1994)
342. Tammer, Chr.: Necessary conditions for approximately efficient solutions of vector approximation. In: Approximation and Optimization in the Caribbean, II (Havana, 1993), 651–663, Approx. Optim., 8, Lang, Frankfurt am Main (1995)
343. Tammer, Chr.: A variational principle and applications for vectorial control approximation problems. Math. J. Univ. Bacau (Romania) (1996)
344. Tammer, Chr.: Approximate solutions of vector-valued control-approximation problems. Studies in Locational Analysis, **10**, 151–162 (1996)

345. Tammer, Chr.: Multiobjective optimal control problems. International Series of Numerical Mathematics 124, Birkhäuser, Basel, pp. 97–106 (1998)

346. Tammer, Chr., Tammer, K.: Generalization and sharpening of some duality ralations for a class of vector optimization problems. ZOR, **35**, 249–265 (1991)

347. Tammer, Chr., Gergele, M., Patz, R., Weinkauf, R.: Standortprobleme in der Landschaftsgestaltung (accepted) (2002)

348. Tammer, K.: Two-level optimization with approximate solutions in the lower level. ZOR, **41**, 231–249 (1995)

349. Tammer, K., Tammer, Chr., Ohlendorf, E.: Multicriterial fractional optimization. In: Guddat, J. et al. (eds.) Parametric Optimization and Related Topics IV. Peter Lang, Frankfurt, pp. 359–370 (1996)

350. Tanaka, T.: Some minimax problems of vector valued functions. J. Optim. Theory Appl., **59**, 505–524 (1988)

351. Tanaka, T.: Existence theorems for cone saddle points of vector-valued functions in infinite-dimensional spaces. J. Optim. Theory Appl., **62**, 127–138 (1989)

352. Tanaka, T.: Two types of minimax theorems for vector-valued functions. J. Optim. Theory Appl., **68**, 321–334 (1991)

353. Tanaka, T.: Generalized quasiconvexities, cone saddle points, and minimax theorem for vector-valued functions. J. Optim. Theory Appl., **81**, 355–377 (1994)

354. Tanino, T.: Sensitivity analysis in multiobjective optimization. J. Optim. Theory Appl., **56**, 479–499 (1988)

355. Tanino, T., Sawaragi, Y.: Duality theory in multiobjective programming. J. Optim. Theory Appl., **30**, 229–253 (1979)

356. Tanino, T., Sawaragi, Y.: Stability of nondominated solutions in multicriteria decision-making. J. Optim. Theory Appl., **30**, 229–253 (1980)

357. Thibault, L.: Sequential convex subdifferential calculus and sequential Lagrange multipliers. SIAM J. Control Optim., **35**, 1434–1444 (1997)

358. Tian, G.Q.: Generalized KKM theorems, minimax inequalities and their applications. J. Optim. Theory Appl., **83**, 375–389 (1994)

359. Turinici, M.: Cone maximal points in topological linear spaces. An. Stiint. Univ. Al.I.Cuza Iasi, N. Ser., Sect. Ia, **37**, 371–390

360. Valadier, M.: Sous-différentiabilité de fonctions convexes à valeurs dans un espace vectoriel ordonné. Math. Scand., **30**, 65–74 (1972)

361. Valyi, I.: On duality theory related to approximate solutions of vector-valued optimization problems. In: Demyanov, V.F., Pallaschke, D. (eds.) Nondifferentiable Optimization: Motivations and Applications. Lecture Notes in Econom. and Math. Systems, 255, pp. 150–162 (1985)

362. Valyi, I.: Approximate saddle-point theorems in vector optimization. J. Optim. Theory Appl., **55**, 435–448 (1987)

363. Vetschera, R.: Visualisierungstechniken in Entscheidungsproblemen bei mehrfacher Zielsetzung. OR Spektrum, **16**, 227–241 (1994)

364. Vetschera, R.: MCVIEW/Windows User Manual. Version 1.11 (1994)

365. Wang, S.: Lagrange conditions in nonsmooth and multiobjective mathematical programming. Mathematics in Economics, **1**, 183–193 (1984)

366. Wanka, G.: Duality in vectorial control approximation problems with inequality restrictions. Optimization, **22**, 755–764 (1991)

367. Wanka, G.: On duality in the vectorial control-approximation problem. ZOR, **35**, 309–320 (1991)

368. Wanka, G.: Characterization of approximately efficient solutions to multiobjective location problems using Ekeland's variational principle. Studies in Locational Analysis, **10**, 163–176 (1996)

369. Ward, J.E., Wendel, R.E.: Characterizing efficient points in location problems under the one-infinity norm. In: Thisse, J.F., Zoller, H.G. (eds.) Locational Analysis of Public Facilities. North Holland, Amsterdam (1983)

370. Wardrop, J.G.: Some theoretical aspects of road traffic research. Proceedings of the Institute of Civil Engineers, Part II, pp. 325–378 (1952)

371. Weber, A.: Über den Standort der Industrien. Tübingen (1909)

372. Weidner, P.: Dominanzmengen und Optimalitätsbegriffe in der Vektoroptimierung. Wissenschaftliche Zeitschrift der TH Ilmenau, **31**, 133–146 (1985)

373. Weidner, P.: Ein Trennungskonzept und seine Anwendungen auf Vektoroptimierungsverfahren. Dissertation B, Martin–Luther Universität Halle–Wittemberg (1991)

374. Weir, T.: A duality theorem for a multiple objective fractional optimization problem. Bull. Austral. Math. Soc., **34**, 415–425 (1986)

375. Wendell, R.E., Hurter, A.P., Lowe, T.J.: Efficient points in location problems. American Institute of Industrial Engineers Transactions, **9**, 238–246 (1977)

376. Wendell, R.E., Peterson, E.L.: A dual approach for obtaining lower bounds to the Weber problem. J. Regional Science, **24**, 219–228 (1984)

377. Whitmore, G.A., Findlay, M.C. (eds.): Stochastic dominance: an approach to decision making under risk. D.C. Heath, Lexington, Mass. (1978)

378. Witzgall, C.: Optimal location of a central facility: Mathematical models and concepts. National Bureau of Standards, Report 8388, Washington (1964)

379. Yang, X.Q.: On some equiivalent conditions of vector variational inequalities. In: Giannessi, F. (ed.) Vector Variational Inequalities and Vector Equilibria. Mathematical Theories. Nonconvex Optimization and its Applications. Kluwer, Dordrecht, pp. 423–432 (2000)

380. Yang, X.Q., Chen, G.Y.: On inverse vector variational inequalities. In: Giannessi, F. (ed.) Vector Variational Inequalities and Vector Equilibria. Mathematical Theories. Nonconvex Optimization and its Applications. Kluwer, Dordrecht, pp. 433–446 (2000)

381. Yang, X.Q., Goh, C.J.: On vector variational inequalities: application to vector equilibria. J. Optim. Theory Appl., **95**, 431–443 (1997)

382. Yang, X.Q., Goh, C.J.: Vector variational inequalities, vector equilibrium flow and vector optimization. In: Giannessi, F. (ed.) Vector Variational Inequalities and Vector Equilibria. Mathematical Theories. Nonconvex Optimization and its Applications. Kluwer, Dordrecht, pp. 447–465 (2000)

383. Yu, S.Y., Yao, J.C.: On vector variational inequalities. J. Optim. Theory Appl., **89**, 749–769 (1996)

384. Yuan, G.X.Z.: The study of minimax inequalities and applications to economies and variational inequalities. Mem. Amer. Math. Soc., **132**, No. 625 (1998)

385. Zălinescu, C.: A generalization of the Farkas lemma and applications to convex programming. J. Math. Anal. Appl., **66**, 651–678 (1978)

386. Zălinescu, C.: Duality for vectorial nonconvex optimization by convexification and applications. An. Univ. "Al. I. Cuza" Iaşi, Sec. Mat. (N.S.), **29** (3), 15–34 (1983)

387. Zălinescu, C.: Solvability results for sublinear functions and operators. Z. Oper. Res. Ser. A-B, **31**, A79–A101 (1987)

388. Zălinescu, C.: On two notions of proper efficiency. In: Brosowski, B., Martensen, E. (eds.) Optimization in Mathematical Physics. Methoden und Verfahren der Mathematischen Physik, **34**, pp. 77–86 (1987)

389. Zălinescu, C.: Stability for a class of nonlinear optimization problems and applications. In: Nonsmooth Optimization and Related Topics (Erice, 1988). Plenum, New York, pp. 437–458 (1989)

390. Zălinescu, C.: On a new stability condition in mathematical programming. In: Giannessi, F. (ed.) Nonsmooth Optimization. Methods and Applications. Gordon and Breach Science Publ., Philadelphia, pp. 429–438 (1992)

391. Zeidler, E.: Nonlinear Functional Analysis and its Applications. Part III: Variational Methods and Optimization. Springer, New York (1986)

392. Zeidler, E.: Nonlinear Functional Analysis and its Applications. Springer, New York (1990)

393. Zhuang, D.: Density results for proper efficiency. SIAM J. Control Optim., **32**, 51–58 (1994)

394. Zowe, J.: Subdifferentiability of convex functions with values in a ordered vector space. Math. Scand., **34**, 69–83 (1974)

395. Zowe, J.: A duality theorem for a convex programming problem in ordered complete vector lattices. J. Math. Anal. Appl., **50**, 273–287 (1975)

Index

algorithm
- approximation
-- Mathematica program   259
- geometric   284
- interactive   264
-- stability results   264
- one-phase vector proximal   270
-- convergence   271
- proximal-point   250
-- convergence results   259
- Spingarn   252, 255
- three-level interactive procedure
   300
- two-phase vector proximal
-- convergence   275
- twophase vector proximal   274
annihilator   165
approximation error   222, 227
associativity   16

base (for a cone)   20
- bounded   36
base of neighborhoods   22
best location   6, 283
bound
- lower   15
- upper   15
Bregman
- distance   267
Bregman function   268
- examples   268

chain   88
closure (of a set)   21

commutativity   16
comparable elements   13
complementary slackness
- approximate   222, 226
concave
- quasi-   177
cone   17
- adjacent   73
- angle property   2
- asymptotic   103
- based   36
- Bouligand tangent   73
- contingent   73
- convex   17
- correct   103
- Daniell   26, 96
-- sequentially   96
- dual   31
- Henig dilating   111
- nontrivial   17
- normal   25, 245
- nuclear   36
- Phelps   197
- (π)-   103
- pointed   17
- proper   17
- reproducing   17
- sequentially C-bound regular   208
- supernormal   36
- Ursescu tangent   73
- well-based   36
cone saddle point   195
cones

– overview 3
control 9
– suboptimal 10
convergence
– Kuratowski–Painlevé 61
– net 22
core (of a set) 27

derivative
– Dini lower 74
– Dini upper 74
Dinkelbach transformation 11, 297
dual pair 30
duality
– axiomatic 154
– conjugation 154
– converse 158, 160
– – for approximation problems 167
– direct 158
– Lagrange 156
– strong 157, 159
– – for approximation problems 166
– weak 156, 159
– – for approximation problems 166
– with scalarization 155
– without scalarization 155

efficiency
– approximate 81
– ε- 81
– εk^0- 81
element
– approximately efficient 81
– efficient 14, 95
– – Benson-proper 141
– – Henig-proper 141
– – strong proper 131
– – weakly 141
– Henig proper maximal 110
– maximal (minimal) 14, 218
– null 16
– properly efficient 159
– properly maximal 110
– unity 16
excess (of A over B) 52
existence
– of equilibria 173, 182
existence of solutions
– of a variation-like inequality 187

– of complementarity problems 192
– of equilibrium problems 4
– of hemivariational inequalities 190
– of vector optimization problems 193

Fan's minimax inequality 180
Fan–KKM lemma 171
function
– B-monotone 40
– – strictly 40
– C-α-convex 49
– C-α-quasiconvex 49
– C-convex 49
– C-lower continuous (C-l.c.) 66
– C-lower semicontinuous 66
– C-mid-convex 49
– C-mid-quasiconvex 49
– C-nearly convex 49
– C-nearly quasiconvex 49
– C-quasiconvex 49
– C-upper semicontinuous 66
– continuous 23
– convex 39
– directional derivative 239
– domain of 39, 49
– epigraph of 39, 49
– lower semicontinuous 39
– marginal 115
– proper 39
– upper continuous (u.c.)
– – C- (C-u.c.) 66
functional
– Minkowski 27
– monotone 28
– positively homogeneous 27
– subadditive 27
– sublinear 27
– symmetric 27

game
– cooperative differential 9
gauge technique 206

hemivariational inequality 190
homeomorphism 24
hull
– affine 17
– convex 17
– – closed 33
– linear 17

infimum of a set   15
interior   21
– algebraic   27
–– relative   75

KKM Lemma   171
KKM-mapping   172
Kolmogorov condition   244
– ε-   240

Lagrange multipliers   212
– existence results   212
Lagrangian   216, 220
– generalized   156, 159
– scalarized   223
limit   23
– inferior   51
– superior   51
lower bound   246, 247

manifold
– linear   16
mapping
– invex   194
– preinvex   194
– pseudomonotone   182
mean-risk approach   11
metric   21
– P-(valued)   208
minimal-point theorem   199
– authentic   202
Minty's linearization   179, 181
multifunction   47
– C-α-quasiconvex, C-mid-quasiconvex,
   C-nearly quasiconvex, C-quasiconvex
   48
– α-concave, C-α-concave   48
– BH-pseudomonotone   189
– C-continuous   61
–– Hausdorff   61
– C-convex, C-α-convex   48
– C-lower continuous (C-l.c.)   61
–– Hausdorff (H-C-l.c.)   61
–– uniformly (at a point on a set)   61
– C-mid-convex, C-nearly convex   48
– C-upper continuous (C-u.c.)   61
–– Hausdorff (H-C-u.c.)   61
– closed   55
–– at a point   55

– closed-valued   55
– compact (at a point)   55
– continuous   51
–– Hausdorff   58
– convex, α-convex   48
– derivative of   141
– domain of   47
– epigraph of   48
– graph of   47
– image of   47
– lower continuous (l.c.)   51
–– Hausdorff (H-l.c.)   58
– lower semicontinuous
–– C- (C-l.s.c.)   63
–– $\mathcal{C}$- ($\mathcal{C}$-l.s.c.)   137
– mid-convex, nearly convex   48
– monotonicity notions   170
– of class $(S)_+$ 189
– optimal-value   122, 138
–– weak   122
– quasi-pseudomonotone   189
– semidifferentiable   75
– solution   138
– sublevel, strict sublevel   48
– upper continuous (u.c.)   51
–– Hausdorff (H-u.c.)   58
– upper Lipschitz   77
– upper semicontinuous
–– C- (C-u.s.c.)   63

neighborhood   22
net   22
– C-decreasing   94
– C-increasing   94
–– strictly   94
– convergent   22
– t-decreasing   88
– t-increasing   88
–– strictly   88
norm   27
– vector-valued   229

operator
– sublinear   50
order
– lexicographic   19
– linear   13
– partial   13
– total   13

Pareto minimum   10
point
- efficient   14, 95
-- Benson-proper   141
-- Henig-proper   141
-- properly   159
-- strong proper   131
-- super   112
-- weakly   141
portfolio optimization problem   318
preorder   13
principle
- Ekeland's variational   196
-- vectorial   197
- Pareto   288
- Pontryagin's minimum   308
-- ε-   304
- Wardrop   5
problem
- L_p-approximation   241
- approximation   161, 245
-- special cases   161
- complementarity   192
- dual
-- strongly   155
-- weakly   155
- equilibrium   3
-- scalar   169
- Fermat–Weber   6, 284
- fractional programming
-- multicriteria   11, 294
-- real-valued   10
-- three-level interactive procedure
    300
- inverse Stefan   242
- linear programming   245
-- perturbed   245
-- surrogate problem   245
- location   245
- multicriteria location   284
-- algorithms   283
-- for a children's playground   289,
    292
-- geometric algorithm   284
-- Mathematica program   284
-- solution set   7
- multicriteria stochastic control   312
- optimal control   245
-- multicriteria   9, 305

-- multiobjective   9
-- vector-valued (vector)   9
- optimal regulator   245
- real-valued approximation   234
- real-valued location   6, 234
- scalar equilibrium   267
- scalarized   84
- vector control approximation   230,
    263
-- finite-dimensional   236, 238
-- interactive algorithm   266
-- necessary conditions   230
-- special cases   234
- vector equilibrium   4, 170
-- discrete case   281
-- penalization   277
-- perturbation   277
-- relaxation   277
- vector location   7
- weak traffic equilibrium   5
- weak vector equilibrium   4, 267
- well-posed   119
-- η-   119
-- weakly   119
process   9
property
- angle (for a cone)   36
- containment   119
-- uniformly around a point   128
- domination   88, 118
proximal-point method   266

quasi-interior   31

relation
- antisymmetric   13
- binary   13
- composition   14
- inverse   14
- reflexive   13
- transitive   13

saddle point
- approximate   220
- (B, e)-   218
- cone   195
- y_0^*-   216
scalarization   84
section

– lower   88
– upper   88
segment parallel to   116
seminorm   27
separation
– of convex sets   32
– of nonconvex sets   44
sequence
– generalized   22
set
– absorbing   24
– affine   16
– α-convex   45
– asymptotically compact (a.c.)   103
– balanced   24
– bounded   34
– – C-upper (lower)   94
– – lower   15
– – upper   15
– boundedly order complete   98
– C-compact   103
– C-complete   99, 123
– – sequentially   99
– C-semicompact   103
– closed   21
– – sequentially   23
– cofinal   22
– connected   56
– convex   17
– directed   22
– epigraph type   39
– full   24
– lower bounded   26
– mid-convex   45
– nearly convex   45
– open   21
– polar   31
– solution   119
– strict sublevel   48
– strict sublevel   49
– sublevel   48, 49
– well-ordered   13, 88
solution
– ε-optimal   119
– optimal   119
space
– Banach   29
– Hilbert   29
– lineality   94

– linear   16
– locally convex   27
– – examples   29
– metric   21
– metrizable   29
– normed   29
– quasi-complete   97
– reflexive   30
– topological   21
– – first-countable   22
– – Hausdorff   23
– – linear   24
– – vector   24
– vector   16
state   9
stochastic dominance   11, 317
structure
– compatible   17
– linear   16
– order   13
subdifferential   50, 230
– of a sublinear multifunction   217
– of norm terms   236
– of the indicator function   238
– of the vector-valued norm   231
subnet   23
subspace
– linear   16
supremum of a set   15

theorem
– Alaoglu–Bourbaki   30
– Fan   4
– – generalized   172
– Hahn–Banach   31
– – geometric form   32
– Kirk–Caristi fixed point   211
– Krein and Rutman   35
– nonconvex separation   44
– Phelps minimal-point   197
– vector minimax   196
topology   20
– compatible   30
– linear   24
– Mackey   30
– product   24
– strong   36
– trace   29
– weak   30

– weak*   30
town planning   283
traffic control   5
transportation network   5

value
– optimal   119

variational inequality
– scalar   185
– vector   185
variational-like inequality   187

Zorn's lemma   16